Aviation Turbulence

Robert Sharman • Todd Lane
Editors

Aviation Turbulence

Processes, Detection, Prediction

Springer

Editors
Robert Sharman
National Center for Atmospheric Research
Boulder, Colorado
USA

Todd Lane
School of Earth Sciences
The University of Melbourne
Melbourne, Victoria
Australia

ISBN 978-3-319-23629-2 ISBN 978-3-319-23630-8 (eBook)
DOI 10.1007/978-3-319-23630-8

Library of Congress Control Number: 2016941512

Printed on acid-free paper

This Springer imprint is published by Springer Nature
The registered company is Springer International Publishing AG Switzerland

Preface

This book provides a compilation of recent work in the area of aviation turbulence research. The motivation for such a compilation was based on two highly successful internationally attended aviation turbulence workshops in the United States in 2013 and 2014 and four aviation meteorology workshops in Seoul, South Korea. Attendees included members of the research community as well as representatives from various government agencies and commercial entities (including airlines). These workshops made it clear that there has not been enough communication between the research community and users who need turbulence information to make operational decisions about where to fly. This book aims to foster that communication, by summarizing recent progress and encouraging future research in this area. In some ways, it is very similar in scope to the Proceedings of a Symposium on Clear Air Turbulence and its Detection held at the Boeing Company in Seattle, Washington, in 1968 edited by Pao and Goldburg (1969). However, there has not been anything similar published since then, which is surprising given the many scientific and operational advances that have taken place in the area of aviation turbulence in the last few decades.

The significant advances have been in the areas of turbulence observations and detection, nowcasting, forecasting and verification, and simulation and modeling studies. Most have been aimed at improving safety for the flying public by application to commercial aircraft in cruise, i.e., in the upper troposphere and lower stratosphere (UTLS). This is in part due to the fact that most injuries to passengers and crew occur in cruise, when they are more likely to be unbuckled. It is also due in part to the interesting research challenges associated with understanding the sources, dynamics, characteristics, and genesis of turbulence at these altitudes and the fact that UTLS turbulence is thought to make a significant contribution to the total dissipation in the atmosphere and may be comparable to that which takes place in the surface boundary layer (e.g., Cadet 1971). Turbulence in the UTLS is also of critical importance for stratosphere–troposphere exchange of chemical constituents. This is not to say there has not been significant progress in understanding of turbulence in the planetary boundary layer (PBL), indeed several recent texts are available on the subject (e.g., Stull 1988; Sorbjan 1989; Kaimal and Finnigan 1994), and in fact some of the recent research in stable boundary layers (SBL) may have relevance to upper level turbulence as well, but PBL turbulence is not the focus of this book.

The book is not intended to be used as a stand-alone textbook for turbulence or even a textbook for aviation turbulence. These are provided in the texts referenced above and many others too numerous to list separately, although mention should be made to those books and scholarly reviews that are particularly relevant to aviation turbulence, including Pao and Goldburg (1969), Burnham (1970), Dutton (1971), Vinnichenko et al. (1980), Lee et al. (1984), Camp and Frost (1987), Bedard (1993), Lester (1994), Knox (1997), Sharman (2005), Wolff and Sharman (2008), Lane et al. (2012), Sharman et al. (2012), and Ellrod et al. (2015). It is, however, intended for a wide range of readers with minimal assumed background knowledge. Some familiarity with meteorology is assumed, but perhaps not an intimate knowledge of turbulence properties. For some readers, the material contained in the chapters would be a review and for others an introduction. We hope the topics covered will encourage interest by the readers and foster further research that is sorely needed to make real progress in these areas.

The book is structured into five main sections providing background material, reviews and applications of detection strategies, forecast and nowcast methods and their verification, observational and modeling studies, and finally discussions of future opportunities and research needs. Each major section contains several chapters for a total of 25 chapters. The authors of each chapter were chosen because of their expertise in their area, and every chapter has been peer reviewed by at least two other experts. We thank all the authors for their contributions, all of the anonymous reviewers who helped considerably with comments and reviews on a number of chapters, and the many sponsors of the work over the years that have led to this compilation. Finally, we pay tribute to our late friend and colleague Rod Frehlich, who inspired us with his enthusiasm for research in this area over the many years we worked together.

Boulder, CO — Robert Sharman
Melbourne, VIC, Australia — Todd Lane

References

Bedard, A.J., Jr.: Atmospheric turbulence aloft: A review of possible methods for detection, warning, and validation of predictive models. In: 31st Aerospace Sciences Meeting and Exhibit, AIAA 93–0847, Reno, NV (1993)

Burnham, J.: Atmospheric gusts—A review of the results of some recent research at the Royal Aircraft Establishment. Mon. Wea. Rev. **98**(10), 723–734 (1970)

Cadet, D.: Energy dissipation within intermittent clear air turbulence patches. J. Atmos. Sci. **34**(1), 137–142 (1977)

Camp, D.W., Frost, W.: Atmospheric turbulence relevant to aviation, Missile, and Space Programs. In: NASA Conf. Pub. 2468 (1987)

Dutton, J.: Clear-air turbulence, aviation, and atmospheric science. Rev. Geophys. Space Phys. **9**, 613–657 (1971). doi:10.1029/RG009i003p00613

Ellrod, G.P., Knox, J.A., Lester, P.F., Ehernberger, L.J.: Clear air turbulence. In: North, G.R. (editor-in-chief), Pyle, J., Zhang, F. (eds.) Encyclopedia of the Atmospheric Sciences, 2nd edn, vol. 1, pp. 177–186 (2015)

Kaimal, J.C., Finnigan, J.J.: Atmospheric Boundary Layer Flows: Their Structure and Measurement. Oxford University Press, New York (1994)

Knox, J.A.: Possible mechanisms of clear-air turbulence in strongly anticyclonic flows. Mon. Wea. Rev. **125**, 1251–1259 (1997)

Lane, T.P., Sharman, R.D., Trier, S.B., Fovell, R.G., Williams, J.K.: Recent advances in the understanding of near-cloud turbulence. Bull. Amer. Meteor. Soc. **93**(4), 499–515 (2012). doi:10.1175/BAMS-D-11-00062.1

Lee, D.R., Stull, R.B., Irvine, W.S.: Clear air turbulence forecasting techniques. In: Air Force Weather Service Report AFGWC/TN-79-001 (REV) (1984)

Lester, P.F.: Turbulence: A New Perspective for Pilots. Jeppesen Sanderson, Inc., Englewood, CO (1993)

Pao, Y.-H., Goldburg, A.: Clear Air Turbulence and Its Detection. Plenum Press, New York (1969)

Sharman, R.: Clear air turbulence. In: Scott, A. (ed.) Encyclopedia of Nonlinear Science. Routledge, New York (2005)

Sharman, R.D., Trier, S.B., Lane, T.P., Doyle, J.D.: Sources and dynamics of turbulence in the upper troposphere and lower stratosphere: A review. Geophys. Res. Lett. **39**, L12803 (2012). doi:10.1029/2012GL051996

Sorbjan, Z.: Structure of the Atmospheric Boundary Layer. Prentice Hall, New Jersey (1989)

Stull, R.: An Introduction to Boundary Layer Meteorology. Springer Netherlands (1988)

Vinnichenko, N.K., Pinus, N.Z., Shmeter, S.M., Shur, G.N.: Turbulence in the Free Atmosphere. Plenum, New York (1980)

Wolff, J.K., Sharman, R.D.: Climatology of upper-level turbulence over the continental Unites States. J. Appl. Meteor. Clim. **47**, 2198–2214 (2008). doi:10.1175/2008JAMC1799.1

Contents

List of Contributors

Alan W. Black IIHR-Hydroscience and Engineering, 100 C. Maxwell Stanley Hydraulics Laboratory, Iowa City, USA

David R. Bright NOAA/National Weather Service, Portland, OR, USA

Piers Buchanan Met Office, Exeter, Devon, UK

P.W. Chan Hong Kong Observatory, Hong Kong, China

William N. Chan NASA Ames Research Center, Moffett Field, CA, USA

Larry B. Cornman Research Applications Laboratory, National Center for Atmospheric Research, Boulder, CO, USA

Wiebke Deierling Research Applications Laboratory, National Center for Atmospheric Research, Boulder, CO, USA

James D. Doyle Naval Research Laboratory, Marine Meteorology Division, Monterey, CA, USA

Corey B. Dunn Department of Geography, University of Georgia, Athens, GA, USA

Tom Fahey Meteorology Department #091, Delta Air Lines, Atlanta, GA, USA

Philip G. Gill Met Office, Exeter, Devon, UK

Robert K. Goodrich Research Applications Laboratory, National Center for Atmospheric Research, Boulder, CO, USA

Jeremiah S. Grant Department of Geography, University of Georgia, Athens, GA, USA

Qingfang Jiang Naval Research Laboratory, Marine Meteorology Division, Monterey, CA, USA

Manoj M. Joshi Centre for Ocean and Atmospheric Sciences, University of East Anglia, Norwich, UK

Michael L. Kaplan Division of Atmospheric Sciences, Desert Research Institute, Reno, NV, USA

Jung-Hoon Kim Colorado State University/Cooperative Institute for Research in the Atmosphere (CSU/CIRA), Fort Collins, CO, USA

National Oceanic and Atmospheric Administration/National Centers for Environmental Prediction/Aviation Weather Center, Kansas City, MO, USA

Stephanie Klipfel Meteorology Department #091, Delta Air Lines, Atlanta, GA, USA

John A. Knox Department of Geography, University of Georgia, Athens, GA, USA

Steve A. Lack NOAA/National Weather Service, Kansas City, MO, USA

Todd Lane School of Earth Sciences, The University of Melbourne, Melbourne, VIC, Australia

Yuh-Lang Lin Department of Physics and Department of Energy and Environmental Systems, North Carolina A&T State University, Greensboro, NC, USA

Bob Lunnon UK Met Office, University of Reading, Reading, UK

Alex Mahalov School of Mathematical and Statistical Sciences and School for Engineering of Matter, Transport and Energy, Arizona State University, Tempe, AZ, USA

Gregory Meymaris Research Applications Laboratory, National Center for Atmospheric Research, Boulder, CO, USA

David S. Nevius Department of Geography, University of Georgia, Athens, GA, USA

Rory O'Loughlin Meteorology Department #091, Delta Air Lines, Atlanta, GA, USA

Stephanie P. Phelps APEX Clean Energy, Charlottesville, VA, USA

Riwal Plougonven Ecole Polytechnique, Laboratoire de Météorologie Dynamique, Palaiseau, France

P. Alex Reinecke Naval Research Laboratory, Marine Meteorology Division, Monterey, CA, USA

Ulrich Schumann Deutsches Zentrum für Luft- und Raumfahrt (DLR), Institut für Physik der Atmosphäre, Oberpfaffenhofen, Germany

Robert Sharman Research Applications Laboratory, National Center for Atmospheric Research, Boulder, CO, USA

Zbigniew Sorbjan Department of Physics, Polish Academy of Sciences, Institute of Geophysics, Warsaw, Poland and Marquette University, Milwaukee, WI, USA

Jesse A. Sparks NOAA/National Weather Service, Kansas City, MO, USA

Banavar Sridhar NASA Ames Research Center, Moffett Field, CA, USA

Melissa Thomas Meteorology Department #091, Delta Air Lines, Atlanta, GA, USA

Stanley B. Trier Mesoscale and Microscale Meteorology Laboratory, National Center for Atmospheric Research, Boulder, CO, USA

Patrick Sergej Vrancken Deutsches Zentrum für Luft- und Raumfahrt (DLR), Institut für Physik der Atmosphäre, Oberpfaffenhofen, Germany

John K. Williams Research Applications Laboratory, National Center for Atmospheric Research, Boulder, CO, USA

The Weather Company, an IBM Business, Andover, MA, USA

Paul D. Williams Department of Meteorology, University of Reading, Reading, UK

Emily N. Wilson Meteorology Department #091, Delta Air Lines, Atlanta, GA, USA

Fuqing Zhang Department of Meteorology, Pennsylvania State University, University Park, PA, USA

Part I
Background

Chapter 1
Nature of Aviation Turbulence

Robert Sharman

Abstract In-flight encounters with turbulence are a well-known hazard to aviation that is responsible for numerous injuries each year, with occasional fatalities and structural damage. Not only are turbulence encounters a safety issue, they also result in millions of dollars in operational costs to airlines and may cause schedule delays and air traffic management problems. For these reasons, pilots, dispatchers, and air traffic controllers attempt to avoid turbulence wherever possible. Despite these motivations, the accuracy of detection and forecasting turbulence for aviation applications is insufficient for acceptable levels of avoidance. This chapter reviews the fundamental problems associated with understanding the properties of aviation-scale turbulence and provides an overview of the advances that are being made in the areas of detection and forecasting. References are included to other chapters in this volume which provide more detail on particular subjects.

1.1 Introduction

Encounters with significant turbulence, or simply "bumpiness" in flight (Lester 1993), are a major concern not only for passenger comfort but also for safe, efficient, and cost-effective aircraft operations. The magnitude of aircraft bumpiness depends on the magnitude and size of the encountered atmospheric turbulent eddies as well as the aircraft response to those eddies. Although there is a large spectrum of eddy sizes in the atmosphere, from macroscale to microscale, aircraft bumpiness is felt mainly for a range of eddy sizes between about 100 m and 1 km; larger eddies cause only slow variations in the flight path while the effect of very small eddies is integrated over the surface of the aircraft. Within this critical range of eddy sizes, the aircraft response to turbulence decreases with wing loading (=W/S, where W is the aircraft weight and S is the wing span) and increases with

R. Sharman (✉)
Research Applications Laboratory, National Center for Atmospheric Research, Boulder, CO, USA
e-mail: sharman@ucar.edu

R. Sharman, T. Lane (eds.), *Aviation Turbulence*,
DOI 10.1007/978-3-319-23630-8_1

altitude and airspeed (e.g., Hoblit (1988), see also Eq. (1.2) below). Since this size range is a subset of the turbulence spectrum that is limited to that which affects aircraft in flight, these turbulence scales are often qualified as "aviation turbulence" or "aviation-scale turbulence."

Some idea of the impact of turbulence on aviation can be obtained through examination of significant turbulence encounters recorded in U. S. National Transportation Safety Board (NTSB) and Federal Aviation Administration (FAA) databases. These databases record "accidents" when "any person suffers death or serious injury or in which the aircraft receives substantial damage." The terms "serious injury" and "substantial damage" are quite narrow in scope (http://www.gpo.gov/fdsys/pkg/CFR-2011-title49-vol7/pdf/CFR-2011-title49-vol7-subtitleB-chapVIII.pdf), and thus many injuries incurred or aircraft damaged may not be officially reported, and termed an "incident," or in many cases not reported at all. According to various analyses of NTSB and FAA databases, for the cases reported, turbulence accounts for about 70 % of all weather-related reported accidents and incidents for commercial aircraft (Part 121) and is the leading cause of injuries to airline passengers and flight attendants with most of these occurring at cruise altitudes even though the seat-belt sign was illuminated. Still, the total number of turbulence events reported is relatively low: according to the FAA (https://www.faa.gov/passengers/fly_safe/turbulence/) from 1980 through 2008, for Part 121 aircraft, there were 234 reported events with 298 serious injuries (including three fatalities). Damage to aircraft can also occur; for example, for smaller general aviation (GA, Part 91) aircraft, significant turbulence encounters lead to three in-flight breakups with seven fatalities over the time period 2000–2013. These numbers are for U. S. airspace; globally of course the numbers are higher. Some examples of commercial aircraft damage due to turbulence encounters are shown in Fig. 1.1. Costs to the airlines are difficult to establish, but one major U.S. air carrier estimated that it pays out "tens of millions" of dollars for customer injuries, and loses about 7000 days in employee injury-related disabilities per year.

Turbulence also impacts air traffic controller workload, since in a known turbulence region, pilots seek deviations, negatively affecting airspace capacity and efficiency. This in turn disrupts schedules of air crews and passengers and uses extra fuel. Finally, it must be admitted that passengers generally do not enjoy a turbulent ride and this can lead to the impression that air travel is unsafe.

The remainder of this chapter provides some background into the nature of atmospheric aviation-scale turbulence and highlights some of the difficulties associated with understanding its properties and observing and forecasting turbulence that readers should be aware of to fully appreciate the following chapters. Some of this information can also be found in previous reviews of the subject (Pao and Goldburg 1969; Burnham 1970; Dutton 1971; Vinnichenko et al. 1980; Camp and Frost 1987; Bedard et al. 1993; Lester 1993; Fedorovich et al. 2004; Lane et al. 2012; Sharman et al. 2012b; Ellrod et al. 2015).

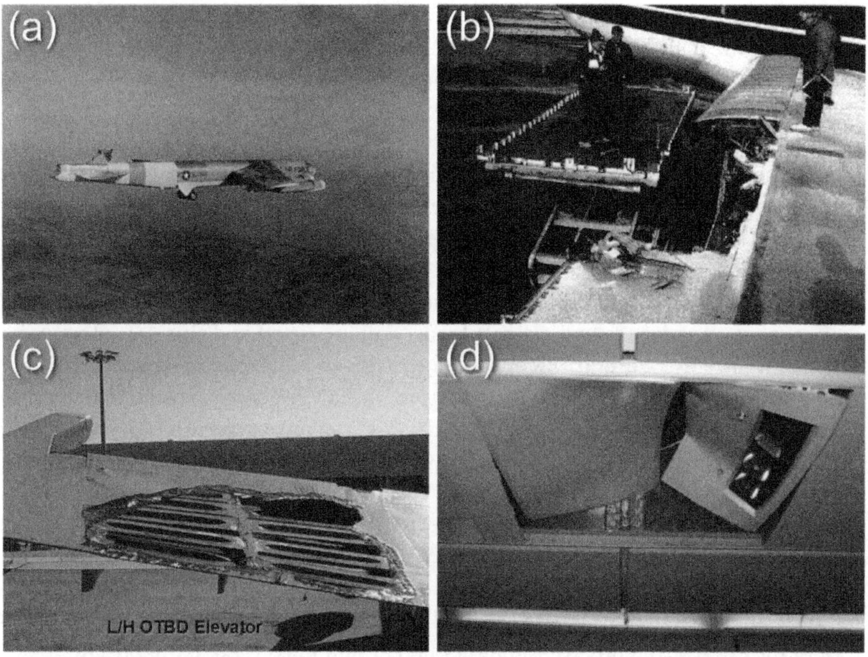

Fig. 1.1 Examples of aircraft damage due to turbulence. (**a**) USAF B-52H tail damage caused by an encounter with severe mountain wave turbulence on January 10, 1964, photo from White Eagle Aerospace History Blog. (**b**) In-flight engine separation of Japan Airlines 747-121 that occurred in extreme turbulence after takeoff from Anchorage AK, March 31, 1993, photo from www.flightsafety.org. (**c**) MD-11 damage to left outboard elevator upper skin due to encounter with severe in-cloud turbulence over the Pacific in January 2004. (**d**) Result of unbuckled passenger damage to overhead in a B737 during severe turbulence event over Korea in April 2006

1.2 Turbulence Intensity Metrics

In engineering applications, transition to turbulence is often identified by exceedance of a critical value of the Reynolds number, $Re = UL/\nu$, where U is a characteristic velocity, L a characteristic length, and ν the kinematic viscosity. In laboratory applications, this critical value is a few thousand. However, application of this concept to the free atmosphere is difficult given the uncertainty in defining a meaningful value of *L;* however if we take U and L to be typical of atmospheric eddy sizes then $Re \sim 10^5$ (e.g., Stewart 1969) which is large enough to suggest that the atmosphere should be almost always turbulent. In the free atmosphere (above the planetary boundary layer) however, the atmosphere is usually stably stratified, meaning that vertically displaced parcels of air will return to their initial position. This has the effect of suppressing turbulent eddies once initiated and therefore of reducing turbulence levels. However stable stratification allows for the production of gravity waves, which may break and lead to turbulence. The degree of stratification is usually quantified in terms of the rate of change of potential temperature θ with height z and for statically stable, neutral, and unstable conditions $\partial\theta/\partial z$ is positive, zero, and negative, respectively (for more details see Chap. 3).

Regardless, the intensity of the turbulent eddies within the range of eddy sizes that affect aircraft is usually small enough to be perceived as "smooth." However, if the eddies within this size range are energetic enough, some level of aircraft bumpiness is perceived by aircraft occupants. Historically, the magnitude of perceived bumpiness is described as "light," "moderate," "severe," and "extreme," but these are obviously aircraft dependent. These terms have common definitions based on occupant experiences and aircraft loads, and are listed in Table 1.1, along with two commonly used aircraft-independent metrics of atmospheric turbulence intensity. From a pilot's perspective, moderate turbulence may require changes in altitude, attitude, or airspeed, but the aircraft remains in positive control at all times. Severe or extreme turbulence causes large, abrupt changes in these parameters, which may cause momentary loss of control. Fortunately, most encounters with moderate or severe turbulence are very short in duration, such as in the example trace shown in Fig. 1.2.

For aviation purposes, the magnitude of atmospheric turbulent eddies by international agreement (ICAO 2013) should be described quantitatively by the cube-root of the energy (or eddy) dissipation rate per unit mass (here termed EDR $= \varepsilon^{1/3}$, $\mathrm{m}^{2/3}\,\mathrm{s}^{-1}$). Physically ε represents the conversion by viscosity of kinetic energy into heat (cf. Eq. 1.4) by the smallest eddies, but $\varepsilon^{1/3}$ is related to the root-mean-square of aircraft vertical acceleration and aircraft loads (see Eq. (1.1) below) and therefore for aviation purposes is the more useful quantity. For aircraft operations $\varepsilon^{1/3}$ can be inferred from the vertical wind velocity (w) variance or aircraft vertical acceleration variance. The EDR metric was originally suggested by MacCready (1964) and must be estimated from time series of vertical wind or acceleration together with assumptions about the spectral nature of the turbulence (see Chap. 5). In the U.S., algorithms to estimate EDR have been implemented on several commercial air carriers and the data is automatically downlinked, removing the pilot from the reporting loop (Sharman et al. 2014). In Table 1.1 EDR thresholds (peak value over 1 min) are given for a medium-weight category transport aircraft (maximum takeoff 15,500–300,000 lbs) flying under typical cruise conditions. Thresholds would be expected to be smaller for light aircraft, and higher for heavy aircraft. Two EDR threshold values are provided in Table 1.1, the first entry is the reference value, and the second is the median value obtained in a recent study by Sharman et al. (2014) where pilot reports of turbulence (PIREPS) were compared to EDR values from the same aircraft.

The relation between EDR and aircraft root-mean-square (RMS) normal acceleration σ_g is given by (e.g., MacCready 1964; Cornman et al. 1995)

$$\sigma_g = V_{\mathrm{T}}^{1/3}\left[\int |A(f)|^2 S_w(f)\ df\right]^{1/2} \varepsilon_w^{1/3} \tag{1.1}$$

Table 1.1 Turbulence reporting criteria and approximate atmospheric turbulence levels

	Aircraft-dependent measures				Atmospheric measures	
Description	Aircraft reaction[a]	Reaction inside aircraft[a]	Peak normal accel (g)[b]	RMS normal accel (g)[c]	U_{de} (m s^{-1})[d]	$\varepsilon^{1/3}$ (m$^{2/3}$ s^{-1})[e]
Light	Turbulence that momentarily causes slight, erratic changes in altitude and/or attitude (pitch, roll, yaw). or Turbulence that causes slight, rapid, and somewhat rhythmic bumpiness without appreciable changes in altitude or attitude ("chop").	Occupants may feel a slight strain against seat belts or shoulder straps. Unsecured objects may be displaced slightly. Food service may be conducted and little or no difficulty is encountered in walking.	0.2–0.5	0.1–0.2	2.0–4.5	0.1–0.39 (0.1–0.21)
Moderate	Turbulence that is similar to light turbulence but of greater intensity. Changes in altitude and/or attitude occur but the aircraft remains in positive control at all times. It usually causes variations in indicated airspeed. or Turbulence that is similar to light chop but of greater intensity. It causes rapid bumps or jolts without appreciable changes in aircraft altitude or attitude.	Occupants feel definite strains against seat belts or shoulder straps. Unsecured objects are dislodged. Food service and walking are difficult.	0.5–1.0	0.2–0.3	4.5–9.0	0.40–0.69 (0.22–0.47)
Severe	Turbulence that causes large, abrupt changes in altitude and/or attitude. It usually causes large variations in indicated airspeed. Aircraft may be momentarily out of control.	Occupants are forced violently against seat belts or shoulder straps. Unsecured objects are tossed about. Food Service and walking are impossible.	1.0–2.0	0.3–0.6	≥9.0	≥0.70 (≥0.48)
Extreme	Turbulence in which the aircraft is violently tossed about and is practically impossible to control. It may cause structural damage.	Not defined	≥2.0	≥0.6		

[a]FAA (2014), Table 7-1-10
[b]E.g., Lee et al. (1984), Lester (1993), Fig. 1-8
[c]Bowles et al. (2009) based on running 5-s window
[d]Gill (2014)
[e]Peak over 1 min. ICAO (2013) definitions, updated values by Sharman et al. (2014) in parentheses

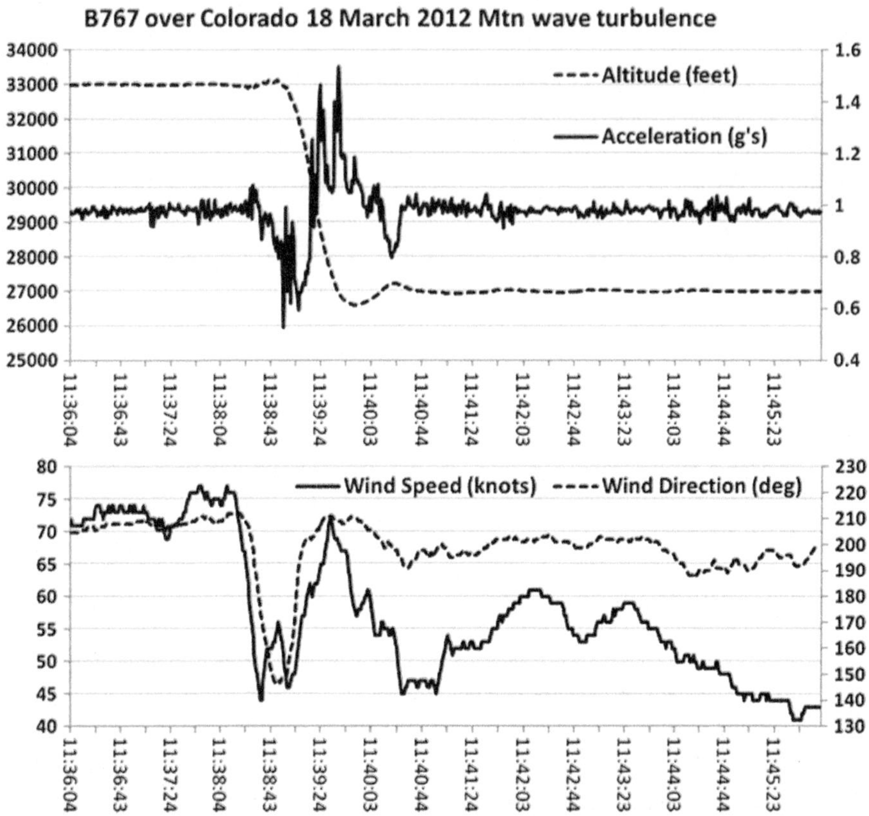

Fig. 1.2 Flight data recorder traces of vertical acceleration and altitude (*upper panel*) and wind speed and direction (*lower panel*) of a B767 encounter with mountain wave turbulence over the Colorado Rockies on 18 March 2012. Time in hh:mm:ss is on the abscissa

where σ_g is the RMS vertical acceleration experienced by the aircraft, $A(f)$ is the aircraft specific response function at frequency f and true airspeed V_T, and S_w is a specified atmospheric spectral model for the vertical wind w with a unit ε_w. Note that for a given atmosphere turbulence intensity ε_w, the aircraft load depends on V_T, so slowing down diminishes the bumpiness experienced by the aircraft.

Also listed in Table 1.1 is another aircraft-independent intensity metric commonly referred to as the Maximum-Derived Equivalent Vertical Gust Velocity (DEVG) or $U_{de} = 10$ DEVG currently implemented on some international B747-400s. U_{de} is defined in, e.g., Hoblit 1988, Eq. 2.1 and can be approximated as (Gill 2014)

$$U_{de} \approx C(M, z) \frac{|\Delta g| M}{V_e} \tag{1.2}$$

where $|\Delta g|$ is the peak modulus value of the deviation of aircraft vertical acceleration from $1\,g$, M is the total aircraft mass, V_e is the equivalent air speed $(= (\rho/\rho_0)^{1/2} V_T$ where ρ is the air density at the aircraft altitude, ρ_0 is the air density at sea level, and V_T is the true air speed), and $C(M, z)$, where z is the aircraft altitude, is an aircraft specific factor. This form has been used by Gill (2014) and others for verification of global turbulence forecasts. However, since U_{de} is really just a gust-loads transfer factor it is not a direct atmospheric turbulence intensity metric and is therefore arguably less useful than EDR for atmospheric turbulence detection, nowcasting and forecasting applications.

Other atmospheric turbulence metrics can be used, such as RMS vertical or horizontal wind speed or turbulent kinetic eddy energy, but these have not found extensive operational use, partly because of their strong dependence on the averaging interval (e.g., Strauss et al. 2015).

1.3 Aviation Turbulence Sources and Climatology

Turbulence or eddies that affect aircraft are initially created by a number of various large-scale forcing mechanisms, and the resulting turbulence is often classified according to its source. Common sources of aviation-scale turbulence are listed below. An easy-to-read more detailed description of these is provided in Lester (1993).

(1) Convective turbulence. This source of turbulence is associated with strong updrafts and downdrafts in dry thermals or convective clouds. Turbulence associated with convective clouds (either in-cloud or near-cloud) is collectively termed convectively induced turbulence or CIT, or to better distinguish in-cloud from out-of-cloud events, turbulence in the clear air just outside the visible cloud boundaries is sometimes referred to as near-cloud turbulence (NCT) (Lane et al. 2012). NCT is commonly caused by convectively induced gravity waves propagating and breaking in the clear air away from the cloud, although other NCT sources are possible, e.g., enhanced shears above the cloud top leading to turbulence (e.g., Lester 1993), cloud interfacial instabilities (e.g., Grabowski and Clark 1991; Lane et al. 2003), and instabilities induced by sublimation of ice underneath anvils (Kudo 2013) and possibly obstacle effects (e.g., Bedard et al. 1993). In all events, since the source of turbulence is directly related to convective cloud structures, the lifetime of CIT is relatively short—typically only a few minutes, but occasionally for mesoscale convective systems (MCS) may be significantly longer. Although data within convective clouds is limited since most pilots try to avoid it, statistical evaluations performed by Honomichl et al. (2013) of penetrations by an armored T-28 over several research programs showed the exceedance frequency of EDRs $> 0.23\ m^{2/3}\ s^{-1}$ was about 2 %, with a maximum occurrence in the upper third of the cloud depth (see also Chaps. 7 and 8).

(2) Low-level turbulence. Low-level turbulence (LLT) can be due to convection and strong winds associated with a surface frontal passage, dry thermals during the daytime over hot surfaces, and mechanical forcing associated with flow over surface obstacles (mountains and hills, trees, buildings) or rough flat terrain. Its intensity is determined by low-level wind speed, stability, height above the ground, and terrain roughness (quantified by the value of the surface roughness parameter z_0). Lee side eddy separation (or "rotor streaming," see, e.g., Gossard and Hooke 1975, Fig. 58.5) can lead to regions of intense LLT. This source of turbulence is typically within the planetary boundary layer and can be a significant safety hazard during takeoffs and landings, especially for small aircraft.

(3) Mountain wave turbulence. Turbulence associated with the large amplitude gravity waves and breaking gravity waves (which can extend to very high levels, through the troposphere, and into the stratosphere and beyond) above and behind mountainous terrain in stably stratified flow is often termed mountain wave turbulence (MWT). It includes gravity (lee) waves at all altitudes and low level effects such as hydraulic jumps and rotors (e.g., Lester 1993). The lee waves themselves have different characteristics depending on the upstream conditions and the modulating effects of the terrain. For conditions favoring vertical propagation of wave energy, amplification and breaking may occur aloft above the mountain; in other situations where wave energy is vertically trapped, wave breaking can occur at lower altitudes downstream from the mountain. Wave amplification and breaking also depends on the character of the underlying terrain, e.g., wave amplitudes tend to be larger over quasi-two-dimensional terrain than over isolated three-dimensional terrain. The slope of the terrain is also important, particularly on the downwind side (Foldvik and Wurtele 1967; Lilly and Klemp 1979).

(4) Clear-air turbulence. Turbulence associated with enhanced wind shears and reduced stabilities in the vicinity of jet streams, the tropopause, and upper-level fronts, is usually termed clear-air turbulence or CAT because it often occurs in clear air (although sometimes in high stratiform clouds). The precise definition of CAT from the National Committee for Clear Air Turbulence is "all turbulence in the free atmosphere of interest in aerospace operations that is not adjacent to visible convective activity (this includes turbulence found in cirrus clouds not in or adjacent to visible convective activity)." This form of turbulence tends to be localized in space and time with the vertical dimensions typically much smaller than the horizontal dimensions and typically occurs in the upper troposphere and lower stratosphere (UTLS). As originally found by Bannon (1952), CAT is favored on the cyclonic side of and above and below the jet stream core. The environmental conditions favorable for CAT can induce Kelvin–Helmholtz instability (KHI) (e.g., Dutton and Panofsky 1970), but gravity wave and inertia gravity wave breaking also contributes to UTLS turbulence (e.g., Lane et al. 2004; Koch et al. 2005; Sharman et al. 2012a; Chaps. 16–20).

(5) Aircraft-induced "turbulence" generated by trailing vortex wakes. This is mainly a concern near airports when a lighter aircraft trails a leading heavier aircraft (e.g., Gerz et al. 2002), but a significant number of encounters can occur at upper levels as well (Schumann and Sharman 2015).

Turbulence climatologies can be constructed and related to these sources through analysis of the temporal and spatial distribution of reported turbulence either from PIREPs, see, e.g., Wolff and Sharman (2008) or in situ estimates of EDR (Sharman et al. 2014); and also by running climate models over long periods of time and analyzing model output fields that are known to be conducive to turbulence (Jaeger and Sprenger 2007; Williams and Joshi 2013; also Chap. 23). Both methods have advantages and disadvantages. Climatologies that rely on observations suffer from their nonuniform spatial and temporal sampling, while climate model-based climatologies depend on accurate representation of the turbulence-related physics. PIREP-derived climatologies over the USA from Wolff and Sharman (2008) indicate increased incidence of Moderate-or-Greater (MoG) turbulence over four major areas, one over the northern middle sections of the country, a second over the eastern seaboard, a third over the Florida peninsula, and a fourth over the Rockies and Sierra Nevada mountain ranges in the West. They attributed the maximum over the eastern seaboard as due to primarily jet stream-related turbulence (CAT), while the maxima over Florida and to some extent the middle western states is associated with the increased frequency of convective storms there (CIT), especially in Summer. The maxima over the Rocky Mountains, Sierra Nevada, and to some extent the Cascades was attributed to MWT. Globally, Jaeger and Sprenger (2007) find upper-level turbulence maxima over the western part of the North Atlantic and North Pacific, the Himalayas, central Europe, and eastern China. Both studies find maxima occur in the Winter due to CAT, while Wolff and Sharman (2008) find a secondary maximum over the USA in the Spring due to CIT. Vertically, both studies find a maximum near the tropopause, while the Wolff and Sharman study also show a maximum near the ground. Williams and Joshi (2013); also Chap. 23 provide evidence that turbulence climatologies may be affected by climate change.

All previous climatological studies show the frequency of MoG encounters is quite small. In situ EDR measurements (Sharman et al. 2014), which don't rely on pilots to report the turbulence, also show that encounters with elevated intensities of turbulence are quite rare. But what is not clear is how much this is due to avoidance bias, since pilots will generally deviate from known regions of severe or even moderate turbulence. This avoidance bias is difficult to estimate, but some estimates are provided in Sharman et al. (2014).

To better quantify turbulence frequency in the free atmosphere, Sharman et al. (2014) constructed probability density functions (PDFs) of recorded in situ peak EDR values over 1 min and found a log-normal distribution fits the data quite well

$$\mathrm{PDF}(x) = \frac{1}{x\left(2\pi\sigma_{\ln x}^2\right)^{1/2}} \exp\left\{ -\frac{\left[\ln(x/\overline{x})\right]^2}{2\sigma_{\ln x}^2} \right\} \tag{1.3a}$$

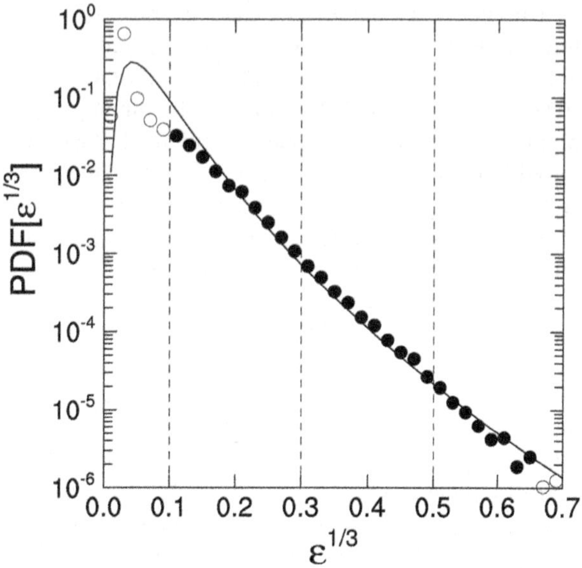

Fig. 1.3 PDF of peak EDR ($\varepsilon^{1/3}$, *circles*) and log-normal fit (*curve*) for DAL 737 data for 2009–2012, for aircraft altitudes >20,000 ft (6.1 km). The *closed circles* indicate data that were used in the fits; the *open circles* were excluded from the fits. The *dashed vertical lines* indicate the ICAO (2001) recommended EDR thresholds values of (0.1, 0.3, 0.5) for ("light", "moderate", "severe"), respectively. From Sharman et al. (2014). © American Meteorological Society. Reprinted with permission

$$\ln(\overline{x}) = \langle \ln(x) \rangle$$
$$\sigma^2_{\ln x} = \left\langle [\ln(x/\overline{x})]^2 \right\rangle \tag{1.3b}$$

where $x =$ peak EDR and the angle brackets indicate the ensemble average. The - log-normal fit to all Delta Air Lines B737 *peak* EDR (over a 1-min sample interval) data for aircraft altitudes >20,000 ft (6.1 km) gave $\langle \ln(\varepsilon^{1/3}) \rangle = -2.85$ and $\sigma_{\ln\varepsilon^{1/3}} = 0.571$, or $<\varepsilon^{1/3}> = 0.068$ m$^{2/3}$ s^{-1} and $<\varepsilon> = 8.3 \times 10^{-4}$ m^2 s^{-3}. Figure 1.3 from Sharman et al. (2014) shows the data and log normal fit. Note the data are available only in linear 0.02 m$^{2/3}$ s^{-1} EDR bins so the abscissa is on a linear scale. According to this, the frequency of occurrence of atmospheric turbulence with EDR > 0.5 m$^{2/3}$ s^{-1} (severe range for medium weight class aircraft, Table 1.1) is only ~10^{-5}. Accurate estimates of the PDF of EDR are important for forecasting applications to ensure that the statistics of the forecasts match the statistics of the observed frequencies.

1.4 Turbulence Characterization

Turbulence is both three-dimensional and nonlinear, making it very difficult to describe; in fact it has been termed "the chief outstanding difficulty" in hydrodynamics (Lamb 1945 and others). Further, the free atmosphere is a stably stratified vertically sheared media, with embedded gravity waves and breaking gravity waves (e.g., Stewart 1969; Gage et al. 2004; Sharman et al. 2012b), and this compounds the theoretical difficulties. Also, stable or laminar waves of wave length short enough to cause significant aircraft response are often reported as "turbulence chop" (Lester 1993; Table 1.1), which extends the definition of aircraft-scale turbulence to explicitly include waves.

As previously stated, using the *Re* as a criterion for the occurrence of instability and turbulence in the atmosphere is probably meaningless. Given these realities, a more useful indicator of instability and turbulence is the Richardson number (Ri), which derives from an turbulent kinetic energy conservation equation for turbulence. For homogeneous flow, this can be written in its simplest form, neglecting transport terms, as (e.g., Schumann and Gerz 1995)

$$\frac{De}{Dt} = P - B - \varepsilon \tag{1.4}$$

where $e = 0.5\left(\overline{u^2} + \overline{v^2} + \overline{w^2}\right)$ is the ensemble averaged (denoted by the overbar) turbulence kinetic energy per unit mass (TKE) of the turbulent velocity components and $\frac{D}{Dt} = \frac{\partial}{\partial t} + \vec{v} \cdot \vec{\nabla}$ is the total or Lagrangian time derivative. In Eq. (1.4) ε is the dissipation of kinetic energy into heat (always positive), P is the shear production, and B is the buoyancy destruction, which are usually assumed to be related to the mean flow gradients through the so-called K-closure assumption,

$$P = -\left(\overline{uw}\frac{\partial \overline{u}}{\partial z} + \overline{vw}\frac{\partial \overline{v}}{\partial z}\right) = K_{\mathrm{M}}S_{\mathrm{V}}^2, B = -\frac{g}{\overline{\theta}}\overline{\theta w} = K_{\mathrm{H}}N^2 \tag{1.5}$$

where θ is the potential temperature, g is the gravitational acceleration, $N^2 = \frac{g}{\theta}\frac{\partial \theta}{\partial z}$ is the Brunt-Väisälä or buoyancy frequency (a measure of the static stability; note the inclusion of moisture considerably complicates this expression, Durran and Klemp 1982), and S_{v} is the vertical shear of the mean horizontal wind. K_{H} and K_{M} are the eddy coefficients of heat and momentum transfer. The ratio of B to P is termed the flux Richardson number $\mathrm{Ri_f}$

$$\mathrm{Ri_f} = \frac{B}{P} = \frac{K_{\mathrm{H}}}{K_{\mathrm{M}}}\frac{N^2}{S_{\mathrm{V}}^2} = \frac{K_{\mathrm{H}}}{K_{\mathrm{M}}}\mathrm{Ri_g} = \frac{1}{\mathrm{Pr_t}}\mathrm{Ri_g} \tag{1.6}$$

$$\mathrm{Ri_g} = N^2/S_{\mathrm{V}}^2 \tag{1.7}$$

where the turbulent Prandtl number, $\mathrm{Pr_t}$ denotes the ratio of K_{H} to K_{M}. Thus, the usual gradient Richardson number $\mathrm{Ri_g}$ is only an approximation to the ratio of buoyancy to shear production sources. However in practice, since K_{H} and K_{M} are difficult to compute, and their ratio $\mathrm{Pr_t}$ is typically O(1), $\mathrm{Ri_g}$ is simply denoted Ri.

Conceptually, to see the effect of the Ri on the production of e, Eq. (1.4) can be rewritten using K closure and assuming $\mathrm{Pr_t} = 1$

$$\frac{De}{Dt} = K_{\mathrm{M}} S_{\mathrm{V}}^2 (1 - \mathrm{Ri}) - \varepsilon \tag{1.8}$$

The effect of Ri is now clearly to increase e only if $\mathrm{Ri} < 1$. This includes convectively unstable conditions where $N^2 < 0$. For convectively stable situations, values of $\mathrm{Ri} > 1$ would decrease e. Of course other terms (advective terms and neglected terms) influence the sign of the tendency of e; however, more complete treatments of this show that $\mathrm{Ri} > 1$ are sufficient to guarantee stability (e.g., Miles 1986). In all events, the exact critical value of Ri is not as important as what the number represents qualitatively and that is the competition between shear which favors instability and buoyancy which favors stability. Thus in the atmosphere, large-scale factors that increase the shear and/or decrease the stability favor the formation of turbulence. These are typically operating at scales larger than those felt by aircraft as bumpiness; thus, aircraft-scale turbulence owes its existence to injection of energy at large scales, which works its way down to aircraft scales through a downscale cascade of larger more energetic eddies to smaller less energetic ones.

Understanding this downscale cascade process is particularly relevant to turbulence forecasting since large-scale numerical weather prediction (NWP) models can only capture the larger scales where energy is injected. One useful description of the downscale cascade process is provided under the admittedly restrictive assumptions of statistical homogeneity (i.e., that the statistical properties of the flow are independent of the location) and statistical isotropy (which requires that the statistics are also invariant under rotations and reflections of the coordinate system). These simplifications allow identification of a region where only inertial forces are acting to transfer energy from larger scales to smaller scales commonly called the inertial subrange. In this range of scales, the spectrum of eddy kinetic energy falls off as $k^{-5/3}$, where k is the horizontal wavenumber (2π/wavelength). In fact, numerous measurements have shown this representation is a very good approximation to the actual atmospheric behavior in both the planetary boundary layer (PBL) and the free atmosphere (e.g., Kaimal and Finnigan 1994; Dutton 1971). Typically, the inertial subrange includes eddy sizes from a few hundred meters to a length scale where viscous effects begin to become important, the so-called inner or Kolmogorov scale $\eta = (\nu^3/\varepsilon)^{1/4}$ which varies with height, but is typically much less than 1 m, i.e., in ranges that are not of interest to aviation turbulence.

In the inertial subrange, the longitudinal one-dimensional one-sided power spectral density level is related to ε through the so-called Kolmogorov spectral form

$$S_u(k) = C_{\mathrm{K}} \varepsilon^{2/3} k^{-5/3} \tag{1.9}$$

where C_{K} is the Kolmogorov constant equal to 0.5–0.6, where u is the longitudinal velocity vector; for aircraft observations, this is the component in the direction of travel. Thus, if the spectrum S is computed over some range of wavenumbers k, $\varepsilon^{2/3}$ may be inferred. Within the constraints of the simplifications used, the transverse spectra $S_v(k)$ and $S_w(k)$ have the same form but theoretically have a level 4/3 higher than the longitudinal spectrum Eq. (1.9). Figure 1.4 from Frehlich and Sharman (2004) shows an example of average longitudinal and transverse velocity and temperature spectra obtained in the mid-troposphere from a research aircraft flight. In this example, the $k^{-5/3}$ slope is a good fit for wavelengths from about 30 m to 6 km for all spectra. At scales smaller than about 30 m, some deviation from the $k^{-5/3}$ slope is apparent due to measuring difficulties and aliasing effects. The w spectrum flattens off at the outer scale (which has various definitions Klipp 2014;

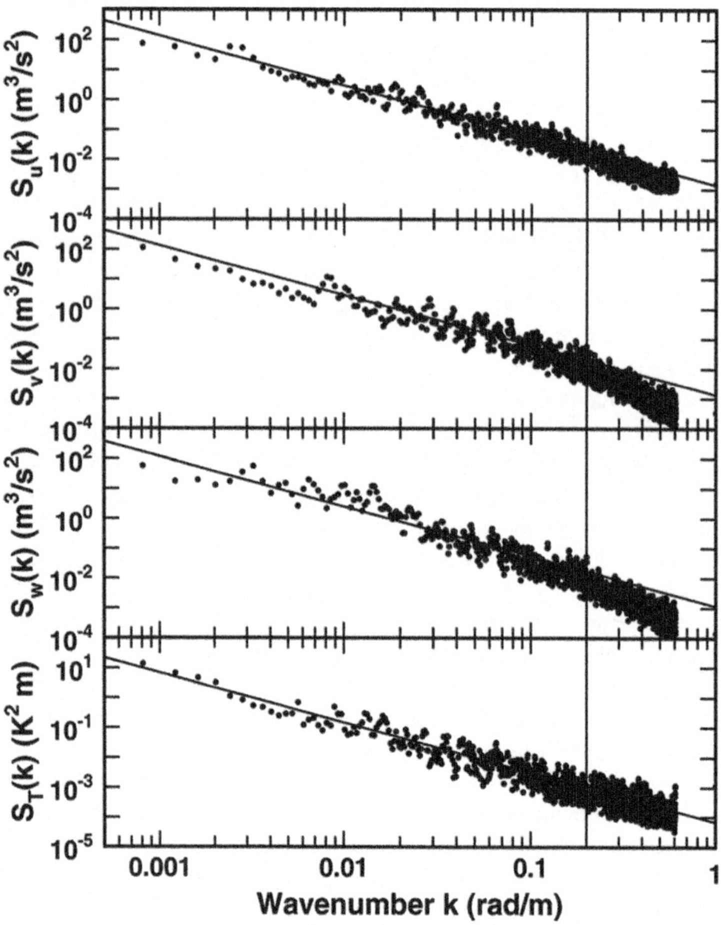

Fig. 1.4 Longitudinal, transverse, vertical velocity, and temperature spectra, denoted $S_u(k)$, $S_y(k)$, $S_w(k)$, and $S_T(k)$, respectively, as derived from INDOEX field campaign data and the best-fit $k^{-5/3}$ line over the interval $k = (0.0005\text{–}0.2\ \text{rad m}^{-1})$. Each point is the average of 40 nonoverlapping spectra produced from a particularly long (600 km) leg of measurements from the NCAR EC-130 aircraft flying above the PBL at 4806-m elevation as part of the Indian Ocean Experiment (INDOEX). From Frehlich and Sharman (2004). © American Meteorological Society. Reprinted with permission

Joseph et al. 2004; Wilson 2004) but is roughly speaking the scale of the largest eddies), and it is therefore usual to modify the Kolmogorov spectral form Eq. (1.9) to include this rolloff. One common spectral model used in aircraft response studies that includes this behavior is the von Kármán form (e.g., Murrow 1987; Sharman et al. 2014)

$$S_w(k) = \frac{9}{110}\alpha\varepsilon^{2/3}L^{5/3}\frac{\left(1+\frac{8}{3}L^2k^2\right)}{\left(1+L^2k^2\right)^{11/6}} \tag{1.10}$$

where L is the assumed outer length scale and $\alpha \approx 1.5$–1.7 is an empirical constant related to C_K (e.g., Kristensen and Lenschow 1987). Note that the horizontal velocity and temperature spectra do not show this rolloff (Fig. 1.4), and in fact the $k^{-5/3}$ spectral behavior for these fields seems to extend to very large scales (a few hundreds of km), consistent with analyses of the Global Atmospheric Sampling Program (GASP) (e.g., Nastrom and Gage 1985) and Measurements of Ozone and Water Vapor by In-Service Airbus Aircraft (MOZAIC) data (e.g., Lindborg 1999). This behavior seems to be duplicated in general circulation and NWP models as well (Koshyk et al. 1999; Skamarock 2004; Frehlich and Sharman 2004). However, these larger scales are outside the inertia subrange, and the explanation of this property of the atmosphere continues to be the subject of research investigations (e.g., Lindborg 2007; Lovejoy et al. 2009).

Although the Kolmogorov form Eq. (1.9) seems to be a good representation of the spatial structure of atmospheric turbulence on average, any individual traverse through turbulence, especially through a discrete event, may not obey Eq. (1.9) or (1.10) very well. An example of a mountain wave turbulence encounter is shown in Fig. 1.5, and although the overall shape shows roughly the $k^{-5/3}$ behavior, it contains quite a variety of energetic wavelengths. Further, the turbulence is often not isotropic (e.g., Lilly et al. 1974; Schumann et al. 1995; Lane and Sharman 2014), and therefore ε derived from longitudinal or transverse horizontal fluctuations may not equal ε_w. In practice, there are certain difficulties in computing reliable spectra and hence ε, due in part to the need for detrending and tapering, and an alternative may be to compute second- or third-order spatial structure functions instead (e.g., Lindborg 1999; Frehlich and Sharman 2004; Lu and Koch 2008; Wroblewski et al. 2010).

For aviation applications, the horizontal spatial structure, whether described in terms of spectra or structure functions, is most relevant since the aircraft are usually moving horizontally through the turbulent atmosphere. In this case, when computing spectra, a time series is obtained and temporal spectra are computed, which can be converted to spatial spectra by invoking Taylor's frozen turbulence hypothesis (this is well satisfied given the high airspeeds of the aircraft relative to the fluctuation velocities, e.g., Bacmeister et al. 1996, Wyngaard and Clifford 1977). But studies of the vertical turbulent structure of the atmosphere are also of interest since they define the vertical extent of turbulent patches. Measurements from high resolution tethered systems (e.g., Frehlich et al. 2004) or radiosondes (e.g., Clayson

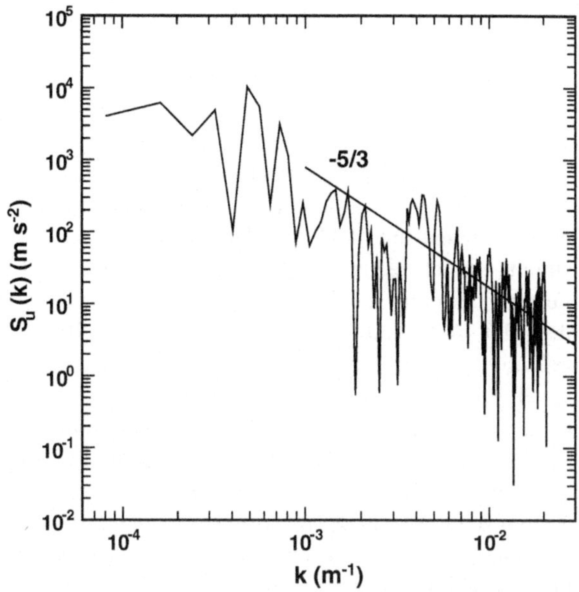

Fig. 1.5 Spectral density of true air speed fluctuations $S_u(k)$ from the flight data recorder derived for a time period of 500 s surrounding a MWT event over Greenland on 25 May 2010. A $k^{-5/3}$ reference line is also shown. From Sharman et al. (2012a). © American Meteorological Society. Reprinted with permission

and Kantha 2008; Haack et al. 2014) show a very rich vertical structure which can have very narrow spikes of turbulence, which probably depends on the local Ri (e.g., Chap. 21). Although some of these are probably too narrow to be felt by the aircraft, commercial aircraft in climbs and descents do show that in the free atmosphere, the vertical extent is much smaller than the horizontal extent, giving rise to pancake-like structures. Typically, the horizontal to vertical aspect ratio is ~100:1 (Vinnichenko et al. 1980; Sharman et al. 2014; Paoli et al. 2014).

1.5 Detection Strategies

PIREPs of turbulence have traditionally been the most common form of in situ turbulence detection. Although quite useful for operational tactical avoidance, PIREPs do have various deficiencies, including subjective assessments of the turbulence experience by the pilot, and spatial and temporal inaccuracies (Sharman et al. 2014). Recently, PIREPs have been augmented by automated in situ EDR reports (Sharman et al. 2014; Chap. 5), and currently several hundred commercial aircraft provide these reports operationally. Reliably estimating EDR from these on-board algorithms does have its difficulties, in part due to the difficulty associated with using Eq. (1.9) or (1.10) in discrete events, and for accelerometer-based methods, the difficulty in obtaining the required aircraft response function (cf. Eq. 1.1). Still, from a forecasting perspective, since it is not practical to predict aircraft-dependent bumpiness, it is only logical to predict atmospheric turbulence levels instead. Since EDR is the ICAO standard for in situ turbulence reports, EDR should be the forecasted turbulence metric of choice, and the in situ EDR reports then provide a consistent verification source for the turbulence forecasts.

Remote detection techniques, either airborne, ground-based, or satellite-based, have been developed, but for the most part are not routinely available to users at large. On-board forward-looking X-band radar algorithms to estimate EDR from spectral width in high reflectivity regions of cloud have been developed (see Chap. 6) and are available on many commercial aircraft. Although this system is useful for real-time CIT avoidance, the on-board radar is limited by power constraints, which leads to significant attenuation in dense cloud and precipitation. Further, currently the data is not transmitted to the ground and is only available to the pilot. A technique to use ground-based NEXRAD (Next-Generation Radar) reflectivity, radial velocity, and spectrum width to estimate EDR has also been developed (termed the NEXRAD Turbulence Detection Algorithm or NTDA; see Chap. 7) and may ultimately become operationally available, providing in-cloud EDR maps over the U. S. airspace. Although radar-based techniques are useful for real-time turbulence assessment and avoidance, their utility is limited spatially to regions of cloud containing sufficiently high reflectivity for reliable EDR estimates and temporally to just a few minutes because of the highly transient nature of CIT. One interesting outcome of NTDA studies is that regions of high reflectivity do not always correlate well with regions of enhanced turbulence; in particular, regions of quite low reflectivity are sometimes associated with moderate or even severe turbulence. However, in-cloud lightning flash density does appear to correlate better with NTDA regions of enhanced turbulence (Chap. 8).

Other ground-based and on-board forward-looking remote sensing techniques have been developed, but for one reason or another have not been practical to implement on a large scale, although these techniques have been invaluable for better understanding the turbulent properties of the atmosphere. The principles behind these are described in Part IV of Pao and Goldburg (1969) and also Chap. 22. Detection by passive infrared and microwave radiation techniques, although once promising, simply does not have adequate resolution to truly detect turbulent fluctuations on scales that affect aircraft. However, recently developed technology using interferometry (Schaffner et al. 2012) may resurrect this detection technique. In any case, microwave temperature profilers (MTP) are useful for retrieval of vertical temperature profiles taken during field campaigns (Haggerty et al. 2014; Mahoney et al. 2009). Ground-based Doppler LIDAR systems have been developed and have been quite successful at identifying turbulence in clear-air regions near the surface (e.g., Frehlich and Cornman 2002; Kühnlein et al. 2013; Chan 2010; Chap. 9), but so far have not had sufficient power to overcome the low aerosol content of the UTLS. Experimental test flights using forward-looking pulsed Doppler LIDAR have successfully monitored MWT ahead of the aircraft (e.g., Hannon et al. 1999), but for operational use, the range and reliability of the instrument would have to be increased substantially, which is not currently cost feasible (Kauffmann 2002). However, research and development in this area continues (Chap. 22). One difficulty with the use of any onboard forward-looking detection device is the fact that longitudinal fluctuations are normally measured, yet the aircraft responds primarily to vertical turbulent fluctuations (Hoblit 1988), and given that the atmosphere may be highly anisotropic at any given time (see e.g., Schumann et al. 1995; Chap. 4) the magnitude of the longitudinal fluctuations may misrepresent the turbulent effect on the aircraft.

Another promising technique that may be useful for aviation applications is the use of high-resolution vertically pointing radars and rawinsondes that can be used to estimate EDR from refractivity measurements, or in the case of radar, also from Doppler spectral width. Currently, these data are not readily available for operational use, but may be in the future, at least for research studies (Love and Geller 2013). The techniques used to derive EDR from the raw data are reviewed in Hocking and Mu (1997), Wilson (2004), and Clayson and Kantha (2008) and require some assumptions to be made about the relation of EDR to the properties actually measured. This introduces the need for some empiricism, although Dehghan et al. (2014) find generally good agreement between radar-derived EDR and EDR estimated from data collected by a coincident research aircraft. Methods to infer turbulence by attaching an accelerometer (Marlton et al. 2015) or a magnetometer (to measure magnetic field fluctuations indicated by turbulence—Harrison and Hogan 2006) to a rawinsonde have been proposed and feasibility studies performed, but so far have not been implemented operationally.

Careful examination of satellite imagery may be useful to infer the presence of CAT, CIT, and MWT. Obviously, the horizontal and temporal resolution of satellite imagery is inadequate to actually detect aviation-scale turbulence, but experience and computer-aided feature identification algorithms may provide a basis for turbulence nowcasts and possible verification of turbulence forecasts. For example, Ellrod (1989) details various infrared, visible, and water vapor satellite image signatures which seem to be well correlated with CAT, including but not limited to, the presence of sharp cloud or moisture edges along the jet stream, transverse cirrus bands on the anticyclonic side of the jet stream axis, billow clouds indicative of KHI, and darkening trends in water vapor imagery. However, the reliability of these indicators is uncertain, they are somewhat human labor intensive to identify, and it is difficult to determine an altitude of occurrence or intensity of the CAT. Wimmers and Moody (2004) have developed an algorithm to infer CAT in the vicinity of tropopause folds. Trapped mountain waves are easily distinguished in satellite imagery by their distinctive banded structure downstream of a mountain ridge (2D) or as "ship wave" (e.g., Sharman 2010) patterns (3D). These may or may not be turbulent, but interference patterns in the waves seem to be indicative of areas of enhanced turbulence (Uhlenbrock et al. 2007). Turbulence can also be inferred in strong leeside downslope wind situations by imagery containing thick cold cirrus plumes with embedded waves and a dark (warm) subsidence zone (föhn gap) in the immediate lee of a large mountain range (e.g., Ellrod 1987). CIT may be inferred from overshooting cloud tops (Bedka et al. 2010; Monette and Sieglaff 2014), gravity wave patterns embedded in the anvil, rapid anvil expansion, but especially banded structures on the outside edge of the anvil (Lenz et al. 2009), which has been confirmed by high-resolution numerical simulations to be related to turbulence (Trier et al. 2010). These banded structures are also present in jet stream-related cirrus (Knox et al. 2010) and may also be turbulent (Kim et al. 2014).

Finally, a more quantitative satellite-based turbulence detection method uses Global Navigation Satellite System (GNSS) occultation data, and seems feasible, but has not yet been implemented on an operational basis (e.g., Cornman et al. 2012).

1.6 Forecasting Methods

Due to the small space and time scales of aviation-scale turbulence, explicit operational forecasting that covers even relatively small regions of the airspace is all but impossible. However, since the energy sources ultimately responsible for the turbulence are at much larger scales, inferences of turbulence likelihood can be made by careful examination of large-scale features known to be conducive to turbulence, especially at upper levels. Some of these large-scale patterns have been reviewed by Hopkins et al. (1977), Lee et al. (1984), and Lester (1993) and in Chaps. 2, 10, 11, and 12. In the past, such patterns were analyzed manually by trained meteorologists at government sponsored weather services or private commercial airlines departments. This has gradually shifted over the years towards the use of automated turbulence prediction algorithms, and there are now only a handful of airline meteorology departments that are actively engaged in turbulence forecasting. Operational turbulence forecasts provided by national and international weather services and the remaining airline meteorology departments now routinely use automated turbulence forecasts as an initial guess which may be modified by a human forecaster to provide the final (or so-called human-over-the-loop) forecast (see Chaps. 10 and 11).

Automated turbulence forecast algorithms are typically derived from large-scale weather forecasts from NWP model output. Almost all these algorithms "diagnose" turbulence by identifying large values in computed horizontal or vertical gradients of different atmospheric state variables (velocity, temperature) from the NWP model output and then threshold these gradients empirically to correspond to light, moderate, and severe turbulence (see reviews by Sharman et al. (2006) and Chap. 12). Because the automated turbulence diagnostics are derived from large-scale NWP models, they implicitly assume a downscale cascade of energy from the larger NWP model resolved scales to the smaller scales that affect aircraft motion as discussed previously in Sect. 1.4. A straightforward approach that should be valid regardless of the source of turbulence would be to use some simplification of the TKE tendency Eq. (1.4), but in practice this has demonstrated only poor performance when compared to observations, especially at upper levels. Therefore, most turbulence prediction algorithms directly or indirectly identify regions where the Ri is relatively small at the larger scales, which through the thermal wind relation can be related to large values of the horizontal temperature gradients. A more physically based approach was introduced by Roach (1970) where an expression for Ri tendency is derived, with the notion that regions where Ri is decreasing over time should also be regions favorable to turbulence. In practice, this has not been a very useful turbulence diagnostic because of the sensitivity to vertical differences, but it does provide a conceptual basis for understanding the origin of many turbulence diagnostics. By taking the total time derivative *D*/*Dt* of *Ri* in Eq. (1.7) and using subscript notation for the derivatives

$$\frac{1}{Ri}\frac{DRi}{Dt} = \frac{1}{Ri}\frac{D}{Dt}\left(\frac{g\theta_z/\bar{\theta}}{u_z^2+v_z^2}\right) = \frac{1}{\theta_z}\frac{D\theta_z}{Dt} - \frac{2}{u_z^2+v_z^2}\left(u_z\frac{Du_z}{Dt} + v_z\frac{Dv_z}{Dt}\right) \tag{1.11}$$

Roach (1970) shows this can be rewritten to a good approximation as

$$\frac{1}{\mathrm{Ri}}\frac{D\mathrm{Ri}}{Dt} = (2\mathrm{Ri}-1)\left(u_z\frac{\theta_x}{\theta_z} + v_z\frac{\theta_y}{\theta_z}\right) + \frac{2u_z v_z}{u_z^2+v_z^2}D_{\mathrm{SH}} - \frac{v_z^2-u_z^2}{u_z^2+v_z^2}D_{\mathrm{ST}} \tag{1.12}$$

where $D_{\mathrm{SH}} = (u_y + v_x)$ is the shearing deformation and $D_{\mathrm{ST}} = (u_x - v_y)$ is the stretching deformation. Simplifications of this lead to Brown's index and the Ellrod index (see Chap. 12). Equation (1.12) shows that Ri tendency is highly dependent on the vertical shear but also on horizontal gradients. The shear terms in Eq. (1.11) can be rewritten in terms of the frontogenesis function *F* using the thermal wind relation

$$\frac{D}{Dt}|\mathbf{u}_z| \approx \frac{g}{f\theta}\frac{D}{Dt}|\nabla_{\mathrm{H}}\theta| = \frac{g}{f\theta}F \tag{1.13}$$

Thus increasing/decreasing *F* (frontogenesis/frontolysis) leads to decreasing/increasing Ri. As with this example, many of the commonly used turbulence diagnostics are related through processes that can ultimately decrease Ri.

In addition to these large-scale effects that reduce Ri, it is now largely accepted that gravity wave (or inertia-gravity wave) breaking is also a contributor to atmospheric turbulence (e.g., Sharman et al. 2012b and references therein), wherein wave-induced temperature and velocity perturbations can lead to local reductions in the environmental Ri. This effect obviously depends on the amplitude of the wave, but even when the wave amplitude is small, if the wave enters a region of already low environmental Ri (perhaps due to large-scale shear effects), the wave induced reduction in Ri may be enough to satisfy local KHI criteria (e.g., Lane et al. 2004; Chap. 16). This considerably complicates our picture of turbulence production sources in the free atmosphere, since now one must consider the atmosphere to be a soup consisting of the superposition of many different gravity waves of different wavelengths and amplitudes generated from many different sources. This also complicates prediction, as the gravity waves are often unresolved or poorly represented by NWP models and because of their transience would normally be expected to have low predictability.

With regard to CIT, identifying large-scale conditions associated with the development of convective storms can be helpful in developing predictive algorithms (Chaps. 15 and 16), and in general, if these storms are of large enough scale to be predicted by NWP models, diagnostics, such as those mentioned above, may successfully predict turbulence due to the strong influence the storms have on gradients of environmental quantities. With ongoing improvements and convection-permitting NWP models, the ability to predict convection and by proxy CIT is rapidly improving.

To be useful for operational forecasting, the automated turbulence diagnostics must be calibrated to some metric of atmospheric turbulence intensity, and probably EDR is as good as any. Although some diagnostics predict turbulence intensity directly (e.g., Frehlich and Sharman 2004; Schumann 2012), most do not (see Chap. 12), and intensity must be derived by correlating diagnostic values to turbulence observations (e.g., Sharman et al. 2006; Kim and Chun 2011; McCann et al. 2012; Kim et al. 2015). Once this is done, an ensemble of diagnostics can be computed and combined (possibly in a weighted manner) to provide a turbulence forecast that takes into account many different potential sources of turbulence. This is the basis of the Graphical Turbulence Guidance (GTG) system (Sharman et al. 2006) as well as some others (e.g., Dutton 1980; Leshkevich 1988; Gill and Stirling 2013).

The highly transient and localized nature of turbulence implies it has low predictability, and therefore deterministic forecasts will remain elusive and to some extent misleading. Ultimately then, the turbulence forecasts should be probabilistic to account for the low predictability of turbulence and the inherent uncertainty in both the NWP model forecast and the diagnostic algorithms. Probabilities of exceeding a certain turbulence intensity could be derived from a suite of turbulence diagnostics, such as used in GTG, possibly driven by an ensemble of NWP models. The percentage agreement among the diagnostics could be inferred as a "probability" (see Chaps. 13 and 14). The use of such probabilistic turbulence forecasts in automated air traffic management systems is discussed in Chap. 24. Of course comprehensive verification of forecast performance is difficult given the lack of reliable, systematic observations. Appropriate verification techniques for this are discussed in Chap. 13.

1.7 Simulation Studies

Over the last 50 years or so, considerable progress has been made in understanding the character and evolution of PBL turbulence (both convective and stably stratified) and to a lesser extent turbulence in the free atmosphere through high-resolution numerical simulation. This simulation capability has been made possible through advances in computing hardware and numerical simulation software. The term "high-resolution" is relative here, and in this context it is high resolution compared to operational NWP model output resolution, which today has grid spacings of ~10 km. The high-resolution simulation models solve numerically the governing time-dependent Navier–Stokes equations, usually with some approximations involved to make the computer resources required manageable. The models are of two types: direct numerical simulation (DNS), in which all turbulent scales are computed, and the so-called large eddy simulation (LES) in which the grid spacings is adequate to only resolve the largest most energetic eddies [recall energy falls off as $k^{-5/3}$, Eq. (1.9)] while the effects of the smaller eddies are parameterized. For a recent review of both the DNS and LES approaches see Kühnlein et al. (2012) and for LES applications see Fedorovich et al. (2004). Since DNS resolves more scales of motion, it is considered the more accurate approach to simulating turbulence flows; however, it is very computationally intensive and so far must be executed on relatively small computational domains. Therefore, most simulation studies relevant to aviation are done in the LES mode. But when applied to realistic aviation turbulence problems, even the LES approach can be computationally intensive, since the simulation model setup must have a domain large enough to capture the large-scale turbulence generating mechanisms i.e., a domain size of ~1000 km horizontally and ~25 km deep, yet have a grid spacing sufficiently small to approach the eddy sizes that affect aircraft (at least a few 100 m), requiring on the order of 10^{10} grid points. In practice then, a nested approach is often used where successively smaller yet higher resolution grids are embedded inside one another. An example is shown in Fig. 1.6. In the outer grid domains, the grid spacing is only adequate to capture the large-scale forcing effects, but the resolution of the inner domain is constructed to at least partially resolve aircraft-scale eddies. The turbulence predicted from the unresolved part is parameterized and is usually represented as a subgrid-scale (SGS) TKE. Most NWP

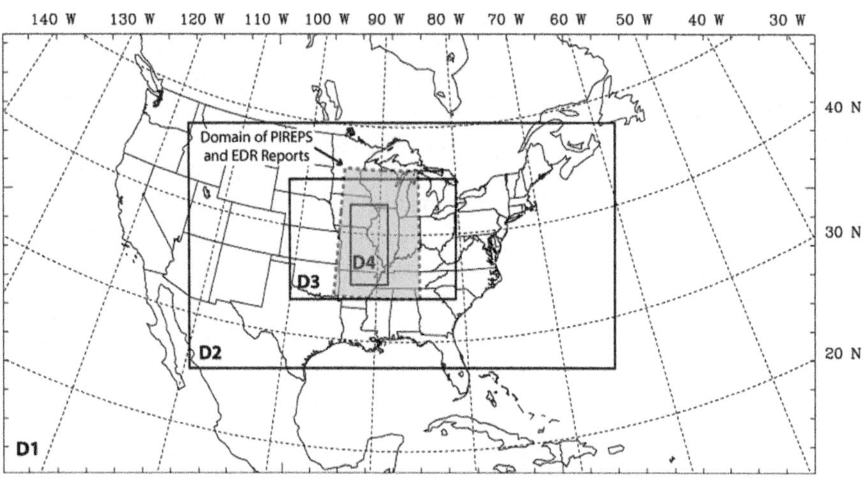

Fig. 1.6 Example of a simulation setup with four nested grids, labeled D1 through D4. The finest grid D4 was set up to reconstruct a turbulence outbreak in the *grey-shaded area*. From Trier et al. (2012). © American Meteorological Society. Reprinted with permission

models produce estimates of SGS TKE; however, these have generally been tuned for best performance in the PBL, and application to upper level turbulence seems to only produce reasonable predictions for relatively high-resolution grids (~1 km or finer).

Of particular relevance to aviation turbulence are simulations that are based on observations of turbulence events (obtained from PIREPs, EDR reports, research aircraft encounters, or other observations). Examples of these include Clark et al. (2000), Kaplan et al. (2005), Lane et al. (2003, 2004, 2006), Doyle et al. (2005), Trier and Sharman (2009), Trier et al. (2010, 2012), Kim and Chun (2010), Kim and Chun (2010), Kim et al. (2014); Kirkwood et al. (2010), Sharman et al. (2012a), and reviews of some of these are provided in Part IV. Three important results have come out of these studies: (1) gravity waves and gravity wave breaking are important contributors to aircraft-scale turbulence. While this has been known for MWT, convective gravity waves and gravity wave breaking appear to also be a major contributor to turbulence in the clear air surrounding cloud and can occur at large distances from visible cloud; (2) often the larger amplitude events are the result of interactions between more than one instability mechanism, and it is often difficult to separate out their individual effects, making forecasting more difficult; (3) executing different numerical simulation models/parameter settings for the same initial conditions can provide substantially different results (see Chap. 18), which again is suggestive of low predictability and underscores the need for probabilistic approaches to turbulence forecasting.

1.8 Closing Comments

Given the complexity of the atmospheric turbulence problem, as noted by Turner (1973, p. 336) "A completely deterministic theory is ... unlikely, and detailed forecasting of clear air turbulence will always be very difficult." Nevertheless, major progress in understanding the causes and life cycles of aviation-scale turbulence has been obtained through denser and more reliable in situ and remote sensing techniques and advances in computing capabilities allowing higher resolution NWP forecasts and still higher resolution simulation case studies. Further high-resolution simulation studies should identify important turbulence production sources and may lead to better turbulence diagnostics, gravity wave parameterizations, and SGS turbulence parameterizations suitable for stratified shear flows, which should lead to better turbulence forecasts for aviation in the future. A combined approach of deriving empirically based turbulence diagnostics, performing analyses of research aircraft turbulence measurements, and investigating turbulence encounters through high-resolution simulation is likely to be the most lucrative strategy for advancing fundamental understanding and ultimately providing reliable turbulence prediction capabilities and real-time avoidance strategies. Immediate research needs to help accomplish these goals are discussed in the final Chap. 25.

References

Bacmeister, J.T., Eckermann, S.D., Newman, P.A., Lait, L., Chan, K.R., Loewenstein, M., Proffitt, M.H., Gary, B.L.: Stratospheric horizontal wavenumber spectra of winds, potential temperature, and atmospheric tracers observed by high-altitude aircraft. J. Geophys. Res. **101**(D5), 9441–9470 (1996)

Bannon, J.K.: Weather systems associated with some occasions of severe turbulence at high altitude. Meteorol. Mag. **81**, 97–101 (1952)

Bedard, A.J. Jr.: Atmospheric turbulence aloft: a review of possible methods for detection, warning, and validation of predictive models. 31st Aerospace Sciences Meeting & Exhibit, AIAA 93-0847, Reno, NV, Jan 11–14 (1993)

Bedka, K.M., Brunner, J., Dworak, R., Feltz, W., Otkin, J., Greenwald, T.: Objective satellite-based overshooting top detection using infrared window channel brightness temperature gradients. J. Appl. Meteorol. Climatol. **49**(2), 181–202 (2010). doi:10.1175/2009JAMC2286.1

Bowles, R.L., Buck, B.K.: A methodology for determining statistical performance compliance for airborne Doppler radar with forward-looking turbulence detection capability. NASA Contractor Report. NASA/CR-2009-215769. http://ntrs.nasa.gov/archive/nasa/casi.ntrs.nasa.gov/20090025483.pdf (2009)

Burnham, J.: Atmospheric gusts—a review of the results of some recent research at the Royal Aircraft Establishment. Mon. Weather Rev. **98**(10), 723–734 (1970)

Camp, D.W., Frost, W.: Atmospheric turbulence relevant to aviation, missile, and space programs. NASA Conf. Pub. 2468 (1987)

Chan, P.W.: LIDAR-based turbulent intensity calculation using glide-path scans of the Doppler Light Detection and Ranging (LIDAR) systems at the Hong Kong International Airport and comparison with flight data and a turbulent alerting system. Meteorol. Z. **19**, 549–562 (2010). doi:10.1127/0941-2948/2010/0471

Clark, T.L., Hall, W.D., Kerr, R.M., Middleton, D., Radke, L., Ralph, F.M., Nieman, P.J., Levinson, D.: Origins of aircraft-damaging clear air turbulence during the 9 December 1992 Colorado downslope windstorm: numerical simulations and comparison to observations. J. Atmos. Sci. **57**(8), 1105–1131 (2000)

Clayson, C.A., Kantha, L.: On turbulence and mixing in the free atmosphere inferred from high-resolution soundings. J. Atmos. Ocean. Technol. **25**(6), 833–852 (2008)

Cornman, L.B., Morse, C.S., Cunning, G.: Real-time estimation of atmospheric turbulence severity from in-situ aircraft measurements. J. Aircr. **32**, 171–177 (1995). doi:10.2514/3.46697

Cornman, L.B., Goodrich, R.K., Axelrad, P., Barlow, E.: Progress in turbulence detection via GNSS occultation data. Atmos. Meas. Tech. **5**, 789–808 (2012). doi:10.5194/amt-5-789-2012

Dehghan, A., Hocking, W.R., Srinivasan, R.: Comparisons between multiple in-situ aircraft measurements and radar in the troposphere. J. Atmos. Sol. Terr. Phys. **118**, 64–77 (2014)

Doyle, J.D., Shapiro, M.A., Jiang, Q., Bartels, D.L.: Large-amplitude mountain wave breaking over Greenland. J. Atmos. Sci. **62**(9), 3106–3126 (2005)

Durran, D.R., Klemp, J.B.: On the effects of moisture on the Brunt-Väisälä frequency. J. Atmos. Sci. **39**(10), 2152–2158 (1982)

Dutton, J.: Clear-air turbulence, aviation, and atmospheric science. Rev. Geophys. Space Phys. **9**, 613–657 (1971). doi:10.1029/RG009i003p00613

Dutton, J., Panofsky, H.A.: Clear air turbulence: a mystery may be unfolding. Science **167**, 937–944 (1970)

Dutton, M.J.O.: Probability forecasts of clear-air turbulence based on numerical output. Meteorol. Mag. **109**, 293–310 (1980)

Ellrod, G.P.: A decision tree approach to clear air turbulence analysis using satellite and upper air data. NOAA Technical Memorandum NESDIS 23. Satellite Applications Laboratory, Washington, DC (1989)

Ellrod, G.P.: Identifying high altitude mountain wave turbulence and strong chinook wind events with satellite imagery. AIAA-87-0183. AIAA 25th Aerospace Sciences Meeting, Reno, NV, Jan 12–15 (1987)

Ellrod, G.P., Knox, J.A., Lester, P.F., Ehernberger, L.J.: In: North, G.R. (editor-in-chief), Pyle, J., Zhang, F. (eds.) Clear Air Turbulence. Encyclopedia of Atmospheric Science, 2nd edn, pp. 177–186. Academic Press/Elsevier (2015)

Federal Aviation Administration (FAA): Aeronautical Information Manual (AIM). http://www.faa.gov/air_traffic/publications/media/AIM_Basic_4-03-14.pdf (2014)

Fedorovich, E., Rotunno, R., Stevens, B. (eds.) Atmospheric Turbulence and Mesoscale Meteorology: Scientific Research Inspired. Cambridge University Press, Cambridge (2004)

Foldvik, A., Wurtele, M.G.: The computation of the transient gravity wave. Geophys. J. R. Astron. Soc. **13**, 167–185 (1967)

Frehlich, R., Cornman, L.: Estimating spatial velocity statistics with coherent Doppler lidar. J. Atmos. Ocean. Technol. **19**(3), 355–366 (2002)

Frehlich, R., Sharman, R.: Estimates of turbulence from numerical weather prediction model output with applications to turbulence diagnosis and data assimilation. Mon. Weather Rev. **132**, 2308–2324 (2004)

Frehlich, R., Meillier, Y., Jensen, M.L., Balsley, B.: A statistical description of small-scale turbulence in the low-level nocturnal jet. J. Atmos. Sci. **61**, 1079–1085 (2004)

Gage, K.S.: Dynamic processes contributing to the mesoscale spectrum of atmospheric motions. In: Lilly, D., Fedorovich, E., Rotunno, R., Stevens, B. (eds.) Atmospheric Turbulence and Mesoscale Meteorology: Scientific Research Inspired, pp. 223–264. Cambridge University Press, Cambridge (2004)

Gerz, T., Holzäpfel, F., Darracq, D.: Commercial aircraft wake vortices. Prog. Aeronaut. Sci. **38**, 181–208 (2002)

Gill, P.G.: Objective verification of World Area Forecast Centre clear air turbulence forecasts. Meteorol. Appl. **21**, 3–11 (2014). doi:10.1002/met.1288

Gill, P.G., Stirling, A.J.: Including convection in global turbulence forecasts. Meteorol. Appl. **20**, 107–114 (2013). doi:10.1002/met.1315

Gossard, E.E., Hooke, W.H.: Waves in the Atmosphere. Elsevier, New York, NY (1975)

Grabowski, W.W., Clark, T.L.: Cloud-environment interface instability: rising thermal calculations in two spatial dimensions. J. Atmos. Sci. **48**(4), 527–546 (1991)

Haack, A., Gerding, M., Lübken, F.-J.: Characteristics of stratospheric turbulent layers measured by LITOS and their relation to the Richardson number. J. Geophys. Res. Atmos. **119**, 10605–10618 (2014). doi:10.1002/2013JD021008

Haggerty, J., Schick, K., Mahoney, M.J., Lim, B.: The NCAR microwave temperature profiler: data applications from recent deployments. In: Microwave Radiometry and Remote Sensing of the Environment (MicroRad), pp. 133–135. 13th Specialist Meeting, Pasadena, CA (2014). doi:10.1109/MicroRad.2014.6878924

Hannon, S.M., Bagley, H.R., Bogue, R.K.: Airborne Doppler lidar turbulence detection: ACLAIM flight test results. Proceedings of SPIE Conference on Laser Radar Technology and Applications IV, SPIE vol. 3707, pp. 234–241. Orlando, FL, April (1999)

Harrison, R.G., Hogan, R.J.: In situ atmospheric turbulence measurement using the terrestrial magnetic field—a compass for a radiosonde. J. Atmos. Ocean. Technol. **23**, 517–523 (2006)

Hoblit, F.M.: Gust Loads on Aircraft: Concepts and Applications. AIAA Education Series. American Institute of Aeronautics and Astronautics, Washington, DC (1988)

Hocking, W.K., Mu, P.K.L.: Upper and middle tropospheric kinetic energy dissipation rates from measurements of C_n^2—review of theories, in-situ investigations, and experimental studies using the Buckland Park atmospheric radar in Australia. J. Atmos. Sol. Terr. Phys. **59**, 1779–1803 (1997)

Honomichl, S.B., Detwiler, A.G., Smith, P.L.: Observed hazards to aircraft in deep summertime convective clouds from 4–7 km. J. Aircr. **50**(3), 926–935 (2013). doi:10.2514/1.C032057

Hopkins, R.H.: Forecasting techniques of clear-air turbulence including that associated with mountain waves. WMO Technical Note No. 155 (1977)
International Civil Aviation Organization (ICAO): Meteorological service for international air navigation. Annex 3 to the Convention on International Civil Aviation, 14th ed. ICAO International Standards and Recommended Practices Tech. Annex, 2001
International Civil Aviation Organization (ICAO): Meteorological service for international air navigation—Annex 3 to the Convention on International Civil Aviation, Appendix 4 Technical Specifications Related to Aircraft Observations and Reports, Section 2.6 Turbulence, 18th Edition, 208 pp. http://store1.icao.int/index.php/publications/annexes/3-meteorological-service-for-international-air-navigation.html (2013)
Jaeger, E.B., Sprenger, M.: A northern-hemispheric climatology of indices for clear air turbulence in the tropopause region derived from ERA40 re-analysis data. J. Geophys. Res. **112**, D20106 (2007). doi:10.1029/2006JD008189
Joseph, B., Mahalov, A., Nicolaenko, B., Tse, K.L.: Variability of turbulence and its outer scales in a model tropopause jet. J. Atmos. Sci. **61**(6), 621–643 (2004)
Kaimal, J.C., Finnigan, J.J.: Atmospheric Boundary Layer Flows: Their Structure and Measurement. Oxford University Press, Oxford (1994)
Kaplan, M.L., Huffman, A.W., Lux, K.M., Cetola, J.D., Charney, J.J., Riordan, A.J., Lin, Y.-L., Waight III, K.T.: Characterizing the severe turbulence environments associated with commercial aviation accidents. Part 2: Hydrostatic mesoscale numerical simulations of supergradient flow and streamwise ageostrophic frontotogenis. Meteorol. Atmos. Phys. **88**, 153–173 (2005)
Kauffmann, P.: The business case for turbulence sensing systems in the US air transport sector. J. Air Transp. Manag. **8**, 99–107 (2002)
Kim, J.-H., Chun, H.-Y.: A numerical study of clear-air turbulence (CAT) encounters over South Korea on 2 April 2007. J. Appl. Meteorol. Climatol. **49**(12), 2381–2403 (2010)
Kim, J.-H., Chun, H.-Y.: Statistics and possible sources of aviation turbulence over South Korea. J. Appl. Meteorol. Climatol. **50**(2), 311–324 (2011)
Kim, J.H., Chun, H.-Y., Sharman, R.D., Trier, S.B.: The role of vertical shear on aviation turbulence within cirrus bands of a simulated western Pacific cyclone. Mon. Weather Rev. **142**(8), 2794–2813 (2014). doi:10.1175/MWR-D-14-00008.1
Kim, J.-H., Chan, W.N., Sridhar, B., Sharman, R.D.: Combined winds and turbulence prediction system for automated air-traffic management application. J. Appl. Meteorol. Clim. **54**(4), 766–784 (2015). doi:10.1175/JAMC-D-14-0216.1
Kirkwood, S., Mihalikova, M., Rao, T.N., Satheesan, K.: Turbulence associated with mountain waves over Northern Scandinavia—a case study using the ESRAD VHF radar and the WRF mesoscale model. Atmos. Chem. Phys. **10**, 3583–3599 (2010)
Klipp, C.: Turbulence anisotropy in the near surface atmosphere and the evaluation of multiple outer length scales. Bound.-Layer Meteorol. **151**, 57–77 (2014). doi:10.1007/s10546-013-9884-0
Knox, J.A., Bachmeier, A.S., Carter, W.M., Tarantino, J.E., Paulik, L.C., Wilson, E.N., Bechdol, G.S., Mays, M.J.: Transverse cirrus bands in weather systems: a grand tour of an enduring enigma. Weather **5**, 35–41 (2010)
Koch, S.E., Jamison, B.D., Lu, C., Smith, T.L., Tollerud, E.L., Girz, C., Wang, N., Lane, T.P., Shapiro, M.A., Parrish, D.D., Cooper, O.R.: Turbulence and gravity waves within an upper-level front. J. Atmos. Sci. **62**(11), 3885–3908 (2005)
Koshyk, J.N., Hamilton, K., Mahlman, J.D.: Simulation of the $k^{-5/3}$ mesoscale spectral regime in the GFDL SKYHI general circulation model. Geophys. Res. Lett. **26**(7), 843–846 (1999)
Kristensen, L., Lenschow, D.H.: An airborne laser air motion sensing system. Part II: Design criteria and measurement possibilities. J. Atmos. Ocean. Technol. **4**, 128–138 (1987)
Kudo, A.: The generation of turbulence below midlevel cloud bases: the effect of cooling due to sublimation of snow. J. Appl. Meteorol. Climatol. **52**(4), 819–833 (2013). doi:10.1175/JAMC-D-12-0232.1

Kühnlein, C., Dörnbrack, A., Gerz, T.: Advanced numerical modeling of turbulent atmospheric flows. In: Schumann (ed.), Atmospheric Physics, Research Topics in Aerospace, U. 529–541. Springer, Berlin (2012). doi: 10.1007/978-3-642-30183-4_39

Kühnlein, C., Dörnbrack, A., Weissmann, M.: High-resolution Doppler lidar observations of transient downslope flows and rotors. Mon. Weather Rev. **141**(10), 3257–3272 (2013)

Lamb, S.H.: Hydrodynamics, 6th edn. Dover Publications, New York, NY (1945)

Lane, T.P., Sharman, R.D.: Intensity of thunderstorm-generated turbulence revealed by large-eddy simulation. Geophys. Res. Lett. **41**, 2221–2227 (2014). doi:10.1002/2014GL059299

Lane, T.P., Sharman, R.D., Clark, T.L., Hsu, H.-M.: An investigation of turbulence generation mechanisms above deep convection. J. Atmos. Sci. **60**(10), 1297–1321 (2003)

Lane, T.P., Doyle, J.D., Plougonven, R., Shapiro, M.A., Sharman, R.D.: Observations and numerical simulations of inertia-gravity waves and shearing instabilities in the vicinity of a jet stream. J. Atmos. Sci. **61**(22), 2692–2706 (2004)

Lane, T.P., Sharman, R., Frehlich, R.G., Brown, J.M.: Numerical simulations of the wake of Kauai with implications for the Helios flights. J. Appl. Meteorol. Climatol. **45**(9), 1313–1331 (2006)

Lane, T.P., Sharman, R.D., Trier, S.B., Fovell, R.G., Williams, J.K.: Recent advances in the understanding of near-cloud turbulence. Bull. Am. Meteorol. Soc. **93**(4), 499–515 (2012). doi:10.1175/BAMS-D-11-00062.1

Lee, D.R., Stull, R.B., Irvine, W.S.: Clear air turbulence forecasting techniques. Air Force Weather Service Report AFGWC/TN-79-001 (REV) (1984)

Lenz, A., Bedka, K., Feltz, W., Ackerman, S.: Convectively-induced transverse band signatures in satellite imagery. Weather Forecast. **24**(5), 1362–1373 (2009)

Leshkevich, T.V.: Automated method of predicting the probability of clear-air turbulence. Meteorol. Gidrol. **10**, 27–33 (1988)

Lester, P.F.: Turbulence: A New Perspective for Pilots. Jeppesen Sanderson, Englewood, CO (1993)

Lilly, D.K., Klemp, J.B.: The effects of terrain shape on nonlinear hydrostatic mountain waves. J. Fluid Mech. **95**(2), 241–261 (1979)

Lilly, D.K., Waco, D.E., Adelfang, S.I.: Stratospheric mixing estimated from high-altitude turbulence measurements. J. Appl. Meteorol. **13**(4), 488–493 (1974). doi:10.1175/1520-0450(1974)013,0488:SMEFHA.2.0.CO;2

Lindborg, E.: Can the atmospheric kinetic energy spectrum be explained by two-dimensional turbulence? J. Fluid Mech. **388**, 259–288 (1999)

Lindborg, E.: Horizontal wavenumber spectra of vertical vorticity and horizontal divergence in the upper troposphere and lower stratosphere. J. Atmos. Sci. **64**(3), 1017–1025 (2007)

Love, P.T., Geller, M.A.: Exploring and improving access to high vertical resolution radiosonde data. Eos Trans. AGU **94**(44), 401 (2013). doi:10.1002/2013EO440006

Lovejoy, S., Tuck, A.F., Schertzer, D., Hovde, S.J.: Reinterpreting aircraft measurements in anisotropic scaling turbulence. Atmos. Chem. Phys. **9**, 5007–5025 (2009)

Lu, C., Koch, S.E.: Interaction of upper-tropospheric turbulence and gravity waves as obtained from spectral and structure function analyses. J. Atmos. Sci. **65**(8), 2676–2690 (2008). doi:10.1175/2007JAS2660.1

MacCready Jr., P.B.: Standardization of gustiness values from aircraft. J. Appl. Meteorol. **3**, 439–449 (1964)

Mahoney, M.J., Denning, R.: A state-of-the-art airborne microwave temperature profiler (MTP). Proceedings of the 33rd International Symposium on the Remote Sensing of the Environment, Stresa, Italy, 4–8 May (2009)

Marlton, G.J., Harrison, R.G., Nicoll, K.A., Williams, P.D.: A balloon-borne accelerometer technique for measuring atmospheric turbulence. Rev. Sci. Instrum. **86**, 016109 (2015). doi:10.1063/1.4905529

McCann, D.W., Knox, J.A., Williams, P.D.: An improvement in clear-air turbulence forecasting based on spontaneous imbalance theory: the ULTURB algorithm. Meteorol. Appl. **19**, 71–78 (2012)

Miles, J.: Richardson's criterion for the stability of stratified shear flow. Phys. Fluids **29**(10), 3470–3471 (1986)

Monette, S.A., Sieglaff, J.M.: Probability of convectively induced turbulence associated with geostationary satellite–inferred cloud-top cooling. J. Appl. Meteorol. Climatol. **53**(2), 429–436 (2014)

Murrow, H.N.: Measurements of atmospheric turbulence. In: Camp, D.W., Frost, W. (eds.) Atmospheric Turbulence Relative to Aviation, Missile, and Space Programs. NASA Conference Publication 2468, pp. 73–92 (1987)

Nastrom, G.D., Gage, K.S.: A climatology of atmospheric wavenumber spectra of wind and temperature observed by commercial aircraft. J. Atmos. Sci. **42**, 950–960 (1985)

Pao, Y.-H., Goldburg, A.: Clear Air Turbulence and Its Detection. Plenum Press, New York, NY (1969)

Paoli, R., Thouron, O., Escobar, J., Picot, J., Cariolle, D.: High-resolution large-eddy simulations of stably stratified flows: application to subkilometer-scale turbulence in the upper troposphere–lower stratosphere. Atmos. Chem. Phys. **14**, 5037–5055 (2014)

Roach, W.T.: On the influence of synoptic development on the production of high level turbulence. Q. J. R. Meteorol. Soc. **96**, 413–429 (1970)

Schaffner, P.R., Daniels, T.S., West, L.L., Gimmestad, G.G., Lane, S.E., Burdette, E.M., Smith, W.L., Kireev, S., Cornman, L., Sharman, R.D.: Experimental validation of a forward looking interferometer for detection of clear air turbulence due to mountain waves. AIAA 2012–2790. 4th AIAA Atmospheric and Space Environments Conference, New Orleans, LA, 25–28 June (2012)

Schumann, U.A.: Contrail cirrus prediction model. Geosci. Model Dev. **5**, 543–580 (2012). doi:10.5194/gmd-5-543-2012

Schumann, U., Gerz, T.: Turbulent mixing in stably stratified shear flows. J. Appl. Meteorol. **34**(1), 33–48 (1995)

Schumann, U., Sharman, R.D.: Aircraft wake-vortex encounter analysis for upper levels. J. Aircr. **52**(4), 1277–1285 (2015). doi:10.2514/1.C032909

Schumann, U., Konopka, P., Baumann, R., Busen, R., Gerz, T., Schlager, H., Schulte, P., Volkert, H.: Estimation of diffusion parameters of aircraft exhaust plumes near the tropopause from nitric oxide and turbulence measurements. J. Geophys. Res. **100**(D7), 14147–14162 (1995). doi:10.1029/95JD01277

Sharman, R.: Ship waves in the atmosphere. In: McGraw-Hill 2010 Yearbook of Science & Technology, pp. 340–342. McGraw-Hill, New York, NY (2010). ISBN: 978-007-163928-6

Sharman, R., Tebaldi, C., Wiener, G., Wolff, J.: An integrated approach to mid- and upper-level turbulence forecasting. Weather Forecast. **21**(3), 268–287 (2006). doi:10.1175/WAF924.1

Sharman, R.D., Doyle, J.D., Shapiro, M.A.: An investigation of a commercial aircraft encounter with severe clear-air turbulence over western Greenland. J. Appl. Meteorol. Climatol. **51**(1), 42–53 (2012a)

Sharman, R.D., Trier, S.B., Lane, T.P., Doyle, J.D.: Sources and dynamics of turbulence in the upper troposphere and lower stratosphere: a review. Geophys. Res. Lett. **39**, L12803 (2012b). doi:10.1029/2012GL051996

Sharman, R.D., Cornman, L.B., Meymaris, G., Pearson, J., Farrar, T.: Description and derived climatologies of automated in situ eddy dissipation rate reports of atmospheric turbulence. J. Appl. Meteorol. Climatol. **53**(6), 1416–1432 (2014). doi:10.1175/JAMC-D-13-0329.1

Skamarock, W.C.: Evaluating mesoscale NWP models using kinetic energy spectra. Mon. Weather Rev. **132**, 3019–3032 (2004)

Stewart, R.W.: Turbulence and waves in a stratified atmosphere. Radio Sci. **4**(12), 1269–1278 (1969)

Strauss, L., Serafin, S., Haimov, S., Grubišić, V.: Turbulence in breaking mountain waves and atmospheric rotors estimated from airborne in situ and Doppler radar measurements. Q. J. R. Meteorol. Soc. **141**(693), 3207–3225 (2015). doi:10.1002/qj.2604

Trier, S.B., Sharman, R.D.: Convection-permitting simulations of the environment supporting widespread turbulence within the upper-level outflow of a Mesoscale Convective System. Mon. Weather Rev. **137**(6), 1972–1990 (2009)

Trier, S.B., Sharman, R.D., Fovell, R.G., Frehlich, R.G.: Numerical simulation of radial cloud bands within the upper-level outflow of an observed mesoscale convective system. J. Atmos. Sci. **67**(9), 2990–2999 (2010)

Trier, S.B., Sharman, R.D., Lane, T.P.: Influences of moist convection on a cold-season outbreak of clear-air turbulence (CAT). Mon. Weather Rev. **140**(8), 2477–2496 (2012). 10:1175/MWR-D-11-00353.1

Turner, J.S.: Buoyancy Effects in Fluids. Cambridge University Press, London (1973)

Uhlenbrock, N.L., Bedka, K.M., Feltz, W.F., Ackerman, S.A.: Mountain wave signatures in MODIS 6.7-μm imagery and their relation to pilot reports of turbulence. Weather Forecast. **22**(3), 662–670 (2007). doi:10.1175/WAF1007.1

Vinnichenko, N.K., Pinus, N.Z., Shmeter, S.M., Shur, G.N.: Turbulence in the Free Atmosphere. Plenum, New York, NY (1980)

Williams, P.D., Joshi, M.M.: Intensification of winter transatlantic aviation turbulence in response to climate change. Nat. Clim. Chang. **3**(7), 644–648 (2013). doi:10.1038/nclimate1866

Wilson, R.: Turbulent diffusivity in the free atmosphere inferred from MST radar measurements: a review. Ann. Geophys. **22**, 3869–3887 (2004)

Wimmers, A.J., Moody, J.L.: Tropopause folding at satellite-observed spatial gradients: 1. Verification of an empirical relationship. J. Geophys. Res. **109**, D19306 (2004)

Wolff, J.K., Sharman, R.D.: Climatology of upper-level turbulence over the continental Unites States. J. Appl. Meteorol. Climatol. **47**(8), 2198–2214 (2008). doi:10.1175/2008JAMC1799.1

Wroblewski, D.E., Coté, O.R., Hacker, J.M., Dobosy, R.J.: Velocity and temperature structure functions in the upper troposphere and lower stratosphere from high-resolution aircraft measurements. J. Atmos. Sci. **67**(4), 1157–1170 (2010)

Wyngaard, J.C., Clifford, S.F.: Taylor's hypothesis and high-frequency turbulence spectra. J. Atmos. Sci. **34**, 922–929 (1977)

Chapter 2
A History of Weather Reporting from Aircraft and Turbulence Forecasting for Commercial Aviation

Tom Fahey, Emily N. Wilson, Rory O'Loughlin, Melissa Thomas, and Stephanie Klipfel

Abstract Manual pilot reporting of weather conditions, including turbulence, has been standard procedure during most of the Twentieth century. By the late 1940s, the International Civil Aviation Organization (ICAO) had developed standards which included aircraft-based weather observations by pilots and subsequent government distribution. In the USA, commercial airlines developed their own collection and distribution methods. This was in response to Federal Aviation Administration (FAA) requirements that pilots should report hazardous weather, and airlines should have ability to monitor the progress of every flight. By the 1980s, flight location as well as wind and temperature were being automatically reported in some areas of the world. But automated turbulence report capability was mainly a research effort with some operational capability introduced during the 1990s and 2000s.

Aviation weather forecasting has seen many government as well as airline efforts over the last 80–100 years. By World War II, military aircraft such as the B-17 were able to reach altitudes of over 30,000 ft. This initiated the need for weather forecasting in the upper troposphere and lower stratosphere. Turbulence and mountain wave forecasting emphasis increased significantly with the transition from propeller to jet powered commercial passenger aircraft in the 1950s and 1960s. By the 2000s, government developed forecasting tools such as the Graphical Turbulence Guidance (GTG) had begun to mature, and airlines continued to focus on minimizing turbulence exposure for passenger and flight attendant safety as well as comfort.

Many US airlines have come and gone, likewise for airline meteorology offices. A brief history of Northwest (NWA), Delta, and Western meteorology; turbulence forecasting, including mountain wave; and the NWA/Delta Turbulence Plot system provide added perspective. Still today, manual and automated aircraft-based

T. Fahey (✉) • E.N. Wilson • R. O'Loughlin • M. Thomas • S. Klipfel
Meteorology Department #091, Delta Air Lines, Inc., P. O. Box 20706, Atlanta, GA 30320-6001, USA
e-mail: tom.fahey@delta.com; emily.n.wilson@delta.com; rory.oloughlin@delta.com; melissa.thomas@delta.com; Stephanie.Klipfel@delta.com

R. Sharman, T. Lane (eds.), *Aviation Turbulence*,
DOI 10.1007/978-3-319-23630-8_2

weather observations as well as government and airline produced turbulence forecasts are very valuable.

2.1 Weather Information from Aircraft

The story of weather reports from aircraft could begin with the first four powered flights by Orville and Wilbur Wright on 17 December 1903. But since no other aircraft were operating and the aircraft's maximum altitude was about 10 ft above ground level (AGL), any Wright brothers' report would have been of little value.

By the 1930s, aircraft were being flown solely for their weather observation capability. In 1931, the predecessor to the National Weather Service (NWS), the Weather Bureau, began regular 5:00 a.m. Eastern Standard Time (EST) aircraft observations at Chicago, Cleveland, Dallas, and Omaha, at altitudes reaching 16,000 ft to replace "kite stations." By the end of the 1930s, the initial Weather Bureau aircraft observation program and all similar military efforts had been replaced by the radiosonde network. But aircraft-based weather reports continued to be important for aviation meteorologists.

By the 1940s, new propeller driven commercial passenger aircraft were routinely reaching en route cruising altitudes over 20,000 ft above sea level (ASL), and use of pilot reports by aviation meteorologists as well as by pilots was a well-established practice. See Table 2.1 for a summary of Certified Service Ceilings (approximate highest usable altitude) for a number of propeller aircraft from the 1930s to 1950s.

From a worldwide perspective, the International Civil Aviation Organization (ICAO) has maintained Standards and Recommended Practices (SARP) relating to meteorology since April 1948 when they were adopted by the ICAO Council. This was done in response to the provisions of the Convention on International Civil Aviation held in 1944. The document containing the meteorology related SARP is designated as "Annex 3—Meteorological Services for International Air Navigation." Aircraft-based weather observations by pilots, as well as collection and distribution by government organizations, is just one of many topics documented in what is usually referred to as "ICAO Annex 3" or simply "Annex 3."

2.1.1 AIREPs: ICAO Requirements

Aviation meteorologists often use the terms PIREP (Pilot Report) and AIREP (air-report) interchangeably. But the official ICAO term is AIREP. Historically, an AIREP has been used to describe a pilot's verbal, radio transmission of the observed weather conditions from an aircraft that has been manually transcribed by an air traffic controller and then distributed by the local meteorology office. The

Table 2.1 Documents and compares altitude capability of some commercial and military propeller aircraft placed into service during the 1930s through 1950s

Selected Propeller Aircraft 1930s–1950s		
Commercial Passenger		
Mftr & Aircraft (& Model)/Name	*Initial Service Year*	*Certified Service Ceiling (ft)*
DC3	1936	23,200
DC4	1942	22,300
Lockheed Constellation	1946	24,000
DC6	1946	–
DC6A (Cargo)	–	21,900
DC6B (passenger)	–	25,000
Convair CV-240	1948	16,000
DC7	1953	28,400
Convair CV-440	–	24,900
Lockheed L-188 Electra	1959	32,000
Military Bombers		
Mftr & Aircraft (& Model)/Name	*Initial Year Flown*	*Certified Service Ceiling (ft)*
Douglas B-23 Dragon	1939	31,600
Boeing B17 Flying Fortress		35,600
Model 299 (B17 test a/c)	1935	(all models)
Boeing B17-(B)	1939	
Boeing B17-(C)	1940	
Boeing B17-(D)	1941 (Feb)	
Boeing B17-(E)	1941 (Sep)	
Boeing B17-(F)	1942	
Boeing B17-(G)	1943	

transcription and distribution of AIREPs from oceanic areas have been very reliable, but over land areas such as Europe, they have been nonexistent. Unfortunately, in the USA the AIREP process has not been much better. PIREPs discussed on an air traffic sector frequency by pilots and controllers often do not make it into the distribution system because it is a cumbersome, manual process. The exception are very significant turbulence encounters.

Most jet aircraft used for commercial aviation and manufactured since the late 1960s have had an on-board computer system with the capability to calculate wind direction and speed as well as the static outside air temperature. By the early 1970s, AIREPs routinely contained these three accurately measured variables. The intensity of turbulence encountered was also often included, and less frequently, turbulence duration as well as cloud, and/or icing information. Unfortunately turbulence reports were based on a pilot's subjective description and were sometimes misrepresented spatially and/or temporarily.

Up until 2010, what ICAO calls "Routine AIREPs," were expected to be made by pilots of all commercial aircraft while en route. It was the responsibility of what

ICAO calls the Meteorological Watch Office (MWO) to distribute the information once they collected it from the associated Air Traffic Service (ATS) unit. The AIREPs were then made immediately available via weather communication circuits. When ICAO Amendment 75 to Annex 3 became applicable on 18 November 2010, routine voice reporting of weather was no longer required (Section 9 AIREP/ AIREP SPECIAL EXCHANGE of the ICAO ROBEX Handbook 12th Edition 2004, 11th Amendment May 2013).

ICAO Annex 3 has made a distinction between AIREPs provided by "routine voice reporting" and those by "data link communications." While routine AIREPs provided by voice have not been required to be distributed since November 2010, the distribution of Routine as well as Special AIREPs provided by datalink continue to be required by Annex 3. The ATS organizations around the world are expected to, "relay them without delay . . .".

ICAO Amendment 77 to Annex 3 planned for 2016 contains draft text reenforcing the importance of AIREPs via datalink. The expectation has been added that AIREPs provided by datalink shall also be relayed to "the centres designated by regional air navigation agreement for the operation of the aeronautical fixed service satellite distribution system and the Internet-based services" [ICAO State Letter, 22 Dec 2014, Subject: Proposals for the amendment of Annex 3 relating to aeronautical meteorology and consequential amendments to Annex 11, PANS-ABC (Doc 8400) and PANS-ATM (Doc 4444)].

In addition, ICAO Annex 3 does continue to include an expectation that special air reports (AIREP SPECIAL) generated via voice be distributed expeditiously. The AIREP SPECIAL report is intended to contain information about a significant weather related aviation hazards such as severe turbulence as well as pilot observed, pre-eruption volcanic activity, or volcanic eruption material in the atmosphere. In the USA, the term and distribution label, Urgent Upper Air (UUA) report, is used rather than AIREP SPECIAL. In some areas of the world, AIREP SPECIALs also include reports of moderate turbulence. In any case, since 2010 the number of AIREPs distributed has decreased significantly.

2.1.2 Airlines' Company Position Reports

Over the years many commercial airlines, especially US carriers, have developed very sophisticated internal processes and infrastructure for producing and collecting what is often called, position reports. This is in addition to ICAO required collection and distribution of reported weather parameters.

2.1.2.1 Transmission of Company Position Reports

In the past Company, Position Reports were made verbally by a member of the flight crew and relayed directly to the airline via VHF or HF radio frequency transmission. Some US airlines developed their own VHF radio infrastructure for domestic US communications that were completely independent of the Federal Aviation Administration (FAA) radio frequencies used for Air Traffic Control. For international operations, HF networks were developed jointly by airlines via the then, airline owned, ARINC organization or other commercial companies. Today, only a few US airlines continue to maintain their own VHF radio network. But, position reports continue to be an important method of obtaining operational information.

2.1.2.2 Federal Air Regulations

There has been a need for US airlines to have communication capabilities between company representatives on the ground and the pilots in the airborne aircraft, in addition to ATC.

Commercial passenger airlines registered in the USA are bound by the requirements in Part 121 of the Federal Aviation Regulations (FAR) which are enforced by the FAA. FAR Part 121 operators, as they are called, are required to maintain capability that "will ensure reliable and rapid communications, under normal operating conditions over the entire route ... between each airplane and the appropriate dispatch office, and between each airplane and the appropriate air traffic control unit" (Source: FAA 2015, FAR Part 121, Section 99 Communication Facilities).

These communication needs have been driven by the FAR's Operational Control model in which, "The pilot in command and the aircraft dispatcher are jointly responsible for the preflight planning, delay, and dispatch release of a flight ..." (Source: FAA 2015, FAR Part 121, Sections 533 and 535, b and c Responsibility for Operational Control). This model traditionally was not used in other parts of the world, but recently has been adopted by other international airlines such as KLM and Air France.

FAR Part 121, Section 561 goes further to require pilots, while in flight to always report "...Potentially Hazardous Meteorological Conditions..." While this responsibility had been fulfilled in the past by a voice radio transmission to the local air traffic facility, it is being addressed more frequently today with what is generically called "data-link" communication capabilities.

Most US airlines have had requirements that every flight transmit a position report on a scheduled basis, at set waypoints or set times. In this way, airlines fulfill both FAA requirements: flight dispatcher ability to monitor the progress of flights and pilot ability to report hazardous weather.

2.1.2.3 Automated Position Reports

As the term implies, a position report includes the flight level as well as the location (by latitude and longitude or by predefined and named waypoints). Airlines usually include fuel remaining onboard as well as weather parameters similar to an AIREP, including turbulence. In the 1980s, the ability to automatically distribute the exact location as well as the calculated static air temperature and wind velocity to an airline ground facility for the purpose of position reporting was adopted by some airlines. In the USA at that time, ARINC, a communications company that was owned in part by the airlines, provided a service called Automated Communications, Addressing And Reporting System (ACARS). This can be thought of as the airlines' version of "texting" between an aircraft and individuals with safety and/or efficiency responsibilities on the ground. ACARS was one of the earliest datalink systems and was widely adopted by airlines. It was designed to use ground-based VHF transmitters/receivers across the USA to send and receive messages to/from en route aircraft. By the early 1990s, airlines were providing approximately 7000 automated reports of wind, temperature, and location information per day over the USA (Benjamin et al. 1991).

By 1996, the volume of automated aircraft reports of wind and temperature reports had more than tripled to approx. 22,000 per day, and a Water Vapor Sensing System (WVSS) was being developed (Fleming 1996). During the 1990s, similar capabilities to automatically report turbulence encountered was developing. By 2002 six domestic airlines, American Airlines (AAL), Delta Air Lines (DAL), Federal Express (FedEx), Northwest Airlines (NWA), United Airlines (UAL), and United Parcel Service (UPS), were providing automated aircraft reports of temperature and wind velocity via ACARS. In addition, some of these six airlines were providing measurements of water vapor or turbulence. By 2002, ACARS provided data exceeded 100,000 reports per day (Moninger et al. 2003).

2.1.3 Aircraft Weather Report Access: Government Systems

2.1.3.1 International Distribution Systems

There are numerous systems for distributing a wide variety of aviation related weather data. But all text weather data types are identified by WMO Bulletin headers for distribution purposes.

ICAO Annex 3 specifies that each country establish and maintain one or more Meteorological (Met) Offices to provide weather service in order to meet the needs of international air navigation as well as flight operations at individual airports. The Met Offices are expected to prepare, and distribute, forecasts and observations. ICAO provides a structure for distribution of what is called Operational Meteorological (OPMET) data worldwide. It includes Terminal Aerodrome Forecasts (TAFs) as well as airport hourly weather observations (METARs) and special between hour observations (SPECIs) which are all airport specific.

ICAO Annex 3 also defines Meteorological Watch Offices which are responsible for monitoring weather conditions over a volume of airspace called a Flight Information Region (FIR). SIGMETs which include turbulence are weather hazard information of Significant Meteorological importance to aviation. SIGMETs are produced and delivered to the air traffic services unit covering the FIR and are included as OPMET data for broader dissemination and access. The previously described AIREPs are also included. But automated aircraft reports of weather variables that expanded in the 1980s (winds, temperatures) and 1990s (turbulence, water vapor) have not been included by ICAO in the list of required OPMET data for international distribution. Fortunately, WMO has developed a loosely organized system for distribution of automated aircraft weather observations called Aircraft Meteorological Data Relay (AMDAR) (see http://www.wmo.int/pages/prog/www/GOS/ABO/AMDAR/resources/index_en.html).

2.1.3.2 WMO's Role

One of the purposes of the WMO is to coordinate the activities of its over 180 member countries in the generation of data and information on weather, based on internationally agreed standards. Technical Regulations are developed which specify the meteorological practices and procedures to be followed by WMO member countries. These Technical Regulations are supplemented by a number of Manuals and Guides which describe in more detail the practices, procedures, and specifications that countries are requested to follow and implement. Manuals contain mandatory practices; Guides such as the WMO *Guide to Meteorological Instruments and Methods of Observation* contain recommended practices. The first edition of this Guide was published in 1954. Since then, the Commission for Instruments and Methods of Observation (CIMO) has periodically reviewed the contents of the Guide, making recommendations for additions and amendments when deemed appropriate.

WMO's AMDAR Program

At the time of writing, June 2015, the present, eighth edition of the WMO Guide to Meteorological Instruments and Methods of Observation is in effect. The purpose of the Guide is to give comprehensive and up-to-date guidance on the most effective practices for carrying out meteorological observations and measurements. Part II of the Guide covers Observing Systems and within Part II, Chap. 3 covers Aircraft Observation. The methods used for automatic meteorological measurements on modern commercial aircraft (i.e., AMDAR) are described (WMO 2012).

Any AMDAR Observing System established by an individual country in coordination with aircraft operators is described in general to include processing to give the following meteorological elements:

1. Pressure altitude, position, and time
2. Static air temperature (SAT)
3. Wind speed
4. Wind direction

On some aircraft, a process to calculate turbulence values is available. Sensors have been installed on some aircraft for measuring ice buildup on the flying surfaces and/or for measuring relative humidity or water vapor mixing ratio. Turbulence measured or calculated by aircraft instrumentation is described in some detail in the WMO Guide including vertical acceleration or "g load," derived equivalent gust velocity, and eddy dissipation rate (EDR) (see Chap. 5).

Comparing US and Other Programs

In 1986, the Australian program became the first operational AMDAR system (Sprinkle 1999). Development was led by the Australian Bureau of Meteorology in conjunction with Qantas, with the bureau paying for the observations and controlling the type of observations, the frequency of reports, the region within which reports are made, data quality, and data distribution, with access free to all users globally. This model was later developed by other new AMDAR systems, with the exception of the US system (Moninger et al. 2003).

In the 1980s, at the same time that Australia was developing a formal system, three US airlines (DAL, NWA, and UAL) began an informal arrangement of sharing automated reports of winds and temperatures with NOAA computer model developers (Benjamin et al. 1991). The US airlines freely shared the data, with the expectation of benefiting from improved computer model weather forecast accuracy. In addition, the participating US airlines were paying for the communication costs to downlink the data. It was not until the mid-2000s that the US government agreed to begin paying for a portion of the airlines' communication costs. As a result US airlines still, today, own and retain control over their aircraft-based weather observations, with the exception being water vapor. SWA and UPS have separate, signed agreements with NOAA for the provision of water vapor observations.

2.2 Turbulence Forecasting

2.2.1 Current International Perspective

ICAO Annex 3 has designated two World Area Forecast Centers (WAFCs) as part of the World Area Forecast System (WAFS). The two centers are located in the USA and in England and as the name implies, they focus on producing and distributing forecast products for international air navigation. The two primary responsibilities of the WAFCs are to produce global forecasts of:

1. Upper level winds and temperatures in grid point format for use in automated flight planning systems.
2. Significant weather (SIGWX) phenomena, including turbulence.

2.2.2 History

2.2.2.1 National Weather Service Organizational Perspective

In 1914, just 11 years after the Wright Brothers' flight at Kitty Hawk, the U.S. Weather Bureau, now the NWS, officially began efforts to support aviation by establishing an aerological section. Immediately after the end of WWI in December 1918, the Weather Bureau issued its first aviation forecast. It was for the Aerial Mail Service route from New York to Chicago (http://celebrating200years.noaa.gov/foundations/aviation_weather/welcome.html).

When the Air Commerce Act of 1926 was established, it included the first official requirements for providing aviation weather services. The act stated in Section 5 Part e:

> It shall be the duty of the Secretary of Commerce ... (a) to furnish such weather reports, forecasts, warnings, and advices as may be required to promote the safety and efficiency of air navigation in the United States and above the high seas, particularly upon the civil airways designated by the Secretary of Commerce under authority of law as routes suitable for air commerce, ...

Lives and property protection for the general public has always been an important focus in addition to the support of commerce. Thunderstorm hazards, which affect the general public as well as aviation, play an important role in the history of support to aviation. Radar entered the observation and forecasting picture in 1942, when the US Navy gave the Weather Bureau 25 surplus aircraft radars. These radars were modified for ground meteorological use, marking the start of a weather radar network and efforts to forecast severe weather.

The predecessor to today's NWS Aviation Weather Center began in the early 1950s with the establishment of a Unit within the Weather Bureau-Army-Navy (WBAN) Analysis Center in 1952; renamed the Severe Local Storms (SELS) Unit in 1953, with a move from the east coast to Kansas City in 1954, and the addition of the Radar Analysis and Development Unit (RADU) soon following in 1956. By 1957, SELS was issuing two categories of forecasts: Aviation and Public (http://www.spc.noaa.gov/history/timeline.html).

Commercial air travel gained momentum in the years following World War II and began to take off in popularity and efficiency with the introduction of jet aircraft such as the B707 in 1958. The ability to consistently cruise at altitudes above 30,000 ft (see Table 2.2) set the stage for a continuous expansion of the commercial passenger and cargo aviation industry. Severe convective weather hazards to aviation as well as turbulence and mountain wave forecasting were just beginning to be recognized as areas needing attention in the late 1950s.

Table 2.2 Documents altitude capability of some commercial jet aircraft placed into service during the late 1950s through the 1960s. Refer to Table 2.1 for a comparison with propeller aircraft

Selected Jet Passenger Aircraft 1958–1974		
Mftr & Aircraft	*Initial Service Year*	*Certified Service Ceiling (ft)*
Boeing 707-120	1958	38,993
DC8 series 10	1959	–
DC8 series 30	–	36,000
Convair 880	1960	41,000
Boeing 727 series 100	1964	36,100
Boeing 727 series 200	1967	42,000
DC-9 series 10	1965	37,000
DC-9 series 30	1967	37,000
Boeing 747-100	1969	45,000

The 1958 Federal Aviation Act established the modern day FAA, initially known as the Federal Aviation Agency. This Act set an important precedent requiring the then named Weather Bureau to work with the FAA to provide the necessary aviation weather products to govern the safety and efficiency of flight. Public Law 85-726 Section 310 states:

> The Administrator is empowered and directed to make recommendations to the Secretary of Commerce for providing meteorological service necessary for the safe and efficient movement of aircraft in air commerce.

By the late 1950s, the Weather Bureau had established dedicated forecast positions that focused on handling aviation products such as the Terminal Aerodrome Forecast (TAF). The Aviation Meteorologist positions in the current 100+ Weather Forecast Offices (WFOs) continue to be tasked today to produce TAFs. But other aviation roles have developed and evolved greatly in the last 50+ years.

By 1961, special training began for Federal Aviation Agency employees to equip them to brief pilots as part of a joint program with the Weather Bureau. During the 1950s and 1960s, the military and airline meteorology offices were focusing a significant amount of resources on aviation forecasting. In fact, the USAF Air Weather Service issued the first official government forecast of clear-air turbulence in 1961 (http://www.weather.gov/timeline).

A number of changes were made within the NWS and the FAA after the Southern Airways, Flight 242 aircraft, crashed near New Hope, Georgia, on 04 April 1977. According to the NTSB Aircraft Accident Report (1978), the aircraft encountered a thunderstorm between 17,000 and 14,000 ft and one of the major contributing factors in the accident was: "... limitations in the FAA's air traffic control system which precluded the timely dissemination of real-time hazardous weather information to the flightcrew."

Changes within the NWS and FAA began in 1978 when the RADU, which was established in 1956, ceased operations and the Convective SIGMET Unit was established in Kansas City, with the responsibility for reporting current locations of convection over the USA. Equally significant for aviation, NWS staffed, Center Weather Service Units (CWSUs) were formed in 1978 at FAA facilities. CWSU Meteorologists continue today to provide Center Weather Advisories (CWAs) and Meteorological Impact Statements (MISs) for the 21 FAA Air Route Traffic Control Centers (ARTCCs) in the United States. The CWSUs are colocated with air traffic operations within each of the ARTCCs to provide support for weather impacting the various centers including turbulence. Organizationally, each CWSU is supervised by the Meteorologist—In-Charge at the nearby WFO.

In 1982, The National Aviation Weather Advisory Unit (NAWAU) was established as part of the National Severe Storms Forecast Center (NSSFC) organization within the NWS in Kansas City. The NSSFC had originally been named in 1966 and existed as predecessor organizations in Kansas City since 1954. In October 1995, the Aviation Weather Center (AWC) was formed in Kansas City, Missouri. It was one part of the then, newly organized National Centers for Environmental Prediction (NCEP). Organizationally, the AWC was a consolidation of the NAWAU and the aviation component of the former National Meteorological Center (NMC) Meteorological Operations Division (MOD) near Washington, DC (see http://www.spc.noaa.gov/history/timeline.html, http://www.srh.weather.gov/topics/html/jan1597.htm).

Ostby (1992) described the role of NAWAU to provide: "alerts, nationally, of hazardous weather, known as AIRMETs and SIGMETs, and convective SIGMETs to aviation interests. NAWAU also issues aviation area forecasts." Today, as the NAUWAU did in the 1980s, the AWC issues area forecasts (FAs), Airman's Meteorological Information statements (AIRMETs), Significant Meteorological Information statements (SIGMETs), and Convective SIGMETs. In addition to the AWC, the Alaska Aviation Weather Unit (AAWU) and the WFO in Honolulu, HI (HFO) contribute to SIGMET issuances. These four products are shared and distributed publicly to pilots via the web and by Flight Service Station Specialists. FAA ARTCC sector controllers also distribute SIGMETs and Convective SIGMETs verbally via their VHF radio frequencies used for air traffic control. SIGMETs are also produced to fulfill ICAO Annex 3—Meteorological Services for International Air Navigation requirements. For a further description of the NWS AWC produced AIRMET and SIGMET product, see Chapter 10.

Today, organizationally the CWSUs (producers of CWAs and AWS) and the 122 WFOs (producers of the TAFs) report to one of six regional NWS offices. The AWC is a separate organization within the NWS. While no organizational structure is perfect, this arrangement has complicated the provision of aviation weather services in the past, including the provision of turbulence related information.

2.2.3 Turbulence Research and Aviation Applications History

2.2.3.1 Clear-Air Turbulence

Turbulence forecasting got its start in the World War II era, when scientists and individuals in the field of aviation first began attempting to correlate observed clear-air turbulence events with large scale synoptic features. Throughout the years, the large scale feature that was most commonly associated with turbulence was frontogenesis. In one of the first published studies to address turbulence, Baughman (1946) used a case study to suggest that clear-air turbulence was occurring over the North Atlantic Ocean in regions where there was a sharp air mass contrast and strong wind shear. Almost a decade later, Chambers (1955) provided several cases of observed moderate to severe turbulence over the North Atlantic Ocean and found that a substantial number of the strong turbulence events occurred in areas of pronounced wind shear. Several years later, Dutton and Panofsky (1970) reported that clear-air turbulence was most likely to occur in areas of strong vertical wind shear and strong horizontal temperature gradients.

While most early studies related turbulence to wind shear and temperature gradients, later studies suggested that one more atmospheric feature, deformation, may play an important role in turbulence production. Mancuso and Endlich (1966) used statistical methods to show that turbulence occurred in areas with large vertical wind shear and deformation values. Roach (1970) also found that large scale shearing and stretching deformation may play a role in turbulence production. Several studies comparing satellite imagery and PIREPs supported the hypothesis that turbulence frequently occurs along deformation zones (Anderson et al. 1982; Ellrod 1985; Ellrod and Knapp 1992).

Continuing the theory that wind shear and deformation are important factors in turbulence production, Ellrod and Knapp (1992) created the Turbulence Index (TI), which forecasts turbulence by highlighting areas that have large vertical wind shear and deformation values. The TI is derived from Petterssen's frontogenetic intensity equation, which relates frontogenesis to an increase in vertical wind shear (Petterssen 1956; Ellrod and Knapp 1992). The forecast technique is supported by earlier research showing that the product of vertical wind shear and deformation resulted in the highest correlation with observed moderate or severe turbulence (Mancuso and Endlich 1966). Using two case studies to illustrate the TI's ability to forecast turbulence, it was determined that the TI correctly forecasted between two-thirds and three-fourths of turbulence events, while maintaining a low false alarm ratio (Ellrod and Knapp 1992).

Although the TI performs well when predicting turbulence in areas where frontogenesis is occurring, it often misses turbulence caused by other atmospheric processes. Knox (1997) compared several different turbulence forecasting methods, including the TI, and found that they may be incorrectly predicting smooth conditions in regions of turbulence caused by processes other than frontogenesis. Ellrod and Knox (2010) attempted to remedy this problem by adding a divergence trend term into the TI, to account for non-frontogenetical situations, creating a divergence-modified turbulence index (DTI). Two case studies of turbulence outbreaks supported the idea that the addition of the divergence trend term to the TI creates a better turbulence diagnostic, as the DTI showed an improvement over the TI in both cases (Ellrod and Knox 2010).

The development of the Graphical Turbulence Guidance (GTG) product led to a significant advancement in the ability to produce automated turbulence forecasts. The GTG takes a different approach to forecasting turbulence: it produces a forecast by using a weighted combination of several individual turbulence diagnostics (Sharman et al. 2006). To test the effectiveness of the GTG, turbulence forecasts were produced from the GTG weighted combination and from each one of the GTG's components, and each forecast was compared to PIREPs for verification. It was determined that the GTG weighted combination performed better than any single turbulence diagnostic. However, in a nod to earlier turbulence research, it was discovered that out of all of the GTG's components, frontogenesis was the single best predictor of turbulence (Sharman et al. 2006). In addition, there are commercial vendors who have begun to produce turbulence forecast algorithms and are now marketing and selling the output.

Forecasted and observed turbulence information has been important for commercial airline operations over the years for a number of reasons. If severe turbulence is encountered, continued use of the aircraft is delayed until a physical inspection can be completed. An aircraft's route and/or altitude will be adjusted preflight, not only to avoid severe turbulence but also for passenger comfort as well as reduction in the risk of passenger and flight attendant injuries. Fahey (1993) reported that one airline was able to realize $700,000 in annual savings in passenger and flight attendant avoided injuries with preflight and en route turbulence products and procedures. By the early 2000s, it was estimated that this same airline was avoiding annual flight attendant turbulence injury costs alone, in comparison to two other airline encounter rates of $0.45 M and $1.33 M (Schultz et al. 2003).

2.2.3.2 Mountain Wave and Turbulence Forecasting

The number and complexity of mountain wave variables has always posed a challenge for developing accurate forecast techniques. Prior to the 1990s, mountain waves were forecast at some airline meteorology offices by correlating surface pressure gradient (thereby assuming wind flow over the mountains), to upper level winds, using a nomogram (see Fig. 2.6 as an example). As computer models have advanced, forecast techniques have become more involved, incorporating more variables into the forecast and leading to more accurate results.

Mountain waves are mostly a winter season phenomena, due to the strong static stability at mountain top and strong jet stream winds above. Air flowing perpendicularly across favorable mountain profiles can displace this stable air at mountain top, thereby doing work on this layer and generating wave energy. This energy then propagates upwards until it reaches the tropopause, unless it encounters a strong layer of stability or wind shear that will act to break the wave prematurely. Since strong jet stream winds and the associated tropopause level are often at the altitudes flown by commercial aircraft, Aviation Meteorologists must monitor the mid-latitude regions around the world where waves tend to develop. Monitoring includes continuous review of current PIREPs and satellite images as well as assessment of both model forecasts and observations of evolving thermal stability profiles, mountain tops, winds, strong surface pressure gradients, and wind shear layers.

2.3 US Airlines' Historical Perspective on Turbulence

2.3.1 Airline Meteorology Offices

In 1927, Pan American Airways (PanAm) became the first U. S. airline to hire meteorologists. Today, DAL, FedEx, UPS, and Southwest Airlines are the only US airlines that run 7 by 24 meteorology offices with airline employed, in-house meteorologists. DAL and two airlines that DAL acquired, Western Airlines in 1987 and NWA in 2008, all have had long traditions of in-house meteorology staffs.

2.3.1.1 Northwest Airlines

Bourke (2007) reports that NWA was the eighth airline to hire meteorologists when it opened its first meteorology office in Spokane, Washington, in 1938. Of those eight airlines in 1938, only two are still operating today, UAL and AAL. By 1952, there were approximately 16 commercial, passenger air carriers in North America with in-house meteorology offices. Bourke (2007) provided a complete list of airlines and dates when they established in-house meteorology office(s). Assuming none of the airlines closed their meteorology office after opening, in 1952 the list would have included: AAL, Braniff, Capital, Chicago and Southern, Continental, DAL, Eastern, Mid-Continent, Northeast, NWA, Panagra, PanAm, Trans Canada (now Air Canada), TWA, UAL, and Western.

In the 1940s and 1950s, NWA staffed meteorology offices at different times at a total of seven different locations. Dan Sowa, who led the NWA weather group for many years, studied meteorology at the Spartan School of Aeronautics during WWII and in 1945 at the age of 22 years, was hired as a NWA meteorologist in the MSP office. In 1947, Dan Sowa was promoted to Area Chief Meteorologist for the Anchorage and Seattle offices and in 1959 was appointed Superintendent Meteorology (Source: RNPA 2000). He then moved back to the Minneapolis-St. Paul (MSP) office in the early 1960s until his retirement in 1985.

2.3.1.2 Delta Air Lines

During the 1930s as the airlines began to hire meteorologists, a monumental initiative was underway. The United States Army had a competition among the airplane manufacturers for a new heavy bomber. The resulting Douglas B-23 Dragon and the Boeing B-17 Flying Fortress were huge modern machines at the time. Both the B-23 and the B-17 had Service Ceilings above 30,000 ft, well within the jet stream levels (see Table 2.1). Unfortunately, both planes' maximum speed was less than 300 kts, and usual cruising speeds were less than 200 kts.

It was found, during WWII with a strong jet stream off the Japanese coast, a bomber may only be making a ground speed of 50–100 kts when approaching Japan from the east. The army needed to understand these very strong winds since they had such a major impact on air operations. Meteorologists were placed onboard to take observations of the impact on the plane performance. The collected data was analyzed, and the "Upper Front Model" was developed. C. L. "Chan" Chandler was one of these US Air Force (USAF) meteorologists.

In 1954, Chan took his expertise in upper winds and turbulence to PanAm. In 1959, as commercial airlines were entering the jet era with DC-8s, Chan joined DAL and led the weather analysis group until his death in 1989. During the three decades of 1960s–1980s, the route and altitude of the flight were selected by the meteorologist based on safety, comfort, and economy. Forecasting upper air turbulence used data from rawinsonde observations and an empirical model using jet cores at FL390, FL340, and FL290 (subtropical, polar, subarctic, by name, respectively). The three jets were associated with 500 mb and 300 mb temperatures, respectively, of −13 C/−39 C (subtropical), −20 C/−46 C (polar), and −27 C/−50 C (subartic). This three-dimensional model of the atmosphere with jet cores, sloping upper front surfaces and associated temperatures was used because observational data was sparse at best. Manual extrapolation of the features 12–24 h in the future based on speed of movement over the past 12–24 h was then used. This was the basis for turbulence forecasting at airline meteorology offices at the time.

Chan also set up an independent company, Southeast Weather, to sell flight plans to other companies and airlines. See Fig. 2.1 for a photo of C. L. Chandler.

2.3.1.3 Western Airlines and Merger with DAL

According to Bourke (2007), in 1929, Western Air Express (later Western Airlines) began employing meteorologists. Western's meteorology office history, prior to mid-1970s, is a little cloudy. But, a portion of the story begins when John Pappas turned 18 during his senior year in 1945. The next day, he quit high school and joined the army. He was assigned to meteorology and learned the Upper Front Model. After retiring from the military, he joined Southeast Weather and was manager there when Western Airlines decided to start up/restart a meteorology department in 1975. Western hired John as manager and John Bordeleau, a former

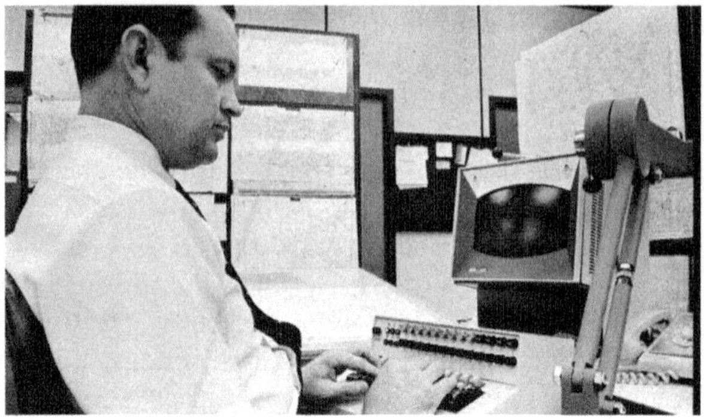

Fig. 2.1 C. L. (Chan) Chandler, Delta superintendent of weather analysis, keys in wind and weather data using IBM 2915 CRT linked to a 7074 computer. *Image and caption source*: Air Transport World magazine, March 1972

manager of Southeast Weather, as assistant manager, as well as three other experienced Southeast Weather employees, when the department was formed in January 1976. In addition, four younger, aspiring meteorologists were added to the team of nine and they started producing flight plans in April 1976 (Mike Heying, personal communication, June 29, 2015, and Conrad Svoboda, July 6, 2015).

The Western Airlines process for producing flight plans and forecasting turbulence in the late 1970s was almost identical to DAL's. Svoboda, (personal communication, July 6, 2015) reports that Chan would fly from DAL's Atlanta office out to Western's office near LAX airport, "2 or 3 times a year and critique our department, sitting down with each of us while on shift—observing our time management, thinking process and offering suggestions."

Svoboda, (personal communication, July 6, 2015) continues by describing the manual route selection, manual wind forecasting, and turbulence forecasting: "It was labor intensive, plotting RAOB data using standard levels: 500–150 mb with 300 mb as the base map, analyzed for ridges, troughs, upper fronts; noting temperature differences, level of maximum winds and the tropopause,. . .. Precedence was given to route selection then upper wind forecasts. Turbulence forecasts were included in every flight plan, but were not stressed."

Brockmeyer (personal communication, June 29, 2015) shared his perspective of Western meteorology just prior to and after the merger with DAL in April 1987, characterizing the process of manual flight planning at the time as "now-casting." Brockmeyer also noted that the model advocated by C. L. Chandler featured a theoretical Z—layer located between the top of the upper front shear and bottom of tropopause shear, which should feature smoother air in the jet core. Chandler (1987) provides a detailed description of the Z—layer concept.

By the 1990s, DAL had introduced computers that produced RAOB plots of observed upper air station data. At this same time, forecast data was computer plotted, providing a layered look at the atmosphere, but was still hand-analyzed for tropopause, troughs, upper front, and other features.

2.3.2 Clear-Air Turbulence Categories: Identification and Forecasting (1960s–1990s)

The level of focus on turbulence reporting and forecasting varied between airlines operating during the 1960s–1990s. Among US Carriers, focus was probably greatest at NWA. This was due to its more northerly routes with greater exposure to both clear-air turbulence and mountain wave activity over the Rockies as well as en route between the northern US and Japan/Asia. But all airlines used similar models to identify a variety of turbulent features in the atmosphere.

2.3.2.1 Upper Front

The most common turbulence events were deemed to be in close proximity to jet streams in an upper-level front, a narrow volume parallel to and below a jet stream. This feature was associated with strong vertical and horizontal shear; horizontal temperature gradient; and a discontinuity (steep decline on synoptic charts) in the tropopause on the cold side of the jet core (Fig. 2.2). From the 1960s well into the 1990s, this synoptic analysis by aviation forecasters was used to locate and forecast upper front turbulence. As described in Sect. 2.3.1.2, temperature associated with the observed jet core types (Subtropical, Polar, and Subarctic) was highlighted on 300 mb charts. And after frontogenesis was identified by noting a trend of increasing horizontal temperature and both horizontal and vertical velocity gradients, forecasts were produced. Initially, forecasts were mostly the product of extrapolation from observed values and "rules of thumb" regarding the movement, development, and increase or decrease of the thermal and velocity gradients. As numerical model output charts increased in accuracy and resolution during the 1970s, 1980s, and 1990s, the procedures for identifying upper level fronts by thermal gradients were adapted for use on various model forecast charts.

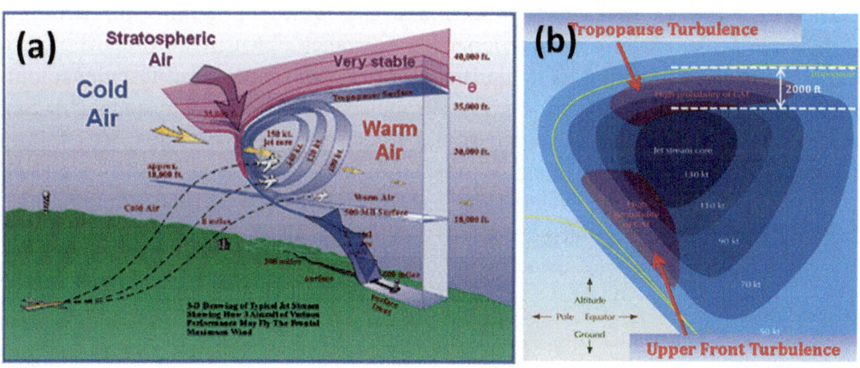

Fig. 2.2 (**a**) Three-dimensional and (**b**) two-dimensional models of upper level front and associated areas of potential turbulence

2.3.2.2 Trough Line

Forecasters also noted numerous turbulence encounters in the vicinity of upper level troughs, near the axis of wind velocity changes. By the 1960s, three rules of thumb were consistently being applied in order to identify which troughs would likely be turbulent. Turbulence was attributed to sharp troughs (a subset being deformation zones known to aviation forecasters as "bottle neck troughs"), troughs moving at speeds at or greater than 30 kts and troughs with marked vertical slope. Troughs were identified by 300 mb analysis and future speed of movement, mainly extrapolation of past movement. By the early 1980s this extrapolation oriented technique had gradually been replaced with upper level model charts from the baroclinic and spectral models. But the turbulence forecast continued to be based on the three troughs model.

2.3.2.3 Tropopause

In the 1960s with passenger aircraft routinely reaching altitudes greater than 35,000 ft (see Table 2.2), turbulence was occasionally being noted in the stratosphere in the absence of the other known causes. By the 1970s, lateral jet stream movements greater than 25 kts was used as a rule of thumb for forecasting moderate of greater turbulence. Today forecasters realize that this turbulence is caused by gravity waves in the tropopause and associated turbulence as the tropopause is abruptly lifted. This type of turbulence is generally associated with rapid cyclogenesis and known to forecasters as an "exploding ridge." In northern latitudes, it is most frequently observed off the east coast of Asia and North America.

2.3.2.4 Convection Induced Turbulence

During the 1980s, forecasters noticed that the "rules of thumb" for avoiding turbulence in the vicinity of thunderstorms (distance downstream proportional to the ambient wind speed) was not sufficient to explain widespread turbulence outbreaks associated with nocturnally formed mesoscale convective systems (MCSs). A related phenomena was the occasional forecast wind bust with flights falling behind or gaining on flight plans. Winds were, at that time, manually forecast and entered into flight planning systems by meteorologist. Forecasters could only react to these events as they occurred.

2.3.2.5 Transverse Band Turbulence

In the 1990s, turbulence that had previously been a mystery began to be recognized with the advent of better satellite imagery. It was not well understood but appeared in extensive cloud plumes usually with some convection at the origin of the plume.

2.3.3 Clear-Air and Convection Induced Turbulence: Identification and Forecasting (2000s)

Over the last 10–15 years, improvements in the resolution of numerical models and a more flexible array of model graphics have allowed for significant improvements in the identification and forecasting of the types of turbulence discussed above. Forecast vertical wind shear profiles are readily available and invaluable in forecasting upper fronts, vertically sloped troughs, and bottle neck troughs. Vertical profiles also have shown that the increased negative shears above the tropopause induced by jet stream movements or building ridges are associated with turbulence reports and are reliably forecast with current model output. Convective Induced Turbulence is still challenging, but improvements in understanding the process of forming the associated radial bands has greatly increased forecasting abilities. Transverse band forecasting is also still challenging but once formed, techniques for forecasting duration have been aided by the ability to display shear vectors and upper level relative humidity charts in addition to upper level advection.

2.3.4 Mountain Wave Turbulence

The seminal mountain wave field research by C. F. Jenkins and J. Kuettner (1953) was conducted in a very turbulent mountain wave region of the USA, the Owens Valley area in east central California.

There was a significant amount of additional, but less known research completed by commercial airline Meteorology offices in the 1950s and 1960s on turbulence as well as on mountain wave activity. UAL began a very comprehensive effort in the 1950s. One of their foci was on the analysis (Harrison 1956) and then on the forecasting (Harrison 1957a) of mountain wave activity over the Denver area. A detailed analysis of mountain wave areas in the entire USA was also published in Harrison (1957b). Documenting the location, frequency and intensity of mountain wave activity in the USA was the purpose of the latter study. It combined a comprehensive analysis of terrain and narratives from over 200 UAL pilots' mountain wave experiences, collected during and after a training module on terrain effects of weather. The terrain analysis applied the findings of Jenkins and Kuettner (1953) that mountain wave intensity is based in part on the height of mountain ridge and steepness of mountain slope on the lee wind side. An example of one of those UAL pilot comments, this one specifically regarding the Sierra Nevada area: "…if you can find an area where the terrain lets down slowly from the top of the ridge instead of abruptly you will find the turbulence reduced by about 90%." (Harrison 1957b, p. 10). The concept noted by another pilot would be developed further as an avoidance technique in the 1960s: "Turbulence of Laramie Valley can still be detoured by a little additional mileage over the town of Medicine Bow and east until north of Cheyenne …" (Harrison 1957b, p. 22).

The introduction of jet aircraft in the late 1950s and early 1960s with Certified Service Ceilings of 36,000–45,000 ft and with the capability of operating day in and day out at altitudes between 31,000 and 41,000 ft resulted in aircraft cruising near or just above the tropopause (see Table 2.2). By 1963, NWA had collected enough empirical data from flight crews to introduce new operating and flight planning procedures. NWA Flight Standard Bulletin 9-63 in December 1963 stated: "The worst turbulence in mountain wave conditions is in or near the tropopause. Although the trop varies about plus or minus 4000 ft in height in such conditions, it is possible to forecast its average height with good accuracy even though intensity and duration of the turbulence cannot be accurately forecasted." (Harrison and Sowa 1966, p. 10).

About this same time, in the early 1960s, NWA proposed a joint study with UAL after noting that in addition to altitude avoidance, preliminary work showed that some wave activity on NWA routes through Montana were confined to a narrow band that could be avoided with small detours. Two routes had been identified, one near Helena, MT, and another east of Whitehall.

Work was done from 1963 to 1965 by both NWA and UAL and was then published (Harrison and Sowa 1966). Their work identified 169 unique wave zones along 31 jet air routes, in the lower 48 states of the USA, west of 100 W longitude and flown by either airline. It included analysis of mountain ridge profiles favorable for triggering wave action, using a variety of data. This included the collection of the observed evidence of wave action and/or turbulence using pilot reports; wave clouds reported by NWS; as well as photos. Calculations of the height of the crest of each of the 169 wave zones as well as the terrain drop from each crest to a lee elevation 15–20 nm downwind were compiled along with estimated activating wind directions based on orientation of the ridge. The observed and calculated data sets were also cross referenced to provide the location along the airway most susceptible to the mountain wave hazards, estimated to within 1 nm. Each of the 169 wave zones were classified as one of two potential intensities: "some wave action" or "strong wave possible." Thirty-seven of the 169 wave zones were classified as "strong wave possible."

2.4 Northwest Airlines Turbulence Plot System

In early June 1965, Paul Soderlind, Director Flight Standards and Dan Sowa, Superintendent Meteorology, proposed an integrated program at NWA, addressing not only mountain wave activity and associated turbulence but also clear-air turbulence, thunderstorms, and wind shear (NWA 1965a, b). And according to Ruble (1986) and Bourke (2007), clear-air turbulence was being forecast at NWA as early as 1957.

By 1968, Sowa and Soderlind had completed all the necessary preparation to address the four aviation hazards identified in the 1965 memos. On 10 October 1968, NWA began a program called the Turbulence Plot (TP) system (NWA 1968).

From a mountain wave perspective, the TP system in 1968 incorporated 29 of the 37 "strong wave possible" zones large mountain wave areas (Fig. 2.3). Mountain Wave Area (MWA) 1 covered the northern Rockies from southern Canada to northern Wyoming. MWA 2 covered the same as 1 except for NW Montana and southern Canada. MWA 3 included Wyoming, Colorado, and a small area of New Mexico. MWA 3 could be activated in its entirety or a subset could be issued that excluded the area over Wyoming and was identified as MWA 3A. The Sierra Nevada range along the California and Nevada border was identified as MWA 4.

The TP system's mountain wave avoidance included both altitude avoidance near the tropopause and horizontal avoidance using "Deviation Routes." Small horizontal route adjustments in order to fly over less steep terrain reduced the risk of turbulence associated with wave activity. For example, the BFF (Scotts Bluff, NE)-MBW (Medicine Bow, WY)—MYTON (Myton, Utah) route is one example of a DAL mountain wave deviation route still used today.

2.4.1 Reporting, Forecasting, and Manual Production: 1970s–1980s

In the 1970s and 1980s, the TP system was much more of a hazard reporting system than a forecasting system. Mountain wave areas were usually activated with a TP after the conditions had developed. Thunderstorm TPs were produced using text summaries of the current detected activity for every radar site in the lower 48 of US. The potential associated convection induced turbulence was implied. Low Altitude Wind shear was limited to areas associated with warm or cold fronts and was estimated using the gradient wind velocity above the elevated frontal surface over the affected airport. All clear-air turbulence TP hazards were usually classified into one of the following causes: Tropopause, Trof Line, or Upper Front and forecasts were inferred with empirical data.

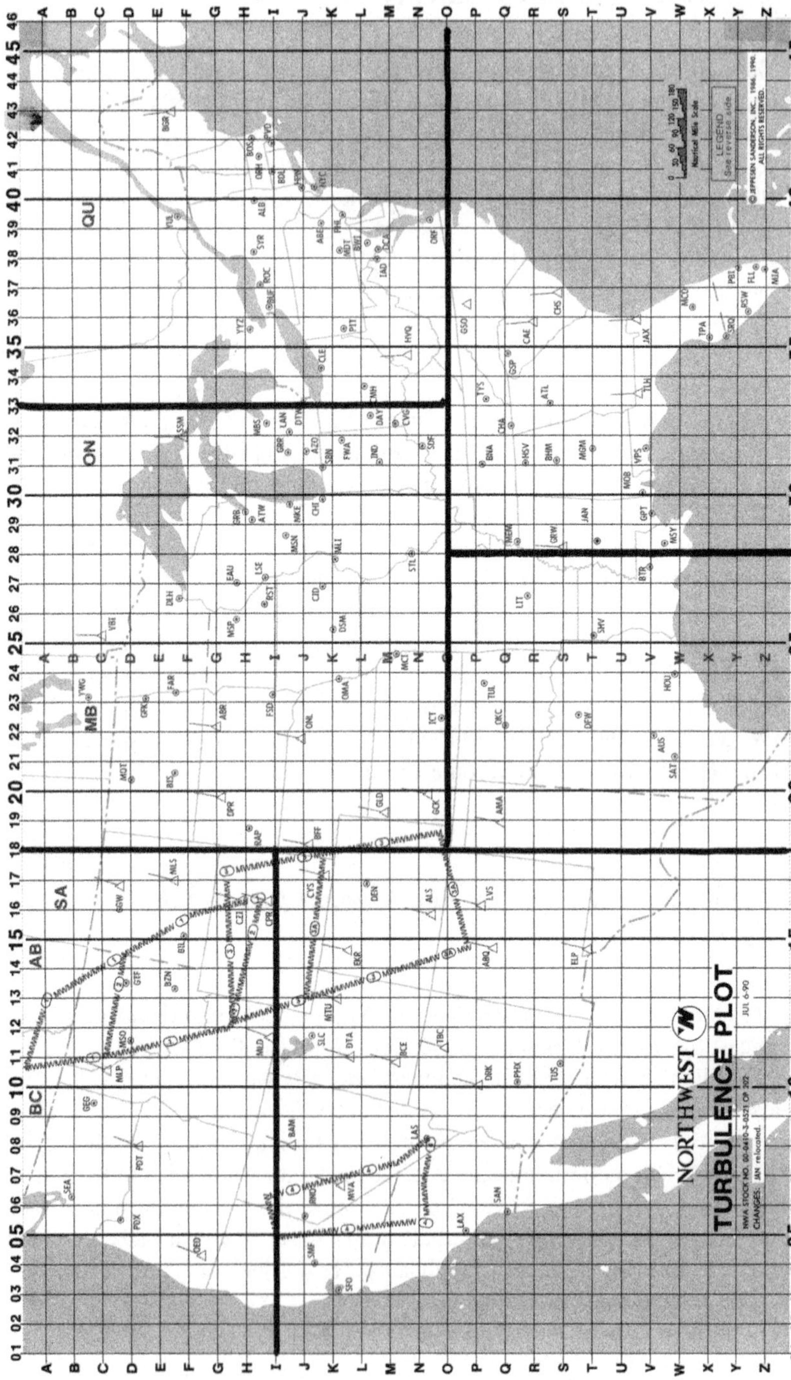

Fig. 2.3 NWA Mountain Wave areas used from 1960s to 1993 and version of the TP form used for plotting from late 1980s to 1993

2.4.2 *TP Distribution and Usage: 1970s–1980s*

In 1970s and 1980s, the TP information was displayed for all dispatchers on a large 2 m by 4 m map of the USA covered in Plexiglas; suspended by counter weights; and able to be raised or lowered using track and wheels similar to those used on a garage door. This large board was located at the Main Headquarters Operations office at MSP airport. The weather hazards, when issued by the NWA meteorologists, were plotted using grease pencils. The meteorologist also used a dumb computer terminal to type the TP message containing hazard description and location. The NWA teletype-based communications system distributed the TP messages to all other NWA airport station operations offices as well as NWA radio operators.

Many of the NWA pilots began their trips in MSP and received paperwork, briefings, and referenced the plotted hazards on the TP board before departing (see Fig. 2.4). At all other stations, NWA pilots received the TP messages that applied to their flight, in their preflight documents, compiled by the NWA airport station agent. In the 1970s, once en route, pilots monitored company radar continuously. All updated TP messages were broadcast by NWA radio operators using specific radio frequencies for separate geographical areas in the USA that matched TP coverage areas. NWA pilots used a specially designed large map, similar in size to an airway chart, to plot the hazard(s) during both their preflight preparation and updates while en route.

By the 1980s, the ACARS system that was being used to send company positions reports from the aircraft to the airline's personnel on the ground (downlinking) was

Fig. 2.4 Turbulence Plot (TP) Board & Pre-Flt prep, Circa 1985. Foreground, NWA First Officer Jon Pendleton reviews preflight paperwork, while on the right, Captain Schellinger consults with Meteorologist, Monique Venne, and in the background, First Officer David Schroeder checks the hazards plotted on the TP Board, and on the left the four flight attendants prep

also beginning to be used for distribution of weather hazard information from the NWA flight dispatch office to the aircraft while en route (uplinking).

2.4.3 TP System 1990s–2009

2.4.3.1 Distribution to Aircraft Preflight and En Route

In the late 1980s, NWA began to supply hand drawn depictions, displaying upper air features such as the jet streams and troughs as well as forecasted areas of turbulence. They were hand delivered to dispatchers and sent via FAX to airport operation personnel, for delivery to flight crews before departure. In the 1990s, turbulence depiction charts began being crafted using computer aided design graphics, and were displayed electronically to be viewed by dispatchers in the operations center and for crews worldwide.

TP messages continued to be plotted by NWA Meteorologist on the same TP board as used in 1970 (see Fig. 2.5), for dispatchers' reference. Flight crews were no longer reporting to the operations center but were still expected to plot the TP information both preflight and while en route. By the early 1990s, all NWA aircraft were able to automatically receive text TP messages in the cockpit sorted for only

Fig. 2.5 Close-up Turbulence Plot (TP) Board. Circa 1970. NWA Meteorology Supervisor Dan Sowa beginning to plot areas of thunderstorm activity on the TP Board, right, while a newly hired NWA pilot provides distance and radial information from individual NWS WSR-57 information off the RAWARC circuit

those that affected the route being flown. Figure 2.3 shows the plotting chart that replaced the large airway map in the 1980s and was used until 1993 when the TP System was revised to use latitude and longitude coordinates.

2.4.3.2 Forecasting Advances

In the late 1980s, NWA began a project to revise both the Mountain Wave areas and the avoidance procedures for North America as well as the mountain wave forecasting procedures. Previously, the very large Mountain Wave Areas (see Fig. 2.3) were defined with particular Mountain Wave Avoidance routes specified from the Rockies all the way to the Sierras. Pilot reports were collected over an 18 months period and plotted on a plastic overlay on top of the USGS topographic maps. It became clear that the avoidance areas could be reduced in size. In addition the forecast tool (surface pressure/wind aloft nomogram) developed by UAL and NWA was deemed too simplistic with too many false alerts (Fig. 2.6).

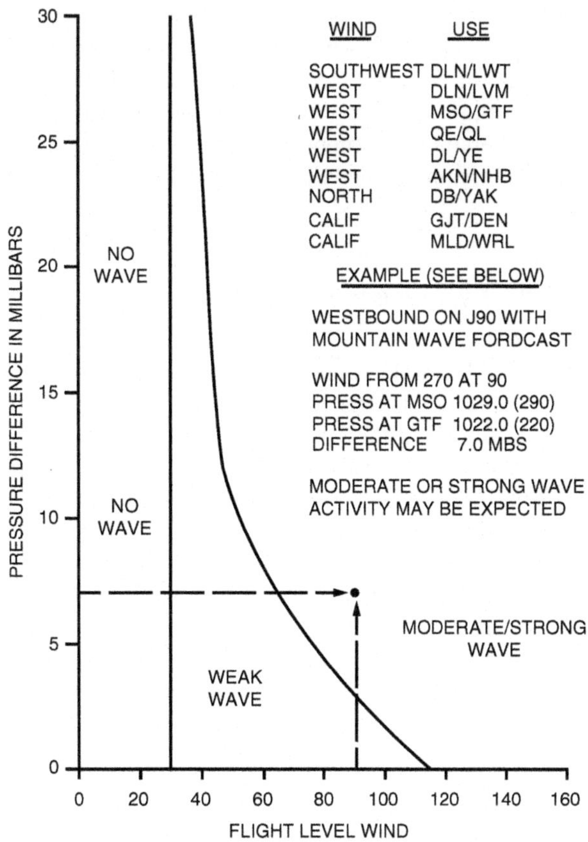

Fig. 2.6 NWA Mountain Wave Forecasting nomogram used during the 1970s to early 1990s

The US mountain ranges were segmented into smaller areas with the wave areas in the USA reduced in size and increased in number—going from 5 large areas to 15 smaller areas. Flights were no longer required to operate on avoidance routes outside of these smaller areas. Various improvements to wave forecasting were also implemented during this period including both identification of favorable synoptic patterns and favorable thermal profiles. A similar project for Alaska Mountain Wave areas was completed in 1995.

After the DAL acquisition of NWA in 2008, operations and procedures were integrated. The TP system including the mountain wave procedures were just one of many DAL or NWA products and procedures that were incorporated into the combined airline operation.

2.5 Conclusions

2.5.1 Weather Information from Aircraft

DAL meteorologists have found turbulence reports less than severe (moderate or light or even smooth) very helpful for nowcasting purposes over the years and still do today. In addition, focus on access to reliable and accurate automated aircraft reports of turbulence as well as other weather variables such as water vapor, winds, and temperatures are important for the benefit of not only airline users but the entire weather community and the general public.

2.5.2 Turbulence Forecasting

The WAFS and individual Meteorology authorities in countries around the world provide SIGWX Charts and SIGMETs, respectively, for airline turbulence awareness. In the USA, the CWSUs produce of CWAs and AWS, and the 122 WFOs produce the TAFs, while the AWC is a separate organization within the NWS and produces SIGMETs. Some airlines choose to provide additional tailoring of products for their internal procedure that are designed to support their unique ways of implementing safe and efficient operations.

With the continued development and anticipated improvement of computer model forecasts of turbulence, it is anticipated that the human role in turbulence forecasting will evolve greatly in the future (see Chap. 11 to read about the Delta Air Lines processes and in-house produced products).

Acknowledgement Thanks to all who shared their expertize. We would like to specifically recognize the following: Bill Watts who provided input regarding the AIREP/PIREP process; David Bright, Steven Lack, and Jesse Spark, from the NWS for their contribution to Sect. 2.2.2.1 "National Weather Service Organizational Perspective"; and to Retired Delta Meteorologist,

Conrad Svoboda, as well as current Delta Meteorologists Joe Brockmeyer and Mike Heying for sharing their recollections of past Delta and Western Meteorology departments and contributions to Sect. 2.3 "US Airlines' Historical Perspective on Turbulence."

References

Anderson, R.K., Gurka, J.J., Steinmetz, S.J.: Application of VAS multispectral imagery to aviation forecasting. Preprints, Ninth Conference on Weather Forecasting and Analysis, American Meteorological Society, Seattle, pp. 227–234 (1982)

Baughman Jr., E.E.: Turbulence with a stable lapse rate. Bull. Am. Meteorol. Soc. **27**, 459–462 (1946)

Benjamin, S.G., Brewster, K.A., Brummer, R.L., Jewett, B.F., Schlatter, T.W., Smith, T.L., Stamus, P.A.: An isentropic 3-hourly data assimilation system using ACARS aircraft observations. Mon. Weather Rev. **119**, 888–906 (1991) (Three airlines identified in the Acknowledgement section)

Bourke, K.: Northwest airlines meteorology department 1938–1982. Personal, Written Communication with Tom Fahey, 27 August 2007. On file in the Delta Air Lines Museum, Atlanta, GA

Chambers, E.: Clear-air turbulence and civil jet operation. J. R. Aeronaut. Soc. **59**, 613–628 (1955)

Chandler, C.L.: Turbulence forecasting. In: Camp, D.W., Frost, W. (eds.) Atmospheric Turbulence Relative to Aviation, Missile, and Space Programs, pp. 137–154. NASA Conference Publication 2468 (1987)

Dutton, J., Panofsky, H.A.: Clear-air turbulence: a mystery may be unfolding. Science **167**, 937–944 (1970)

Ellrod, G.P.: Detection of high level turbulence using satellite imagery and upper air data. NOAA Technical Memorandum NESDIS 10, U.S. Department of Commerce, Washington, DC (1985)

Ellrod, G.P., Knapp, D.L.: An objective clear-air turbulence forecasting technique: verification and operational use. Weather Forecast. **7**, 150–165 (1992)

Ellrod, G.P., Knox, J.A.: Improvements to an operational clear-air turbulence diagnostic index by addition of a divergence trend term. Weather Forecast. **25**, 789–798 (2010)

FAA: FAR (Federal Aviation Regulations) Part 121 (2015). Available online at: http://www.ecfr.gov/cgi-bin/text-idx?SID=222a0cb93d8edc53cbda769499d67d1f&mc=true&node=pt14.3.121&rgn=div5

Fahey III, T.H.: The northwest airlines atmospheric hazards advisory & avoidance system. Aviation Range & Aerospace Meteorology, 8th International Conference on Aviation Weather Systems, American Meteorological Society, Vienna, VA (1993)

Fleming, R.: The Use of commercial aircraft as platforms for environmental measurements. Bull. Am. Meteorol. Soc. **77**, 2229–2242 (1996)

Harrison, H.T.: Synoptic features of the mountain wave at Denver, Colorado. UAL Meteorology Circular No. 41 (1956)

Harrison, H.T.: Forecasting the mountain wave at Denver, Colorado. UAL Meteorology Circular No. 42 (1957a)

Harrison, H.T.: Mountain wave zones in the United States. UAL Meteorology Circular No. 43 (1957b)

Harrison, H.T., Sowa, D.F.: Mountain wave exposure on jet routes of Northwest Airlines and United Air Lines. UAL Meteorology Circular No. 60 (1966)

ICAO: Annex 3, 18th edn., 2013, 76th Amendment, Meteorological services for international air navigation

ICAO: ROBEX Handbook, 12th edn., 11th Amendment May 2013, Section 9 AIREP/AIREP SPECIAL EXCHANGE (2004)

Jenkins, C.F., Kuettner, J.: Flight aspects of the mountain wave. Air Force Surveys in Geophysics No. 35 (1953)
Knox, J.A.: Possible mechanism of clear-air turbulence in strongly anticyclonic flows. Mon. Weather Rev. **125**, 1251–1259 (1997)
Mancuso, R.L., Endlich, R.M.: Clear air turbulence frequency as a function of wind shear and deformation. Mon. Weather Rev. **94**, 581–585 (1966)
Moninger, W.R., Mamrosh, R.D., Pauley, P.M.: Automated meteorological reports from commercial aircraft. Bull. Am. Meteorol. Soc. **84**, 203–216 (2003)
NTSB: Aircraft Accident Report, Southern Airways, Inc., DC-9-31, N1335U, Report# AAR-78-3, 26 Jan (1978)
NWA: Interoffice Communication, from D. F. Sowa to P. A. Soderlind; 02 June 1965. On file in the Delta Air Lines Museum, Atlanta, GA (1965a)
NWA: Interoffice Communication, from P. A. Soderlind to Benjamin G. Griggs; 04 June 1965. On file in the Delta Air Lines Museum, Atlanta, GA (1965b)
NWA: Ground Services Memo No. 128-68; 07 Oct 1968. On file in the Delta Air Lines Museum, Atlanta, GA (1968)
Ostby, F.P.: Operations of the National Severe Storms Forecast Center. Weather Forecast. **7**, 546–563 (1992)
Petterssen, S.: Weather Analysis and Forecasting, vol. 1. McGraw-Hill, New York, NY (1956)
Retired Northwest Airlines Pilots' Association: Newsletter #134, May 2000, p. 19. On file in the Delta Air Lines Museum, Atlanta, GA (2000)
Roach, W.T.: On the influence of synoptic development on the production of high level turbulence. Q. J. R. Meteorol. Soc. **96**, 413–429 (1970)
Ruble, K.: Flight to the Top. Viking Press, New York, NY (1986)
Schultz, A.W., Fahey, T.H., Zimmerman, J.D.: Mountain Wave and Clear Air Turbulence Forecasting at Northwest Airlines. On file in the Delta Air Lines Museum, Atlanta, GA (2003)
Sharman, R., Tebaldi, C., Wiener, G., Wolff, J.: An integrated approach to mid and upper-level turbulence forecasting. Weather Forecast. **21**, 268–287 (2006)
Sprinkle, C.H.: The AMDAR program. Preprints, Eighth Conference on Aviation, Range, and Aerospace Meteorology, American Meteorological Society, Dallas, TX, pp. 552–556 (1999)
WMO: WMO Guide to Meteorological Instruments and Methods of Observation, 8th edn. (2012)

Chapter 3
Instabilities Conducive to Aviation Turbulence

Yuh-Lang Lin

Abstract Atmospheric instabilities, including static, Kelvin–Helmholtz (KHI), inertial, conditional, potential (convective), symmetric, conditional symmetric, and potential symmetric instabilities, might trigger aviation-scale turbulence such as the clear-air turbulence (CAT), mountain wave turbulence (MWT), and near-cloud turbulence (NCT) that are reviewed in this chapter. Energy conversion associated with several atmospheric instabilities is reviewed through an equation governing energy transfer in an inviscid, stably stratified shear flow, and the Taylor–Goldstein equation is used to prove a corollary of Howard's semicircle theorem and the Miles–Howard Theorem. The criteria for static (gravitational) instability, conditional instability, potential instability, inertial instability, symmetric instability, conditional symmetric instability, and potential symmetric instability are discussed and summarized in a table. Their implications for the development of turbulence diagnostics or indices for aviation turbulence forecasting are also discussed.

3.1 Introduction

One major source of aviation hazard is the turbulence that occurs at the typical commercial aircraft cruise level in the upper troposphere and lower stratosphere (i.e., approximately 8–14 km above the flat terrain (Lane et al. 2012; Sharman et al. 2012). It is well known that aviation-scale turbulence often has its origins in mesoscale instabilities. For example, the clear air turbulence originated from Kelvin–Helmholtz instability may be induced by enhanced shears and reduced Richardson numbers associated with the jet stream and upper level fronts. Thus, in order to understand the formation of the aviation-scale turbulence, it is important

Y.-L. Lin (✉)
Department of Physics, North Carolina A&T State University, Greensboro, NC, USA

Department of Energy and Environmental Systems, North Carolina A&T State University, Greensboro, NC, USA
e-mail: ylin@ncat.edu

R. Sharman, T. Lane (eds.), *Aviation Turbulence*,
DOI 10.1007/978-3-319-23630-8_3

to understand the energy transfer associated with atmospheric instabilities and integral theorems of stratified flow which will be discussed in Sects. 3.2 and 3.3, respectively. Following Sects. 3.2 and 3.3, several mesoscale instabilities which might induce aviation-scale turbulence, such as static instability, conditional instability, potential instability, Kelvin–Helmholtz instability, inertial instability, and symmetric instability, will be presented in subsequent sections of this chapter.

One of the most well-known categories of aviation turbulence is *clear air turbulence* (CAT), which is often generated by the Kelvin–Helmholtz instability (KHI) over a flat surface (Lane et al. 2012; Sharman et al. 2012; Trier et al. 2012) or mountains through the amplification, steepening, overturning, and breaking of mountain waves (e.g., Nastrom and Fritts 1992; Lester 1993; Wurtele et al. 1996; Smith 2002; Wolff and Sharman 2008; Kim and Chun 2010, 2011; Doyle et al. 2011). The latter is also known as *mountain wave turbulence* (MWT). When propagating energy upward, mountain waves may amplify in the vicinity of (a) weaker stratification; (b) stronger wind, such as a jet stream or jet streak; see, e.g., Lin (2007, Eq. (5.2.32)); (c) a critical level where the wind speed approaches zero (see also Sect. 3.3 below); and (d) a decrease in air density, leading to an increase of $\sqrt{\rho_s/\rho(z)}$ where ρ_s and $\rho(z)$ are the air density at surface and a certain height z, respectively, all of which make prediction of MWT difficult in practice. Wave breaking or overturning tends to reduce stratification and increase the vertical wind shear locally, which may potentially lead to *static instability*, *KHI*, or both. Similarly, inertia-gravity waves over either flat terrain or mountains may steepen, leading to inertial instability which can induce CAT (e.g., Mahalov et al. 2007; Knox 1997).

In addition to CAT and MWT, turbulence may also occur near a convective cloud and is known as *near cloud turbulence* (NCT) (Lane et al. 2012; Sharman et al. 2012; Trier et al. 2012). Unlike convectively induced turbulence (CIT), which is produced within the cloud by moist instabilities and can be detected effectively through remote sensing instruments, NCT occurs in the vicinity of convective clouds. NCT is weaker, more difficult to detect, and its formation mechanisms are less well understood (Lane et al. 2012). Possible mechanisms leading to the NCT include the entrainment/detrainment of cloudy air below the cloud top, convective outflows at the level of the cloud top, or gravity waves above the convective cloud top, which may be triggered by shear instability, KHI, and/or static instability directly and moist instabilities indirectly.

Atmospheric instabilities, which may trigger directly or indirectly the above-mentioned turbulence leading to an aviation hazard, include static, Kelvin–Helmholtz (KHI), inertial, conditional (CI), potential (convective; PI), symmetric (SI), conditional symmetric (CSI), and potential symmetric (PSI) instabilities. In fact, CAT may also be generalized, in addition to MWT and NCT, to include the turbulence that interacts with the aircraft wake vortices leading to their breakup into vortex rings, known as the Crow instability (e.g., Crow 1970; Spalart 1998; Han et al. 2000; Raab and Foster 2011).

In aviation turbulence forecasting, turbulence diagnostics or indices have been proposed for identifying areas of turbulence likelihood. For example, the Graphical Turbulence Guidance (GTG; Sharman et al. 2006) is based on dynamic weighting of a number of dimensional or nondimensional parameters, such as the *Colson-Panofsky* (1965) *index*, *Richardson number* (*Ri*), *Ri tendency*, diagnostic *turbulence kinetic energy* (*TKE*), *frontogenesis function*, *unbalanced flow diagnostic*, *horizontal temperature gradient*, *turbulence index 1*, *NCSU index 1* (*NCSU1*), *eddy dissipation rate* (*EDR*), and *isentropic vertical motion*. Most of these indices or parameters are associated with one or more atmospheric instabilities explicitly or implicitly, such as *Ri* and a curvature measure of strongly anticyclonic flows, which have been used to represent partially or fully for *KHI* and inertial instability, respectively. Thus, a basic understanding of these instabilities is essential to improving the understanding of aviation turbulence and its forecasting. In this chapter, we will review some basic dynamics of mesoscale atmospheric instabilities that might lead to aviation turbulence. Part of the text is based on Lin (2007).

3.2 Energy Conversion Associated With Atmospheric Instabilities

When atmospheric instability occurs, one type of energy is converted into kinetic energy, which leads to the acceleration of fluid particles away from their origins. The energy conversion associated with atmospheric instabilities can be understood by considering the following equation, which governs the energy transfer in an inviscid, Boussinesq atmosphere with a basic flow, $U(y, z)$, on a planetary f-plane similar to Eq. (7.1.2) of Lin (2007):

$$\left(\frac{\partial}{\partial t}+U\frac{\partial}{\partial x}\right)E+\rho_o u'w'U_z+\rho_o u'v'U_y-\left(\frac{\rho_o g f}{N^2\theta_o}\right)v'\theta' U_z+\nabla\cdot(p'\mathbf{v}') = \left(\frac{\rho_o g^2}{c_p T_o N^2\theta_o}\right)\theta' q', \quad (3.1)$$

where

$$E=\frac{\rho_o}{2}\left[\left(u'^2+v^2+w'^2\right)+\left(\frac{g}{N\theta_o}\right)^2\theta'^2\right] \quad (3.2)$$

is the total perturbation energy, (u', v', w') is the perturbation wind velocity, θ' the perturbation potential temperature, ρ_o, θ_o, and T_o are constant reference density, potential temperature, and temperature, respectively, f the Coriolis parameter, $U(y, z)$ the basic wind velocity, N the Brunt–Väisälä frequency, p' the perturbation pressure, g the gravitational acceleration, c_p the heat capacity at constant pressure, and q' the diabatic heating. Equation (3.1) contains potential generation mechanisms of: (a) pure and inertia-gravity waves, (b) static instability, (c) conditional instability, (d) potential (convective) instability, (e) shear (Kelvin–Helmholtz) instability, (f) symmetric instability, (g) inertial instability, and (h) moist absolute instability.

The total perturbation energy consists of the perturbation kinetic energy and the perturbation potential energy, which are, respectively, represented by the first and second terms inside the square bracket of Eq. (3.2). Integrating Eq. (3.1) horizontally from $-\infty$ to $+\infty$ for a localized disturbance or over a single wavelength for a periodic disturbance in both x and y directions and vertically from $z = z_0$ to the top of either the physical or numerical model domain, $z = z_T$, yields (with primes dropped):

$$\underset{\text{A}}{\frac{\partial E_T}{\partial t}} = \underset{\text{B}}{-\rho_o \int_0^{z_T} \overline{uw}\, U_z dz} \underset{\text{C}}{- \rho_o \int_0^{z_T} \overline{uv}\, U_y dz} + \underset{\text{D}}{\left(\frac{\rho_o g f}{N^2 \theta_o}\right) \int_0^{z_T} \overline{v\theta}\, U_z dz} \\ \underset{\text{E}}{-\overline{pw}(z_T)} \underset{\text{F}}{+\overline{pw}(z_0)} + \underset{\text{G}}{\left(\frac{\rho_o g^2}{c_p T_o N^2 \theta_o}\right) \int_0^{z_T} \overline{\theta q}\, dz}, \tag{3.3}$$

where E_T is the domain-integrated total perturbation energy.

In Eq. (3.3), Term A represents the local rate of change of E_T, and Term B represents the vertical momentum flux transfer between the kinetic energy of the basic current and the perturbation energy. When *shear instability* occurs, the energy is transferred from the basic state shear flow to the perturbation as represented by this term. The phase relationship of the basic shear and the perturbation velocities can be represented by

$$\overline{uw}U_z = -\left[\frac{\partial \psi}{\partial x} / \frac{\partial \psi}{\partial z}\right] \left(\frac{\partial \psi}{\partial z}\right)^2 U_z = \left[\frac{\partial z}{\partial x}\right]_\psi \left(\frac{\partial \psi}{\partial z}\right)^2 U_z, \tag{3.4}$$

where ψ is the *perturbation streamfunction* defined as $u = \partial\psi/\partial z$ and $w = -\partial\psi/\partial x$. The above equation implies that if *shear instability* exists (i.e., term A > 0) and $U_z > 0$, then on average it requires

$$\left[\frac{\partial z}{\partial x}\right]_\psi < 0. \tag{3.5}$$

Therefore, the growing wave in a stably stratified flow must have an upshear phase tilt, as depicted in Fig. 3.1.

Term C of Eq. (3.3) represents the horizontal momentum flux transfer between the kinetic energy of the basic state horizontal shear and the perturbation wave energy. When *inertial instability* occurs, the energy stored in the basic state horizontal shear is transferred to horizontal perturbation momentum flux. The horizontal upshear phase tilt is analogous to that presented in Fig. 3.1 above for

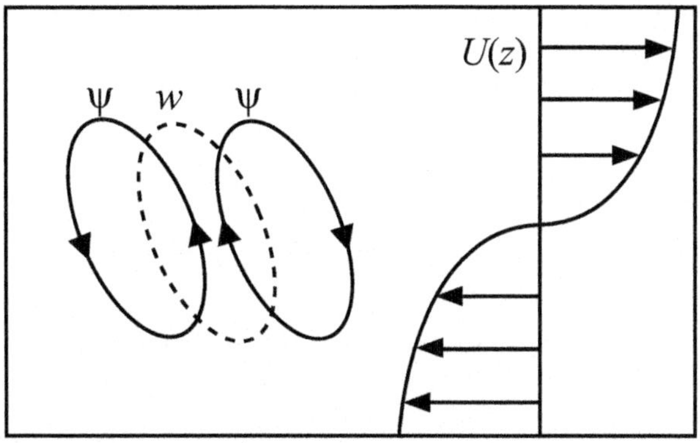

Fig. 3.1 A sketch of the basic wind profile and the upshear tilt of the perturbation streamfunctions (*solid*) and updraft (*dashed*) associated with an unstable growing gravity wave in a stably stratified flow. The perturbation wave energy is converted from the basic flow shear (after Lin and Chun 1993; Lin 2007)

shear instability. Term D represents the energy exchange between the basic state vertical shear and the perturbation heat flux associated with the *baroclinicity* (i.e., the horizontal temperature gradient through thermal wind balance). When baroclinic instability occurs, the *available potential energy* stored in the system is transferred to perturbation kinetic energy. Terms E and F represent the forcing from the upper and lower boundaries and are specified by the respective boundary conditions. Term G represents the contribution from diabatic heating or cooling to the system.

3.3 Integral Theorems of Stratified Flow

The governing equation of vertical motion for a small-amplitude, two-dimensional, inviscid, nonrotating, Boussinesq fluid flow system in the absence of thermal and momentum forcing can be derived as (e.g., see Lin 2007)

$$\left(\frac{\partial}{\partial t}+U\frac{\partial}{\partial x}\right)^2\left(\frac{\partial^2 w'}{\partial x^2}+\frac{\partial^2 w'}{\partial z^2}\right)-U_{zz}\left(\frac{\partial}{\partial t}+U\frac{\partial}{\partial x}\right)\frac{\partial w'}{\partial x}+N^2\frac{\partial^2 w'}{\partial x^2}=0. \quad (3.6)$$

Substituting $w' = \hat{w}\, e^{ik(x-ct)}$ into Eq. (3.6) leads to the *Taylor–Goldstein equation*

$$\frac{\partial^2 \hat{w}}{\partial z^2}+m^2\hat{w} = 0, \quad (3.7)$$

where

$$m^2 = \frac{N^2}{(U-c)^2}-\frac{U_{zz}}{U-c}-k^2. \quad (3.8)$$

Introducing a new variable $h \equiv \hat{w}/(c-U)$, and letting $c = c_r + ic_i$, multiplying the complex conjugate of h, and integrating from $z=0$ to ∞ yields

$$\int_0^\infty c_i(c_r - U)\left(\left|\frac{\partial h}{\partial z}\right|^2 + k^2|h|^2\right)dz = 0. \tag{3.9}$$

Since $w' = \hat{w}\exp(ik(x - c_r t))\exp(kc_i t)$, for instability to occur, c_i must be greater than 0.

Based on Eq. (3.9), the term $(c_r - U)$ must change sign at some level between $z = 0$ and ∞. In other words, *in order for instability to occur in a two-dimensional, nonrotating flow requires the existence of a critical level at which* $U = c_r$. Note that this is a corollary of the *Howard's semicircle theorem* (Howard 1961): *The complex phase speed, c, of an unstable normal mode must lie within the semicircle enclosed by* $U_{\min}$ *and* $U_{\max}$,

$$[c_r - (U_{\min} + U_{\max})/2]^2 + c_i^2 \leq [(U_{\min} - U_{\max})/2]^2$$

Equations (3.7) and (3.8) may be rearranged as

$$\frac{\partial}{\partial z}\left[(U-c)^2\frac{\partial h}{\partial z}\right] + \left(N^2 - k^2(U-c)^2\right)h = 0, \tag{3.10}$$

where $h(z_1) = h(z_2) = 0$. For an unstable flow, $c_i > 0$. Then c becomes a complex number giving $U - c \neq 0$ for any z. Equation (3.10) may be rewritten by defining a new function G as $h \equiv G/\sqrt{U-c}$ and substituting h into Eq. (3.10); then integrating Eq. (3.10) from z_1 to z_2 yields the imaginary part (Miles 1966):

$$c_i\left(\int_{z_1}^{z_2}\left[\left|\frac{\partial G}{\partial z}\right|^2 + k^2|G|^2\right]dz + \int_{z_1}^{z_2}\left|\frac{G}{U-c}\right|^2 U_z^2\left(Ri - \frac{1}{4}\right)dz\right) = 0, \tag{3.11}$$

where $Ri \equiv N^2/U_z^2$ is the *Richardson number*, which is also called the *gradient Richardson number*. For instability to occur, Eq. (3.11) requires $Ri < 1/4$ at some level between z_1 and z_2 or as stated in *Miles–Howard Theorem*: A *sufficient condition for a stratified flow to be stable is* $Ri \geq 1/4$ *everywhere* (Miles 1966; Howard 1961).

3.4 Static Instability

Static instability, also known as *buoyant* or *gravitational instability*, can be understood through the vertical momentum equation,

$$\frac{Dw}{Dt} = -\frac{1}{\rho}\frac{\partial p}{\partial z} - g, \tag{3.12}$$

where ρ and p are the density and pressure of the air parcel, respectively. The air parcel follows the *parcel theory*:

1. The pressure of the air parcel adjusts immediately to that of its environment ($\overline{p}$), i.e., $p = \overline{p}$, when it moves up or down from its initial level.
2. The air parcel's environment is in hydrostatic balance.
3. No compensating motions are allowed in the parcel's environment.
4. No mixing between the air parcel and its environment is allowed.

Applying Eq. (3.1) to the air parcel leads to

$$\frac{Dw}{Dt} = g\left(\frac{\overline{\rho} - \rho}{\rho}\right) \equiv b, \tag{3.13}$$

where b is the *buoyancy*. For an infinitesimal vertical displacement η, Eq. (3.13) becomes

$$\frac{D^2\eta}{Dt^2} + N^2\eta = 0, \tag{3.14}$$

where

$$N^2 = -\frac{\partial b}{\partial z} = \frac{g}{\overline{\theta}}\frac{\partial \overline{\theta}}{\partial z} = \frac{g}{T}(\Gamma_d - \gamma). \tag{3.15}$$

Here, $\gamma \equiv -\partial T/\partial z$ is the *environmental lapse rate* observed by the sounding and $\Gamma_d = g/c_p$ is the *dry adiabatic lapse rate* associated with the air parcel, which is approximately 9.76×10^{-3} km^{-1}. Assuming $\eta(t) = A\exp(iNt) + B\exp(-iNt)$ and substituting it into Eq. (3.15) leads to the criteria for static stability, static neutrality, and static instability:

$$\begin{array}{lrl} (a) & \text{Absolutely stable}: & N^2 > 0,\ \gamma < \Gamma_\text{d},\ \text{or}\ \partial\overline{\theta}/\partial z > 0 \\ (b) & \text{Dry neutral}: & N^2 = 0,\ \gamma = \Gamma_\text{d},\ \text{or}\ \partial\overline{\theta}/\partial z = 0\,. \\ (c) & \text{Dry absolutely unstable}: & N^2 < 0,\ \gamma > \Gamma_\text{d},\ \text{or}\ \partial\overline{\theta}/\partial z < 0 \end{array} \tag{3.16}$$

Note that this type of *Archimedean buoyancy force* derives its buoyancy from the density or temperature difference between the air parcel and its surrounding air, based on the parcel theory. The vertical perturbation pressure gradient force may be added (i.e., a non-Archimedean approach; Das 1979; Davies-Jones 2003) leading to the more complete vertical momentum equation (Doswell and Markowski 2004),

$$\frac{Dw}{Dt} = -\frac{1}{\bar{\rho}}\frac{\partial p_d'}{\partial z} + \left(b - \frac{1}{\bar{\rho}}\frac{\partial p_b'}{\partial z}\right), \tag{3.17}$$

where p_d' and p_b' are the *dynamic pressure* and *buoyancy pressure*, respectively. The dynamic pressure arises from the flow field differences created by the fluid motion, while the buoyancy pressure is generated by the vertical buoyancy gradient.

In saturated air, the Brunt–Väisälä frequency (N) in Eqs. (3.14) and (3.15) should be replaced by the *saturated moist Brunt*–Väisälä frequency (N_m; Lalas and Einaudi 1974),

$$N_m^2 = \frac{g}{\bar{\bar{T}}}\left(\frac{\partial \bar{T}}{\partial z} + \Gamma_s\right)\left(1 + \frac{Lq_{vs}}{R_d\bar{T}}\right) - \left(\frac{g}{1+q_w}\right)\frac{\partial q_w}{\partial z}, \tag{3.18}$$

or (Emanuel 1994)

$$N_m^2 = \left(\frac{1}{1+q_w}\right)\left\{\left[\frac{\Gamma_s(c_p + c_l q_w)}{\bar{\theta}_e}\right]\frac{\partial \bar{\theta}_e}{\partial z} - \left[g - \Gamma_s c_l \ln\left(\frac{\bar{\theta}_e}{\bar{\bar{T}}}\right)\right]\frac{\partial q_w}{\partial z}\right\}, \tag{3.19}$$

where Γ_s is the *moist adiabatic lapse rate*, q_{vs} the *saturation mixing ratio of water vapor*, R_d the *ideal gas constant for dry air*, L the *latent heat of condensation* or *evaporation*, q_w the *total water mixing ratio*, c_l the specific heat capacity at constant pressure of liquid water and $\bar{\theta}_e$ the *equivalent potential temperature* of the saturated environmental air.

Based on Eq. (3.14) and Eqs. (3.18) or (3.19), the cloudy air is *moist statically stable* or *moist absolutely unstable* if $N_m^2 > 0$ or $N_m^2 < 0$, respectively (Durran and Klemp 1982). *The criterion for moist absolute instability* (*MAI*) is then $\Gamma_s < \gamma_s$, where Γ_s and γ_s are the moist and environmental saturated lapse rate, respectively. MAI can be created and maintained by a *moist absolutely unstable layer* (MAUL) formed ahead of a density current (Bryan and Fritsch 2000).

The static instability is closely related to KHI, to be discussed later; thus, the Brunt–Väisälä frequency, as part of the Richardson number, has been used implicitly in the prediction of aviation turbulence, such as the Graphical Turbulene Guidance (GTG; Sharman et al. 2006) for detecting aviation turbulence.

3.5 Conditional Instability

If the environmental lapse rate (γ) is between the moist adiabatic lapse rate (Γ_s) and the *dry lapse rate* (Γ_d), i.e., $\Gamma_s < \gamma < \Gamma_d$, and an unsaturated air parcel is lifted up to its *lifting condensation level* (*LCL*), further lifting will result in condensation and may reach a level where the temperatures of the air parcel and the environment are equal, i.e., the *level of free convection* (*LFC*). Above the LFC, the air parcel will accelerate upward freely without any further lifting. This type of instability is referred to as *conditional instability*, and the necessary conditions for it to occur are: (a) $\Gamma_s < \gamma < \Gamma_d$ and (b) a lifting of the air parcel past its *LFC*.

The criterion for conditional instability can also be determined by the vertical gradient of the *saturation equivalent potential temperature* (θ_e^*), which is defined as

$$\theta_e^* = \theta e^{Lq_{vs}/c_pT}. \tag{3.20}$$

Thus, θ_e^* is the equivalent potential temperature that the air parcel would have if it were saturated initially at the same temperature and pressure. The criterion for conditional instability can then be derived by considering an air parcel lifted from $z_o - \delta z$ to z_o (e.g., see Lin (2007) for details):

$$\frac{D^2\eta}{Dt^2} + \left(\frac{g}{\overline{\theta}_e^*}\frac{\partial\overline{\theta}_e^*}{\partial z}\right)\eta = 0. \tag{3.21}$$

Therefore, *the conditional instability criterion for a saturated air becomes*

$$\frac{\partial\overline{\theta}_e^*}{\partial z} < 0. \tag{3.22}$$

Equation (3.22) is often used to show the possible existence of conditional instability from observed or model-simulated soundings. However, the air parcels in general are not saturated at their initial lifting levels (Schultz et al. 2000; Sherwood 2000); thus, *the criteria for conditional instability should be modified to that of* $\partial\overline{\theta}_e^*/\partial z < 0$ *when the LFC has been reached.* A strong, mesoscale, mechanically forced ascent along density currents, elevated terrain, or fronts may bring a conditionally unstable environmental layer to saturation.

The *convective available potential energy* (*CAPE*) is defined as the energy available for free convection and can be calculated by

$$CAPE = \int_{z_{LFC}}^{z_{LNB}} b\,dz = \int_{z_{LFC}}^{z_{LNB}} g\left(\frac{\overline{\rho}-\rho}{\rho}\right)dz = \int_{z_{LFC}}^{z_{LNB}} g\left(\frac{T-\overline{T}}{\overline{T}}\right)dz$$
$$= \int_{z_{LFC}}^{z_{LNB}} g\left(\frac{\theta-\overline{\theta}}{\overline{\theta}}\right)dz. \tag{3.23}$$

Thus, *CAPE* is the work done by the buoyancy force in lifting an air parcel from its *LFC* to *LNB*. In a thermodynamic diagram, *CAPE* is proportional to the area enclosed by the environmental temperature curve, and the moist adiabat of the air parcel in between the *LFC* and *LNB* (Fig. 3.2), thus, may be represented by the *positive area* (*PA*) if the initial level of lifting of the air parcel (z_i) is assumed to be at *LFC*. Similarly, a *negative area* (*NA*) may be defined by the area confined by the dry adiabat (below *LCL*) or the moist adiabat (above *LCL*) to the left, and the

Fig. 3.2 An idealized sounding with conditional instability displayed on a skew-T log-p thermodynamic diagram. The isotherms, saturation mixing ratio, and dry sadiabat are denoted by *slanted solid*, *dotted*, and *dashed curves*. The environmental temperature (*T*) and dew point profile (Td) are denoted by *thick solid* and *dashed curves*, respectively. The lifting condensation level (LCL), level of free convection (LFC), level of neutral buoyancy (LNB), convective available potential energy (CAPE), and convective inhibition (CIN) for the air parcel originated at A are denoted in the figure (after Lin 2007)

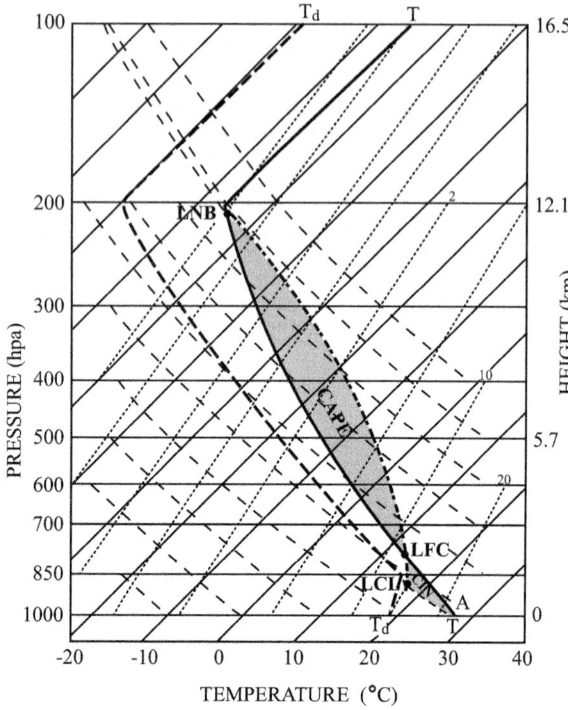

sounding to the right, from the initial level to the *LFC*. In other words, the negative area represents the energy needed to lift an air parcel vertically and dry adiabatically or pseudoadiabatically to its *LFC* and is also known as the *convective inhibition* (*CIN*), which is defined as

$$CIN = \int_{z_i}^{z_{LFC}} g\frac{\overline{T} - T}{\overline{T}}dz = \int_{z_i}^{z_{LFC}} g\frac{\overline{\theta} - \theta}{\overline{\theta}}dz. \tag{3.24}$$

Without horizontal advection, the maximum vertical velocity can be estimated by converting all the potential energy into kinetic energy, i.e., $w_{\max} = \sqrt{2CAPE}$. The maximum downdraft generated by evaporative or melting cooling can be estimated as $-w_{\max} = \sqrt{2DCAPE_i}$, where $DCAPE_i$ is the *downdraft convective available potential energy* and is defined as

$$DCAPE_i = \int_{z_s}^{z_i} g\left(\frac{\overline{T} - T}{\overline{T}}\right) dz, \tag{3.25}$$

where the air parcel descends from its initial level z_i to z_s.

In summary (also see Table 3.1), there are six static stabilities:

Table 3.1 Criteria for different types of instabilities (from Lin 2007; adapted after Schultz and Schumacher 1999)

	Static (gravitational)	Inertial	Symmetric
Dry	Absolute instability	Inertial instability	Symmetric instability
	$\partial\overline{\theta}/\partial z < 0$ $\gamma > \Gamma_d$	$\partial\overline{M}/\partial x < 0$ $\zeta_{ga} + f < 0$	$(\partial\overline{\theta}/\partial z)_{\overline{M}} < 0$; $(\partial\overline{M}/\partial x)_{\overline{\theta}} < 0$ $\Gamma_d\vert_{\overline{M}} < \gamma\vert_{\overline{M}}$ $\vert\partial z/\partial x\vert_{\overline{M}} < \vert\partial z/\partial x\vert_{\overline{\theta}}$ $PV_g < 0$
Moist	Moist Absolute Instability (MAI)	N/A	N/A
	$\Gamma_s < \gamma_s$		
Conditional[a]	Conditional Instability (CI)	N/A	Conditional Symmetric Instability (CSI)
	$\partial\overline{\theta}_e^*/\partial z < 0$ $\Gamma_s < \gamma < \Gamma_d$ (parcel lifted above LFC)		$\partial\overline{\theta}_e^*/\partial z\vert_{\overline{M}} < 0$; $\partial\overline{M}/\partial z\vert_{\overline{\theta}_e^*} < 0$ $\Gamma_s\vert_{\overline{M}} < \gamma\vert_{\overline{M}} < \Gamma_d\vert_{\overline{M}}$ $\vert\partial z/\partial x\vert_{\overline{M}} < \vert\partial z/\partial x\vert_{\overline{\theta}_e^*}$ $MPV_g^* < 0$ (parcel lifted above LFC)
Potential[b]	Potential Instability (PI)	N/A	Potential Symmetric Instability (PSI)
	$\partial\overline{\theta}_e/\partial z < 0$		$\partial\overline{\theta}_e/\partial z\vert_{\overline{M}} < 0$; $\partial\overline{M}/\partial z\vert_{\overline{\theta}_e} < 0$ $\vert\partial z/\partial x\vert_{\overline{M}} < \vert\partial z/\partial x\vert_{\overline{\theta}_e}$ $MPV_g < 0$

[a]At saturation, $\overline{\theta}_e = \overline{\theta}_e^*$

[b]$\overline{\theta}_w$ can be used equivalently to $\overline{\theta}_e$

Meanings of symbols: (1) γ: observed environmental lapse rate; (2) Γ_d: dry lapse rate; (3) Γ_s: moist lapse rate; (4) γ_s: observed environmental saturated lapse rate; (5) $\overline{\theta}$: environmental potential temperature; (6) $\overline{\theta}_e$: environmental equivalent potential temperature; (7) $\overline{\theta}_e^*$: environmental saturation equivalent potential temperature; (8) $\overline{M}$: environmental geostrophic absolute momentum; (9) PV_g: geostrophic potential vorticity (PV); (10) MPV_g: moist geostrophic PV; and (11) MPV_g^*: saturated geostrophic PV

$$
\begin{array}{lcc}
(1) & \text{Absolutely stable} & \gamma < \Gamma_s \\
(2) & \text{Saturated neutral} & \gamma = \Gamma_s \\
(3) & \text{Conditionally unstable} & \Gamma_s < \gamma < \Gamma_d \\
(4) & \text{Dry neutral} & \gamma = \Gamma_d \\
(5) & \text{Dry absolute unstable} & \gamma > \Gamma_d \\
(6) & \text{Moist absolutely unstable} & \gamma_s > \Gamma_s
\end{array}, \tag{3.26}
$$

where γ_s is the saturated lapse rate of the environmental air.

Note that the above instabilities were based on parcel theory, which is developed in terms of buoyancy, and thus do not take into account instabilities that depend on horizontal pressure gradients, such as baroclinic and barotropic instabilities (Emanuel 1994). The parcel theory may be improved to include the entrainment and detrainment of the environmental air and the compensating downward motion of the environmental air associated with convective clouds and the effect of condensate on buoyancy.

3.6 Potential Instability

When an atmospheric layer composed of a much drier air on top of a moist layer is lifted up, such as going over a broad mountain range, a frontal surface, density current, or a cyclone, *potential* (*convective*) *instability* may occur. Under this situation, it is more appropriate to apply the *layer theory* instead of the *parcel theory*, which assumes that (a) the mass is conserved within the layer, and (b) the atmosphere is in hydrostatic balance. Figure 3.3 illustrates the process leading to potential instability by lifting an initially absolutely stable layer AB with $\partial\overline{\theta}_e/\partial z < 0$, where the *equivalent potential temperature* (θ_e) can be approximately calculated by

$$\theta_e = \theta e^{Lq_v/c_pT_{LCL}}. \tag{3.27}$$

The top of the layer (B) cools faster following a dry adiabat to saturation at B', while the bottom of the layer reaches saturation earlier at its LCL and then cools at a slower rate following a moist adiabat to A′. Eventually, the final saturated layer

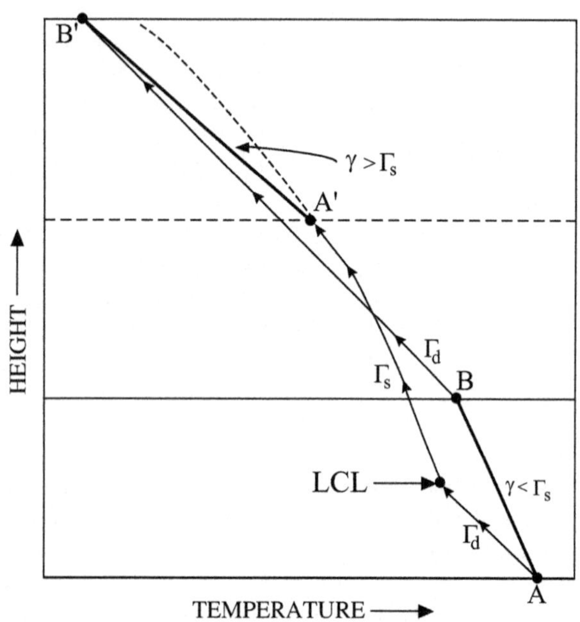

Fig. 3.3 Illustration of potential (convective) instability by lifting an initially absolutely stable layer AB with $\partial\overline{\theta}_e/\partial z < 0$. The *top* of the layer (B) follows a dry adiabat to saturation at B′, while the *bottom* of the layer becomes saturated earlier (at LCL) and then follows moist adiabat to A′. The lapse rate of the final saturated layer (A′B′) is greater than the moist adiabat, thus is unstable (after Lin 2007, adapted after Darkow 1986)

(A′B′) is moist unstable, i.e., $\gamma > \Gamma_s$. Therefore, the criteria for potential (convective) stability, neutrality, and instability are:

$$\frac{\partial \overline{\theta}_e}{\partial z} \begin{cases} > 0 & \text{potentially stable} \\ = 0 & \text{potentially neutral} \\ < 0 & \text{potentially unstable} \end{cases}. \tag{3.28}$$

Note that a continuous supply of forcing is required to realize the potential instability because there is no reservoir of potential energy (Emanuel 1994), and the *wet-bulb potential temperature* ($\overline{\theta}_w$) has also been used to replace $\overline{\theta}_e$ in Eq. (3.28) for determining potential instability. Based on Eqs. (3.22) and (3.28), potential instability is equivalent to conditional instability in a saturated atmosphere because the vertical gradients of $\overline{\theta}_e$ and $\overline{\theta}_e^*$ are equivalent (Schultz and Schumacher 1999). It has been suggested that an embedded convective cellular convection might result from potential instability through layer lifting, while a single convective cell might result from conditional instability through parcel lifting (Lin 2007). The exact relationship between these two instabilities is not quite clear and deserves further investigation.

3.7 Kelvin–Helmholtz Instability

The instability induced by shear in a stratified fluid flow is referred to as the *Kelvin–Helmholtz instability* (*KHI*). It is also referred to as *shear instability* in the literature, although strictly speaking the shear instability includes KHI and the instability induced by shear alone in a homogeneous fluid. KHI may produce large-amplitude gravity waves and clear air turbulence (*CAT*). CAT is a major cause of aviation turbulence and an extraordinarily challenging subject to aviators because of the significant impact of this process on aviation safety. When the atmosphere is moist and air parcels are lifted above their lifting condensation levels, KHI may become visible as *K-H waves* or *billow clouds*. The criterion for KHI can be derived along Miles–Howard Theorem (Miles 1966; Howard 1961). As demonstrated in Eq. (3.11), KHI requires $Ri < 1/4$ at some level of the fluid, where Ri is the Richardson number and is defined as $Ri \equiv N^2/U_z^2$.

The criterion $Ri < 1/4$ requires a strong vertical shear and/or a weak thermal stratification for the KHI to occur. When $N^2 < 0$, the criterion $Ri < 0$ is automatically satisfied, which means the KHI may be induced by either the buoyancy effect (static instability) alone or a mixture of shear and static instabilities. Based on numerically simulated severe downslope winds, Scinocca and Peltier (1993) found that a local static instability develops first when internal gravity waves steepen and overturn, which produces a well-mixed layer, and then generates a local region of enhanced shear over the lee slope by the large-amplitude disturbance leading to the development of small-scale secondary shear instability or KHI before the development of severe downslope wind. They then proposed that when $N^2 < 0$, KHI is dominated by static instability through the buoyancy effect when Ri is less negative and shear instability through the shear effect when Ri is more negative. Whether this argument applies to the KHI induced CAT or MWT in the upper troposphere or lower stratosphere remains to be investigated. In addition, the interaction of turbulence, KHI, and static instability may come into play when $N^2 < 0$ and needs to be studied.

During the wave-overturning process, a local critical level is induced for a stationary internal gravity wave, such as a mountain wave or a upward propagating gravity wave induced by a stationary convective cloud or system, because $U(z) = c$ (=0) there. This may lead to a region of wave breaking and turbulent mixing, which is characterized by $Ri < 1/4$ (Clark and Peltier 1984; Teixeira et al. 2005; Wang and Lin 1999). In the vicinity of a critical level, nonlinearity needs to be considered since the flow is highly nonlinear there. Based on numerical experiments, it was proposed (Breeding 1971) that when a wave approaches a critical level, (a) for $Ri > 2.0$, most of the wave energy cannot penetrate through it and is absorbed, as described by the linear theory; (b) for $0.25 < Ri < 2.0$, some wave energy is absorbed or transmitted through the critical level while the rest of the wave energy is reflected; and (c) for $Ri < 0.25$, wave overreflection may occur. In regime (c), shear instability may occur and the incident wave is able to extract energy and momentum from the basic flow leading to wave overreflection (Lindzen and Rosenthal 1983). When the overreflected waves are in phase with the incident waves, waves may grow exponentially with time by resonance leading to the *normal mode instability*. On the other hand, when the overreflected waves are only partially in phase with the incident waves, waves may grow algebraically with time by partial resonance leading to the *algebraic mode instability*. Research is needed by using more state-of-the-art and sophisticated numerical models to revisit the problem of interaction of gravity waves and critical level, such as more accurately identifying the regime boundaries of wave absorption; wave transmission and reflection; and wave overreflection, normal mode, and algebraic mode growth of KHI. Large eddy simulation (*LES*) may be adopted to explore the interaction of turbulence, RHI, and static instability in this situation as well.

Note that in the literature, the *Richardson number* is also referred to as the *gradient Richardson number*, which is approximated by the *bulk Richardson number* in moist convection (AMS 2014)

$$R_B = \frac{(g/\overline{T}_v)\Delta\theta_v\Delta z}{(\Delta U)^2 + (\Delta V)^2} \tag{3.29}$$

or

$$R_B = \frac{CAPE}{(\Delta U)^2 + (\Delta V)^2}, \tag{3.30}$$

where Eq. (3.30) is a three-dimensional extension from Weisman and Klemp's (1982) definition. The gradient or bulk Richardson number (e.g., Dutton and Panofsky 1970; Endlich 1964; Kronebach 1964),

$$R_i = N^2 / \sqrt{|u_z|^2 + |v_z|^2},$$

where $N^2 = (g/\theta)\partial\theta/\partial z$ or $(g/\theta_e)\partial\theta_e/\partial z$ has been adopted fully or partially, such as Richardson number tendency, *dRi/dt* (Keller 1990; Roach 1970) and *NCSU1* (Kaplan et al. 2005a, b) in the Graphical Turbulene Guidance (GTG; Sharman et al. 2006) for forecasting aviation turbulence.

3.8 Inertial Instability

The criterion of inertial instability may be derived by considering a horizontally displaced air parcel embedded in a geostrophically balanced, inviscid shear flow on an f-plane, $(u, v) = (0, v_g(x))$. The horizontal momentum equations may be approximated by

$$\frac{Du}{Dt} = f(v - v_g) = f(M - \overline{M}), \tag{3.31}$$

$$\frac{Dv}{Dt} = -fu - \frac{1}{\rho}\frac{\partial p}{\partial y} = -fu = -f\frac{Dx}{Dt}, \tag{3.32}$$

where $M = v + fx$ and $\overline{M} = v_g + fx$ are the total *absolute momentum* and the basic-state *geostrophic absolute momentum* of the fluid, respectively. If an air parcel is displaced from its initial position $x = x_o$ to $x = x_o + \delta x$, integration of Eq. (3.32) leads to

$$v(x_o + \delta x) = v_g(x_o) - f\delta x \tag{3.33}$$

assuming the air parcel moves with the basic flow initially. The geostrophic wind at $x = x_o + \delta x$ may also be approximated by

$$v_g(x_o + \delta x) = v_g(x_o) + \frac{\partial v_g}{\partial x}\delta x. \tag{3.34}$$

Substituting Eqs. (3.33) and (3.34) into Eq. (3.31) at $x_o + \delta x$ yields

$$\frac{D^2\delta x}{Dt^2}+f\left(\frac{\partial v_g}{\partial x}+f\right)\delta x=0. \tag{3.35}$$

Equation (3.35) leads to the criteria for inertial stability, neutrality, and instability:

$$f\frac{\partial \overline{M}}{\partial x}=f\left(\frac{\partial v_g}{\partial x}+f\right)\begin{cases}>0 & \text{inertially stable}\\ =0 & \text{inertially neutral}\\ <0 & \text{inertially unstable}\end{cases} \tag{3.36}$$

corresponding to $Du/Dt=D^2\delta x/Dt^2<0$, $=0$, and >0, respectively, for an eastward displacement $\delta x>0$ in the Northern Hemisphere ($f>0$). Physically, the inertial stability can be explained by the fact that conservation of absolute momentum forces a laterally displaced air parcel to return to its initial position when an imbalance between the Coriolis force and the pressure gradient force is created. The above inertial instability criterion can be generalized to $\partial v_g/\partial x-\partial u_g/\partial y+f=\zeta_{ga}<0$, where ζ_{ga} is the absolute geostrophic vorticity for a three-dimensional shear flow. The above argument is independent of horizontal coordinates and can thus be applied to geostrophically balanced, inviscid zonal flow, $(u,v)=(u_g(y),0)$. In this case, the absolute momentum is defined as $M=fy-u$, which gives the inertial instability criterion: $f\,\partial\overline{M}/\partial y=f\,(f-\partial u_g/\partial y)<0$. Note that the absolute momentum corresponds to angular momentum in a circular motion and that the *inertial oscillation* is a special case of an inertially stable flow in which a parcel follows an anticyclonic circular trajectory in a quiescent atmosphere.

In the atmosphere, inertial instability occurs on meso-α/β or large scales and may be released by a strong divergent anticyclone flow which possesses either strong anticyclonic shear or strong anticyclonic curvature, which may be located equatorward of a westerly jet streak or in subsynoptic-scale ridges in the midlatitudes. Flow with low potential vorticity and low Richardson number, which are typical of anticyclonically shearing jet streams embedded within strong frontal systems, is directly related to inertial instability. For extratropical synoptic-scale systems, flows are nearly always inertially stable. On the other hand, inertially unstable conditions are often found in the subtropical upper troposphere (Schumacher and Schultz 2001). In addition, it has been found that (1) Rossby wave breaking in the tropics may trigger equatorial inertial instability; (2) inertial instability can enhance the outflow from mesoscale convective systems such as thunderstorms, tropical plumes, and hurricanes; and (3) the divergence-convergence couplets of inertial instability appear to determine the location of near-equatorial convection and the mean latitude of the ITCZ (Knox 2003).

Inertial instability thus provides a potential mechanism for turbulence development in the vicinity of jet entrance regions, fronts, regions of low Richardson number, regions with strong deformation, and regions of low upper-tropospheric absolute vorticity (e.g., Arakawa 1952; MacDonald 1977; Stone 1966) and could trigger CAT by promoting gravity wave genesis and breaking (Knox 1997). In fact, measures related to inertial instability in strongly anticyclonic flows (e.g., Arakawa 1952; Kaplan et al. 2005a, b; Knox 1997; Stone 1966) or related parameters, such as negative vorticity advection (Bluestein 1992), unbalanced flow (Knox 1997; Koch and Caracena 2002; McCann 2001; O'Sullivan and Dunkerton 1995), and curvature (Hopkins 1977; Lester 1993), have been incorporated fully or partially in the Graphical Turbulence Guidance (GTG; Sharman et al. 2006) for forecasting aviation turbulence.

3.9 Symmetric Instability

In a statically stable ($N^2 > 0$) and inertially stable ($f\,\partial\overline{M}/\partial x = f\,(\partial v_g/\partial x + f)$ > 0) atmosphere, under certain conditions, a slantwise displaced air parcel may accelerate away from its origin leading to *symmetric instability*. It was named symmetric because the argument is independent of the horizontal coordinates. Thus in the following, we will assume $\partial/\partial y = 0$ and thus defined $M = v + fx$ since the argument can be easily applied to cases with $\partial/\partial x = 0$. Dynamically, static, potential, inertial, and symmetric instabilities are very closely related in a way that each of them can be thought of as resulting from an unstable distribution of body forces acting on a fluid element, i.e., the gravitational force for static and potential instabilities, the Coriolis force for inertial instability, and a combination of gravitational and Coriolis forces for symmetric instability (e.g., Emanuel 1983; Ooyama 1966; Xu and Clark 1985).

Assuming a Boussinesq, hydrostatic, and geostrophic basic flow, the thermal wind relation may be written as

$$f\frac{\partial v_g}{\partial z} = \frac{g}{\theta_o}\frac{\partial\overline{\theta}}{\partial x}, \tag{3.37}$$

where $\overline{\theta}(x)$ and θ_o are the mean and constant reference potential temperature, respectively. Considering an idealized front aligned with the y-axis and located between the cold and warm regions (Fig. 3.4a), v_g becomes the along-front wind. Since $\partial\overline{\theta}/\partial z > 0$ and $\partial\overline{M}/\partial x > 0$ (Fig. 3.4), the basic flow is statically and inertially stable. For an air parcel displaced slantwise upward anywhere within the shaded wedge in Fig. 3.4b, however, it will be subjected to an upward and leftward acceleration in the direction of its initial displacement since $\overline{\theta}$ decreases and $\overline{M}$ increases along the displacement AB. Thus, the air parcel will accelerate away from A leading to *symmetric instability* (SI), which can be easily explained by viewing it as either static instability on a $\overline{M}$ surface (i.e., $\partial\overline{\theta}/\partial z|_{\overline{M}} < 0$) or inertial instability on an isentropic surface (i.e., $\partial\overline{M}/\partial x|_{\overline{\theta}} < 0$). Thus, any slantwise displacement within the shaded wedge between $\overline{M}$ and $\overline{\theta}$ may trigger the symmetric instability.

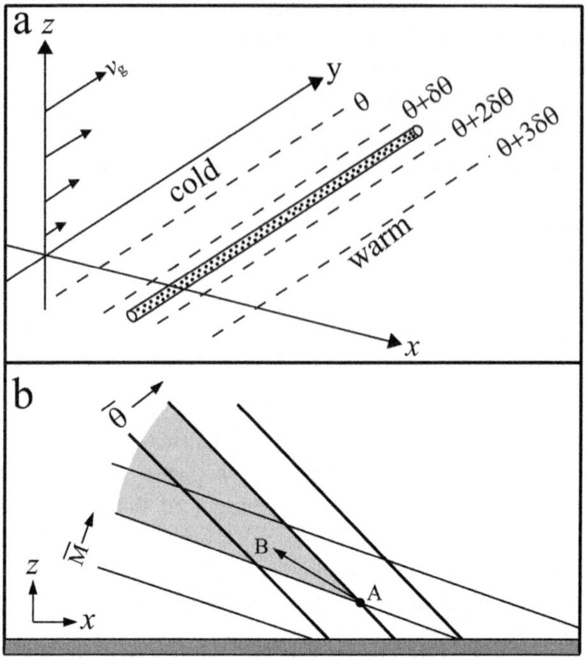

Fig. 3.4 Schematic of a mean state with symmetric instability. (**a**) A meridional, steady baroclinic flow. The unidirectional geostrophic wind v_g is in thermal wind balance. One may imagine an idealized broad front (*dotted tube*) that is aligned with the y-axis and located in between the cold and warm region. In this situation, v_g is the along-front wind (from Lin 2007, adapted after Emanuel 1994). (**b**) $\overline{M}$ and $\overline{\theta}$ surfaces on the x–z plane are tilted such that both of them increase upward and eastward. Displacement of an air parcel upward anywhere within the shaded area from point A is symmetrically unstable (after Lin 2007)

Based on Fig. 3.4 and the above analysis, the necessary condition for the SI to occur is that the slope of the $\overline{M}$ surfaces must be less than that of the $\overline{\theta}$ surfaces, i.e.,

$$-\left.\frac{\partial z}{\partial x}\right|_{\overline{\theta}} > -\left.\frac{\partial z}{\partial x}\right|_{\overline{M}}. \tag{3.38}$$

Substituting relations of $\delta\overline{\theta} = (\partial\overline{\theta}/\partial x)\delta x + (\partial\overline{\theta}/\partial z)\delta z = 0$ on $\overline{\theta}$ surfaces and $\delta\overline{M} = (\partial\overline{M}/\partial x)\delta x + (\partial\overline{M}/\partial z)\delta z = 0$ on $\overline{M}$ surfaces into Eq. (3.38), and applying Eq. (3.37) and $N^2 = (g/\theta_o)(\partial\overline{\theta}/\partial z)$, the necessary condition for the SI can be obtained

$$\frac{Ri}{f}\frac{\partial\overline{M}}{\partial x} < 1, \quad \text{or } Ri < \frac{f}{\zeta_g + f}, \tag{3.39}$$

where $Ri = N^2/(\partial v_g/\partial z)^2$ is the Richardson number. Therefore, symmetric instability is favored by low static stability, strong basic state vertical wind shear, and anticyclonic relative vorticity. If no horizontal wind shear exists, the above criterion is reduced to $Ri < 1$. An alternative method to derive the criterion for SI is to consider a small displacement δx along a constant θ surface (i.e., $(\delta x)_\theta$) and to then use Eqs. (3.32) and (3.33) with thermal wind relation, which gives

$$\frac{D^2(\delta x)_\theta}{Dt^2} + f^2\left[\frac{\zeta_g + f}{f} - \frac{1}{Ri}\right](\delta x)_\theta = 0. \tag{3.40}$$

Thus for SI to occur, it requires the term in square brackets of (3.40) to become negative, which leads to the criterion of Eq. (3.39). It can be shown that the critical (maximum) Richardson number for symmetric instability is $f/(\zeta_g + f)$.

It also can be shown that the *geostrophic potential vorticity*,

$$q_g \equiv \frac{g}{\theta_o}(\boldsymbol{\omega}_g \cdot \nabla\theta) = (\zeta_g + f)N^2\left(1 - \frac{f}{\zeta_g + f}\frac{1}{Ri}\right), \tag{3.41}$$

becomes 0 when Ri is critical, i.e., $Ri = f/(\zeta_g + f)$. Hence, symmetric instability occurs when $q_g < 0$. In (3.41), $\boldsymbol{\omega}$ is the total vorticity vector and the subscript g denotes the geostrophic value. The second equality is derived by expanding the first equality, dropping the y derivatives, and using the thermal wind balance (Emanuel and Raymond 1984).

Applying the parcel and layer theories to symmetric instability in a moist atmosphere (i.e., *moist symmetric instability*) leads to the necessary conditions for *conditional symmetric instability* (*CSI*) and *potential symmetric instability* (*PSI*), i.e., $\partial\overline{\theta}_e^*/\partial z\big|_{\overline{M}} < 0$ and $\partial\overline{\theta}_e/\partial z\big|_{\overline{M}} < 0$, respectively (Bennetts and Hoskins 1979; Ooyama 1966). The necessary condition for dry symmetric instability in terms of potential vorticity, $q_g < 0$ can be extended (e.g., see Lin 2007 for a brief review) to:

$$q_g^* = MPV_g^* = \frac{g}{\theta_o}(\boldsymbol{\omega}_{ga} \cdot \nabla\theta_e^*) < 0, \tag{3.42}$$

where $q_g^* \left(= MPV_g^*\right)$ is the *saturated geostrophic potential vorticity*. In summary, based on parcel theory, the necessary conditions for CSI to occur is

$$\partial\overline{\theta}_e^*/\partial z\big|_{\overline{M}} < 0, \tag{3.43a}$$

$$\partial\overline{M}/\partial z\big|_{\overline{\theta}_e^*} < 0, \tag{3.43b}$$

$$\Gamma_s\big|_{\overline{M}} < \gamma\big|_{\overline{M}} < \Gamma_d\big|_{\overline{M}}, \tag{3.43c}$$

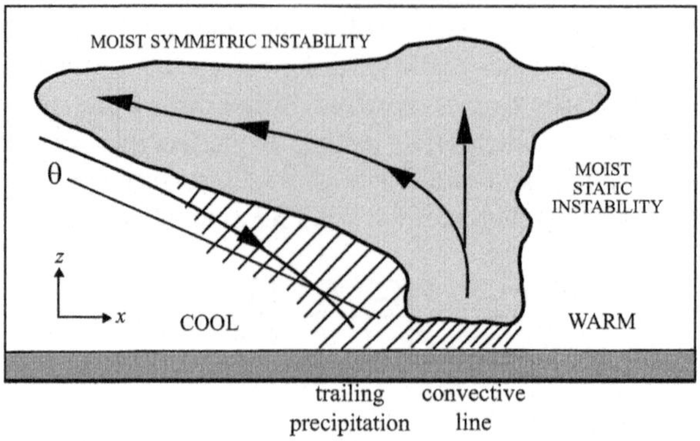

Fig. 3.5 A conceptual model of upscale convective-symmetric instability in a midlatitude mesoscale convective system. The *bold curve* encloses the cloud (*shaded*). The *arrows* represent the direction of circulation. The *labeled solid line* represents the orientation of typical potential-temperature contours in the cool air. The *hatched area* is proportional to precipitation intensity. The upright updraft caused by the release of static (gravitational) instability (CI or PI) is followed by slantwise convection, caused by the release of symmetric instability (CSI or PSI), which then produces downdrafts that descend following sloping (dry and/or moist) isentropes (adapted after Lin 2007, Seman et al. 1992 and Schultz and Schumacher 1999)

$$|\partial z/\partial x|_{\overline{M}} < |\partial z/\partial x|_{\overline{\theta}_e^*} \quad \text{or} \tag{3.43d}$$

$$q_g^* \equiv MPV_g^* = g\left(\boldsymbol{\omega}_{ga} \cdot \nabla\theta_e^*\right) < 0 \tag{3.43e}$$

provided that the air tube has been lifted above its LFC and the environment is conditionally and inertially stable.

Figure 3.5 shows a conceptual model of upscale convective-symmetric instability in a midlatitude mesoscale convective system. CSI and PSI have also been misused in the literature on several occasions, so readers should exercise caution when using these terminologies (Schultz and Schumacher 1999). It is important to distinguish between CSI and PSI, even though both tend to produce *slantwise convection* when conditions are met. In the real atmosphere, CI/PI and CSI/PSI may coexist, creating *convective-symmetric instability*, which is responsible for producing some mesoscale convective systems (e.g., Jascourt et al. 1988).

The criteria for static (gravitational) instability, conditional instability, potential instability, inertial instability, symmetric instability, conditional symmetric instability, and potential symmetric instability are summarized in Table 3.1.

3.10 Summary and Discussions

Atmospheric instabilities, including static, Kelvin–Helmholtz (KHI), inertial, conditional (CI), potential (convective; PI), symmetric (SI), conditional symmetric (CSI), and potential symmetric (PSI) instabilities, might trigger aviation-scale turbulence, such as the clear-air turbulence (CAT), mountain wave turbulence (MWT), and near-cloud turbulence (NCT), are reviewed in this chapter. Some of the turbulence diagnostics or indices that have been proposed for aviation turbulence forecasting are also discussed. Energy conversion associated with several atmospheric instabilities is reviewed in Sect. 3.2 through an equation governing energy transfer in an inviscid, stably stratified shear fluid flow. When *shear instability* occurs, the energy is transferred from the basic state shear to perturbation energy giving an upshear phase tilt in a stably stratified fluid flow. When *inertial instability* occurs, the energy stored in the basic state horizontal shear is transferred to horizontal perturbation momentum flux giving a horizontal upshear phase tilt. When baroclinic instability occurs, the *available potential energy* stored in the system is transferred to perturbation kinetic energy, as represented by the energy exchange between the basic state vertical shear and the perturbation heat flux. In Sect. 3.3, the Taylor–Goldstein equation is used to prove that in order for instability to occur in a two-dimensional, nonrotating flow, the existence of a critical level is required, as a corollary of the Howard's semicircle theorem and the Miles–Howard Theorem.

In Sect. 3.4, the criteria for static instability in dry and saturated atmospheres is reviewed. The static instability is closely related to KHI; thus, the Brunt–Väisälä frequency has been used implicitly in the prediction of aviation turbulence, such as the Graphical Turbulene Guidance (GTG) for detecting aviation turbulence. The criteria for *conditional instability* to occur are derived in Sect. 3.5. In practice, CAPE and CIN are used to examine the likelihood of conditional instability. Six static and conditional instabilities are given in Table 3.1. It is shown in Sect. 3.6 that when an atmospheric layer composed of a much drier air on top of a moist layer is lifted up, the atmosphere is potentially (convectively) unstable. The exact relationship between potential and conditional instabilities is still not quite clear and deserves further investigation.

KHI may produce large-amplitude gravity waves and CAT leading to aviation turbulence. The criterion for KHI is $Ri < 1/4$ at some level of the fluid, which requires strong vertical shear and/or weak thermal stratification. It has been proposed that when $N^2 < 0$, the KHI might be more dominated by static instability when Ri is less negative and shear instability when Ri is more negative. Whether this argument applies to the KHI induced CAT or MWT in the upper troposphere or lower stratosphere remains to be investigated. In addition, the interaction of turbulence, KHI, and static instability may come into play when $N^2 < 0$ and thus requires more investigation. The existence of a critical level may make the situation even more complicated and deserves further investigation, especially for $Ri < 1/4$.

The criterion for inertial instability is discussed in Sect. 3.8. In the atmosphere, inertial instability may be released by a strong divergent anticyclone flow equatorward of a westerly jet streak or in subsynoptic-scale ridges in the midlatitudes. Inertial instability is often found in the subtropical upper troposphere, thus providing a potential mechanism for turbulence development in the vicinity of jet entrance regions, fronts, regions of low Richardson number, regions with strong deformation, and regions of low upper-tropospheric absolute vorticity, and could trigger

CAT by promoting gravity wave genesis and breaking. In fact, measures related to inertial instability in strongly anticyclonic flows or related parameters, such as negative vorticity advection, unbalanced flow, and curvature, have been incorporated fully or partially in the GTG for forecasting aviation turbulence.

In Sect. 3.9, it is shown that in a statically stable ($N^2 > 0$) and inertially stable ($f\left(\partial v_g/\partial x + f\right) > 0$) atmosphere, a slantwise displaced air parcel may accelerate away from its origin leading to *symmetric instability* if $Ri < f/\left(\zeta_g + f\right)$. Alternatively, it can be shown that symmetric instability occurs when the geostrophic potential vorticity is negative, as shown in Eq. (3.14).

All these various instability mechanisms are summarized in Table 3.1.

Acknowledgments This research was partially supported by the National Science Foundation Award AGS-1265783, HRD-1036563, OCI-1126543, and CNS-1429464.

References

AMS.: "Bulk Richardson Number". Glossary of Meteorology, Amer. Meteor. Soc. Available online at http://glossary.ametsoc.org/wiki/bulk_richardson_number (2014)

Arakawa, H.: Severe turbulence resulting from excessive wind-shear in tropical cyclones. J. Meteor. **9**, 221–223 (1952)

Bennetts, D.A., Hoskins, B.J.: Conditional symmetric instability—a possible explanation for frontal rainbands. Quart. J. R. Meteor. Soc. **105**, 945–962 (1979)

Bluestein, H.B.: Principles of Kinematics and Dynamics. Vol. I. Synoptic-Dynamic Meteorology in Midlatitudes. Oxford University Press, New York (1992)

Breeding, R.J.: A nonlinear investigation of critical levels for internal atmospheric gravity waves. J. Fluid Mech. **50**, 545–63 (1971)

Bryan, G.H., Fritsch, J.M.: Moist absolute instability: The sixth static stability state. Bull. Amer. Meteor. Soc. **81**, 1207–1230 (2000)

Clark, T.L., Peltier, W.R.: Critical level reflection and the resonant growth of nonlinear mountain waves. J. Atmos. Sci. **41**, 3122–3134 (1984)

Colson, D., Panofsky, H.A.: An index of clear-air turbulence. Quart. J. R. Meteor. Soc. **91**, 507–513 (1965)

Crow, S.C.: Stability theory for a pair of trailing vortices. AIAA J. **8**, 2172–2179 (1970)

Darkow, G. L.: Basic thunderstorm energetics and thermodynamics. In: Kessler, E. (ed.) Thunderstorm Morphology and Dynamics, pp. 59–72. Univ. of Oklahoma Press (1986)

Das, P.: A non-Archimedean approach to the equation of convection dynamics. J. Atmos. Sci. **36**, 2183–2190 (1979)

Davies-Jones, R.: An expression for effective buoyancy in surroundings with horizontal density gradients. J. Atmos. Sci. **60**, 2922–2925 (2003)

Doswell, C.A., Markowski, P.M.: Is buoyancy a relative quantity? Mon. Wea. Rev. **132**, 853–863 (2004)

Doyle, J.D., Jiang, Q., Smith, R.B., Grubišić, V.: Three-dimensional characteristics of stratospheric mountain waves during T-REX. Mon. Wea. Rev. **139**, 3–23 (2011)

Durran, D.R., Klemp, J.B.: On the effects of moisture on the Brunt-Väisälä frequency. J. Atmos. Sci. **39**(10), 2152–2158 (1982)

Dutton, J., Panofsky, H.A.: Clear air turbulence: A mystery may be unfolding. Science **167**, 937–944 (1970)

Emanuel, K.A.: On assessing local conditional symmetric instability from atmospheric soundings. Mon. Wea. Rev. **111**, 2016–2033 (1983)

Emanuel, K.A.: Atmospheric Convection. Oxford University Press, New York (1994)

Emanuel, K.A., Raymond, D.J.: In: Klemp, J.B.(eds.) Dynamics of Mesoscale Weather Systems. NCAR (1984)

Endlich, R.M.: The mesoscale structure of some regions of clear-air turbulence. J. Appl. Meteor. **3**, 261–276 (1964)

Han, J., Lin, Y.-L., Schowalter, D.G., Arya, S.P.: Large eddy simulation of aircraft wake vortices within homogeneous turbulence: Crow instability. AIAA J. **38**, 292–300 (2000)

Hopkins, R.H.: Forecasting techniques of clear-air turbulence including that associated with mountain waves. WMO Tech. Note 155 (1977)

Howard, L.N.: Note on a paper of John W. Miles. J. Fluid Mech. **10**, 509–512 (1961)

Jascourt, S.D., Lindstrom, S.S., Seman, C.J., Houghton, D.D.: An observation of banded convective development in the presence of weak symmetric stability. Mon. Wea. Rev. **116**, 175–191 (1988)

Kaplan, M.L., Huffman, A.W., Lux, K.M., Charney, J.J., Riordan, A.J., Lin, Y.-L.: Characterizing the severe turbulence environments associated with commercial aviation accidents. A 44 case study synoptic observational analyses. Meteor. Atmos. Phys. **88**, 129–152 (2005a)

Kaplan, M.L., Huffman, A.W., Lux, K.M., Cetola, J.D., Charney, J.J., Riordan, A.J., Lin, Y.-L., Waight III, K.T.: Characterizing the severe turbulence environments associated with commercial aviation accidents. Hydrostatic mesobeta scale numerical simulations of supergradient wind flow and streamwise ageostrophic frontogenesis. Meteor. Atmos. Phys. **88**, 153–174 (2005b)

Keller, J.L.: Clear air turbulence as a response to meso- and synoptic-scale dynamic processes. Mon. Wea. Rev. **118**, 2228–2242 (1990)

Kim, J.-H., Chun, H.-Y.: A numerical study of clear-air turbulence (CAT) encounters over South Korea on 2 April 2007. J. Appl. Meteor. Clim. **49**, 2381–2403 (2010)

Kim, J.-H., Chun, H.-Y.: Statistics and possible sources of aviation turbulence over South Korea. J. Appl. Meteor. Clim. **50**, 311–324 (2011)

Knox, J.A.: Possible mechanism of clear-air turbulence in stronglyanticyclonic flows. Mon. Wea. Rev. **125**, 1251–1259 (1997)

Knox, J.A.: Inertial instability. In: Holton, J.R., et al. (eds.) Encyclopedia of Atmospheric Sciences, pp. 1004–1013. Academic (2003)

Koch, S.E., Caracena, F.: Predicting clear-air turbulence from diagnosis of unbalance flow. In: Preprints, 10th Conference on Aviation, Range, and Aero. Meteor., Portland, OR, Amer. Meteor. Soc., pp. 359–363 (2002)

Kronebach, G.W.: An automated procedure for forecasting clear-air turbulence. J. Appl. Meteor. **3**, 119–125 (1964)

Lalas, D.P., Einaudi, F.: On the correct use of the wet adiabatic lapse rate in the stability criteria of a saturated atmosphere. J. Appl. Meteor. **13**, 318–324 (1974)

Lane, T.P., Sharman, R.D., Trier, S.B., Fovell, R.G., Williams, J.K.: Recent advances in the understanding of near-cloud turbulence. Bull. Amer. Meteor. Soc. **93**(4), 499–515 (2012)

Lester, P.F.: Turbulence: A new perspective for pilots. Jeppesen Sanderson, Inc., Englewood, CO (1993)

Lin, Y.-L.: Mesoscale Dynamics. Cambridge University Press, New York (2007)

Lin, Y.-L., Chun, H.-Y.: Structures of dynamically unstable shear flows and their implications for shallow internal gravity waves. Meteor. Atmos. Phys. **52**, 59–68 (1993)

Lindzen, R.S., Rosenthal, A.J.: Instabilities in a stratified fluid having one critical level. Part III: Kelvin-Helmholtz instabilities as overreflected waves. J. Atmos. Sci. **40**, 530–542 (1983)

MacDonald, A.E.: On a type of strongly divergent steady state. Mon. Wea. Rev. **105**, 771–785 (1977)

Mahalov, A., Moustaoui, M., Nicolaenko, B., Tse, K.L.: Computational studies of inertia-gravity waves radiated by upper tropospheric jets. Theor. Comput. Fluid Dyn. **21**, 399–422 (2007)

McCann, D.W.: Gravity waves, unbalanced flow, and aircraft clear air turbulence. Nat'l Wea. Dig. **25**, 3–14 (2001)
Miles, J.W.: On the stability of heterogeneous shear flow. J. Fluid Mech. **10**, 496–508 (1966)
Nastrom, G.D., Fritts, D.C.: Sources of mesoscale variability of gravity waves. Part I: Topographic excitation. J. Atmos. Sci. **49**, 101–110 (1992)
O'Sullivan, D., Dunkerton, T.J.: Generation of inertia-gravity waves in a simulated life cycle of baroclinic instability. J. Atmos. Sci. **52**, 3695–3716 (1995)
Ooyama, K.: On the instability of baroclinic circular vortex: a sufficient criterion for instability. J. Atmos. Sci. **23**, 43–5 (1966)
Raab, H., Foster, J.: Crow instability in contrails. Earth Sci. Picture of the Day, Aug. 09, 2011, USRA. http://epod.usra.edu/blog/2011/08/crow-instability-in-contrails.html%20 (2011)
Roach, R.T.: On the influence of synoptic development on the production of high level turbulence. Quart. J. R. Meteor. Soc. **96**, 413–429 (1970)
Schultz, D.M., Schumacher, P.N., Doswell III, C.A.: The intricacies of instabilities. Mon. Wea. Rev. **128**, 4143–4148 (2000)
Schultz, D.M., Schumacher, P.N.: The use and misuse of conditional symmetric instability. Mon. Wea. Rev. **127**, 2709–2732; Corrigendum, **128**, 1573 (1999)
Schumacher, R.S., Schultz, D.M.: Inertial instability: climatology and possible relationship to severe weather predictability. In: 9th Conference on Mesoscale Processes, Amer. Meteor. Soc., Fort Lauderdale, FL, 30 July–2 August (2001)
Scinocca, J.F., Peltier, W.R.: The instability of Long's stationary solution and the evolution toward severe downslope windstorm flow. Part I: Nested grid numerical simulations. J. Atmos. Sci. **50**, 2245–2263 (1993)
Seman, C.J.: On the role of nonlinear convective-symmetric instability in the evolution of a numerically simulated mesoscale convective system. In: Preprints, 5th Conference on Mesoscale Processes, Atlanta, GA, Amer. Meteor. Soc. pp. 282–287 (1992)
Sharman, R., Tebaldi, C., Wiener, G., Wolff, J.: An integrated approach to mid- and upper-level turbulence forecasting. Wea. Forecasting **21**(3), 268–287 (2006)
Sharman, R.D., Trier, S.B., Lane, T.P., Doyle, J.D.: Sources and dynamics of turbulence in the upper troposphere and lower stratosphere: A review. Geophys. Res. Lett. **39**, L12803 (2012). doi:10.1029/2012GL051996
Sherwood, S.C.: On moist instability. Mon. Wea. Rev. **128**, 4139–4142 (2000)
Smith, R.B.: Stratified flow over topography. In: Grimshaw, R. (ed.) Environmental Stratified Flows, pp. 121–159. Kluwer Acad, Boston, MA (2002)
Spalart, P.R.: Airplane trailing vortices. Ann. Rev. Fluid Mech. **30**, 1–35 (1998)
Stone, P.H.: On non-geostrophic baroclinic instability. J. Atmos. Sci. **23**, 390–400 (1966)
Teixeira, M.A.C., Miranda, P.M.A., Argain, J.L., Valente, M.A.: Resonant gravity-wave drag enhancement in linear stratified flow over mountains. Q. J. R. Meteor. Soc. **131**, 1795–1814 (2005)
Trier, S.B., Sharman, R.D., Lane, T.P.: Influences of moist convection on a cold-season outbreak of clear-air turbulence (CAT). Mon. Wea. Rev. **140**(8), 2477–2496 (2012)
Wang, T.-A., Lin, Y.-L.: Wave ducting in a stratified shear flow over a two-dimensional mountain. Part II: Implication for the development of high-drag states for severe downslope windstorms. J. Atmos. Sci. **56**, 437–452 (1999)
Weisman, M.L., Klemp, J.B.: The dependence of numerically simulated convective storms on vertical wind shear and buoyancy. Mon. Wea. Rev. **110**, 504–20 (1982)
Wolff, J.K., Sharman, R.D.: Climatology of upper-level turbulence over the continental Unites States. J. Appl. Meteor. Clim. **47**, 2198–2214 (2008)
Wurtele, M.G., Sharman, R.D., Datta, A.: Atmospheric lee waves. Annu. Rev. Fluid Mech. **28**, 429–476 (1996)
Xu, Q., Clark, J.H.E.: The nature of symmetric instability and its similarity to convective and inertial instability. J. Atmos. Sci. **42**, 2880–2883 (1985)

Chapter 4
Turbulence Events Interpreted by Vortex Rolls

Bob Lunnon

Abstract The severity of an aircraft encounter with turbulence is most commonly quantified by reference to the aircraft normal acceleration. For an encounter between an idealised aircraft and an idealised vortex, the acceleration vector is equal to the cross product of half the vorticity vector with air velocity vector. More realistic expressions for the aircraft acceleration have been derived in the context of aircraft encounters with wake vortices, and, although the scale of wake vortices is smaller than that of naturally occurring vortices, the broad methodology is broadly applicable.

This chapter looks both at these idealised cases and more practical issues. Many of the issues touched on are dealt with more comprehensively elsewhere. Two issues are addressed which are particular to the interpretation of turbulence using vortex constructs. Firstly, the feasibility of forecasting vortex characteristics is addressed. Secondly, a method of representing the significant statistical properties of anisotropic turbulence is suggested.

4.1 Introduction

Aircraft trailing wake vortices are a clear example of well-defined vortices in the atmosphere, that are a known hazard to aviation. But previous studies have suggested that enhanced levels of turbulence may also be represented by a series of vortices. For example, both Förchtott (1957) and Gerbier and Berenger (1961) represent upper level turbulence associated with mountain waves as a series of moving or standing rotors aligned horizontally with the mountain ridge. Also, Parks et al. (1985) showed two cases where it was possible to represent the wind field that caused severe turbulence encounters by a sequence of vortices to imitate the roll-up region of Kelvin–Helmholtz billows. Figure 4.1 shows such a representation. As would be expected, there were violent normal accelerations stemming from the two

B. Lunnon (✉)
UK Met Office (retired), University of Reading, Reading, UK
e-mail: bob.lunnon@cantab.net

R. Sharman, T. Lane (eds.), *Aviation Turbulence*,
DOI 10.1007/978-3-319-23630-8_4

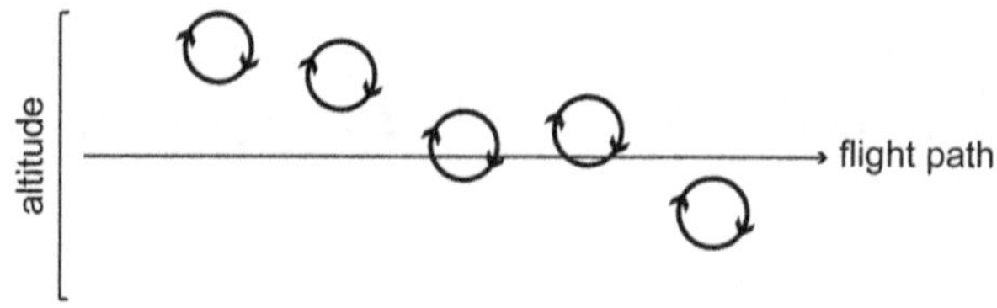

Fig. 4.1 Schematic of vortex array determined from estimated winds for case 1 from Parks et al. (1985). Adapted from Parks et al. (1985)

vortices directly intersected by the flight path in Fig. 4.1, but not for the other three vortices in the figure. This clearly indicates that for this case, the sequence of vortices may be a useful way to represent the atmospheric conditions. The vortex representation has been used subsequently, e.g. by Wingrove and Bach (1994), Spilman and Stengel (1995), Mehta (1987), Misaka et al. (2008) for aircraft response studies. In the Misaka et al. study, a 4-D variational method was used to generate the wind vector field which was consistent with measurements of aircraft-measured wind vector and potential temperature. The diagnosed wind field in one of the two cases can be described as a pair of vortices. The evolving wind field generated by these simulations is in broad terms consistent with the equations governing atmospheric motion, as well as with the aircraft measurements.

The well-known turbulence encounter reported in Clark et al. (2000) was associated with horizontal vortices, in that case parallel to the prevailing wind. Although a single encounter does not constitute rigorous confirmation that this is a useful general approach, it encourages further research on the question. Although vortices are a useful concept, as explained in Sect. 4.8, one term in the expression for vorticity has much more general application.

4.2 Normal Acceleration of Aircraft Flying Perpendicular to Sequence of Vortices

There is considerable interest in normal acceleration in particular (this is vertical acceleration when the aircraft is flying straight and level). Large normal acceleration will result in considerable stress to the wings and other aircraft components which ultimately can lead to damage to the aircraft superstructure. However from the perspective of the effects on passengers, horizontal components of acceleration can be at least as important. This issue is addressed in Jacobson et al. (1978).

The way atmospheric motion causes aircraft acceleration is illustrated in Fig. 4.2. In the scenario illustrated, it is assumed that there is a sequence of vortices, all with the same physical characteristics including sense of rotation. Implicit in this is that there will be areas of vorticity of the opposite sign between the explicitly represented vortices. The aircraft experiences acceleration from each of the explicit vortices as well as from the implicit areas of counter-rotating vorticity. The initial trajectory of the aircraft (before encountering the first vortex) is set so that the combined effects of all the vortices is to return the aircraft to its original trajectory.

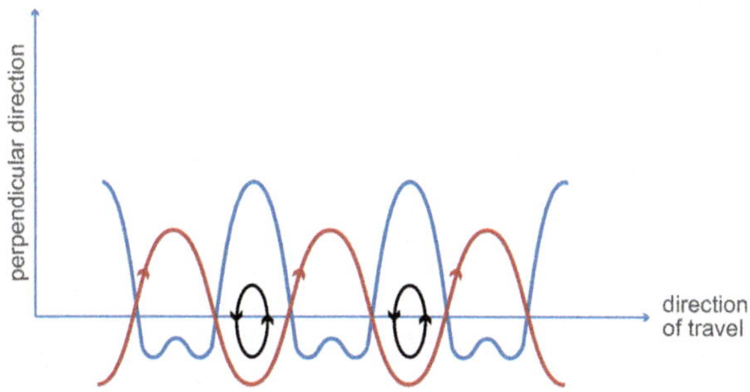

Fig. 4.2 Aircraft position and acceleration in response to a sequence of vortices. The figure illustrates the effect of a sequence of vortices on the path of an aircraft. The *red line* indicates the position of the aircraft while the *blue line* indicates the acceleration of the aircraft. The *black ellipses* with *arrows* indicate the positions of vortices and the sense of rotation. There are additional vortices (not shown) to the *left* and *right* of the diagram. Although the diagram can be interpreted as a sequence of vortices with horizontal axes (as in Kelvin Helmholtz instability), it can also be interpreted as a sequence of vortices with vertical axes (as in inertial instability)

However, in between the aircraft experiences accelerations and which in turn result in oscillatory position changes. Figure 4.2 shows the position of the vortices and both the aircraft position and the aircraft acceleration. The figure is constructed with the assumption that the acceleration is proportional to the second derivative of the position with respect to distance travelled in the direction of the initial trajectory. The representation of the vortices is schematic.

4.3 Normal Acceleration of Aircraft Encountering an Isolated Vortex

The severity of an aircraft encounter with turbulence is most commonly quantified by reference to the aircraft normal acceleration. For an encounter between an idealised aircraft and an idealised vortex, the acceleration vector is equal to the cross product of half the vorticity vector with air velocity vector. More realistic expressions for the aircraft acceleration have been derived in the context of aircraft encounters with wake vortices, and, although the scale of wake vortices is smaller than that of naturally occurring vortices, the broad methodology is broadly applicable. Figure 4.3 shows the dependency of normal acceleration on heading angle for a simulated wake vortex. As can be seen, there is enormous sensitivity to heading angle (relative to the orientation of the vortex) and the dependency is approximately sinusoidal. This suggests that forecasting the orientation of vortices will be very helpful in predicting normal acceleration.

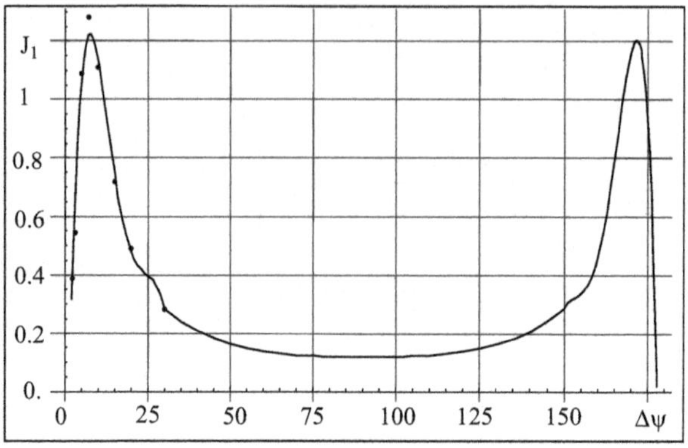

Fig. 4.3 Load factor (normalised acceleration, ordinate) plotted against heading angle (abscissa). The response of an aircraft to a vortex also depends on aircraft specific factors, of which airspeed has already been mentioned. In the RECAT II project, which is categorising aircraft response to wake vortices, aircraft wingspan, maximum take-off weight and maximum landing weight are also identified as major factors. Figure generated by TsAGI (Russia) as part of the FLYSAFE project

4.4 Aircraft Response to Other Forms of Turbulence

Aircraft in general avoid the adverse effect of wake vortices by using separation distances which reflect the time for a vortex to decay to a state in which it does not constitute a significant hazard. The decay process may lead to a vortex with greatly reduced vorticity or to what might be described as 3-D turbulence. Figure 4.4 shows a numerically simulated field of flow around a decaying wake vortex. In Fig. 4.4, the flow some distance away from the vortices is not considered to pose a threat to a following aircraft.

The success of the application of the aircraft separations described earlier implies that wake vortices, which consist of very well defined vortices, pose a threat of a significant normal acceleration but the more 3-D turbulence does not pose such a significant threat. The conclusion is that it is plausible that 2-D turbulence is more likely on average to constitute a threat than 3-D turbulence and that representing turbulent flows by sequences of vortices may well provide useful guidance of the threat posed by the turbulence.

4.5 Causes of Turbulence

Causes of turbulence can be summarised as falling into three categories:

(a) Hydrodynamic instability
(b) Obstacles
(c) Gravity waves

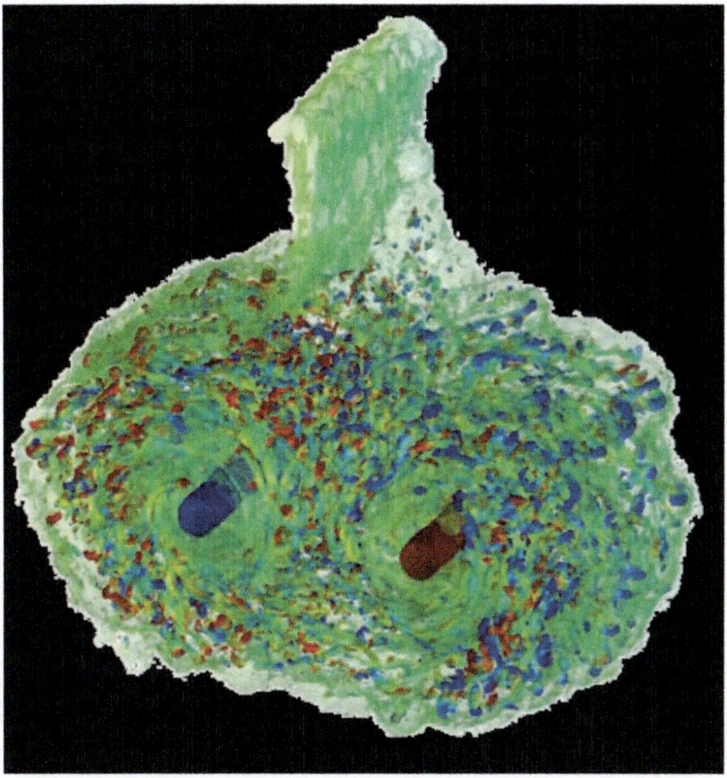

Fig. 4.4 Flow field round two decaying wake vortices indicating turbulence which is quasi-3-D. Figure generated by UCL (Belgium) as part of the CREDOS project

Hydrodynamic instabilities include Kelvin Helmholtz instability and convective instability. A generic approach to these instabilities was attempted by Lighthill (1952) which in turn led to the more recent analysis by Knox et al. (2008). Note that in Lighthill's original analysis, the mechanism giving rise to turbulence is highly anisotropic and it follows that the turbulence resulting directly from the instability will be anisotropic (subsequent processes will tend to make the turbulence more isotropic). Naturally occurring obstacles are likely to give rise to anisotropic turbulence, the rotors stemming from flow over mountains [see, e.g. Gerbier and Berenger (1961)]. The wing of an aircraft, which can hardly be considered a naturally occurring obstacle, will give rise to wake vortices (if the wing is developing lift), and these are self-evidently anisotropic. Imposing a grid within a wind tunnel will generate turbulence which, a short distance downstream, can be considered to be isotropic, but this can be considered to be outside the scope of turbulence as it affects aviation. Gravity waves are also anisotropic. An isolated updraft (which could be a result of convective instability) can be considered to be an isotropic form of turbulence—a ring vortex—although one with a more structure than the turbulence illustrated in Fig. 4.4. Thus most forms of turbulence which affect aviation can be considered to be, at least initially, anisotropic.

A question which is linked to the feasibility of representing turbulence by sequences of vortices is: is the turbulence which causes highest amplitude aircraft accelerations quasi 2-D? In general it can be difficult from, say, aircraft data alone, to determine how 3-D turbulence is. It is noted that in Doyle et al. (2005) turbulence is characterised as being 3-D on the grounds that the three components of the TKE are the same order of magnitude. However note that if an aircraft encounters horizontal vortices as shown in Fig. 4.1 at an angle of 45°, then the three components of TKE will be the same order of magnitude.

This interpretation is consistent with Clodman (1957) where repeated flights of an aircraft through patches of turbulence, travelling in different directions, were described. The encounters reported show that in most cases the aircraft response to the turbulence is strongly dependent on the direction of motion, indicating that in these cases the turbulence was anisotropic.

4.6 Feasibility of Forecasting Vortex Characteristics

A seminal paper on prediction of clear-air turbulence (CAT) was Roach (1970), which focussed on Kelvin Helmholtz instability. The derived CAT predictor was essentially the time rate of change of the Richardson Number, Ri, and can be written as (Dixon and Roach 1970)

$$\Phi \equiv -\frac{D}{Dt}\ln \mathrm{Ri} = \text{temperature term} + \cos 2\alpha D_{\mathrm{ST}} - \sin 2\alpha D_{\mathrm{SH}} \qquad (4.1)$$

where $D_{\mathrm{SH}} = (u_y + v_x)$ is the shearing deformation, $D_{\mathrm{ST}} = (u_x - v_y)$ is the stretching deformation, and α is the direction of the shear vector $\partial \mathbf{v}/\partial p$. If the last two terms in Eq. (4.1) are examined, it can be shown that the terms represent the effect of large scale deformation on vorticity about a horizontal axis, as illustrated in Fig. 4.5. These figures represent vortex stretching effects. The classic 3-D vorticity equation contains a vortex stretching term which is the only process by which mean squared vorticity is increased (e.g. Tennekes and Lumley 1972). Imagine one has a pre-existing small scale vortex with the axis of rotation parallel to the x axis. In Fig. 4.5a, a large scale horizontal deformation field is applied such that the x-axis is being expanded and the y-axis is being compressed [$u_x > 0$, $v_y < 0$, $u_y = v_x = 0$]. This deformation field will increase the magnitude of the x-vorticity in the pre-existing vortex. (A common analogy is an ice skater changing his/her moment of inertia and consequently changing his/her rate of rotation). In Fig. 4.5b, the large scale horizontal deformation field is applied such that fluid elements along the x-axis are compressed leading to a decrease in the x-vorticity. The x-vorticity (axis of rotation parallel to the x-axis) is $\xi = w_x - u_z$. In a case of strong vertical wind shear but no small scale horizontal variability, the u_z term is

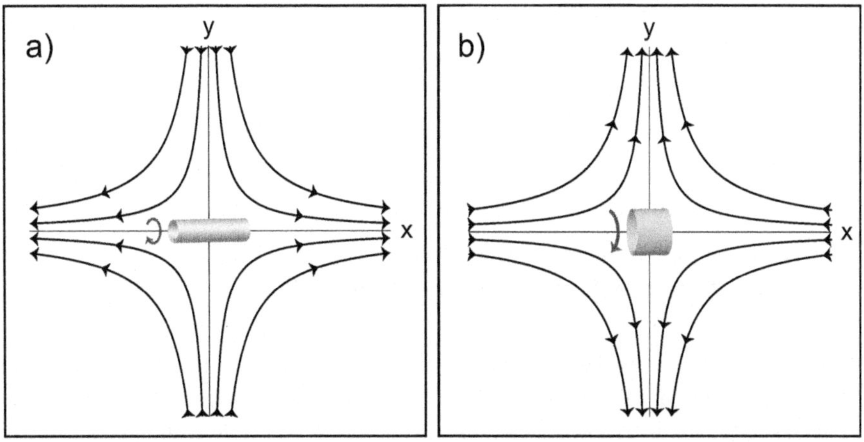

Fig. 4.5 Effect of large scale horizontal deformation field (*streamlines* with *arrows*) in the x–y plane on a single horizontal vortex tube (as represented by a *shaded cylinder* with an *arrow*). (**a**) Case when the deformation field stretches the vortex and increases the x-vorticity. (**b**) as in (**a**) but in this case the deformation field compresses the vortex and decreases the x-vorticity

dominant. The nature of Kelvin Helmholtz instability is such that rolls will develop, i.e. the w_x term increases in magnitude, locally. However, such turbulence is likely to destroy itself quite quickly. The deformation term means that once a roll had been created, it will be subjected to one process which will reduce the magnitude of the vorticity (through dissipation) and a second process which will increase the magnitude of the vorticity (vortex stretching). The lifespan of the roll will depend on the balance between these processes.

A CAT predictor based on these principals, called the "Lunnon index", is discussed in Gill and Buchanan (2014). It is contrasted with the Ellrod TI1 predictor (Ellrod and Knapp 1992). Both Ellrod and Lunnon make reference to the horizontal deformation field and the vertical wind shear field, but whereas Ellrod uses a scalar product between the two fields, Lunnon uses a vector product. Gill and Buchanan show that the Lunnon predictor shows improved skill over the Ellrod TI1 predictor. This implies that NWP models have useful skill in predicting not only the magnitude of deformation but also the orientation of the deformation field. It is plausible that including the direction of travel of the aircraft in the prediction will improve skill further, but this has yet to be demonstrated.

Note that the Lunnon predictor is based on the physical principal that deformation can act on a field of vertical wind shear linked to Kelvin Helmholtz instability. The deformation increases horizontal vorticity by vortex stretching. However, this process will occur whatever the cause of the horizontal vorticity, which could be mountain wave processes or convection.

The formal definition of the Lunnon predictor is

$$\Phi = \left[\left(\frac{\partial v}{\partial p}\right)^2 - \left(\frac{\partial u}{\partial p}\right)^2\right]D_{\mathrm{ST}} - \left(2\frac{\partial u}{\partial p}\frac{\partial v}{\partial p}\right)D_{\mathrm{SH}} \tag{4.2}$$

4.6.1 Technical Advances Which Will Help

Although the direct observation and prediction of vortices on the global scale will be impracticable for the foreseeable future, technical advances will greatly help our understanding of the issues. Computer power will allow the implementation of regional models with sufficiently high resolution to predict vortices or the relevant scales—this has already happened in the T-REX programme (Doyle et al. 2009). In that programme, high resolution lidars were operated giving relatively routine observations of detail (albeit in a local area) which could previously be obtained only from research aircraft. As far as modelling is concerned, one could anticipate a scenario in which a global model is run routinely and, if/when severe turbulence is predicted in a local area, a much higher resolution model is run covering that local area. The frequency of severe turbulence encounters affecting aircraft overflying Greenland (see, e.g. Sharman et al. 2012) suggests that routine running of a high resolution model covering that area might could be beneficial, and there are other areas of the world where this would apply.

More generally, high resolution models could be used to develop techniques for diagnosing the parameters in lower resolution models which characterise the vortices which cannot be explicitly represented in the low resolution models. In general, the lower resolution models will provide boundary conditions for the higher resolution models. One can construct a situation in which the low resolution model provides all the information required by the high resolution model (i.e. no information passes from the high resolution model to the low resolution model), and the orography data sets used by the two models are the same. In these conditions, the vortices in the high resolution model will be a function of the information supplied by the low resolution model, and it should be possible to develop an approach for diagnosing the broad characteristics of the vortices from the information passed from the low resolution model.

Note that it is accepted that, for the foreseeable future, it will not be possible to explicitly forecast individual vortices that pose a threat to aircraft, at least not on the global scale. However as shown in Sect. 4.1, there is some evidence that it is already possible to forecast forcing mechanisms, and this is one of the grounds for optimism. It is anticipated that it will be possible to forecast discrete volumes of the atmosphere in which there can be expected to be a finite number of vortices with certain characteristics, e.g. orientation, vorticity, separation. The high resolution observed data will be useful for verifying that the high resolution model is broadly behaving correctly.

4.6.2 Preliminary Work Which Should Be Undertaken

An obvious preliminary step is to assess whether, for turbulence encounters which are strongly suspected of being Kelvin Helmholtz instability, forecast skill is improved by taking the direction of movement of the aircraft into account. Determining the direction of movement of aircraft is straightforward from aircraft datasets where there are regular reports from aircraft whether or not they encounter turbulence, e.g. the Global Atmospheric Data Set (GADS) data as used by Gill and Buchanan (2014).

Other work should address the questions

1. To what extent is a sequence of vortices an accurate representation of turbulence in the atmosphere?
2. How well can we predict the significant properties of vortices?
3. How well can we predict aircraft behaviour given the characteristics of the vortices?

Question 1 can be clarified by considering what parameters might be useful for representing vortex characteristics. For a numerically simulated flow field, e.g. from a large eddy simulation (LES) model, it is possible to derive the vorticity vector at every grid point. For an appropriate domain one could then derive average statistics.

4.7 Possible Approach to Providing Statistical Information on Anisotropic Turbulence

Assuming that the aircraft normal acceleration can be derived from the aircraft velocity and the vorticity vector, it follows that a probabilistic distribution of normal acceleration can be derived from a probabilistic distribution of the vorticity vector (given knowledge of the aircraft velocity). Therefore, there is value in a forecasting organisation providing a probabilistic distribution of the vorticity vector and then the airline combining that information with knowledge of the aircraft velocity to generate a probabilistic distribution of normal acceleration for that aircraft for an individual promulgation grid box.

The current World Area Forecast System (WAFS) provides products to end users on a $1.25° \times 1.25°$ grid (~140 km × 140 km). For turbulence, there is a vertical resolution of 50 hPa. Given that the UK Met Office runs a 25 km resolution global model, each promulgation grid box contains many model grid boxes. Therefore, there

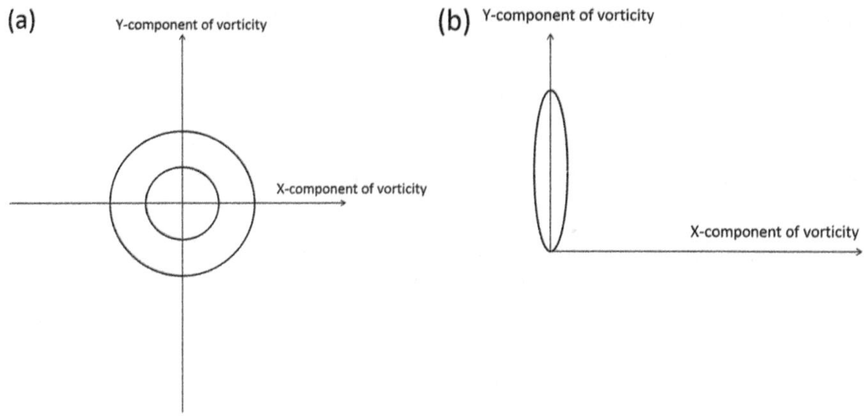

Fig. 4.6 (**a**) Contours indicating a pdf of vorticity in which the coordinates are the x component of vorticity and the y component of vorticity. The contours indicate isotropic turbulence. (**b**) As in (**a**) but for anisotropic turbulence

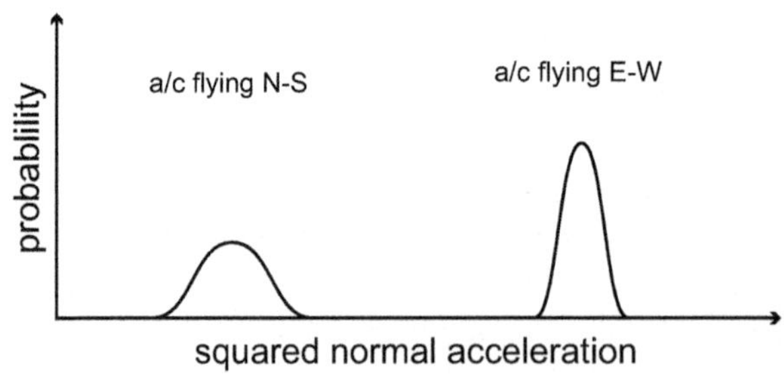

Fig. 4.7 For the anisotropic case (Fig. 4.6b), the PDFs of squared normal acceleration are shown, in the left pane for an aircraft flying N-S and in the right pane for an aircraft flying E-W

is scope for providing statistical information for each promulgation grid box—at present (2014) mean and maximum turbulence indicators are provided.

From high resolution simulations, it would be possible to generate the statistical distribution of the horizontal vorticity vector (magnitude and direction) for individual promulgation grid boxes. For each point in this distribution, the aircraft normal acceleration could be (pre-)calculated. For this promulgation grid box, the statistical distribution of aircraft normal acceleration could be generated. The risk of large normal acceleration could be derived for relevant promulgation grid boxes (e.g. those just above/below preferred route) and well informed route choices made.

This is illustrated in Fig. 4.6. Figure 4.6a shows the probability distribution function (pdf) for the vorticity vector in a case of isotropic turbulence. In this case, the contours of probability are circles centred on the point having zero vorticity. Figure 4.6b shows the pdf for the vorticity vector in the case of anisotropic turbulence. In this case, the contours of probability are approximately elliptical. Figure 4.7 shows the pdfs for two cases—one of an aircraft flying N-S and the other for an aircraft flying E-W. In the E-W case, the maximum probability occurs at a much higher normal acceleration value than for the N-S case, indicating that for the aircraft flying E-W, there is a very high probability of a very high normal acceleration.

4.8 Generating statistics on horizontal gradient of vertical motion

The y component of horizontal vorticity is given by $\partial u/\partial z - \partial w/\partial x$. In the case of symmetrical vortices, these two terms are in general equal. More generally, the term which determines the aircraft normal acceleration is the $\partial w/\partial x$ term. From high resolution NWP output, it is possible to derive this vector, and this should be used in probability distributions.

An issue is the dependency of aircraft response to different scales of turbulence. An aircraft will be relatively unaffected by vortices having a characteristic scale much smaller than the aircraft dimensions, because, assuming the aircraft is essentially rigid, it will integrate out the effect of these wind variations. Intentional movement of aircraft control surfaces will enable the aircraft to mitigate the effects of relatively large scale wind fluctuations. Therefore between these two scales, there exists a scale where the effect of, say, a single isolated vortex on the aircraft motion will be maximum. It is potentially very useful to say something about the scale of vortices.

4.9 Summary and Conclusions

I have shown considerable, largely theoretical evidence that the most severe forms of turbulence are anisotropic and that it is helpful to represent the flow by vortex rolls. Assuming this to be correct, it follows that for such turbulent events, the aircraft direction of motion should be taken into account when quantifying the effect of the turbulence on the aircraft. Methods for doing this in practice have been recommended. The aviation community may be slow to become convinced of the usefulness of this approach. The scientific community may benefit from the inclusion of anisotropic effects in their verification schemes because this should allow more effective tuning of the prediction algorithms.

References

Clark, T.L., Hall, W.D., Kerr, R.M., Middleton, D., Radke, L., Ralph, F.M., Nieman, P.J., Levinson, D.: Origins of aircraft-damaging clear air turbulence during the 9 December 1992 Colorado downslope windstorm: numerical simulations and comparison to observations. J. Atmos. Sci. **57**(8), 1105–1131 (2000)

Clodman, J.: Anisotropic high-level turbulence. Q. J. R. Meterol. Soc. **83**(355), 116–120 (1957). doi:10.1002/qj.49708335511

Dixon, R., Roach, W.T.: A note on the paper "On the influence of synoptic development on the production of high level turbulence" and Reply. Q. J. R. Meterol. Soc. **96**(410), 758–760 (1970). doi:10.1002/qj.49709641019

Doyle, J.D., Shapiro, M.A., Jiang, Q., Bartels, D.L.: Large amplitude mountain wave breaking over Greenland. J. Atmos. Sci. **62**, 3106–3126 (2005)

Doyle, J.D., Grubišić, V., Brown, W.O.J., De Wekker, S.F.J., Dörnbrack, A., Jiang, Q., Mayor, S. D., Weissmann, M.: Observations and numerical simulations of subrotor vortices during T-REX. J. Atmos. Sci. **66**, 1229–1249 (2009)

Ellrod, G.P., Knapp, D.I.: An objective clear air turbulence forecasting technique: verification and operational use. Weather Forecast. **7**, 150–165 (1992)

Förchtott, J.: Active turbulent layer downwind of mountain ridges. Stud. Geophys. Geod. **1**, 173–181 (1957)

Gerbier, N., Berenger, M.: Experimental studies of lee waves in the French Alps. Q. J. R. Meteorol. Soc. **87**, 13–23 (1961)

Gill, P.G., Buchanan, P.: An ensemble based turbulence forecasting system. Meteorol. Appl. **21** (1), 12–19 (2014)

Jacobson, I., Kulthau, A., Richards, L., Conner, D.: Passenger ride quality in transport aircraft. J. Aircr. **15**(11), 724–730 (1978)

Knox, J.A., McCann, D.W., Williams, P.D.: Application of the Lighthill–Ford theory of spontaneous imbalance to clear-air turbulence forecasting. J. Atmos. Sci. **65**, 3292–3304 (2008)

Lighthill, M.J.: On sound generated aerodynamically. I. General theory. Proc. R. Soc. Lond. **211A**, 564–587 (1952)

Mehta, R.S.: Modeling clear-air turbulence with vortices using parameter-identification techniques. J. Guid. Control Dyn. **10**(1), 27–31 (1987)

Misaka, T., Obayashi, S., Endo, E.: Measurement-integrated simulation of clear air turbulence using a four-dimensional variational method. J. Aircr. **45**(4), 1217–1229 (2008)

Parks, E.P., Wingrove, R.C., Bach, R.E., Mehta, R.S.: Identification of vortex-induced clear air turbulence using airline flight records. J. Aircr. **22**(2), 124–129 (1985)

Roach, W.T.: On the influence of synoptic development on the production of high level turbulence. Q. J. R. Meterol. Soc. **96**(410), 413–429 (1970)

Sharman, R.D., Doyle, J.D., Shapiro, M.A.: An investigation of a commercial aircraft encounter with severe clear-air turbulence over western Greenland. J. Appl. Meteorol. Climatol. **51**(1), 42–53 (2012)

Spilman, D.R., Stengel, R.F.: Jet transport response to a horizontal wind vortex. J. Aircr. **32**(3), 480–485 (1995)

Tennekes, H., Lumley, J.L.: A First Course in Turbulence. MIT Press, Cambridge, MA (1972). 300 p

Wingrove, R.C., Bach, R.E.: Severe turbulence and maneuvering from airline flight records. J. Aircr. **31**(4), 753–760 (1994)

Part II
Turbulence Detection Methods and Applications

Chapter 5
Airborne In Situ Measurements of Turbulence

Larry B. Cornman

Abstract Airborne in situ measurements of turbulence play a key role in mitigating the operational turbulence problem for aviation. These measurements can be used for warning following aircraft, used in turbulence forecasting and nowcasting algorithms, and assisting in the development and verification of those algorithms. Furthermore, in situ measurements are useful in furthering the phenomenological understanding of turbulence in terms of climatologies and case studies. In this chapter, we focus on energy dissipation rate (EDR) calculations for both homogeneous and inhomogeneous turbulence and describe the connection between EDR and aircraft vertical acceleration response. The chapter concludes with a simulation analysis comparing EDR estimates from vertical velocity, from scaling the root mean square of the velocities and accelerations, and from scaling peak accelerations.

5.1 Introduction

Encounters with turbulence continue to be a significant problem for aviation, from both safety and efficiency perspectives. One of the key factors in helping to mitigate this problem is airborne measurements of turbulence. The purpose of this chapter is to study various aspects of in situ calculations of turbulence. We will use the terms, "measurement" and "calculation," synonymously, even though technically one calculates, as opposed to measures turbulence. This is because turbulence intensity metrics, such as the turbulent kinetic energy or energy dissipation rate, are functions of measurable fluid state variables such as velocities, pressure, and temperature, rather than being measurable quantities themselves. Furthermore, we distinguish in situ measurements from onboard forward-looking ones, since the former are typically used in different ways from the latter. Forward-looking measurements, such as from Doppler radars or lidars, can be used for real-time warning of impending

L.B. Cornman (✉)
Research Applications Laboratory, National Center for Atmospheric Research, Boulder, CO, USA
e-mail: cornman@ucar.edu

R. Sharman, T. Lane (eds.), *Aviation Turbulence*,
DOI 10.1007/978-3-319-23630-8_5

encounters, whereas in situ measurements can be used for warning following aircraft, used in turbulence forecasting and nowcasting algorithms (Sharman et al. 2014), as well as supporting the development and verification of those algorithms. Furthermore, in situ measurements are useful in furthering the phenomenological understanding of turbulence, in terms of climatologies and case studies. In this chapter, we will focus on in situ measurements in an operational setting as opposed to those in a research one, though many of the concepts and techniques discussed below are applicable equally to both types of measurement platforms.

It has been widely accepted that the energy dissipation rate, ε, is a useful measure of turbulence intensity (Cornman et al. 1995; Sharman et al. 2014). Likewise, the absolute value of the peak, $|\ddot{z}_{\text{peak}}|$, and the root mean square $\sigma_{\ddot{z}}$, of vertical acceleration are operationally meaningful measures of aircraft response to turbulence. We note that the common terminology for the energy dissipation rate is "EDR," which refers to $\varepsilon^{1/3}$, and the root mean square of vertical acceleration, divided by one-g, is "RMS-g." It is important to note that for a correlated random process, the statistics of the sample standard deviation are different from those of the sample root mean square. This is because, even for a zero-mean process, the sample mean will not necessarily be zero. This can have a substantial effect for small sample sizes; hence, we use the root mean square herein. One must be careful when consulting the literature, since oftentimes authors are not careful in making this distinction and use the phrase, "RMS," when in fact they are computing a sample standard deviation. Another measure of turbulence is the so-called derived equivalent gust velocity or U_{de} (Hoblit 1988). This is an indirect atmospheric measurement in that it estimates the amplitude of an assumed 1-cosine gust profile from the measurement of the absolute value of peak vertical acceleration. As U_{de} is directly proportional to the peak vertical acceleration, we shall just consider the latter quantity in this chapter.

It is important to distinguish between atmospheric measures of turbulence and aircraft response metrics. The key difference is that the former can be calculated independent of the latter, whereas the opposite is not true. That is, measurements of atmospheric turbulence intensity can be made independent of aircraft response; but the aircraft response is inherently dependent on the atmosphere as the forcing mechanism. As we shall discuss below, mathematically this is the consequence of a linear system input–output relationship, with the input being the atmospheric turbulence and the output being the aircraft vertical acceleration response. For linear systems, this also means that one can go back and forth between the two—up to the accuracy in the input and the characterization of the system response function. This has practical implications, since measurements of atmospheric turbulence can be used for all the applications mentioned above, both meteorological and operational, and it does not require knowledge of the source of the information (i.e., the aircraft) for its use. To make this concept more concrete, consider the following example. Aircraft "A" makes a measurement of EDR and transmits it to aircraft "B." Aircraft B can use that information as is, or with knowledge of its own response function, the EDR can be translated into a response metric such as $\sigma_{\ddot{z}}$. On the other hand, if aircraft A measures $(\sigma_{\ddot{z}})_{\text{A}}$ and transmits it to aircraft B, and the latter wanted to convert that into EDR or $(\sigma_{\ddot{z}})_B$, it would also require knowledge of the aircraft response function of aircraft A. This means that both $(\sigma_{\ddot{z}})_A$ and information about aircraft A need to be transmitted. So while both atmospheric and aircraft response metrics are useful, the wider and simpler application of the former means that it is the preferred information to be calculated and transmitted from aircraft.

Atmospheric turbulence can manifest in myriad forms, but to simplify things, it is typically characterized as being either continuous or discrete. Metrics such as EDR or $\sigma_{\ddot{z}}$ are usually related to the former and $|\ddot{z}_{peak}|$ or U_{de} to the latter. More precisely, the term "continuous" refers to homogeneous turbulence, i.e., the condition such that the statistical properties of the field are independent of translations of the coordinate system and the associated basis vectors. (We are considering vector velocity fields, as opposed to scalar fields such as temperature.) Isotropy is another simplification in modeling turbulence and is a subset of homogenous fields. An isotropic field is one that is homogeneous and, further, whose statistical properties are invariant under rotation and/or reflection of the components of the field and the associated basis vectors. By "discrete" turbulence, we are referring to inhomogeneous fields in the classical turbulence sense, as opposed to coherent structures, such as vortices or small-scale shears. That is, we are considering locally homogeneous fields, whose intensity varies smoothly—though possibly rapidly—as a function of position. In this limited view of inhomogeneous fields, we can consider so-called uniformly modulated homogeneous fields (Mark and Fischer 1976) as a useful model. Note that we are using spatial characteristics of the velocity field as opposed to temporal ones, even though aircraft measurements are made as a function of time. In the turbulence literature, the term stationarity typically refers to invariance of the statistical properties at a point, as a function of time. However, since the aircraft usually transits the spatial turbulent field much more rapidly than the field is evolving or advecting, we can invoke Taylor's hypothesis (Hinze 1975), essentially $x = Vt$ (where V is the aircraft airspeed), to study the spatial properties of the velocity field from the time series of aircraft measurements.

In the following sections, we will focus on measurements of EDR from homogenous and isotropic velocity fields. Using a modulated homogeneous turbulence model, we also consider an empirical approach to EDR measurements from inhomogeneous fields. This is followed by a discussion of the vertical acceleration response of aircraft to turbulence. We conclude the chapter with a simulation analysis comparing EDR measurements from vertical velocity time series, from scaling the root mean square of the velocities and accelerations, as well as from scaling peak accelerations.

5.2 Energy Dissipation Rate for Homogeneous Wind Fields

In notational form, the equation describing the temporal evolution of (mechanical) turbulent kinetic energy (TKE) per unit mass, E, is given by

$$\frac{dE}{dt} = P + A + D - \varepsilon \tag{5.1}$$

where P is the TKE production due to gradients in the mean flow, A is the TKE change due to advection in the mean flow (can be positive or negative), D is the diffusion of TKE due to small-scale motions in the flow, and ε, the energy dissipation rate, is the TKE lost to heating due to viscous forces. Equation (5.1) is a consequence of the Navier–Stokes equation. For an incompressible fluid, ε at a given measurement point is defined by Hinze (1975)

$$\varepsilon = \upsilon \sum_{i,j} \left\langle \frac{\partial u_i}{\partial x_j} \left(\frac{\partial u_i}{\partial x_j} + \frac{\partial u_j}{\partial x_i} \right) \right\rangle \tag{5.2}$$

where u_α is the α-component of the velocity field, υ is the kinematic viscosity, and the angled brackets refer to an ensemble average. If the turbulent velocity field is homogeneous and isotropic, Eq. (5.2) can be written as

$$\varepsilon_u = c \left\langle \frac{\partial u_i}{\partial x_j} \right\rangle \tag{5.3}$$

where $c = 15\upsilon$ if $i = j$ (i.e., for a longitudinal component), and $c = (15/2)\upsilon$ if $i \neq j$ (i.e., for a transverse component). As the calculation of the energy dissipation rate in Eqs. (5.2) or (5.3) is given by derivatives of the velocity field, these are not practical definitions unless the measurement scales are such that finite-difference approximations to the derivatives are accurate. For most airborne applications, the measurement scales are on the order of meters, which does not meet this requirement; hence, we look to empirical approaches to calculate EDR. For a homogeneous and isotropic field, the energy dissipation rate is given by

$$\varepsilon_u = \upsilon \int_0^\infty k^2 E(k) dk \tag{5.4}$$

where k (rad m^{-1}) is the magnitude of the wave number vector, and $E(k)$ is the three-dimensional energy spectrum for isotropic turbulence (Hinze 1975). For a von Kármán energy spectrum, the asymptotic behavior for large wave numbers matches that for the Kolmogorov energy spectrum, i.e., it has a $k^{-5/3}$ form. It can be seen that using this functional form in Eq. (5.4) gives an infinite result, i.e., the integral of $k^{2-5/3} = k^{1/3}$ does not converge as $k \to \infty$. This is because the Kolmogorov spectrum is valid only in a specific range of wave numbers, the so-called inertial subrange, not out to arbitrarily large ones. In order to resolve this problem, an energy spectrum that accounts for the largest wave numbers, i.e., a so-called dissipation range spectrum, must be used. Without resorting to that approach, the energy dissipation rate is calculated from the von Kármán spectrum by requiring that the large wave number regime of the von Kármán spectrum matches the Kolmogorov spectrum. Specifically, we have for the von Kármán energy spectrum, E_{vK} (Hinze 1975)

$$E_{vK}(k) = \frac{55}{9\pi}\sigma_u^2 L \frac{(k/k_0)^4}{\left(1 + (k/k_0)^2\right)^{17/6}} \tag{5.5}$$

where $k_0 = \sqrt{\pi}\Gamma(5/6)/(\Gamma(1/3)L)$, and Γ is the gamma function. The velocity variance for a single velocity component and the integral length scale, L, are given by

$$\sigma_u^2 = \frac{2}{3}\int_0^\infty E(k)dk, \quad L = \frac{\pi}{2\sigma_u^2}\int_0^\infty k^{-1}E(k)dk \tag{5.6}$$

Expanding Eq. (5.5) in a power series for large k gives

$$E_{vK}(k) \rightarrow \frac{55\,\sigma_u^2}{9\pi^{1/6}L^{2/3}}\left[\frac{\Gamma(5/6)}{\Gamma(1/3)}\right]^{5/3} k^{-5/3} \quad \text{large}\, k \tag{5.7}$$

The Kolmogorov energy spectrum is

$$E_K(k) = A\,\varepsilon^{2/3}k^{-5/3} \tag{5.8}$$

where A is a constant approximately equal to 1.5. This formula says that for fully developed turbulence in the inertial subrange, i.e., for spatial scales well separated from those of the large-scale forcing mechanisms, and those at the smallest, viscous dissipation scales, the turbulent kinetic energy is solely dependent on a single parameter, ε. The typical spatial scales for which the Kolmogorov form is valid are commensurate with scales that affect aircraft (Cornman et al. 1995). These two facts are the rationale for using EDR as an atmospheric turbulence measurement parameter for aviation. Equating Eqs. (5.7) and (5.8) gives

$$\varepsilon^{1/3} = \left\{\frac{55}{9A\,\pi^{1/6}}\left[\frac{\Gamma(5/6)}{\Gamma(1/3)}\right]^{5/3}\right\}^{1/2} \frac{\sigma_u}{L^{1/3}} \tag{5.9}$$

Therefore, the energy dissipation rate used with the von Kármán spectrum is not a consequence of the Navier–Stokes equation, but rather it is given by matching the asymptotic forms of the von Kármán and Kolmogorov energy spectra. We shall use a similar approach in defining the energy dissipation rate for modulated homogeneous turbulence in the next section. Note that for a given length scale, Eq. (5.9) shows that $\varepsilon^{1/3}$, i.e., EDR, is directly proportional to the root mean square of the turbulence velocities. Furthermore, as we shall show later, for linear aircraft response to turbulence, the root mean square of the turbulence velocities is directly proportional to the root mean square of aircraft vertical accelerations. These are further reasons that EDR is used as a turbulence metric for airborne measurements.

5.3 Calculation of EDR for Modulated Homogeneous Turbulence

This section describes a rationale and methodology for calculating EDR from a specific class of inhomogeneous turbulence: modulated homogeneous turbulence. By "modulated," we mean that the inhomogeneous data is the product of a deterministic function and the homogeneous field. In the first part, we will concentrate on continuous spatial sampling—but over a finite window—and in the second part, we will extend the results to accommodate discrete sampling.

Consider the one-dimensional velocity power spectrum along a specified direction, e.g., from the transverse velocity component along the x-direction. For example, from the vertical velocity component measured along the flight path, the one-dimensional spectrum is given by the Fourier transform of the associated velocity autocorrelation function. This latter function is given by the ensemble average,

$$R_u(\rho) = \langle u(x)u(x+\rho)\rangle \tag{5.10}$$

where, by definition of a homogeneous and isotropic field, this autocorrelation is solely a function of the length of the displacement vector, here, $\rho = \boldsymbol{\rho} \cdot \mathbf{e}_x$, where $\mathbf{e}_x$ is the unit vector in the x-direction. The spectrum of u is given by the one-dimensional Fourier transform of the autocorrelation function,

$$\Phi_u(k_x) = \frac{1}{2\pi} \int_{-\infty}^{\infty} R_u(\rho) \mathrm{e}^{ik_x\rho} d\rho \tag{5.11}$$

where $k_x = \mathbf{k} \cdot \mathbf{e}_x$ is the component of the wave number vector in the x-direction. Note that this is a two-sided spectrum, such that

$$\sigma_u^2 = \int_{-\infty}^{\infty} \Phi_u(k_x) dk_x \tag{5.12}$$

For large wave numbers (but still in the inertial subrange), the one-dimensional von Kármán spectrum of the homogeneous and isotropic wind field approaches a Kolmogorov form,

$$\Phi_u(k_x) \to \frac{12}{55} A \varepsilon_u^{2/3} k_x^{-5/3} \quad \text{large } k_x \tag{5.13}$$

It will be shown that for large k_x, the one-dimensional spectrum for modulated homogeneous turbulence will also follow a $k_x^{-5/3}$ power law, though shifted in magnitude. In the following, we will motivate this "spectral scaling" for modulated homogeneous turbulence and determine the specific form of the scaling function.

5.3.1 Continuous Spatial Domain Theory

In practical applications, one deals with a finite sequence of data, and to reduce the induced bias and variance in the spectral estimates, a window function is often employed. Papoulis (1973) and Mark and Fischer (1976) are good references for truncated-windowed data and modulated turbulence data, respectively. In the following, we combine the two approaches; in fact, we shall see that much of the concepts inherent in the spectral analysis of truncated-windowed data carry over to modulated data. That is, mathematically we can consider the modulation to be an extended form of windowing. We shall defer a discussion of discrete sampling until later. The truncated, modulated, windowed data is defined by

$$v(x) = \begin{cases} w(x)\, m(x)\, u(x) & |x| \le X \\ 0 & |x| > X \end{cases} \tag{5.14}$$

where $w(x)$ is the window function, $m(x)$ is the modulation function, $u(x)$ is the homogeneous turbulence data, and X is the length of the truncation window. In general, the window function does not necessarily have to go to zero at the boundaries of the truncation window, but we will assume this in the subsequent analysis. Hence, when we refer to "windowed data," we mean "windowed, truncated data." By default, we are assuming that both the modulation and window functions are centered in the window. (The methods presented below can be generalized for modulation functions not centered in the window, but that is beyond the scope of this chapter.) The autocorrelation function for the modulated, windowed data is defined by

$$R_v(\rho,X)=\left\langle \frac{1}{X}\int_{x=-\left(X-\frac{|\rho|}{2}\right)}^{X-\frac{|\rho|}{2}} w\left(x-\frac{\rho}{2}\right)w\left(x+\frac{\rho}{2}\right)m\left(x-\frac{\rho}{2}\right)m\left(x+\frac{\rho}{2}\right)u\left(x-\frac{\rho}{2}\right)u\left(x+\frac{\rho}{2}\right)dx\right\rangle$$

$$=R_u(\rho)\frac{1}{X}\int_{x=-\left(X-\frac{|\rho|}{2}\right)}^{X-\frac{|\rho|}{2}} w\left(x-\frac{\rho}{2}\right)w\left(x+\frac{\rho}{2}\right)m\left(x-\frac{\rho}{2}\right)m\left(x+\frac{\rho}{2}\right)dx$$

$$=R_u(\rho)R_{w,m}(\rho,X) \tag{5.15}$$

where $R_{w,m}(\rho,X)$ is the correlation of the combined modulation and window functions. It is assumed that the window and modulation functions are deterministic. From the convolution theorem, the Fourier transform of a product is equal to the convolution of their Fourier transforms. Hence, the Fourier transform of Eq. (5.15) gives the one-dimensional power spectrum of the windowed, modulated data

$$\Phi_v(k_x,X)=\frac{1}{2\pi}\int_{-\infty}^{\infty} R_v(\rho,X)e^{ik_x\rho}d\rho=\int_{-\infty}^{\infty}\Phi_{w,m}(\alpha,X)\Phi_u(k_x-\alpha)d\alpha \tag{5.16}$$

$\Phi_{w,m}(\alpha,X)$ will typically go to zero rapidly away from $\alpha=0$, and this will occur more quickly for larger X. Since we are interested in the behavior of $\Phi_v(k_x,X)$ for large k_x, we write

$$\Phi_u(k_x-\alpha)\approx\Phi_u(k_x)\left\{1+O\left[\left(\frac{\alpha}{k_x}\right)^n\right]\right\}\quad \text{large } k_x \tag{5.17}$$

where $n\geq 1$ is an integer, and $O[x]$ refers to "order of x." If, over the domain where $\Phi_{w,m}(\alpha,X)$ has significant values, the correction term in Eq. (5.17) is small, i.e., $\Phi_u(k_x-\alpha)$ is approximately independent of α over that domain, then we can write Eq. (5.16) as

$$\Phi_v(k_x,X)\approx\Phi_u(k_x)\int_{-\infty}^{\infty}\Phi_{w,m}(\alpha,X)d\alpha=g_{w,m}(X)\Phi_u(k_x)\quad \text{large } k_x \tag{5.18}$$

Note that there are two intertwined criteria that must be met for Eq. (5.18) to be an accurate approximation: large enough X and large enough k_x. Reducing X will mean typically that k_x must be increased to maintain the same level of approximation. Since $\Phi_{w,m}(\alpha,X)$ is the Fourier transform of $R_{w,m}(\rho,X)$, its integral over α shows that $g_{w,m}(X)=R_{w,m}(0,X)$, and from Eq. (5.15)

$$R_{w,m}(0,X) = \frac{1}{X} \int_{x=-X}^{X} m^2(x)w^2(x)dx \tag{5.19}$$

Using Eqs. (5.13) and (5.18) gives

$$\Phi_v(k_x, X) \approx R_{w,m}(0,X)\frac{12}{55}A\,\varepsilon_u^{2/3}k_x^{-5/3} \quad \text{large}\, k_x \tag{5.20}$$

and we see that the spectrum of the modulated and windowed homogeneous field is a scaled version of the homogeneous spectrum, with the scale factor being $R_{w,m}(0, X)$. We can then define EDR for modulated turbulence via

$$\varepsilon_v^{1/3}(X) = \sqrt{R_{w,m}(0,X)}\,\varepsilon_u^{1/3} \quad \text{large } k_x \tag{5.21}$$

so that the asymptotic forms of the one-dimensional von Kármán homogeneous spectrum and the modulated turbulence spectrum have the same functional form. In essence, the spectral scaling approach assumes that for large wave numbers, the spectrum of the modulated homogeneous turbulence looks identical to the spectrum from a larger-intensity homogenous velocity field. And since we are assuming that for large wave numbers the homogenous and isotropic velocity spectrum is directly proportional to $\varepsilon_u^{2/3}$, Eq. (5.21) provides a rational definition for EDR from modulated homogeneous and isotropic turbulence. It is important to realize that since this EDR calculation for inhomogeneous turbulence occurs over a finite window, the resultant value will be a function of the window length, i.e., via $R_{w,m}(0,X)$. Nevertheless, it is a deterministic relationship. That is, every homogenous velocity field whose expected EDR value is $\varepsilon_u^{1/3}$, and is acted upon with a window and modulation function whose correlation function is given by $R_{w,m}(0,X)$, will have an expected EDR value determined by Eq. (5.21). In order to calculate an EDR which is window length independent, one would have to resort to the Navier–Stokes definitions, e.g., Eqs. (5.2) or (5.3), which, as we have discussed, are not suited for practical applications.

5.3.2 Discrete Time Domain Calculations

In distinction with the continuous spatial domain analysis presented above, in an airborne application, we deal with discrete sampling and finite window lengths in the time domain. As mentioned above, a typical aircraft transverses the spatial field more rapidly than the turbulence evolves temporally, so we can invoke Taylor's hypothesis, which for our application has the simple effect of setting $\Delta x = V\Delta T$, where V is the average true airspeed over the sampling window and ΔT is the discrete sampling interval. An example modulation function in the time domain is

$$m(q) = 1 + A_1 \mathrm{e}^{-\pi\left(\frac{q\Delta T - t_0}{T_\sigma}\right)^2} \tag{5.22}$$

where t_0 is the center of the window, A_1 is a dimensionless amplitude parameter, $T_\sigma = L_\sigma/V$, and L_σ is a spatial width parameter for the modulation. Mark and Fischer (1976) use a similar, though spatial domain, model but without the additive "1." However, the additive "1" is required to have the correct limit to homogeneous turbulence when $A_1 \to 0$ and/or $L_\sigma \to \infty$, i.e., we require that $m(q) \to 1$ in the limit of vanishing modulation. Since we are dealing with discrete and finite sampling, the power spectrum of the modulated and windowed homogeneous velocities will be given by the expected periodogram:

$$S_v(p) = 2\Delta T N_w \sum_{q=-(M-1)}^{M-1} R_{w,m}(q) R_u(q) \mathrm{e}^{2\pi i q p/M} \tag{5.23}$$

where $p = 0, \ldots, M-1$, and N_w is a normalization factor for the window function. Note that this is the discrete-time and finite window length analogue of Eq. (5.16), excepting that this is a one-sided spectrum (hence, the factor of 2 in front). That is,

$$\sigma_v^2 = \Delta f \sum_{p=0}^{M-1} S_v(p) \tag{5.24}$$

where the linear frequency is given by $f = p\Delta f$, with $\Delta f = 1/(M\Delta T)$. The normalization factor for the window function is given by

$$N_\mathrm{w} = \left[\frac{1}{M}\sum_{q=0}^{M-1} w^2(q)\right]^{-1} \tag{5.25}$$

Figure 5.1 illustrates the spectral scaling concept. The solid gray curve is the power spectrum of the windowed homogeneous data, the solid black curve is that of the windowed and modulated homogeneous data, and the dashed gray curve is that of the scaled, windowed homogeneous data. The parameters used are $\varepsilon_u^{1/3} = 0.05\ \mathrm{m}^{2/3}\ \mathrm{s}^{-1}$, $L = 500$ m, $A_1 = 20$, $L_\sigma = 500$ m, $V = 200\ \mathrm{m\ s}^{-1}$, $\Delta T = 0.125$ s, and $M\Delta T = 60$ s. A Hann window function (Harris 1978) was used in each case. Clearly, for large frequencies the scaled spectrum aligns with that of the modulated data. Due to the interaction between the modulation and window functions, things are more complicated for small window lengths, but that is beyond the scope of this chapter. Note that the modulation introduces a "hump" from around 0.1 to 0.5 Hz but then dips below the scaled homogenous spectrum at smaller frequencies. Integrating Eq. (5.16) over all wave numbers, and using Eq. (5.19), it can be shown that the total power for the scaled homogeneous and modulated homogeneous data is the same, i.e., $\sigma_v^2(X) = R_{\mathrm{w,m}}(0, X)\sigma_u^2$. Hence, the modulation function is redistributing the power in the signal into different frequency bands.

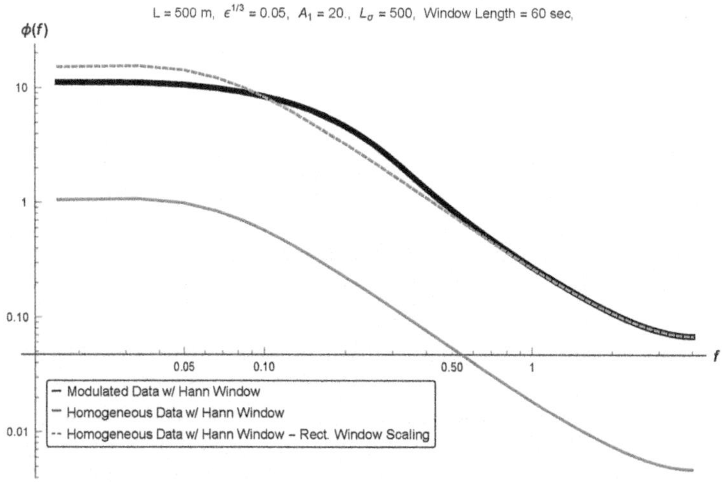

Fig. 5.1 *Curves* illustrating the spectral scaling method. The *solid gray curve* is the windowed homogeneous spectrum, the *solid black curve* is the windowed and modulated spectrum, and the *dashed gray curve* is the scaled spectrum

A standard approach for calculating EDR from measured power spectra is the maximum likelihood (ML) method (Sharman et al. 2014),

$$\varepsilon^{1/3} = \left[\frac{1}{p_2 - p_1 + 1}\sum_{p=p_1}^{p_2}\frac{\hat{S}_v(p)}{S_v(p)}\right]^{1/2} \tag{5.26}$$

where $\hat{S}_v(p)$ is the power spectral estimate, $S_v(p)$ is the expected periodogram with $\varepsilon = 1$, and p_2 and p_1 are upper and lower cutoff frequency indices, respectively. It should be pointed out that Eq. (5.26) is an approximation, valid for large sample sizes or uncorrelated data. (Specifically, when the spectral bins are uncorrelated.) Nevertheless, it provides reasonable results for window lengths that are larger than the correlation scale of the process. Note that if the modulation function was known and the expected periodogram included it, then Eq. (5.26) would provide an estimate of $\varepsilon_u^{1/3}$. However, since we are interested in estimating $\varepsilon_v^{1/3}$, the expected periodogram should represent just the windowed homogeneous data—not that with the modulation function. Furthermore, the lower cutoff frequency should be chosen such that the spectrum from the modulated data and the scaled homogeneous data do not differ greatly. Using Fig. 5.1, one might choose the lower cutoff frequency to be 0.5 Hz or slightly larger for this example case. Note that this same rationale applies to homogeneous turbulence, since the form of the von Kármán velocity spectrum at low frequencies should not be considered universal. That is, the shape of the spectrum at lower frequencies is connected to the specific larger-scale forcing mechanisms. On the other hand, at higher frequencies, the $f^{-5/3}$ form can be considered to be universal. That is, for typical homogeneous atmospheric turbulence, the averaged spectrum at large frequencies usually exhibits the Kolmogorov form.

5.4 Vertical Acceleration Response to Homogeneous Turbulence

For most aircraft operations, the primary acceleration response due to turbulence is in the body-axis vertical direction and due to the vertical component of the wind. In this section, we will elucidate this relationship and describe a scaling method to estimate EDR from the root mean square of the vertical accelerations. We make the fundamental assumption that to a good approximation, the aircraft/autopilot system is a linear one. This allows for the use of all the machinery of linear system theory.

For a one-dimensional linear system, the system output spectrum in the frequency domain, $\Phi_o(f)$, is related to the system input spectrum, $\Phi_i(f)$, via the modulus square of the response function, $H_i^o(f)$ (Bendat and Piersol 2011). This comes from taking the Fourier transform of the correlation function of the accelerations. The accelerations themselves are given by the convolution of the aircraft impulse response function, $h(t)$ (inverse Laplace transform of the response function for zero initial conditions),

$$\ddot{z}(t) = \int_0^\infty h(t-\tau)u(\tau)d\tau \tag{5.27}$$

With the input field being the vertical wind and the output being the vertical acceleration, we have

$$\Phi_{\ddot{z}}(f) = \left|H_w^{\ddot{z}}(f)\right|^2 \Phi_w(f) \tag{5.28}$$

Note that the subscript "*w*" here refers to the vertical component of the wind, not a window function. We assume that the wind spectrum can be expressed as

$$\Phi_w(k) = \varepsilon^{2/3}\Psi_w(k) \tag{5.29}$$

Since we are analyzing things in the time domain, the wind spectrum needs to be converted from wave number to linear frequency, f. This is done using the relationship $k = 2\pi f/V$, where V is the true airspeed, and gives $\Phi_{\ddot{z}}(f) = (2\pi/V)\Phi_{\ddot{z}}(k)$. Using one-sided spectra, the variance of the vertical accelerations is given by

$$\begin{aligned}\hat{\sigma}_{\ddot{z}}^2 &= \int_0^\infty \Phi_{\ddot{z}}(f)df = \int_0^\infty \left|H_w^{\ddot{z}}(f)\right|^2 \Phi_w(f)df = \varepsilon^{2/3}\int_0^\infty \left|H_w^{\ddot{z}}(f)\right|^2 \Psi_w(f)df \\ &= \varepsilon^{2/3} I\end{aligned} \tag{5.30}$$

where the integral factor, I, contains all the information about the aircraft response and the turbulence wind spectrum (besides the intensity level). Taking the square root of both sides of Eq. (5.30) gives the desired relationship between EDR ($\varepsilon^{1/3}$) and the root mean square of the vertical accelerations. It is important to note that the integral factor is a function of the specific aircraft and the operating condition (e.g., altitude, weight, airspeed, autopilot mode, etc.). The factor I can also be used as a scaling factor to estimate EDR from measured root mean squares of vertical acceleration,

$$\hat{\varepsilon}^{1/3} = I^{-1/2}\hat{\sigma}_{\ddot{z}} \tag{5.31}$$

Dividing the left-hand side of Eq. (5.31) by g, the acceleration due to gravity, gives the relationship between the root mean square of the g-values and EDR. With the assumptions made above, we see that there is a simple linear relationship between EDR and $\sigma_{\ddot{z}}$. Note that for two aircraft flying through turbulence with the same EDR level, we can write

$$\left(\frac{\sigma_{\ddot{z}}}{I^{1/2}}\right)_1 = \varepsilon^{1/3} = \left(\frac{\sigma_{\ddot{z}}}{I^{1/2}}\right)_2 \tag{5.32}$$

or

$$\frac{(\sigma_{\ddot{z}})_1}{(\sigma_{\ddot{z}})_2} = \frac{\left(I^{1/2}\right)_1}{\left(I^{1/2}\right)_2} \tag{5.33}$$

Therefore, we see that the relative aircraft response is related directly to the ratio of the integral factors. Note that via the factors I, this relationship is dependent on the (unit-EDR) spectrum of the vertical velocities, and so it is not simply a relationship between aircraft characteristics.

5.5 2-DOF Aircraft Vertical Acceleration Frequency Response Function

This section describes a stability derivative version of the modulus square of the rigid-body aircraft frequency response function for input vertical wind and output center of gravity body-axis vertical acceleration, i.e., $\left|H_{\mathrm{w}}^{\ddot{z}}(f)\right|^2$ in Eq. (5.28). Contributions due to unsteady aerodynamics, i.e., the lag in lift due to step changes in aircraft motion and/or vertical gust velocity, can be accommodated in the analysis but are beyond the scope of this first-order exposition. Hoblit (1988) is a good introductory reference for unsteady aerodynamics and aircraft response. In matrix form, the Laplace-transformed equations of motion for aircraft vertical velocity, w, and pitch angle, θ, in the stability axes are given by

$$\mathbf{A}\begin{pmatrix} w \\ \theta \end{pmatrix} = \mathbf{b}w_{\mathrm{g}} + \mathbf{c}\delta_{\mathrm{e}} \tag{5.34}$$

where w_{g} is the gust velocity, and δ_{e} is the elevator deflection angle. For the short-period motion of interest (i.e., pitching and vertical displacements, without airspeed variations), the matrix $\mathbf{A}$ is given by [see Eq. (5.14) in McRuer et al. (1973)]

$$\mathbf{A} = \begin{pmatrix} s - Z_{\mathrm{w}} & -Vs \\ -(sM_{\dot{\mathrm{w}}} + M_{\mathrm{w}}) & s(s - M_{\mathrm{q}}) \end{pmatrix} \tag{5.35}$$

where $s = 2\pi i f$ is the Laplace transform variable. The other quantities are dimensional stability derivatives. The vectors $\mathbf{b}$ and $\mathbf{c}$ are given by

$$\mathbf{b} = -\begin{pmatrix} Z_{\mathrm{w}} \\ \left(M_{\dot{\mathrm{w}}} - \frac{M_{\mathrm{q}}}{V}\right)s + M_{\mathrm{w}} \end{pmatrix}, \quad \mathbf{c} = \begin{pmatrix} Z_{\delta} \\ M_{\delta} \end{pmatrix} \tag{5.36}$$

From Eq. (5.34) we can write

$$w = \left(\mathbf{A}^{-1}\mathbf{b}\right)_1 w_{\mathrm{g}} + \left(\mathbf{A}^{-1}\mathbf{c}\right)_1 G_{\theta}\theta \tag{5.37}$$

where the subscript refers to the "1" component of the associated vector. Letting $\delta_{\mathrm{e}} = G_{\theta}\theta$, we take the pitch feedback to have a simple PID (proportional, integral, derivative) form,

$$G_{\theta} = K_{\theta} + \frac{K_{\bar{\theta}}}{s} + K_{\dot{\theta}}s \tag{5.38}$$

Since the stability axis is rotating relative to the body axis, the vertical acceleration in the latter is given approximately by

$$\ddot{z} \approx s(w - V\theta) \tag{5.39}$$

The desired frequency response function is then given by

$$\begin{aligned} H^{\ddot{z}}_{w_{\mathrm{g}}}(s) &= \frac{\ddot{z}(s)}{w_{\mathrm{g}}(s)} \\ &= \frac{Z_{\mathrm{w}}s^3 - 2M_{\mathrm{q}}Z_{\mathrm{w}}s^2 + G_{\theta}s\left(-Z_{\mathrm{w}}M_{\delta} + \left(-\frac{M_{\mathrm{q}}}{V}s + (M_{\mathrm{w}} + M_{\dot{\mathrm{w}}}\,s)\right)Z_{\delta}\right)}{-s^3 + s^2\left(M_{\mathrm{q}} + M_{\dot{\mathrm{w}}}V + Z_{\mathrm{w}}\right) + s\left(M_{\mathrm{w}}V - M_{\mathrm{q}}Z_{\mathrm{w}}\right) + G_{\theta}(M_{\delta}(s - Z_{\mathrm{w}}) + (M_{\mathrm{w}} + M_{\dot{\mathrm{w}}}\,s)Z_{\delta})} \end{aligned} \tag{5.40}$$

The response function without autopilot can be written in the more standard form,

$$\left[H_{w_g}^{\ddot{z}}(s)\right]_{noAP} = -\frac{2\xi\omega_0}{\gamma}\left[\frac{2\xi\omega_0\left(1-\frac{1}{\gamma}\right)s+s^2}{\omega_0^2+2\xi\omega_0 s+s^2}\right] \tag{5.41}$$

where

$$\omega_0^2 = -\left(M_{\rm w}V - M_{\rm q}Z_{\rm w}\right), \quad 2\xi\omega_0 = -\left(M_{\rm q} + M_{\dot{\rm w}}V + Z_{\rm w}\right) \tag{5.42}$$

$$2\xi\omega_0/\gamma = \frac{V}{\mu c}, \quad \mu = \frac{2m}{\rho ScC_{L_\alpha}} \tag{5.43}$$

and m is the aircraft mass, c is the mean aerodynamic chord, S is the reference area, ρ is the air density, C_{L_α} is the lift curve slope, and μ is the so-called mass parameter.

Figure 5.2 shows a measured and simulated vertical acceleration time series from the NASA B757 aircraft flying through severe turbulence. The simulation used a simple autopilot model, but one can see that the main aspects of the measured data are captured in the simulation [see Buck and Newman (2006), for more details on the NASA B757 aircraft and the turbulence flight program].

Equation (5.31) shows that the scale factor between the root mean square of vertical acceleration and EDR is $I^{-1/2}$. As a function of altitude, this factor has a complicated shape, and so we look for a scaling approach that simultaneously simplifies the functional form and collapses different aircraft types onto a single curve. The approach taken here is to multiply the scale factor $I^{-1/2}$ by $V/c^{2/3}$ and then parameterize this scaled function by the mass parameter, μ [see Eq. (5.43)], which in turn is a function of altitude. Figure 5.3 shows the result of this approach for four different aircraft types: B737, B747, B757, and a small business jet (SBJ, Hull 2007). The aircraft response functions were calculated for nominal flight

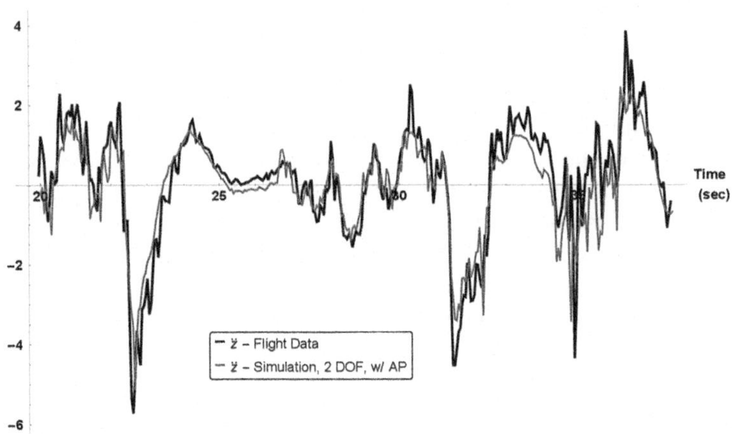

Fig. 5.2 Measured (*black*) and simulated (*gray*) vertical accelerations for NASA 757 aircraft flight through severe turbulence

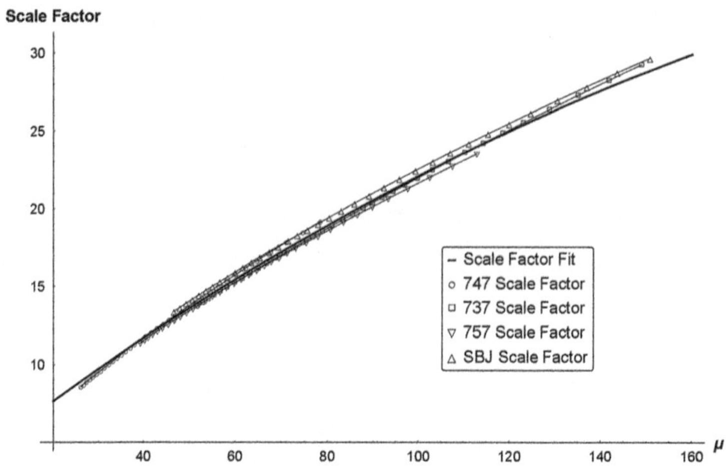

Fig. 5.3 RMS-$\ddot{z}$ to EDR scale factor for four aircraft types, along with functional fit (*solid black curve*)

conditions over the altitude range from 5000 to 40,000 feet. The solid black curve in the figure is a functional approximation, $\eta(\mu)$,

$$\eta(\mu) = \frac{V}{c^{2/3} I^{1/2}}(\mu) = a_0 + a_1(\mu - \mu_0) + a_2(\mu - \mu_0)^2 \tag{5.44}$$

where μ_0 is a reference value of the mass parameter (taken to be 40 here), and the polynomial coefficients are calculated by curve fitting. It can be seen that the scaling and parameterization by the mass parameter has had the desired effect of simplifying and mapping the different aircraft types to a simple functional form. A similar approach can be found in Coupry (1991). Over the entire range of flight conditions, and for all four aircraft, the maximum error between the actual curves and the approximate model is approximately 3.5 %.

Next, we consider the acceleration spectrum for homogeneous and modulated homogeneous turbulence. This spectrum is the product of the von Kármán lateral velocity spectrum (homogeneous or modulated) and the modulus square of the aircraft vertical acceleration frequency response function [see Eq. (5.28)]. Figure 5.4 shows the acceleration spectrum for the B757 at the flight condition for the data shown in Fig. 5.2. For this figure, the unsteady aerodynamic function, "f_1," described in Coupry (1991) and used in Cornman et al. (1995), has been applied. The wind field and modulation parameters are the same as used for the data in Fig. 5.1, i.e., $\varepsilon^{1/3} = 0.05\ \mathrm{m}^{2/3}\ \mathrm{s}^{-1}, L = 500\,\mathrm{m}, A_1 = 20$, and $L_\sigma = 500$ m. The plot uses log–linear scaling to better illustrate how the acceleration power is proportioned over frequency. This scaling comes from the relationship $f\phi(f)d[\log(f)] = \phi(f)df$; hence, if the spectrum is multiplied by f and plotted versus log (f), then the area under the curve for any infinitesimal log-frequency band is the

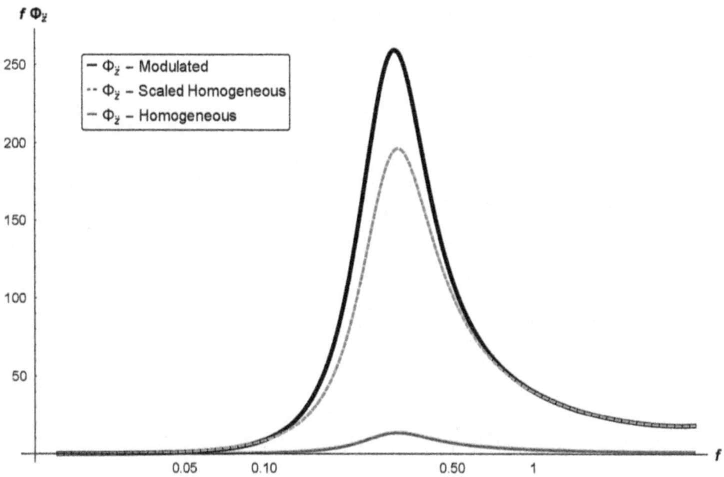

Fig. 5.4 Log–linear plot of the vertical acceleration spectra for the B757, for homogenous von Kármán turbulence (*solid gray*), modulated homogeneous (*solid black*), and scaled homogeneous (*dashed gray*) von Kármán turbulence

same as that for the spectrum itself in linear–linear scale. The solid gray curve in the figure is the acceleration spectrum from the homogeneous input wind field. The solid black curve is the acceleration spectrum from the modulated homogeneous field, and the dashed gray curve is for the scaled homogeneous field. Note that the spectral scaling method described above applies equally well to the acceleration spectrum as it did for the wind spectrum. That is, the scaled homogeneous acceleration spectrum aligns with the modulated homogeneous acceleration spectrum at large frequencies. The spectral scaling method is more complicated with the acceleration time series than with just the wind. The spectral scaling method is based on the application of Eq. (5.15), and to accommodate the acceleration data, the modulated homogenous wind field (transformed to the time domain), $m(t)u(t)$, is replaced by the accelerations, $\ddot{z}(t)$, due to the modulated homogenous field. The window function (and truncation) is then applied to the accelerations. The accelerations are given by the convolution of the modulated wind and the acceleration impulse response function [see Eq. (5.27) with $u(t) \rightarrow m(t)u(t)$], and their autocorrelation function will be given by a double convolution (Bendat and Piersol 2011). Nevertheless, it can be shown that the spectral scaling method is applicable if a condition similar to, but much more complicated than that given in Eq. (5.17) is postulated.

From Fig. 5.4, it can be seen that the acceleration power for both the homogeneous and modulated field is concentrated around the short-period undamped natural frequency for the aircraft at these flight conditions (approximately 0.3 Hz). The "inverted V" shape is due to two counteracting factors. From Fig. 5.1, we see that the von Kármán velocity spectrum is essentially flat at low frequencies and decays as $f^{-5/3}$ for high frequencies (excepting for aliasing and spectral leakage effects). On the other hand, the modulus square of the acceleration response function is almost the opposite, essentially flat at high frequencies (or decaying proportional to something like f^{-2} with the inclusion of unsteady aerodynamic effects) and proportional to f^2 at low frequencies. Therefore, multiplying these two functions together results in the inverted V shape. Note that at very low frequencies, the two-degree-of-freedom, short-period model used here breaks down and must be replaced by a model that takes into account slow changes in the airspeed, i.e., the so-called phugoid mode. However, for the analysis of the vertical acceleration response to turbulence, the short-period model accounts for most of the important effects.

5.6 Simulation Results for Homogenous Turbulence

In this section, we consider EDR estimation from simulated homogeneous, isotropic wind fields. A three-dimensional, three-component von Kármán wind field simulation method is described in Frehlich et al. (2001). For a single dimension and single wind field component, i.e., the vertical component along the x-axis, $w(n\Delta x)$, the method is straightforward. Since the simulated wind field is calculated via a DFT, the resulting velocity field will (correctly) exhibit the artifacts of discrete and finite sampling, i.e., aliasing and spectral leakage. Typically, a long sequence of samples is simulated, so that the effect due to leakage will be minimal. The black curve in Fig. 5.5 is the average of over 1300 spectra, taken over 48 km windows, and the expected periodogram [Eq. (5.23), with rectangular window function and

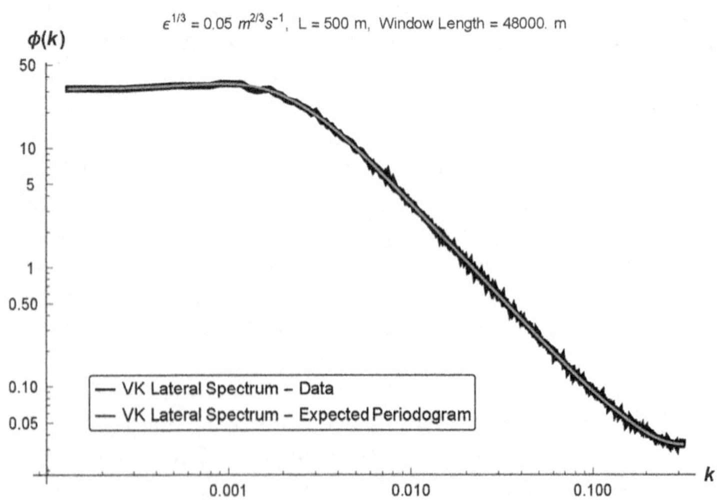

Fig. 5.5 Averaged spectrum from simulated von Kármán lateral velocity component (*black*) and expected periodogram (*gray*)

no modulation] is shown in gray. It can be seen that the simulation does an excellent job of replicating the correct spatial structure of the discretely sampled turbulent field—including aliasing.

For the EDR analysis presented below, the wind field simulation was performed in the spatial domain and then interpolated (via nearest neighbor) for a given airspeed to the time domain. The spatial resolution was ten meters, and 1,310,720 grid points were generated for each EDR level. The aircraft response simulation was performed in the time domain using these interpolated wind field values. A constant airspeed of 200 m s^{-1} was employed, and the sampling rate was $\Delta T = 0.125\ s$ (8 Hz). The EDR levels ranged from 0.05 to 0.5 m$^{2/3}$ s^{-1} ("none" to "severe") in steps of 0.05, and the integral length scale was fixed at $L = 500$ m. EDRs were calculated in four ways: from the vertical wind using the maximum likelihood (ML) method [Eq. (5.26)], scaling the root mean square of the vertical wind (RMS-W) and the vertical acceleration (RMS-$\ddot{Z}$), and scaling the peak absolute value acceleration. The results are presented for ten- and sixty-second sampling windows. The scaling of the root mean squares follows Eq. (5.9) for the wind and Eq. (5.31) for the vertical accelerations, excepting that the scale factors are modified to take into account the discrete and finite sampling. That is, the spectra used in the scaling are expected periodograms, not the theoretical ones. Since the expected periodograms are dependent on the window lengths, the scaling factors will also be dependent on them. The scaling of the peak absolute value acceleration is performed in two steps: a scaling between the peak and root mean square of the accelerations is calculated, and then the scaling between the root mean square of the accelerations and EDR is applied, as just discussed above. The relationship between the peak and root mean square accelerations is dependent on the window lengths and is determined by computing a best-fit line between the two data fields.

Figure 5.6 shows the results of calculating EDR from the root mean square scaling method (vertical axes) versus the vertical wind ML method (horizontal axes). The upper row shows the EDRs from $RMS - W$, and the lower row is the EDRs from $RMS - \ddot{Z}$. The left-hand column is from ten-second windows, and the right-hand column is the result from sixty-second windows. In all cases, the correlation is good and improves with increasing window length. It can also be seen that for larger EDR levels, the values from the scaled root mean squares tend to be larger than those from the ML method. This is because the root mean square is equal to the square root of the area under the spectral curve; and for a power-law spectrum, the area will be dominated by the lower-frequency spectral values. This result is a consequence of the discrete form of Parseval's theorem for a real-time series (Bendat and Piersol 2011),

$$\sum_{n=0}^{M-1} x_{\mathrm{T}}^2(n\Delta t) = \frac{2}{M}\sum_{m=0}^{M/2} |X_{\mathrm{T}}(m\Delta f)|^2 \tag{5.45}$$

where x_{T} is the truncated time series, and X_{T} is the associated DFT. Taking the expected value of each side, dividing by M, and taking the square root give

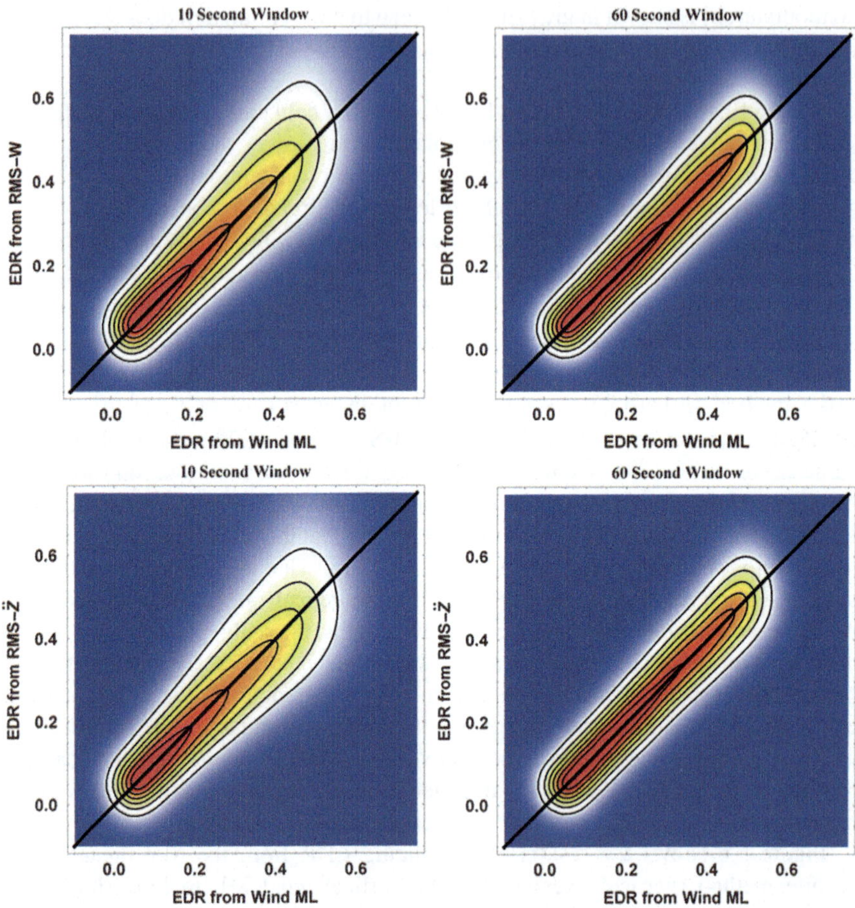

Fig. 5.6 *Upper row*: EDR calculated from the root mean square of the vertical wind (RMS-W) versus EDR from vertical wind maximum likelihood method. *Lower row*: EDR from aircraft accelerations (RMS-$\ddot{Z}$) versus EDR from vertical wind maximum likelihood method

$$\sigma_{x_T} = \sqrt{\Delta f \sum_{m=1}^{M/2-1} S_{x_T}(m\Delta f) + \frac{\Delta f}{2}\left[S_{x_T}(0) + S_{x_T}\left(\frac{M}{2}\Delta f\right)\right]} \tag{5.46}$$

where we have used the appropriate versions of Eqs. (5.23) and (5.24). Equation (5.46) relates the root mean square of the truncated time series to the square root of the power in the data (or equivalently, the square root of the area under the spectral curve in the sense of a Riemann sum). If the underlying data, e.g., the turbulence velocity field, is from a Gaussian probability distribution, then for large M the calculated power spectral values at each spectral bin (excepting for the zero and the Nyquist frequencies), $\hat{S}_{x_T}(m\Delta f)$, will tend to an exponential distribution. The standard deviation of an exponential random variable is equal to its mean value, i.e., the expected value of the spectrum at that frequency, $S_{x_T}(m\Delta f) = \langle \hat{S}_{x_T}(m\Delta f) \rangle$. (Note that this standard deviation is for the distribution of spectral values at each frequency, not that of the time series.) Therefore, for a power-law spectrum, the random error in the root mean square of the time series will be dominated by the lower-frequency spectral values. From Fig. 5.5, we see that the von Kármán lateral velocity spectrum flattens out at lower frequencies, so it is not technically a power-law spectrum over all frequencies. However, the power at lower frequencies is larger than those in the power-law region. On the other hand, the ML EDR method as used herein limits the frequency values away from both the lowest and highest frequencies; and hence, with an appropriate choice of cutoff frequencies, it mitigates much of this problem. A similar issue occurs for the vertical acceleration data. Equation (5.46) is modified to accommodate the vertical acceleration frequency response function,

$$\sigma_{y_T} = \sqrt{\Delta f \sum_{m=1}^{M/2-1} \left|H_{x_T}^{y_T}(m\Delta f)\right|^2 S_{x_T}(m\Delta f) + \frac{\Delta f}{2}\left[\left|H_{x_T}^{y_T}(0)\right|^2 S_{x_T}(0) + \left|H_{x_T}^{y_T}\left(\frac{M}{2}\Delta f\right)\right|^2 S_{x_T}\left(\frac{M}{2}\Delta f\right)\right]} \tag{5.47}$$

where y_T is the response (e.g., vertical acceleration) to the input variable x_T (e.g., vertical wind component), and $\left|H_{x_T}^{y_T}\right|^2$ is the modulus square of the frequency response function. As discussed above, the modulus square of the vertical acceleration frequency response function tends to damp the lower frequencies from the input wind spectrum; and hence, the vertical acceleration spectrum has an "inverted V" shape, with a peak near the short-period undamped natural frequency of the aircraft (see Fig. 5.4). Hence, the area under the spectral curve—and the random error in the time series root mean square—will be dominated by those values near the peak.

Figure 5.7 illustrates the simulation results for the peak acceleration to EDR scaling method. The upper row shows the relationship between peak absolute value acceleration and the root mean square of the accelerations over ten-second (left-hand column) and sixty-second (right-hand column) windows. The solid black line shows the best-fit line, $\left|\ddot{z}_{\text{peak}}\right| = a\sigma_{\ddot{z}}$. Note that the slope of the line is larger for the longer window length. For homogeneous random data, as the window length increases, the root mean square estimates will asymptote to their expected value. On the other hand, the probability of finding a larger peak value will increase with increasing window length. These two factors result in the slope increasing for longer window lengths. As mentioned above, the scaling between the peak acceleration and EDR occurs in two steps. In the first step, the peak absolute value of the accelerations over the window is scaled into root mean square values by using the slope parameter from the best-fit line (i.e., dividing the peak by a); this is followed by using the same scaling method that takes the acceleration root mean square values into EDRs. From the figure, it can be seen that the results for the peak

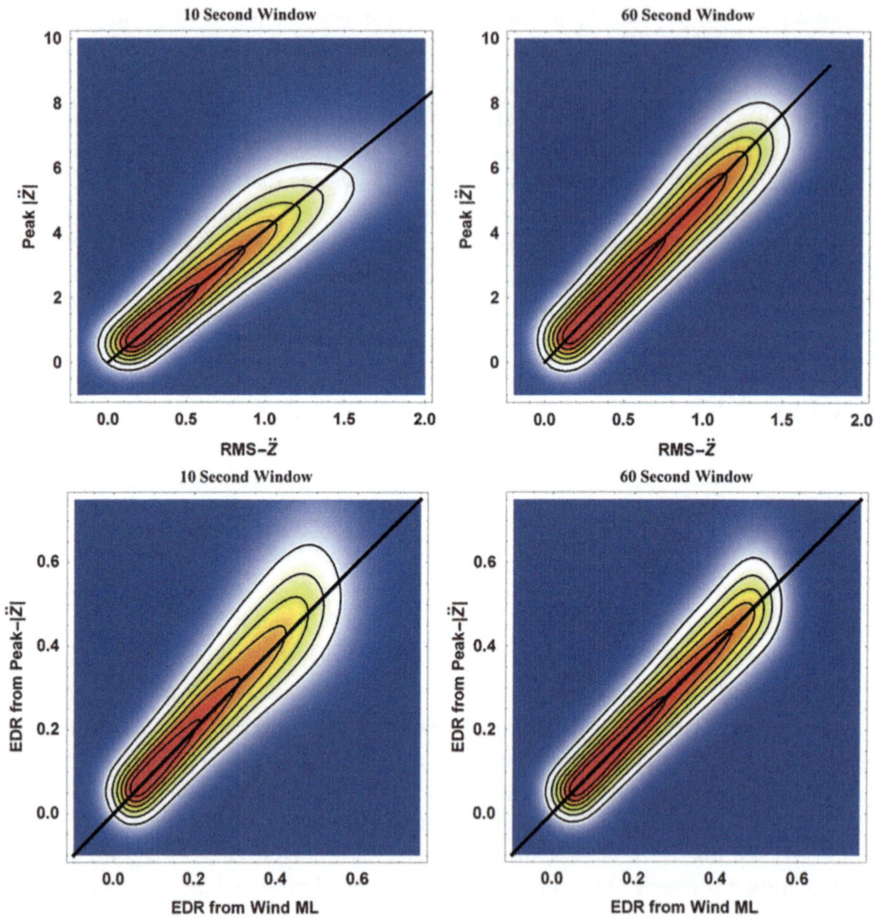

Fig. 5.7 *Upper row*: Peak absolute value versus root mean square of vertical accelerations. *Lower row*: EDR from peak absolute accelerations versus EDR from vertical wind (ML method)

acceleration to EDR scaling look very similar to those for the root mean square to EDR scaling. It is not obvious from the figure, but there are actually more outliers in the peak acceleration scaling method. This is not surprising, as the random errors introduced in the first scaling step propagate into the second one.

Three other EDR estimation methods were analyzed, but the results are not shown, as they were similar to the wind-based ML method, albeit with more outliers. These methods included an ML approach using vertical accelerations [i.e., Eq. (5.26), but the measured and model spectra are for accelerations] and an "area method" for both wind and accelerations, which, instead of using the root mean square values computed over the entire frequency band, just uses those over the range of cutoff frequencies used in the ML approach, that is, using Eq. (5.46), but with a restricted set of frequency indices. It should be noted that the results presented in the two figures above should be considered best-case scenarios. That is, while they elucidate the essential statistical and physical processes described by the theory, they do not take into account certain real-world aspects. These attributes include onboard sensor errors and noise (in computing vertical wind and measuring vertical accelerations) and mismatches between model assumptions and actual conditions. In this latter category, there are two subcategories: the atmosphere and the aircraft response. We have modeled the turbulence via a von Kármán velocity spectrum, and while there is ample real-world verification of the $k^{-5/3}$ large wave number aspect, the lower wave number portion of the spectrum is tied to the specific environmental conditions, i.e., the large-scale turbulence-forcing mechanisms. Hence, the small wave number regime for the von Kármán velocity spectrum cannot be considered universal in form. However, as mentioned above, typical EDR algorithms choose a lower cutoff frequency to mitigate errors due to the precise shape of the velocity spectrum at the smaller wave numbers. In this chapter, we have depicted the aircraft vertical acceleration response function with a two-degree-of-freedom, rigid-body model—augmented with a simple autopilot model. However, in practice, the aircraft has six degrees of freedom (three translational and three rotational velocities), as well as numerous flexible modes, and sophisticated autopilot systems. For large transport aircraft encountering large-amplitude turbulent structures, the vertical acceleration will have contributions, to a greater or lesser extent, from many of these factors. Even for the two-degree-of-freedom, rigid-body model, there are numerous parameters that vary as a function of flight condition, including the stability derivatives, the mass, and the pitching moment of inertia. Nevertheless, we have shown that this method did a reasonably good job of modeling the real-world encounter depicted in Fig. 5.2. Of course, this is just a single case from a single aircraft type, not a statistical analysis, nor does it predict how accurate EDR estimation can be for an onboard implementation. Due to the simplicity and more direct nature of the wind-based EDR method and its ability to minimize the potential sources for error in real-world applications, it is the preferred method.

5.7 Conclusion

In this chapter, a review of airborne measurements of turbulence has been presented, with a focus on EDR measurements from homogenous turbulence. Since it requires accurate small-scale measurements of the velocity field, we discussed how the calculation of EDR from first principles, i.e., from the Navier–Stokes TKE equation, is impractical for most applications. Therefore, we defined EDR in an empirical fashion, by matching the large wave number portions of the von Kármán and Kolmogorov energy spectra for homogeneous and isotropic turbulence. A similar, though more complicated, approach was used to define EDR for modulated homogeneous turbulence, leading to the "spectral scaling" method. To motivate this technique, we first showed how it applied to finite data windows but with continuous spatial (or temporal) sampling. This method was then extended to accommodate the real-world scenario of discrete-time and finite window sampling. The acceleration response of aircraft to turbulence was then described via standard linear input–output relationships, and it was shown how EDR is related to the root mean square of the vertical accelerations. A two-degree-of-freedom stability derivative model, including a simple PID autopilot model, was used to calculate the frequency response function for a rigid-body aircraft, driven by vertical wind turbulence. A brief discussion was presented on how the spectral scaling method can be applied to the accelerations due to aircraft response to modulated homogeneous turbulence. The chapter concluded with a simulation study comparing EDR calculations from homogeneous turbulence, via scaling the root mean square of the velocities and accelerations, as well as by scaling peak accelerations. It was seen that with scaling functions that incorporated the effects of discrete and finite sampling, the correlation between the various methods was very good.

References

Bendat, J.S., Piersol, A.G.: Random Data. Wiley Series in Probability and Statistics, 4th edn. Wiley, Hoboken, NJ (2011)

Buck, B.K., Newman, B.A.: Aircraft acceleration prediction due to atmospheric disturbances with flight data validation. J. Aircraft **43**(1), 72–81 (2006)

Cornman, L.B., Morse, C.S., Cunning, G.: Real-time estimation of atmospheric turbulence severity from in-situ aircraft measurements. J. Aircraft **32**, 171–177 (1995)

Coupry, G.: Improved reduction of gust loads data for gust intensity. In: Houbolt, J. (ed.) Manual on the flight of flexible aircraft in turbulence. AGARD Report: AGARD-AG-317 (1991)

Frehlich, R., Cornman, L., Sharman, R.: Simulation of three-dimensional turbulence velocity fields. J. Appl. Meteorol. **40**, 246–258 (2001)

Harris, F.J.: On the use of windows for harmonic analysis with the discrete Fourier transform. Proc. IEEE **66**(1), 51–83 (1978)

Hinze, J.O.: Turbulence, 2nd edn. McGraw-Hill, New York (1975)

Hoblit, F.M.: Gust Loads on Aircraft: Concepts and Applications. AIAA Education Series. AIAA (1988)

Hull, D.G.: Fundamentals of Airplane Flight Mechanics. Springer, Berlin Heidelberg (2007)

Mark, W.D., Fischer, R.W.: Investigation of the effects of nonhomogenous (or nonstationary) behavior on the spectra of atmospheric turbulence. NASA Contractor Report NASA-CR-2745 (1976)

McRuer, D., Ashkenas, I., Graham, D.: Aircraft Dynamics and Automatic Control. Princeton University Press, Princeton, NJ (1973)

Papoulis, A.: Minimum-bias windows for high-resolution spectral estimates. IEEE Trans. Inform. Theory **IT-19**(1), 9–12 (1973)

Sharman, R.D., Cornman, L.B., Meymaris, G., Pearson, J., Farrar, T.: Description and derived climatologies of automated in situ eddy-dissipation-rate reports of atmospheric turbulence. J. Appl. Meteorol. Climatol. **53**(6), 1416–1432 (2014)

Chapter 6
Doppler Radar Measurements of Turbulence

Larry B. Cornman and Robert K. Goodrich

Abstract A first-principle analysis of Doppler radar measurements of turbulence is presented. A set of limited, though practical, conditions are assumed to make the problem more tractable, with the primary condition being the discrete and finite temporal sampling of the radar signals. A theoretical derivation of the Doppler spectrum under these conditions is performed, and the distinction between what the radar actually measures and what results after a theoretical ensemble averaging is delineated. This is an important consideration, as all of the theoretical development in the literature is based upon ensemble averaging. It is shown that in the limit of an infinite number of samples, and after ensemble averaging, the Doppler spectrum can be represented by a sum of Dirac delta distributions and furthermore that the normalized spectrum will equal the probability distribution of the scatterer velocities. We show that the correlation structure of the velocity field manifests itself primarily in the square of the first moment. That is, a correlated field will have a certain degree of patchiness, which leads to variations in the first moments from realization to realization. These theoretical considerations are then studied via simulation. Parameters for a typical airborne X-band Doppler radar are used, and correlated von Kármán and uncorrelated random fields are employed. Energy dissipation rate estimates are calculated from the simulated Doppler spectra, and the performance based on moment averaging and spectral averaging is presented. Real-world application of the turbulence measurement methods is then shown with airborne X-band detection of convective turbulence.

L.B. Cornman (✉) • R.K. Goodrich
Research Applications Laboratory, National Center for Atmospheric Research, Boulder, CO, USA
e-mail: cornman@ucar.edu; goodrich@ucar.edu

R. Sharman, T. Lane (eds.), *Aviation Turbulence*,
DOI 10.1007/978-3-319-23630-8_6

6.1 Introduction

Encounters with convective turbulence continue to be a key factor in commercial transport accidents. Providing accurate short-term forecasts of these turbulence events is not yet a reality; hence, the ability to detect and warn pilots of an impending encounter is important. One important tool in providing these warnings is ground-based and airborne Doppler radar. There is a long and rich history in the literature regarding the detection of turbulence from Doppler radars. Key references include Rodgers and Tripp (1964), Srivastava and Atlas (1974), Frisch and Clifford (1974), Bohne (1982), and Doviak and Zrnic (1993). Each of these works makes a singular assumption in the analysis that an ensemble average over the random scatterer positions and velocities can be made. Due to practical considerations inherent with actual radar measurements, however, only limited spatial and/or temporal averaging can be accomplished. The purpose of this chapter is to investigate some of the essential considerations in radar measurements, specifically related to averaging and the resultant estimates of turbulence intensity. To accomplish this, we strip the problem down to certain basic elements: the form of the Doppler spectrum for finite and discrete temporal sampling and how this spectrum, its moments, and their averages relate to turbulence measurements. This is done via theoretical analysis and numerical simulation. We conclude the chapter by illustrating some real-world results from airborne radar measurements.

The analysis of radar measurements from random fields is a daunting problem. This is the result of a number of factors, including, but not limited to the following items. Atmospheric turbulence and collisional processes give rise to random positions, velocities, orientations, and shapes (of liquid drops) of the scatterers, which then results in random amplitudes and phases of the scattered electric fields. Due to scatterer inertia, their motion does not necessarily equate identically to the motion of the atmosphere. The radar is typically scanning, and furthermore, with an airborne platform, the location of the radar is changing as a function of time. In fact, if the aircraft is flying through turbulence itself—even with motion compensation—the position and orientation of the radar can be perturbed. Further complications to the analysis include signal processing methods, e.g., pulse compression and Doppler moment estimation algorithms. Furthermore, real-world turbulence never fully satisfies the idealized assumptions of stationarity, homogeneity, and/or isotropy. Finally, larger-scale, deterministic variations in quantities (e.g., advection or gradients in the velocity and reflectivity fields) can also be present, further complicating the analysis.

The purpose of this chapter is to analyze the process of radar measurements of turbulence by considering a limited, though substantive and practical set of conditions. These conditions include discrete and finite temporal sampling, homogeneous and isotropic turbulence, negligible scatterer inertia, uniform spatial and temporal scatterer cross sections, and infinite signal to noise. We note that, besides the first condition, these assumptions are commensurate with those in the analyses referenced above. Furthermore, it will be seen that even with these assumptions, the problem is still formidable.

6.2 The Doppler Velocity Spectrum

Based on first-order scattering theory, the electric field returned from a collection of N scatterers and measured by a monostatic radar at a time mT_s, $(m = 0, \ldots, M - 1)$ is given by

$$E(mT_s) = \sum_{j=1}^{N} A_j(mT_s)\text{Exp}\left[-i\frac{4\pi}{\lambda}r_j(mT_s)\right] \tag{6.1}$$

where T_s is the pulse repetition period, λ is the radar transmitter wavelength, $r_j(mT_s)$ is the distance to the receiver of the jth scatterer at time mT_s, and $A_j(mT_s)$ is the scattering amplitude of the jth scatterer at time mT_s. The connection between the scatterer position and its radial velocity is typically not discussed in the literature, so we will describe this important relationship in some detail. If we assume that the velocities of the scatterers are constant over a short temporal interval $T = (0, (M - 1)T_s)$, i.e., $\mathbf{V}_j\left[\mathbf{r}_j(t)\right] \approx \mathbf{V}_j\left[\mathbf{r}_j(0)\right]$, then the vector position of the jth scatter at time mT_s relative to the radar is given by

$$\mathbf{r}_j(mT_s) = \mathbf{r}_j(0) + \int_0^{mT_s} \mathbf{V}_j\left[\mathbf{r}_j(t)\right] dt \approx \mathbf{r}_j(0) + \mathbf{V}_j\left[\mathbf{r}_j(0)\right] mT_s \tag{6.2}$$

For notational clarity, we just write $\mathbf{V}_j = \mathbf{V}_j\left[\mathbf{r}_j(0)\right]$ in what follows. Furthermore, it is the velocity of the scatterer, not necessarily that of the atmosphere, with which we are dealing. For scatterers with small inertia, such as ice crystals, these quantities will be close, but for large raindrops or hail, the scatterer velocity will not equal that of the atmospheric velocity. However, in the current analysis, we will assume that the scatterer has negligible inertia and, hence, is a perfect tracer of the wind. The scatterer position r_j at time mT_s in Eq. (6.1) is given by the magnitude of the position vector, i.e., the radial distance to the radar, $r_j(mT_s) = \left\|\mathbf{r}_j(mT_s)\right\| \approx \left\|\mathbf{r}_j(0) + \mathbf{V}_j mT_s\right\|$. In order to show the relationship between this expression and the radial velocity, we write

$$\begin{aligned} r_j(mT_s) &= \left\|\mathbf{r}_j(0) + \mathbf{V}_j mT_s\right\| \\ &= \left\|\mathbf{r}_j(0)\right\| \left[1 + \frac{\left\|\mathbf{V}_j\right\|^2 (mT_s)^2}{\left\|\mathbf{r}_j(0)\right\|^2} + \frac{2mT_s}{\left\|\mathbf{r}_j(0)\right\|} \frac{\mathbf{r}_j(0)}{\left\|\mathbf{r}_j(0)\right\|} \cdot \mathbf{V}_j\right]^{1/2} \end{aligned} \tag{6.3}$$

Note that $\mathbf{r}_j(0)/\|\mathbf{r}_j(0)\| = \mathbf{e}_r[\mathbf{r}_j(0)]$ is the radial unit vector to the point $\mathbf{r}_j(0)$, and so $(\mathbf{r}_j(0)/\|\mathbf{r}_j(0)\|) \cdot \mathbf{V}_j$ is the radial velocity of scatterer at the point $\mathbf{r}_j(0)$, which we will denote as v_j. Since the magnitude of the scatterer position vector will typically be much larger than the magnitude of the velocity vector (and for short time periods), the second term in brackets in Eq. (6.3) will be much less than one. Therefore, applying a first-order binomial expansion to Eq. (6.3), we have the approximate formula

$$r_j(mT_\mathrm{s}) \approx \|\mathbf{r}_j(0)\| + mT_\mathrm{s}v_j = r_j(0) + mT_\mathrm{s}v_j \tag{6.4}$$

This gives the desired relationship between the scatterer range as a function of time and its radial velocity.

In order to focus on the effects of the velocity turbulence, we assume that the amplitudes in Eq. (6.1) are deterministic and constant over the sampling interval. This is consistent with assuming that the scatterers are neither moving an appreciable proportion of their distance to the radar nor are they changing shape or orientation over the sampling time. Inserting Eq. (6.4) into Eq. (6.1), and since $r_j(0)$ is much larger than $\lambda/2$ and the complex exponential is periodic in 2π, the first term in the square brackets in Eq. (6.1) acts like a uniform random variable over the interval $(0, 2\pi)$. Denoting $\phi_j = 4\pi r_j(0)/\lambda$, Eq. (6.1) can be written as

$$E(mT_\mathrm{s}) = \sum_{j=1}^{N} A_j \mathrm{Exp}\left[-\mathrm{i}\left(\phi_j + \frac{4\pi}{\lambda} mT_\mathrm{s}v_j\right)\right] \tag{6.5}$$

The discrete Fourier transform (DFT) of the electric field is given by

$$E(q\Delta f) = \sum_{m=0}^{M-1} E(mT_\mathrm{s})\,\mathrm{e}^{-2\pi imq/M} \tag{6.6}$$

and the periodogram form of the Doppler frequency spectrum is given by

$$S(q\Delta f) = \frac{T_\mathrm{s}}{M}|E(q\Delta f)|^2 \tag{6.7}$$

where $\Delta f = 1/MT_\mathrm{s}$ is the fundamental Doppler frequency. Using Eq. (6.5) and Eq. (6.6) in Eq. (6.7) gives double sums over both temporal and scatterer indices. A representative sum at the frequency index q is

$$\sum_{m=0}^{M-1} \mathrm{Exp}\left[-2\pi\mathrm{i}\left(\frac{mq}{M} + \frac{2}{\lambda}mT_\mathrm{s}v_j\right)\right] = \mathrm{Exp}\left[-\frac{2\pi\mathrm{i}}{\lambda}(M-1)T_\mathrm{s}v_j + \mathrm{i}\pi\frac{q}{M}\right]\frac{\mathrm{Sin}\left(\frac{2\pi}{\lambda}MT_\mathrm{s}v_j\right)}{\mathrm{Sin}\left(\pi\frac{q}{M} + \frac{2\pi}{\lambda}T_\mathrm{s}v_j\right)} \tag{6.8}$$

where the summation has been evaluated by using the geometric sum formula. Equation (6.7) can then be written as

$$S(q\Delta f) = \frac{T_s}{M}\sum_{j,k=1}^{N} A_j A_k \mathrm{Exp}\left[-\mathrm{i}(\phi_j - \phi_k)\right]\mathrm{Exp}\left[-\frac{2\pi \mathrm{i}}{\lambda}(M-1)T_s(v_j - v_k)\right] \frac{\mathrm{Sin}\left(\frac{2\pi}{\lambda}MT_s v_j\right)}{\mathrm{Sin}\left(\pi\frac{q}{M}+\frac{2\pi}{\lambda}T_s v_j\right)} \frac{\mathrm{Sin}\left(\frac{2\pi}{\lambda}MT_s v_k\right)}{\mathrm{Sin}\left(\pi\frac{q}{M}+\frac{2\pi}{\lambda}T_s v_k\right)} \tag{6.9}$$

We have now expressed the Doppler spectrum solely in terms of sums over the scatterers. This approach to radar signal analysis is useful when one is only interested in the properties of the Doppler spectrum and its moments, i.e., time series processing considerations are not important. In order to transform Eq. (6.9) from (discrete) Doppler frequency (f_D) to (discrete) Doppler velocity (u_D), we use the relationship $u_D = -(\lambda/2)f_D = -(\lambda/2)(q/MT_s)$, or $q = -(2/\lambda)MT_s u_D$, and $S(u_D) = (2/\lambda)S(f_D)$. The sine terms in the numerators in Eq. (6.9) can be written as

$$\mathrm{Sin}\left(\frac{2\pi M}{\lambda}T_s v_j\right) = (-1)^{-(q+1)}\mathrm{Sin}\left[\frac{2\pi M}{\lambda}T_s(u_D - v_j)\right] \tag{6.10}$$

Furthermore, we can write

$$\mathrm{Exp}\left[-\mathrm{i}(\phi_j - \phi_k)\right]\mathrm{Exp}\left[-\frac{2\pi \mathrm{i}}{\lambda}(M-1)T_s(v_j - v_k)\right] = \mathrm{Exp}\left[-\mathrm{i}(\theta_j - \theta_k)\right] \tag{6.11}$$

where $\theta_j = \left[4\pi r_j(M-1)/(2\lambda T_s)\right]$ is the phase associated with the location of the jth scatterer at the midpoint of the sampling period, and as with ϕ_j, it is a uniform random variable in the interval $(0, 2\pi)$. Using Eqs. (6.10) and (6.11) in Eq. (6.9) gives

$$S(u_D) = \frac{2T_s}{\lambda M}\sum_{j,k=1}^{N} A_j A_k \mathrm{Exp}\left[-\mathrm{i}(\theta_j - \theta_k)\right] \frac{\mathrm{Sin}\left[\frac{2\pi M}{\lambda}T_s(u_D - v_j)\right]}{\mathrm{Sin}\left[\frac{2\pi}{\lambda}T_s(u_D - v_j)\right]} \frac{\mathrm{Sin}\left[\frac{2\pi M}{\lambda}T_s(u_D - v_k)\right]}{\mathrm{Sin}\left[\frac{2\pi}{\lambda}T_s(u_D - v_k)\right]} \tag{6.12}$$

From Eq. (6.11) it can be seen that the random phases are independent of the Doppler velocities, u_D, so that their effect is identical across the spectral bins. The ratio of the sine terms are so-called Dirichlet kernels and in the limit of large M approach Dirac distributions. It is important to note that the product of two Dirichlet kernels (for $j \neq k$) does not act like a single Dirac distribution, and it is the interaction of these pairs of kernels that produce a substantial part of the random fluctuations in the Doppler spectrum. Furthermore, the amplitude of the product of the two Dirichlet kernels will be largest when v_j and v_k are close to each other and maximum when they are equal (i.e., when $j = k$) and equal to u_D. In that scenario the ratio of the sine terms will equal M^2. Because of this localized effect, one way to mitigate the random variations in the Doppler spectrum is to average across spectral bins. We will revisit this concept later.

Since the probability density function (pdf) of the sum of two independent random variables is given by the convolution of their pdfs and since both θ_j and θ_k are uniform random variables over $(0, 2\pi)$, the pdf for their difference (for $j \neq k$) will also be that of a uniform random variable over $(0, 2\pi)$. The expected value of the complex exponential of a uniform random variable over $(0, 2\pi)$ is zero. Thus, the expected value of the expression in Eq. (6.11) for $j \neq k$ will be zero. In the literature, this fact is typically invoked to reduce the double sum in Eq. (6.12) to the single sum over $j = k$. However, for a single realization, or averaging over a small number of realizations, the $j \neq k$ terms will not sum to zero, and so we cannot ignore them.

Next, we express Eq. (6.12) in an explicitly real form. Letting

$$z_j = A_j \frac{\mathrm{Sin}\left[\frac{2\pi M}{\lambda} T_s\left(u_D - v_j\right)\right]}{\mathrm{Sin}\left[\frac{2\pi}{\lambda} T_s\left(u_D - v_j\right)\right]} \tag{6.13}$$

Since z_j and θ_j are real functions, it can be shown that the double sum in Eq. (6.12) can be written as

$$\sum_{j,k=1}^{N} z_j z_k \mathrm{Exp}\left[-\mathrm{i}\left(\theta_j - \theta_k\right)\right] = \left(\sum_{j=1}^{N} z_j \mathrm{Cos}\theta_j\right)^2 + \left(\sum_{j=1}^{N} z_j \mathrm{Sin}\,\theta_j\right)^2 \tag{6.14}$$

Using the definition of z_j, we can write Eq. (6.12) as

$$S(u_D) = \frac{2T_s}{\lambda M}\left\{\left(\sum_{j=1}^{N} A_j \frac{\mathrm{Sin}\left[\frac{2\pi M}{\lambda}T_s\left(u_D - v_j\right)\right]}{\mathrm{Sin}\left[\frac{2\pi}{\lambda}T_s\left(u_D - v_j\right)\right]}\mathrm{Cos}\theta_j\right)^2 + \left(\sum_{j=1}^{N} A_j \frac{\mathrm{Sin}\left[\frac{2\pi M}{\lambda}T_s\left(u_D - v_j\right)\right]}{\mathrm{Sin}\left[\frac{2\pi}{\lambda}T_s\left(u_D - v_j\right)\right]}\mathrm{Sin}\,\theta_j\right)^2\right\} \tag{6.15}$$

where u_D ranges from $-\lambda/(4T_s)$ to $\lambda/(4T_s)$, in steps of $\lambda/(2MT_s)$. We note that the main characteristics of discrete sampling and finite time intervals (e.g., aliasing and periodogram bias) have been retained in this formulation. Implicit in the analysis is a rectangular window function, which leads to the ratio of sine terms. Nevertheless, using a non-rectangular window function would result in a similar form, albeit with a more complicated function replacing the ratio of sines.

From Eq. (6.12), it can be seen that the $j = k$ part of the Doppler velocity spectrum is given by

$$S_{j=k}(u_{\mathrm{D}}) = \frac{2T_{\mathrm{s}}}{\lambda M}\sum_{j=1}^{N} A_j^2 \frac{\mathrm{Sin}^2\left[\frac{2\pi M}{\lambda}T_{\mathrm{s}}(u_{\mathrm{D}} - v_j)\right]}{\mathrm{Sin}^2\left[\frac{2\pi}{\lambda}T_{\mathrm{s}}(u_{\mathrm{D}} - v_j)\right]} \tag{6.16}$$

Note that

$$\lim_{M\to\infty}\frac{2T_{\mathrm{s}}}{\lambda M}\frac{\mathrm{Sin}^2\left[\frac{2\pi M}{\lambda}T_{\mathrm{s}}(u_{\mathrm{D}} - v_j)\right]}{\mathrm{Sin}^2\left[\frac{2\pi}{\lambda}T_s(u_{\mathrm{D}} - v_j)\right]} = \delta(u_{\mathrm{D}} - v_j) \tag{6.17}$$

where the Dirac delta function on the right-hand side is understood in the distributional sense. The ratio of the sine-squared terms on the left-hand side is the well-known Fejer kernel from Fourier analysis. Using Eq. (6.17) in Eq. (6.16), and dropping the "$j = k$" subscript, we have for large M:

$$S(u_{\mathrm{D}}) = \sum_{j=1}^{N} A_j^2\delta(u_{\mathrm{D}} - v_j) \tag{6.18}$$

Taking the expected value of this expression over the probability density function of the random radial velocity field, $P(v_j)$, gives

$$\langle S(u_{\mathrm{D}})\rangle = \sum_{j=1}^{N} A_j^2\int_{-\infty}^{\infty}\delta(u_{\mathrm{D}} - v_j)P(v_j)dv_j = \sum_{j=1}^{N} A_j^2 P(u_{\mathrm{D}}) \tag{6.19}$$

(Recall that we have assumed that the amplitudes are deterministic quantities.) Normalizing by $\sum_{j=1}^{N} A_j^2$ gives

$$\langle S_N(u_{\mathrm{D}})\rangle = \frac{\langle S(u_{\mathrm{D}})\rangle}{\sum_{j=1}^{N} A_j^2} = P(u_{\mathrm{D}}) \tag{6.20}$$

Therefore, we see that for large M the expected value of the normalized Doppler velocity spectrum (the $j = k$ terms) is just the pdf of the Doppler velocities. Another way of showing this, using the correlation function of the electric field, is presented in the next section.

As mentioned above, the $j = k$ part of the Doppler spectrum comes about from taking the expected value of the complex exponential terms in Eq. (6.12). However, since the random phases, θ_j, are functions of the random radial velocities of the scatterers, technically these quantities cannot be isolated from the expected value operation (i.e., they are not independent random variables). Nevertheless, if the pulse volume over which the summation occurs is much larger than the correlation length of the turbulence (or more specifically, that between the scatterers), then one can consider "clumps" of scatterers that are uncorrelated with other clumps. In this case, we can consider the expected value of the random phases going to zero in an approximate sense. This is a form of a central limit theorem for correlated samples.

We will also need to consider the cross-correlation of the expected value of the Doppler spectrum (for large M) at two Doppler velocities:

$$\begin{aligned}\langle S_N(u_\mathrm{D})S_N(v_\mathrm{D})\rangle &= \int S_N(u_\mathrm{D})S_N(v_\mathrm{D})P(u_j,v_k)\,du_jdv_k\\ &= \frac{1}{\left(\sum_j A_j^2\right)^2}\int\sum_{j,k} A_j^2A_k^2\,\delta(u_\mathrm{D}-u_k)\delta(v_\mathrm{D}-v_k)P(u_j,v_k)du_jdv_k = P(u_\mathrm{D},v_\mathrm{D})\end{aligned} \tag{6.21}$$

where $P(u_j, v_k)$ is the joint pdf of the scatterer radial velocities. So, as with the expected value of the Doppler spectrum itself, under the assumptions used above, the cross-correlation of the Doppler spectrum at two Doppler velocities gives the joint pdf of the Doppler velocities. If the scatterer velocities are independent of each other, $P(u_j, v_k) = P(u_j)P(v_k)$ and hence $P(u_\mathrm{D}, v_\mathrm{D}) = P(u_\mathrm{D})P(v_\mathrm{D})$. Note that for a joint-Gaussian field, this is equivalent to the scatterer velocities being uncorrelated.

6.3 Doppler Spectral Moments and Turbulence

Next, consider the first three Doppler moments: M_0, the total power; M_1, the pulse volume-weighted mean radial velocity; and M_2, the pulse volume-weighted variance of the radial velocities. The total power is given by

$$M_0 = \int_{u=-\lambda/(4T_s)}^{\lambda/(4T_s)} S(u)du \tag{6.22}$$

In practice, we would use discrete sums instead of continuous integrals. We define the normalized Doppler spectrum as $S_N(u_D) = S(u_D)/M_0$. Note that this normalization is different from that used in Eq. (6.20) above. This is because technically M_0 is a random variable, so what is used in Eq. (6.20) is akin to normalizing by $\langle M_0\rangle$. Since we have assumed that the electric field amplitudes, A, are deterministic and constant over time, M_0 will not vary too much from realization to realization.

Hence, we will assume that it is a deterministic quantity. The first and second Doppler moments are then given by

$$M_1 = \int_{u=-\lambda/(4T_s)}^{\lambda/(4T_s)} u S_N(u) du \tag{6.23}$$

$$\begin{aligned} M_2 &= \int_{u=-\lambda/(4T_s)}^{\lambda/(4T_s)} (u - M_1)^2 S_N(u) du = \int_{u=-\lambda/(4T_s)}^{\lambda/(4T_s)} u^2 S_N(u) du - M_1^2 \\ &= \int_{u=-\lambda/(4T_s)}^{\lambda/(4T_s)} u^2 S_N(u) du - \int_u \int_v uv S_N(u) S_N(v) du dv \end{aligned} \tag{6.24}$$

Taking the expected value of Eqs. (6.23) and (6.24) gives

$$\langle M_1 \rangle = \int_{u=-\lambda/(4Ts)}^{\lambda/(4Ts)} u \langle S_N(u) \rangle du = \langle u \rangle \tag{6.25}$$

and

$$\langle M_2 \rangle = \int_{u=-\lambda/(4Ts)}^{\lambda/(4Ts)} u^2 \langle S_N(u) \rangle du - \int_u \int_v uv \langle S_N(u) S_N(v) \rangle du dv = \langle u^2 \rangle - \langle uv \rangle \tag{6.26}$$

where we have used Eqs. (6.20) and (6.21) (and the assumptions inherent in those equations). Note that if the scatterer velocities are taken from a zero-mean random process, then $\langle M_1 \rangle = 0$, and furthermore if the scatterer velocities are uncorrelated, the cross-correlation terms in Eq. (6.26) will be zero [see the discussion following Eq. (6.21)]. Considering a uniform continuum approximation of the scatterer positions in Eq. (6.18), we can write

$$S(u_D) du_D \approx A^2(\mathbf{x}) \delta(u_D - v(\mathbf{x})) d\mathbf{x} \tag{6.27}$$

And further we set $A^2(\mathbf{x}) = \eta(\mathbf{x}) I(\mathbf{x}, \mathbf{x}_0)$ as the returned power at the spatial position $\mathbf{x}$, where η is the reflectivity function and I is the pulse volume illumination function centered at the position $\mathbf{x}_0$ (Doviak and Zrnic 1993). We assume that the reflectivities are constant over the given pulse volume. This means that M_0 is constant. Writing

$$I_N(\mathbf{x},\mathbf{x}_0) = \frac{I(\mathbf{x},\mathbf{x}_0)}{\int I(\mathbf{x},\mathbf{x}_0)d\mathbf{x}} \tag{6.28}$$

Equations (6.25) and (6.26) can then be written as

$$\langle M_1 \rangle \approx \int \langle v(\mathbf{x}) \rangle I_N(\mathbf{x},\mathbf{x}_0)d\mathbf{x} \tag{6.29}$$

and

$$\langle M_2(\mathbf{x}_0) \rangle \approx \int \langle v^2(\mathbf{x}) \rangle I_N(\mathbf{x},\mathbf{x}_0)d\mathbf{x} - \int\int \langle v(\mathbf{x})v(\mathbf{y}) \rangle I_N(\mathbf{x},\mathbf{x}_0) I_N(\mathbf{y},\mathbf{x}_0)d\mathbf{x}d\mathbf{y} \tag{6.30}$$

If we assume that the turbulence is homogeneous and isotropic, then $\langle v(\mathbf{x}) \rangle = 0$, $\left\langle v^2(\mathbf{x}) \right\rangle = \sigma_r^2$, and $\langle v(\mathbf{x})v(\mathbf{y}) \rangle = R_{rr}(\boldsymbol{\rho})$, where σ_r^2 is the variance and $R_{rr}(\boldsymbol{\rho})$ is the autocorrelation of the radial velocities, with $\boldsymbol{\rho} = \mathbf{y} - \mathbf{x}$. This means that for a purely turbulent field (i.e., no ambient velocity field), $\langle M_1 \rangle = 0$. Note that since we are dealing with radial velocities, we have to assume that the field is both homogeneous and isotropic to remove the dependence on the position $\mathbf{x}$. We choose a Cartesian coordinate system with the z-direction along the line of sight from the radar to the center of the pulse volume, and the x- and y-directions perpendicular to each other and z. If we further assume that the pulse volume weighting function is sufficiently narrow in the crossbeam directions, we can take the turbulence quantities as those of a longitudinal Cartesian component of the velocity vector field, i.e., the z-component $v(\mathbf{x}) = \mathbf{u}(\mathbf{x}) \cdot \mathbf{x}/\|\mathbf{x}\| \approx u_z(\mathbf{x})$. That is, over the entire pulse volume of interest, the radial unit vector is approximately the same as the unit vector along the Cartesian z-axis, $\mathbf{x}/\|\mathbf{x}\| \approx \mathbf{e}_z$. This allows us to write the variance, $\left\langle v^2(\mathbf{x}) \right\rangle \approx \left\langle u_z^2(\mathbf{x}) \right\rangle = \sigma_z^2$ and the correlation function, $\langle v(\mathbf{x})v(\mathbf{y}) \rangle \approx \langle u_z(\mathbf{x})u_z(\mathbf{y}) \rangle = R_{zz}(\rho)$. Equation (6.30) can then be written as

$$\langle M_2(\mathbf{x}_0) \rangle \approx \sigma_z^2 - \langle M_1^2(\mathbf{x}_0) \rangle = \sigma_z^2 - \int\int R_{zz}(\rho) I_N(\mathbf{x},\mathbf{x}_0) I_N(\mathbf{x}+\rho,\mathbf{x}_0)d\mathbf{x}d\rho \tag{6.31}$$

The standard approach in the literature [e.g., Srivastava and Atlas (1974) or Frisch and Clifford (1974)] is to consider the Fourier space version of Eq. (6.31). However, we will show that the same result can be obtained in physical space. The advantage of this approach is that we do not need to assume that the turbulence is homogeneous over all space, as is needed in the Fourier approach. This can be seen from Eq. (6.31). The integral in $\boldsymbol{\rho}$ is over all space; however, in practice the correlation function (and the pulse volume illumination functions) will be non-negligible over a finite spatial domain; hence, we only need to assume that the velocity field is homogeneous (and/or isotropic) over that domain.

For isotropic turbulence, the components of the correlation tensor can be written as (Hinze 1975)

$$R_{\alpha\beta}(\boldsymbol{\rho}) = (R_{LL}(\rho) - R_{NN},(\rho))\frac{\rho_\alpha \rho_\beta}{\rho^2} + R_{NN}(\rho)\delta_{\alpha\beta} \tag{6.32}$$

where $\rho_\alpha = \boldsymbol{\rho} \cdot \mathbf{e}_\alpha$ is the component of the displacement vector along the α-coordinate, $\rho = \| \boldsymbol{\rho} \|$ is the length of the displacement vector, and R_{LL} and R_{NN} are the longitudinal and normal correlation functions. These correlation functions are only dependent on ρ. For the $\alpha = \beta = z$ components, we have

$$R_{zz}(\boldsymbol{\rho}) = R_{LL}(\rho)\cos^2\phi + R_{NN}(\rho)\sin^2\phi \tag{6.33}$$

where we have used spherical coordinates such that $\rho_z = \rho\cos\phi$. Using a von Kármán form for the correlation functions, we can write $R_{LL}(\rho) = B_1(\rho)$ and $R_{NN}(\rho) = B_1(\rho) + B_2(\rho)$, such that Eq. (6.33) can be written as

$$R_{zz}(\boldsymbol{\rho}) = B_1(\rho) + B_2(\rho)\sin^2\phi \tag{6.34}$$

where ϕ is the polar angle from the z-axis to the plane containing the x- and y-axes and the functions B_1 and B_2 are given in terms of modified Bessel functions of the second kind (Cornman and Goodrich 1996). We use a Gaussian model for the normalized pulse volume illumination function:

$$I_n(\mathbf{x}, \mathbf{x}_0) = \frac{1}{a^2 b (2\pi)^{3/2}} Exp\left[-\frac{(x - x_0)^2 + (y - y_0)^2}{2a^2} - \frac{(z - z_0)^2}{2b^2} \right] \tag{6.35}$$

where $a = R_0 \Delta\phi / (4\sqrt{ln2})$ and $b = \Delta R / (4\sqrt{ln2})$, with $\Delta\phi$ and ΔR as the polar and radial full-width at half-max (FWHM) points for the two-way electric field patterns, respectively. Note that the radar parameter a is a function of the range to the pulse volume center, $R_0 \approx z_0$. Using this form of the pulse volume illumination function, integrating over $\mathbf{x}$, converting to spherical coordinates for $\boldsymbol{\rho}$, and using Eq. (6.34), gives for the integral in Eq. (6.31)

$$\frac{1}{8a^2 b \pi^{3/2}} \int_0^{2\pi}\int_0^{\pi}\int_0^{\infty} \left[B_1(\rho) + B_2(\rho)\sin^2\phi \right] Exp\left\{ -\frac{\rho^2 \left[a^2 + b^2 + (a^2 - b^2)\cos 2\phi \right]}{8a^2 b} \right\} \times \rho^2 \sin\phi d\rho d\phi d\theta \tag{6.36}$$

The integral over θ gives 2π and the integral over ϕ can be calculated analytically. Using the explicit forms for B_1 and B_2, Eq. (6.36) can then be written as a one-dimensional integral over ρ:

$$\frac{9A\pi\varepsilon^{2/3}}{55Lk_0^{5/3}}\frac{L\,k_0^{4/3}}{2^{1/3}\sqrt{\pi}\Gamma(5/6)a\sqrt{a^2-b^2}}\int_0^\infty \rho^{4/3}\left\{-\frac{k_0}{\sqrt{\pi}}\frac{ab}{\sqrt{a^2-b^2}}K_{2/3}(k_0\rho)e^{-\rho^2/4b^2}\right.$$
$$\left.+\left[K_{1/3}(k_0\rho)-\frac{k_0}{2\rho}\left(\rho^2-\frac{2a^2b^2}{a^2-b^2}\right)K_{2/3}(k_0\rho)\right]e^{-\rho^2/4a^2}Erf\left(\frac{\sqrt{a^2-b^2}}{2ab}\rho\right)\right\}d\rho$$
$$=\frac{9A\pi}{55Lk_0^{5/3}}G(a,b,L)\,\varepsilon^{2/3} \tag{6.37}$$

where $k_0=\sqrt{\pi}\Gamma(5/6)/(\Gamma(1/3)L)$, Γ is the gamma function, K_ν is the modified Bessel function of the second kind with order ν, *Erf* is the error function (for complex argument), A is a constant on the order of 1.5, and ε is the energy dissipation rate. L is the turbulence integral length scale, defined by

$$L=\frac{1}{\sigma_v^2}\int_0^\infty R_{LL}(\rho)d\rho \tag{6.38}$$

Given a length scale, L, and the radar parameters a and b, the integral can then be evaluated numerically.

For the von Karman model, the variance of the velocities, σ_z^2, in Eq. (6.31) can also be written in terms of the energy dissipation rate (Cornman et al. 1995):

$$\sigma_z^2=\frac{9A\pi}{55Lk_0^{5/3}}\varepsilon^{2/3} \tag{6.39}$$

Therefore, we can write the expected value of the Doppler second moment in terms of the atmospheric turbulence intensity parameter, ε,

$$\langle M_2(\mathbf{x}_0)\rangle\approx(1-G[a(R_0),b,L])\frac{9A\pi}{55Lk_0^{5/3}}\varepsilon^{2/3} \tag{6.40}$$

where the function G is given by Eq. (6.37), and we have explicitly indicated the functional dependence of the radar parameter a on the radar range, R_0. Note that in the limit that L goes to zero, i.e., the velocity field is uncorrelated spatially, the function G goes to zero, so that $\langle M_1^2(\mathbf{x}_0)\rangle=0$ and hence $\langle M_2(\mathbf{x}_0)\rangle=\sigma_z^2$. This corresponds to the results presented in the discussion following Eq. (6.26) above. On the other hand, in the limit that L goes to infinity, i.e., the velocity field is perfectly correlated (i.e., the spatial correlation coefficient of the joint pdf is one), the function G goes to one, and $\langle M_2(\mathbf{x}_0)\rangle=0$.

Solving Eq. (6.40) for $\varepsilon^{1/3}$ gives

$$\varepsilon^{1/3} = \left[\langle M_2(\mathbf{x}_0) \rangle \frac{55 L k_0^{5/3}}{9 A \pi (1 - G[a(R_0), b, L])} \right]^{1/2} \tag{6.41}$$

Instead of using Eq. (6.31), we could also use Eqs. (6.26) and (6.39) to give

$$\varepsilon^{1/3} = \left[\frac{55 L k_0^{5/3}}{9 A \pi} \left(\langle M_2(\mathbf{x}_0) \rangle + \left\langle M_1^2(\mathbf{x}_0) \right\rangle \right) \right]^{1/2} \tag{6.42}$$

That is, we use an average of measured first moments, as opposed to the theoretical expected value.

Next, consider the moments calculated from the expected value of the Doppler spectrum, denoted with an over bar:

$$\overline{M}_0 = \int_{u=-\lambda/(4Ts)}^{\lambda/(4Ts)} \langle S(u) \rangle du \tag{6.43}$$

$$\overline{M}_1 = \int_{u=-\lambda/(4Ts)}^{\lambda/(4Ts)} u \langle S_N(u) \rangle du \tag{6.44}$$

$$\overline{M}_2 = \int_{u=-\lambda/(4Ts)}^{\lambda/(4Ts)} \left(u - \overline{M}_1\right)^2 \langle S_N(u) \rangle du = \sigma_r^2 - \left(\overline{M}_1\right)^2 \tag{6.45}$$

Comparing with the expected values of the moments, Eqs. (6.25) and (6.26), we see that $\overline{M}_1 = \langle M_1 \rangle$ and $\overline{M}_2 = \langle M_2 \rangle + \left\langle M_1^2 \right\rangle - \langle M_1 \rangle^2$. As mentioned above, for a purely turbulent field, $\langle M_1 \rangle = 0$, and hence, $\overline{M}_1 = 0$ and $\overline{M}_2 = \sigma_r^2 \approx \sigma_z^2$. This means that ensemble averaging the Doppler spectra removes all information regarding the correlation structure of the wind field—at least from the perspective of the first and second moments. We can also see this by assuming a white noise correlation function (i.e., setting B_1 and B_2 proportional to delta functions) in Eq. (6.36). That results in $\left\langle M_1^2 \right\rangle = 0$, and hence for an uncorrelated velocity field, $\langle M_2 \rangle = \sigma_r^2 = \overline{M}_2$.

For a purely turbulent velocity field, Eq. (6.45), along with Eq. (6.39), gives

$$\overline{M}_2 \approx \sigma_z^2 = \frac{9 A \pi}{55 L k_0^{5/3}} \varepsilon^{2/3} \tag{6.46}$$

Solving for $\varepsilon^{1/3}$ gives

$$\varepsilon^{1/3} = \left[\overline{M}_2 \frac{55Lk_0^{5/3}}{9A\pi}\right]^{1/2} \tag{6.47}$$

Therefore, we see that this equation has the same form as Eq. (6.41) with the function G set to zero. In a similar fashion to Eq. (6.42), we can also write

$$\varepsilon^{1/3} = \left[\frac{55Lk_0^{5/3}}{9A\pi}\left(\overline{M}_2(\mathbf{x}_0) + \overline{M}_1^{\,2}(\mathbf{x}_0)\right)\right]^{1/2} \tag{6.48}$$

where we do not a priori assume that $\overline{M}_1 = 0$, for example, in the practical situation when a limited number of spectra are averaged—as opposed to a theoretical ensemble.

6.4 Simulation Results

In this section, we use a simulation to analyze many of the concepts presented above. The radar parameters are that of a typical airborne X-band Doppler radar: $\lambda = 0.032$ m, $\tau = 1.5$ μs, $T_s = 333$ μs, $P_t = 19$ dB, $G = 35$ dB, and $M = 64$. A constant reflectivity value of $\eta = -92$ dB was used. There is no additive (e.g., receiver) noise. The amplitudes, $A_j = A(\mathbf{r}_j)$ are given by the square root of the radar equation for a point scatterer,

$$A(\mathbf{r}_j) = \left[\frac{P_t G^2 \lambda^2 \eta}{(4\pi)^3 \|\mathbf{x}_0\|^4} f^4(\mathbf{r}_j, \mathbf{x}_0) g^2(\mathbf{r}_j, \mathbf{x}_0) \Delta x \Delta y \Delta z\right]^{1/2} \tag{6.49}$$

where the location of the pulse volume center is $\mathbf{x}_0$, f^4 and g^2 are the two-way azimuthal and range intensity patterns, respectively,

$$f^4(\mathbf{r}_j, \mathbf{x}_0) g^2(\mathbf{r}_j, \mathbf{x}_0) = \mathrm{Exp}\left\{-8\ln 2\left[\left(\frac{\sin\alpha}{\Delta\alpha}\right)^2 + \left(\frac{r_j - R_0}{\Delta R}\right)^2\right]\right\} \tag{6.50}$$

where α is the polar angle from the beam center to the given scatterer, $r_j = \|\mathbf{r}_j\|$, and $\Delta\alpha = 0.061$ rad (i.e., 3.5°) and $\Delta R = 263$ m are as described in the text below Eq. (6.35) [note that Eq. (6.35) is a normalized Cartesian approximation to Eq. (6.50)]. The simulation was based on a grid with 50 m resolution, so that $\Delta x = \Delta y = \Delta z = 50$ m. The number of grid points, N, was determined by finding those points in the grid where both f^4 and g^2 had dropped to 0.01 (from a maximum value of 1.0 at the pulse volume center). For $R_0 = \|\mathbf{x}_0\| = 21{,}600$ m and $\Delta\alpha$ and ΔR as described above, $N = 22{,}360$. The velocity field was generated using a 64^3

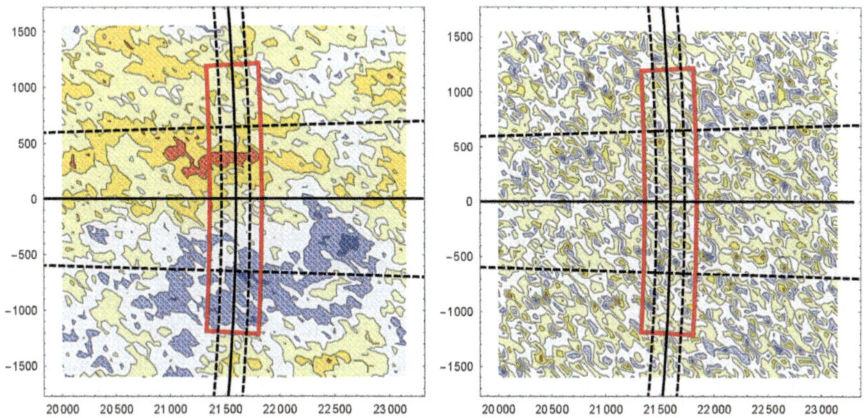

Fig. 6.1 Two-dimensional slice of simulated von Kármán turbulence field (*left-hand side*) and uncorrelated Gaussian field (*right-hand side*)

discrete simulation of a von Kármán turbulence field, as described in Frehlich et al. (2001). The integral length scale is $L = 500$ m and the energy dissipation rate is $\varepsilon = 0.064$ m^2 s^{-3} ($\varepsilon^{1/3} = 0.4$ m$^{2/3}$ s^{-1}). The left-hand side of Fig. 6.1 is a 2-D slice of the z-component of a single realization of the von Kármán field (recall that z in this case is along the bore sight of the radar). The right-hand side shows a zero-mean uncorrelated random field with variance given by Eq. (6.39) using the ε and L values as used with the von Kármán field. The colors range from -12 m/s (blue) to +12 m/s (red). The horizontal black line is the beam center, and the solid black arc indicates a range of 21,600 m (the radar is off to the left). The dashed black lines bracketing the solid black lines indicate the $\pm\Delta\alpha/2$ and $\pm\Delta R/2$ points. The red outline indicates the region of the grid used in the radar simulation (i.e., based on the 0.01 level as described above). Comparing the left- and right-hand sides of the figure, the difference between a correlated and uncorrelated field is obvious. The correlated field is "clumpy" whereas the uncorrelated field is "well-mixed." This is a key factor in what the radar measures—especially for small pulse volumes. That is, for each realization, a correlated field means that the radar sees a preferential distribution of scatterer radial velocities and the variation over different realizations results in variations in the first moments, as described by Eqs. (6.26) and (6.31). On the other hand, each realization of an uncorrelated field will look very much the same, and so the first and second moment variations will be much smaller. It is important to realize that the von Kármán and uncorrelated random velocity fields are both derived from Gaussian pdfs and that their ensemble-averaged distributions will be identical. However, due to the correlation structure, an averaged distribution from the von Kármán field will converge more slowly to the ensemble average one.

The correlation structure we have been referring to above is a spatial one, but there are also temporal correlations that are pertinent. For turbulence, there are two correlation times that are important here, τ_T, a correlation time of the larger-scale turbulence field, and τ_E, a correlation time of the smaller-scale turbulence, and hence for the electric field. The von Kármán wind field simulation used herein does not contain any temporal correlation, and so when we talk about realizations, we are considering them in the context of temporal intervals, τ_T, large enough such that all of the realizations are statistically independent from each other. Contrast this scale to a correlation scale of the electric field, i.e., the time it takes the scatterers to move an appreciable percentage of a wavelength. Two Doppler spectra that are calculated at times greater than τ_E will be statistically independent in the sense that the random fluctuations at each Doppler velocity bin—primarily due to the $j \neq k$ terms—will be uncorrelated. The time scale for the large-scale properties of the turbulence to change appreciably will typically be much greater than that for the electric field, i.e., $\tau_T \gg \tau_E$. Hence, any variations in the first moments of the Doppler spectrum over the time scale τ_E will occur primarily due to small-scale properties of the turbulence via the interaction of the $j \neq k$ terms, as opposed to coming from the larger-scale properties of the turbulence (we are not considering advection here). For times greater than the correlation time of the turbulence field, the first moments are affected by both phenomena, but in very different ways. We shall revisit these concepts below in the context of the simulation results.

In order to illustrate some of the Doppler spectrum concepts discussed above, consider Fig. 6.2. The radar parameters and scatterer geometry are as presented above. The scatterer radial velocities, v_j, are sampled randomly from a spatially uncorrelated, skew-Gaussian pdf, $P_{sG}(v_j)$:

$$P_{sG}\left(v_j;\beta,\mu,\sigma\right) = \frac{1}{\sqrt{2\pi}\sigma} e^{-\frac{\left(v_j-\mu\right)^2}{2\sigma^2}} Erfc\left[-\frac{\beta\left(v_j-\mu\right)}{\sqrt{2}\sigma}\right] \tag{6.51}$$

where *Erfc* is the complimentary error function, μ is the mean value of the distribution, σ is the standard deviation, and the parameter β determines the amount of skewness. Note that since $Erfc(0) = 1$, setting $\beta = 0$ gives the usual Gaussian pdf. The rationale for using this pdf is not necessarily that it reflects a physical scenario, but to verify Eq. (6.20), i.e., that the ensemble-averaged Doppler spectrum (and for a large number of pulses) reflects the velocity pdf of the scatterers, not a Gaussian form a priori. For illustrative purposes, the following pdf characteristics were used: $\mu = 18$ m/s, $\sigma = 5$ m/s, and $\beta = 8$. The electric field amplitudes were calculated using Eq. (6.49). The solid points connected by the thin black curve in the figure come from Eq. (6.15), and the triangles connected by the dashed gray curve come from Eq. (6.16) (the "$j = k$" or "diagonal" terms); the thick black curve is the pdf, Eq. (6.51), multiplied by the sum of the squares of the amplitudes. The vertical dashed lines are at the unambiguous velocities (i.e., Nyquist Doppler velocities), $\pm\lambda/(4T_s)$. It is clear that the correct aliasing structure—wrapping of the Doppler spectrum around the unambiguous velocity on the right-hand side of Fig. 6.2—occurs naturally with the formulation given by Eq. (6.15). By comparing the full spectrum versus the diagonal terms, it can be seen that the $j \neq k$ terms dominate the randomness in the Doppler spectrum. As mentioned above, this is

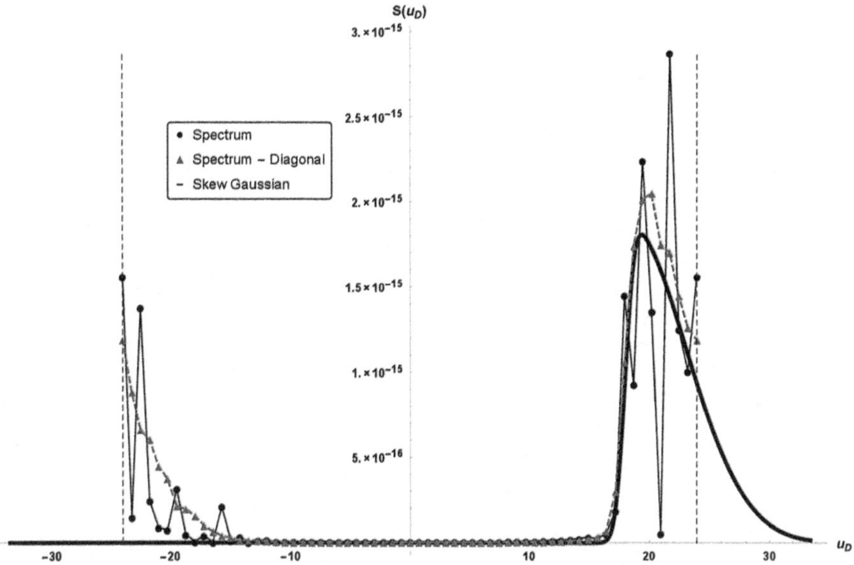

Fig. 6.2 Simulated Doppler velocity spectrum. The *thick black curve* is the pdf [Eq. (6.51)], the *dashed gray curve* is the full Doppler spectrum [Eq. (6.15)], and the *thin black curve* is the "$j = k$" part of the Doppler spectrum [Eq. (6.16)]

mainly due to the interaction of the Dirichlet kernel terms in Eq. (6.12). Unlike the Fejer kernel in the diagonal terms, the Dirichlet kernels oscillate in sign, and so the product of two of them manifests constructive and destructive interference effects. Figure 6.2 is from a single realization of the random wind field. When averaging over many realizations, the Doppler spectrum (not shown) converges to the pdf—with the spectrum from just the diagonal terms converging much more rapidly than the spectrum using all the terms. It is important to note that the spectra in Fig. 6.2 result from an uncorrelated random scatterer velocity field, showing that these interaction effects, as opposed to correlated velocity structures, dominate the fluctuations in a single Doppler spectrum.

Figure 6.3 shows the same quantities as in Fig. 6.2, but now using a single realization of the correlated von Kármán turbulence field. The mean and standard deviation for the Gaussian curve [Eq. (6.51) with $\beta = 0$] come from the sampled von Kármán field. The inset in Fig. 6.3 shows the beneficial effect of averaging the Doppler spectrum along spectral bins, which in this case was implemented with a running five-point average. This is shown in the inset via the curve with the solid squares. It can be seen that, at least for this example, the bin-averaged data is close to the Doppler spectrum from the diagonal terms and both are close to the Gaussian curve. It is important to realize that this is not always the case and there is a significant amount of variability realization to realization—though not nearly as much as with the full Doppler spectrum. In the results presented below, the spectral

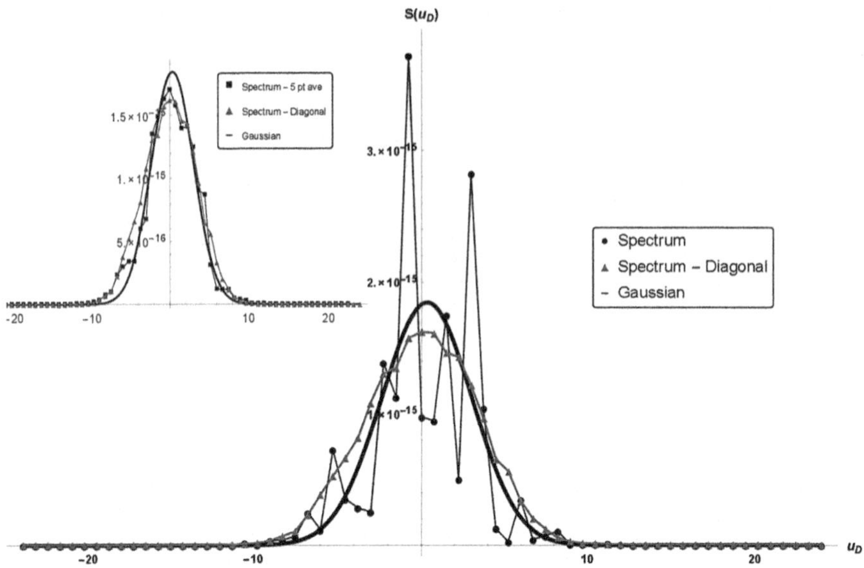

Fig. 6.3 Same as Fig. 6.2, except that the scatterer velocities are from a simulated von Kármán turbulence field. The *inset* shows the spectrum from the diagonal terms along with the five point bin-averaged spectrum

bin averaging is applied to both the von Kármán and uncorrelated Gaussian data, as well as for averaging moments and averaging spectra.

Figure 6.4 (in log-linear format) shows the effect of averaging 200 realizations of the Doppler spectrum from von Kármán fields. The solid black curve is the Gaussian pdf [multiplied by the sum of the squares of the amplitudes from Eq. (6.49)], the curve with the solid triangles is the averaged spectrum, Eq. (6.16), and the curve with the solid squares is the expected periodogram of the Doppler spectrum (described below). It can be seen that the averaged spectrum and the expected periodogram are very close to each other and that for values surrounding the peak value, they are both very close to the Gaussian curve. Away from the peak, both are biased high relative to the Gaussian curve. As is well known, this bias is due to the finite time interval and discrete sampling. The expected periodogram for the finite time, discrete sampling Doppler spectrum is essentially the expected value of Eq. (6.7) and is given by Zrnic (1979)

$$\langle S(p\Delta u_{\mathrm{D}})\rangle = \frac{2}{\lambda}T_{\mathrm{s}} \sum_{q=-(M-1)}^{M-1} \left(1-\frac{|q|}{M}\right) R(qT_{\mathrm{s}})\mathrm{Exp}\left(2\pi \mathrm{i}\frac{qp}{M}\right) \tag{6.52}$$

where $p = \left\{-\frac{M}{2}, \ldots, \frac{M}{2}\right\}$, $\Delta u_D = \lambda/(2MT_s)$, and $R(qT_s)$ is a discrete sampling of the expected value of the electric field correlation function. Note that the term

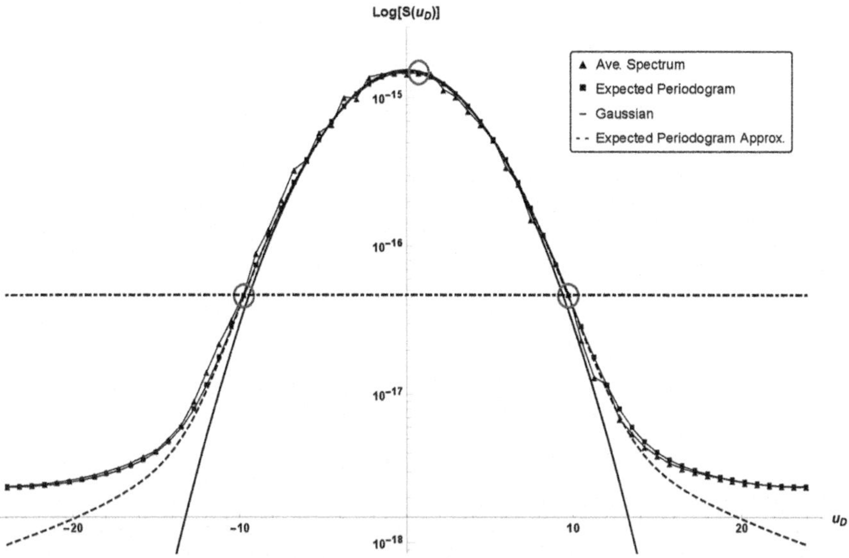

Fig. 6.4 Log-linear plot of the averaged Doppler velocity spectrum from von Kármán turbulence. The *solid black curve* is the (scaled) Gaussian pdf; the *curve* with *solid triangles* is the averaged spectrum, Eq. (6.16); the *curve* with the *solid squares* is the discrete expected periodogram, Eq. (6.52); the *dashed curve* is the approximation to the expected periodogram, Eq. (6.55); and the *dot-dashed horizontal line* indicates the cutoff power level. The *gray circles* indicate the midpoint and cutoff Doppler velocities used in the moment calculations

$1 - |q|/M$ is the correlation function of a rectangular window of length M and that the DFT of this window function (without the factor $2T_s/\lambda$) is the Fejer kernel from Eq. (6.17). From the discrete convolution theorem, Eq. (6.52) is equivalent to the convolution of the Fejer kernel and the DFT of the correlation function, R. Since the Fejer kernel is always nonnegative, this convolution will result in a high bias in the expected periodogram away from the center of the DFT of the correlation function. Assuming that the electric field, Eq. (6.5), is stationary over the M samples, the expected value of the correlation function of the electric field (after averaging over the random initial positions) is

$$R(qT_s) = \langle E(mT_s)E^*(nT_s)\rangle = \sum_{j=1}^{N} A_j^2 \int_{-\infty}^{\infty} \mathrm{Exp}\left(-2\pi \mathrm{i}\frac{2v_j}{\lambda}qT_s\right)P(v_j)\,dv_j \quad (6.53)$$

where $q = m - n$. It is assumed that the velocities for all the scatterers are sampled from the same pdf, that is, $P(v_j) = P(v_k)$, for all scatterers j and k. Note that in the limit $T_s \to 0$ and $M \to \infty$ such that $T = MT_s$, and with $qT_s \to \tau$ and $p\Delta u_D \to u_D$, both continuous variables, Eq. (6.52) becomes a continuous-time Fourier transform over the finite interval $(-T, T)$. Dividing by $\sum_{j=1}^{N} A_j^2$ to give the normalized Doppler spectrum results in

$$\begin{aligned}\langle \hat{S}_N(u_\mathrm{D})\rangle &= \frac{2}{\lambda}\int_{-T}^{T}\int_{-\infty}^{\infty} Exp\left(-2\pi i\frac{2v_j}{\lambda}\tau\right)P(v_j)\left(1-\frac{|\tau|}{T}\right)\mathrm{Exp}\left(2\pi i\frac{2u_\mathrm{D}}{\lambda}\tau\right)dv_j\,d\tau \\ &= \frac{2}{\lambda}\int_{-T}^{T}\mathrm{Exp}\left(-\frac{4\pi i\mu}{\lambda}\tau-\frac{8\pi^2\sigma_r^2}{\lambda^2}\tau^2\right)\left(1-\frac{|\tau|}{T}\right)Exp\left(2\pi i\frac{2u_\mathrm{D}}{\lambda}\tau\right)d\tau\end{aligned} \tag{6.54}$$

where we have used the over-caret on S to denote this continuous variable version of the expected periodogram, and in going from the first to second line, a Gaussian pdf [i.e., Eq. (6.51), with $\beta = 0$] was used. Note that the transform variables are time lag, τ, and Doppler frequency, $f_\mathrm{D} = -(2/\lambda)u_\mathrm{D}$, and recall that $S(u_\mathrm{D}) = (2/\lambda)S(f_\mathrm{D})$. The integral can be evaluated in closed form; however, an accurate approximation for most practical situations is given by

$$\langle \hat{S}_N(u_\mathrm{D})\rangle \approx \frac{1}{\sqrt{2\pi}\sigma_r}\,\mathrm{Exp}\left[-\frac{(u_\mathrm{D}-\mu)^2}{2\sigma_r^2}\right]\left[1+\frac{\lambda(u_\mathrm{D}-\mu)}{4\pi T\sigma_r^2}Erfi\left(\frac{u_\mathrm{D}-\mu}{\sqrt{2}\sigma_r}\right)\right]-\frac{\lambda}{4\pi^2 T\sigma_r^2} \tag{6.55}$$

where *Erfi* is the imaginary error function (which is real-valued for real arguments). Note that taking the limit as $T \to \infty$ shows that the normalized expected value Doppler velocity spectrum reduces to a Gaussian form—which is another way of proving Eq. (6.20), at least for a Gaussian pdf. However, using the same assumptions, we can generalize this to arbitrary pdfs. That is, using the $T \to \infty$ limit of (the first line of) Eq. (6.54) for continuous temporal lag τ, we can write

$$\begin{aligned}\langle \hat{S}_N(u_\mathrm{D})\rangle &= \int_{-\infty}^{\infty} P(v_j)\left\{\frac{2}{\lambda}\int_{-\infty}^{\infty}\mathrm{Exp}\left[2\pi i\left(\frac{2u_\mathrm{D}}{\lambda}-\frac{2v_j}{\lambda}\right)\tau\right]d\tau\right\}dv_j \\ &= \int_{-\infty}^{\infty} P(v_j)\delta(u_\mathrm{D}-v_j)dv_j = P(u_\mathrm{D})\end{aligned} \tag{6.56}$$

where we have used the relationship, $\delta(\alpha x) = \delta(x)/|\alpha|$. Comparing this analysis to that which led to Eq. (6.20) above, we see that the approaches are consistent, i.e., both methods result in a delta function Doppler spectrum. A related derivation can be found in Doviak and Zrnic (1993).

Equation (6.54) accommodates one of the real-world aspects embodied in Eq. (6.52): a finite time sampling interval. Its approximation, Eq. (6.55), is shown by the dashed curve in Fig. 6.4. By using a continuous-time variable, this formulation does not take into account aliasing effects, as seen via the lack of power in the tails of that curve. This can be handled by adding the aliased power to Eq. (6.55) via

$$\langle S(u_{\mathrm{D}})\rangle \approx \langle \hat{S}\,(u_{\mathrm{D}})\rangle + \sum_{n=1}^{N}\left[\left\langle \hat{S}\left(2n\frac{\lambda}{4T_{\mathrm{s}}} - u_{\mathrm{D}}\right)\right\rangle + \left\langle \hat{S}\left(2n\frac{\lambda}{4T_{\mathrm{s}}} + u_{\mathrm{D}}\right)\right\rangle\right] \quad (6.57)$$

where $\lambda/(4T_{\mathrm{s}})$ is the unambiguous Doppler velocity (i.e., Nyquist velocity), $u_{\mathrm{D}} = \{-\lambda/(4T_{\mathrm{s}}), \lambda/(4T_{\mathrm{s}})\}$, and N, the number of Nyquist interval pairs [note the factor of $2n$ in Eq. (6.57)], is on the order of 10–100, depending on the desired level of approximation. With a value of $N = 40$, the curves from Eqs. (6.57) and (6.52) are indistinguishable for this case (not shown in Fig. 6.4).

As mentioned above, and clearly seen in Fig. 6.4, the discrete sampling and finite interval effects will produce a high bias of the expected periodogram relative to the limiting-case Gaussian spectrum. Note that the effect is inherent in the use of discrete sampling and a finite interval and not a function of the random processes involved. That is, no amount of averaging can reduce the bias, it just reduces the fluctuations relative to the expected periodogram. Clearly, the further out in Doppler velocity away from the peak of the Doppler spectrum, the more substantial the bias effect. Hence, we would like to compute the Doppler moments in a way that accommodates or mitigates this bias. As mentioned above, one could use a window function other than a rectangular one to mitigate some of these effects, but for simplicity, we will stick with the rectangular window in our analysis. First, consider the bias in computing the second moment over a truncated Doppler velocity range, $(-a, a)$ [not to be confused with the radar pulse volume parameter, a, in Eq. (6.35) above],

$$\hat{M}_2(a) = \frac{\int_{-a}^{a} (u_{\mathrm{D}} - \mu)^2 S(u_{\mathrm{D}})du_{\mathrm{D}}}{\int_{-a}^{a} S(u_{\mathrm{D}})du_{\mathrm{D}}} \quad (6.58)$$

For a Gaussian spectrum, we have

$$\hat{M}_2(a) = \sigma_r^2\left\{1 - \frac{\sigma_r\sqrt{\frac{2}{\pi}}\mathrm{Exp}\left[-\frac{(a+\mu)^2}{2\sigma_r^2}\right]\left[a + \mu + (a-\mu)\mathrm{Exp}\left[-\frac{2a\mu}{\sigma_r^2}\right]\right]}{Erf\left(\frac{a-\mu}{\sqrt{2}\sigma_r}\right) + Erf\left(\frac{a+\mu}{\sqrt{2}\sigma_r}\right)}\right\} \quad (6.59)$$

Note that as $a \to \infty$, the second term in curly brackets goes to zero ($\lim_{z\to\infty} Erf(z) = 1$), so $\lim_{a\to\infty}\hat{M}_2(a) = \sigma_r^2$, but otherwise $\hat{M}_2(a)$ will have a

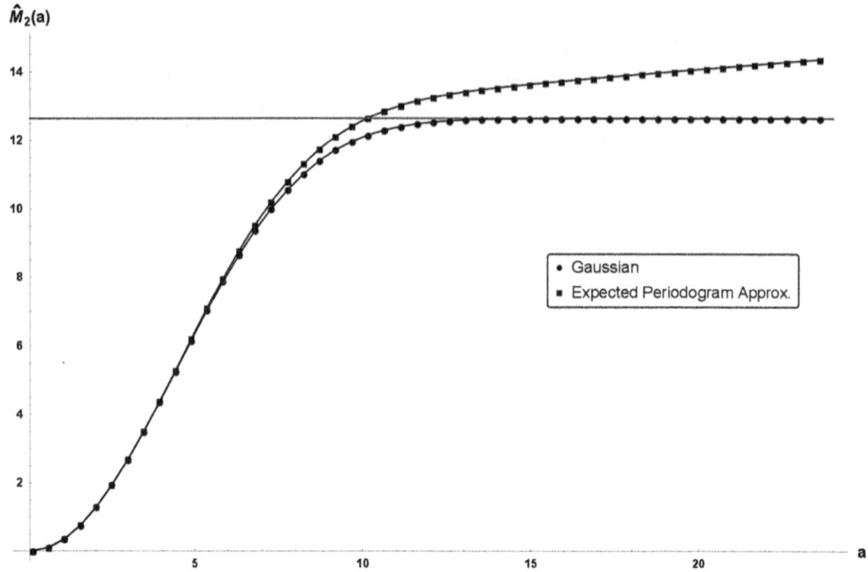

Fig. 6.5 Second moment as a function of cutoff velocity, for Gaussian spectrum (*filled circles*) and the approximation to the expected periodogram (*filled squares*), Eq. (6.55). The *horizontal line* shows the true value of the second moment for this zero-mean case, 12.64

negative bias (for $|a| > |\mu|$). On the other hand, using the expected periodogram approximation, Eq. (6.55), [or Eq. (6.57)], can produce a high, low, or no bias, depending on the choice of the parameter a. In fact, one can choose a to minimize the bias, by solving numerically the equation $\hat{M}_2(a) = \sigma_r^2$ for a, giving a "cutoff velocity" $\hat{a}$. A "cutoff power" can then be computed by inserting $\hat{a}$ into Eq. (6.55) [or Eq. (6.57)]. Figure 6.5 illustrates these concepts. The curve with filled circles is the second moment as a function of cutoff velocity for the Gaussian spectrum [Eq. (6.59)], the curve with filled squares is the second moment for the approximation of the expected periodogram, Eq. (6.58) using Eq. (6.55), and the horizontal line is the true second moment for $\mu = 0$ and $\sigma_r^2 = 12.64\ \text{m}^2/\text{s}^2$ [see Eq. (6.39), with $L = 500$ m and $\varepsilon^{1/3} = 0.4$]. The parameter $\hat{a}$ is given by the intersection of the latter two curves. Returning to Fig. 6.4, the peak and cutoff velocities are indicated by the gray circles, and the horizontal dashed line shows the cutoff power level. It is important to realize that this analysis has been performed for an infinite SNR scenario and that in practice one would choose the cutoff to be the larger of the noise power and the cutoff power computed in this fashion. If the noise power is greater than the cutoff power, then Eq. (6.58) [with Eq. (6.55) or Eq. (6.57)] could be used to estimate the bias in the second moment (see also Zrnic 1979).

As mentioned above, one of the key differences between Doppler radar measurements of an uncorrelated random velocity field and a correlated one is revealed in the quantity $\langle M_1^2 \rangle$. That is, in the former case, $\langle M_1^2 \rangle = 0$, whereas in the latter

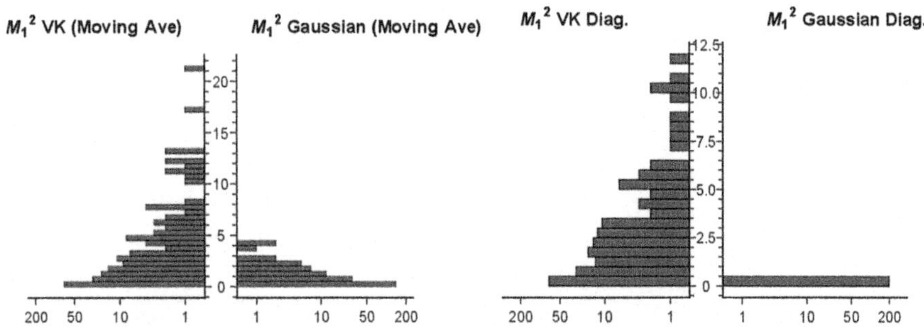

Fig. 6.6 Distributions, in log-counts, of M_1^2 (diagonal terms) for von Kármán velocity field and Gaussian velocity field. The *left-hand side* shows the distributions for the bin-averaged spectra and the *right-hand side* shows those for just the diagonal terms

case, it is given—at least theoretically—by the integral in Eq. (6.31). Figure 6.6 shows distributions (in log counts) of M_1^2 from the 200 realizations of the von Kármán and uncorrelated Gaussian velocity fields. The left-hand side panel shows the results from the five spectral bin moving average method and the right-hand side panel shows those for the Doppler spectra from the diagonal terms. It is clear that the M_1^2 values from the uncorrelated Gaussian field are more clustered around zero than for the von Kármán field. From Eq. (6.31) it can be seen that for an uncorrelated field, the expected value of the second moment will equal the variance of the radial velocities and hence will be larger than that for a correlated field, the difference of course being the $\langle M_1^2 \rangle$ factor. This is an important consideration in simulating Doppler radar performance for turbulence detection.

Table 6.1 shows summary statistics for $\varepsilon^{1/3}$ estimation via moment and spectral averaging. By "spectral averaging," we mean that the spectra are first averaged across realizations and then the moments are calculated from these averages. Note that in Table 6.1, the notations $\langle x \rangle$ and $\overline{x}$ refer to sample means over a given set of realizations, not the ensemble averages used in Eqs. (6.25) and (6.26) for moment averaging or Eqs. (6.44) and (6.45) for spectral averaging, respectively. The "% Error" is given by

$$\%\,\text{Error} = 100\left(\frac{\frac{1}{N}\sum_{i=1}^{N} d_i}{E(d)} - 1\right) \tag{6.60}$$

where d is the data element (i.e., $\varepsilon^{1/3}$ estimate), the numerator in the ratio is the sample mean, and $E(d)$ is the expected value (0.4 in this case). The "RMSE" is the root mean square error, given by

Table 6.1 Summary statistics for $\varepsilon^{1/3}$ from averaging moments or spectra

$\varepsilon^{1/3}$ est. method	Vel. type	# Ave	% Error	RMSE
$\langle M_2\rangle$	VK	1	−2.42	0.062
$\langle M_2\rangle + \langle M_1^2\rangle$	VK	1	−1.67	0.072
$\langle M_2\rangle + \langle M_1^2\rangle$	Uncorrelated	1	−0.62	0.045
$\langle M_2\rangle$	VK	10	−1.35	0.019
$\langle M_2\rangle + \langle M_1^2\rangle$	VK	10	−0.19	0.023
$\langle M_2\rangle + \langle M_1^2\rangle$	Uncorrelated	10	−0.05	0.014
$\overline{M}_2 + \overline{M}_1^2$	VK	10	0.15	0.021
$\overline{M}_2 + \overline{M}_1^2$	Uncorrelated	10	0.68	0.013
$\langle M_2\rangle$	VK	100	−1.25	0.007
$\langle M_2\rangle + \langle M_1^2\rangle$	VK	100	−0.04	0.006
$\langle M_2\rangle + \langle M_1^2\rangle$	Uncorrelated	100	0.01	0.003
$\overline{M}_2 + \overline{M}_1^2$	VK	100	−0.03	0.006
$\overline{M}_2 + \overline{M}_1^2$	Uncorrelated	100	0.04	0.004

$$\text{RMSE} = \sqrt{\frac{1}{N}\sum_{i=1}^{N}\left[d_i - E(d)\right]^2} \tag{6.61}$$

where $N = 200$. In each scenario, bin averaging was applied to the individual spectra prior to moment calculation or averaging. Since there were only 200 realizations, a randomized sampling method was used to calculate the "10" and "100" averages scenarios. That is, a random selection (though minimizing overlap) of the 200 realizations was used to generate the 200 averages, resulting in a certain amount of correlation between the estimates. If the data were uncorrelated, one would expect a $1/\sqrt{M}$ decrease in the RMSE as a function of the number of averages, M. From the table, we can see that this holds approximately, so the random sampling does not overly affect the general nature of the results. Note that for $M = 1$, moment averaging and spectral averaging are the same, and hence, only three of the five $\varepsilon^{1/3}$ estimation methods are shown. From the table, we can see that the moment averaging and spectral averaging produce similar results. Recall that these are infinite SNR results, so spectral averaging should be better in lower SNR scenarios. However, bin averaging would also help in that regard. The effect of bin averaging is substantial, even in this infinite SNR case. Without this method, the percent errors are approximately five times larger and the RMSE values are almost doubled (results not shown in the table). The bin averaging helps reduce the errors in the moment estimates. This can be understood by considering Figs. 6.3 and 6.4. The noisier the spectrum, there is a higher likelihood that the spectral points that go into the moment calculation, i.e., those points above the cutoff power level, will vary from realization to realization. Smoothing the spectra helps to mitigate this problem, resulting in more accurate and stable moment estimates. Obviously, averaging the spectra over different realizations also has this beneficial effect.

We also note that using Eq. (6.42) instead of the Eq. (6.41), i.e., using the calculated instead of theoretical value of $\langle M_1^2 \rangle$, improves the percent errors, while minimally decreasing the RMSEs. Not surprisingly, the results with the uncorrelated velocity data are generally better than the results with the von Kármán data—especially for the RMSE values. This is due to the fact mentioned above that the velocity distributions for the uncorrelated random data converge more rapidly to the theoretical ones, than for the correlated data. It is important to realize that the values in the table are for infinite SNR and are also dependent on the tuning of the cutoff velocities; nevertheless, they do give insight into the key characteristics and methods.

6.5 Flight Test Results

Many commercial transport aircraft are outfitted with predictive windshear radar systems. Therefore, as part of the US National Aeronautics and Space Administration's (NASA) Aviation Safety Program (AvSP), the Turbulence Prediction and Warning System (TPAWS) team developed real-time detection algorithms that can provide timely and accurate warnings of convective turbulence in relatively low reflectivity situations (<30 dBZ). As part of the these R&D activities, the National Center for Atmospheric Research (NCAR) developed new turbulence detection and data quality control algorithms for use with airborne Doppler radars. The detection algorithm was based on a version of Eq. (6.41), but with the moments coming from spectral averaging in both range and azimuth (Cornman et al. 2000). The energy dissipation rate values were then scaled into RMS vertical accelerations using an aircraft response function. A thorough verification exercise was performed in 2001 using NASA's B-757 research aircraft. A team of human experts analyzed 55 cases. Forty-two of these cases contained moderate or greater turbulence encounters. A comparison of the radar-based turbulence algorithm and in situ aircraft data shows that the algorithm provided reliable detection well in front of the aircraft, even at very low reflectivities.

In this scoring exercise, NCAR's post-analysis of the flight test data produced plots depicting the predictions made by the detection algorithm with the actual RMS g-load experienced by the aircraft overlaid along its track. Human experts considered the magnitude, persistence, and proximity to the aircraft encounter of the radar turbulence diagnoses in order to perform the scoring. Figure 6.7 illustrates the situation for Flight 232, event 10, a case in which consensus among the experts participating in the manual scoring exercise was easily achieved. This event consisted of a rapidly building thunderstorm, with the aircraft encountering severe turbulence in a very low reflectivity region above the cell. The event consists of a time segment in which the aircraft experienced a sequence of turbulence encounters deemed to have a common operational impact. A sample of three sequential hazard diagnoses produced for this event are displayed as color-scaled, contoured images,

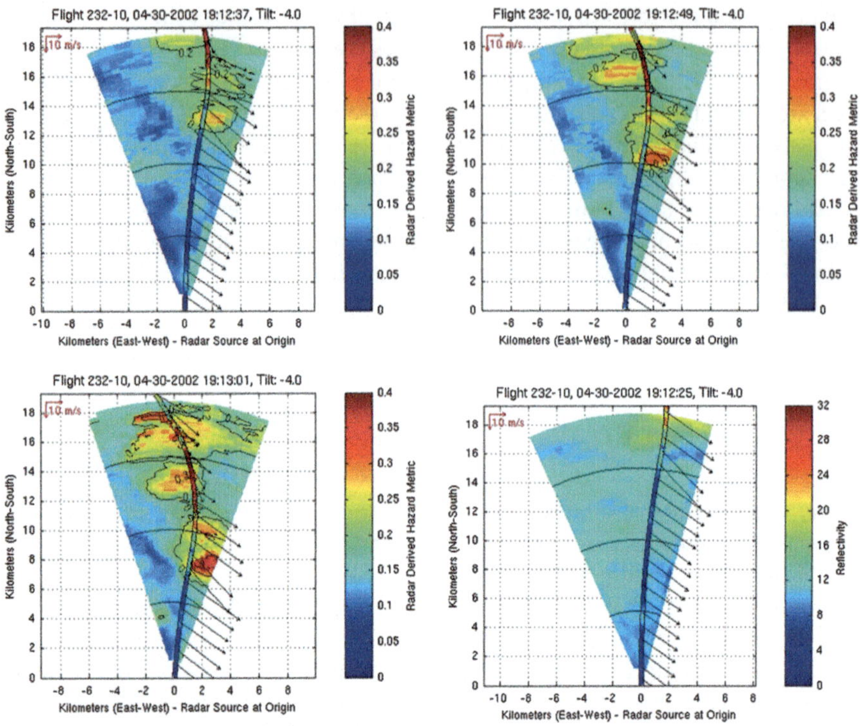

Fig. 6.7 NASA TPAWS Event 232-10. From *upper left* to *lower right*: radar-derived hazard metric for three successive radar sweeps, radar reflectivity (*lower right*). Aircraft flight track with hazard metric color-coding overlaid

with the over-plotted RMS g-load from the aircraft's subsequent penetration representing "truth." These are from the $-4°$ elevation scans, approximately 12 s apart, and are shown in the upper row and bottom left-hand side of the figure. Moderate or greater (MOG) turbulence, defined here as a predicted RMS load ≥ 0.2 g RMS (with the RMS computed over 1 km of the flight path) is displayed as light-green to red values, and a contour is drawn at 0.2 g on the radar hazard plot. Although a precise spatial and quantitative match of the turbulence algorithm and aircraft-measured values was not achieved for this case, the algorithm was judged by the human experts to have correctly diagnosed the MOG aircraft turbulence well in advance of 30 s before the encounter. In particular, the ≥ 0.2 g interval on the aircraft track persistently overlaps the 0.2 g contoured regions on all three radar hazard plots. The lower right-hand image in Fig. 6.7 illustrates the reflectivities for the $-4°$ elevation scan 12 s before the time period shown in the hazard plots. Note that the maximum values are all below 25 dBZ, that is, a standard aircraft radar would be displaying black (<20 dBZ) or barely green (20–30 dBZ). At the zero degree elevation angle, the reflectivities were even lower. This case clearly illustrates a scenario that could have resulted in a severe encounter for a commercial transport aircraft. That is, severe turbulence embedded within a low reflectivity region—a region that a typical pilot would not consider hazardous. In all, the human experts analyzed 52 cases, 42 of which contained MOG turbulence. The Probability of Detection (POD) was 81 % and the False Alarm Ratio (FAR) was 11 %. In fact, from an operational point of view, the performance numbers were actually better than these values. Of the eight cases that were deemed to be missed detections, five were just marginally so. That is, the aircraft-measured values were just above the detection threshold while the radar predicted values were just below. Similarly, three of the four false detections were marginal.

6.6 Conclusion

A first-principle analysis of Doppler radar measurements of turbulence has been presented. A set of limited, though practical, conditions has been assumed to make the problem more tractable, with the primary condition being the discrete and finite temporal sampling of the radar signals. A theoretical derivation of the Doppler spectrum under these conditions was performed, and the distinction between what the radar actually measures and what results after a theoretical ensemble averaging was delineated. This is an important consideration, as all of the theoretical development in the literature is based upon ensemble averaging. It was shown that in the limit of an infinite number of samples, and after ensemble averaging, the Doppler spectrum can be represented by a sum of Dirac delta distributions and furthermore that the normalized spectrum will equal the probability distribution of the scatterer velocities. We have seen that the interaction between the off-diagonal terms produces the majority of the fluctuations in the Doppler spectrum and that these terms are present irrespective of the correlation structure in the scatterer velocity field. These terms vanish under ensemble averaging but are clearly an issue for real-world analysis, where only a limited amount of averaging is possible. The correlation structure of the velocity field manifests itself primarily in the square of the first moment. That is, a correlated field will have a certain degree of patchiness, which leads to variations in the first moments from realization to realization. These theoretical considerations were then studied via simulation. Parameters for a typical airborne X-band Doppler radar were used, and correlated von Kármán and uncorrelated random fields were employed. Simulated Doppler spectra were shown for selected cases to illustrate many of the concepts described above. Two approaches to minimizing the effect of the cross-terms were investigated: averaging across spectral bins for a single spectrum and averaging at a given spectral bin across realizations. Both were seen to have beneficial effects in reducing the bias and random error in turbulence estimates. It was also shown that the inclusion of the measured squares of the first moments—instead of the theoretical value—also improved the performance of the turbulence calculations. Real-world application of the turbulence measurement methods was then shown with airborne X-band detection of convective turbulence.

References

Bohne, A.R.: Radar detection of turbulence in precipitation environments. J. Atmos. Sci. **39**, 1819–1837 (1982)

Cornman, L.B., Morse, C.S., Cunning, G.: Real-time estimation of atmospheric turbulence severity from in-situ aircraft measurements. J. Aircraft **32**(1), 171–177 (1995)

Cornman, L.B., Goodrich, R.K.: The detection of atmospheric turbulence using Doppler radars. In: Proceedings, Workshop on Wind Shear and Wind Shear Alert Systems, American Meteorological Society, Oklahoma City, OK, 12–15 November 1996

Cornman, L.B., Williams, J.K., Goodrich, R.K.: The detection of convective turbulence using airborne Doppler radars. In: Proceedings, Ninth Conference on Aviation, Range, and Aerospace Meteorology, American Meteorological Society, Orlando, FL, 11–15 September 2000

Doviak, R. J., Zrnic, D.S.: Doppler radar and weather observations, 2nd ed., Academic Press San Diego (1993)

Frehlich, R., Cornman, L., Sharman, R.: Simulation of three dimensional turbulent velocity fields. J. Appl. Meteorol. **40**, 246–258 (2001)

Frisch, A.S., Clifford, S.F.: A study of convection capped by a stable layer using Doppler radar and acoustic echo sounders. J. Atmos. Sci. **31**, 1622–1628 (1974)

Hinze, J.O.: Turbulence, 2nd edn. McGraw-Hill, New York (1975)

Rodgers, R.R., Tripp, B.R.: Some radar measurements of turbulence in snow. J. Appl. Meteorol. **3**, 603–610 (1964)

Srivastava, R.C., Atlas, D.: Effect of finite radar pulse volume on turbulence measurements. J. Appl. Meteorol. **13**, 472–480 (1974)

Zrnic, D.S.: Estimation of spectral moments for weather echoes. IEEE Trans. Geosci Electron **GE-17**(4), 113–128 (1979)

Chapter 7
Remote Turbulence Detection Using Ground-Based Doppler Weather Radar

John K. Williams and Gregory Meymaris

Abstract Turbulence in and around clouds can pose a significant hazard to aviation, with convective turbulence identified as being responsible for a majority of all turbulence-related aircraft accidents. Regions of convective turbulence may be small (~1 km) and highly transient (~few minutes). Graphical Turbulence Guidance, an operational NWP-based turbulence forecast, produces hourly forecasts at a relatively coarse spatial resolution and does not explicitly forecast convective turbulence. High-resolution storm data from radar reflectivity or satellites may provide some indication of the likelihood of convective turbulence development, but cannot pinpoint its location or severity. The NCAR/NEXRAD Turbulence Detection Algorithm (NTDA) uses ground-based Doppler weather radar data to measure in-cloud turbulence, with a focus on identifying convective turbulence hazards. NTDA utilizes Level II data from the U.S. network of WSR-88Ds (NEXRADs) to produce real-time, rapid-update, three-dimensional mosaics of in-cloud turbulence. An NTDA product is also produced operationally in the Taiwan Advanced Operational Aviation Weather System using NEXRAD and Gematronik radar data. NTDA turbulence maps are suitable for tactical use by pilots and airline dispatchers and for providing input to comprehensive turbulence nowcasts. They also provide information about storm evolution useful for studying the relationship of turbulence production to thunderstorm dynamics and kinematics. This chapter motivates and describes the NTDA and discusses its performance.

J.K. Williams (✉)
National Center for Atmospheric Research, Research Applications Laboratory, Boulder, CO, USA

The Weather Company, an IBM Business, Andover, MA, USA
e-mail: john.williams@weather.com

G. Meymaris
National Center for Atmospheric Research, Research Applications Laboratory, Boulder, CO, USA
e-mail: meymaris@ucar.edu

R. Sharman, T. Lane (eds.), *Aviation Turbulence*,
DOI 10.1007/978-3-319-23630-8_7

7.1 Introduction

Convective and convectively induced turbulence (CIT) pose a significant hazard to en route aircraft. While Federal Aviation Administration (FAA) guidelines prescribe that aircraft should circumnavigate thunderstorms by wide margins horizontally and/or vertically, this is often not practical given fuel costs and traffic density constraints, and aircraft routinely penetrate airspace in and near storms where turbulence may be moderate or even severe. Several researchers have suggested that such turbulence may be responsible for a majority of turbulence-related aircraft accidents. Cornman and Carmichael (1993) estimated the fraction at 60 %, and of the 44 aviation accident cases analyzed by Kaplan et al. (2005), 86 % were found to be within 100 km of convection. Figure 3 in Lane et al. (2012) indicates that aircraft are 15 times more likely to encounter moderate-or-greater turbulence when within 15 km of convection, and nearly 90 times more likely if flying within 0.8 km above a radar echo top. Even non-convective clouds are often associated with greater likelihood of turbulence, as most air travelers can attest. To assist pilots, dispatchers, and air traffic controllers in identifying safe routes, it is desirable to discriminate which areas in and around clouds are hazardous and which are benign. Both unnecessary airspace closures and potentially hazardous turbulence encounters by "pathfinder" aircraft can be minimized when turbulence can be remotely detected. The NCAR/NEXRAD Turbulence Detection Algorithm (NTDA) described in this chapter accomplishes this with an accuracy, timeliness, and broad-area coverage that no other currently available technology provides.

Convective turbulence can appear and disappear over spatial and temporal scales on the order of kilometers and minutes within dynamically evolving storms and is notoriously difficult to diagnose or forecast (Lane et al. 2012). Fine-scale numerical modeling studies resolve turbulence in and around convection, and they often compare well with observed turbulence in case study investigations. However, operational numerical weather prediction (NWP) does not resolve storms and their associated wind fields with the temporal and spatial resolution, accuracy, and timeliness needed to predict turbulence impacts on en route aircraft. Two-dimensional maps of radar reflectivity and echo tops have long been the primary source of real-time, fine-scale information relating to storms' potential turbulence, hail, lightning, and icing hazards. However, radar reflectivity is determined by hydrometeor distributions, not turbulence, and is often a poor indicator of the location and severity of CIT. Convective SIGMET polygons are large, long-lived, infrequently updated, and do not pinpoint the location and nature of the potential aviation hazards. Approaches have been developed to use geostationary satellite imagery to identify cloud characteristics associated with convective turbulence, e.g., the overshooting tops algorithm developed by Bedka et al. (2010). Deierling et al. (2011) and Al-Momar et al. (2015) have shown that lightning density may be associated with the volume of in-cloud convective turbulence. However, such satellite and lightning products offer only indirect inferences of convective turbulence and have difficulty locating it in three dimensions (3D).

Given that altitude changes are often pilots' preferred method for avoiding turbulence, this is a significant limitation.

The NTDA provides the prospect of remotely pinpointing 3-D turbulence and making the data available to aviation users within minutes, often in time to influence routing and cabin management decisions. Airborne LIDAR and SODAR have been proposed for this purpose (e.g., Chap. 22), but have various limitations that have prevented their adoption in commercial aircraft. For instance, LIDAR units are heavy and expensive and suffer severe attenuation in areas of high water vapor or hydrometeors. In contrast, turbulence detection using airborne Doppler weather radars has become routine in commercial aircraft, using an approach similar to the NTDA's. Airborne radar provides rapid updates, minimal latency, and the ability to scan the airspace directly ahead of an aircraft, making it very useful for avoiding convective turbulence. However, these systems are limited by power constraints, significant attenuation of the X-band frequency in dense cloud and precipitation that limits deep signal penetration of large storms, and the fact that they afford only a limited view of the turbulence ahead of an aircraft. Furthermore, because onboard radar information is typically only displayed within the cockpit, it does not promote a shared situational awareness with dispatchers and air traffic controllers. Ground-based Doppler weather radars typically operate at the longer C- and S-band wavelengths, affording them the ability to deeply penetrate storms. This allows them to provide a broad area view of the turbulence hazard within clouds and precipitation that can facilitate a common situational awareness among various aviation stakeholders and be provided to general aviation users without onboard radar.

The NTDA is designed to routinely measure turbulence (eddy dissipation rate, EDR) in clouds and precipitation using ground-based Doppler weather radar data. The Doppler spectrum width (SW) represents the variability of the wind's radial velocity—i.e., its speed toward or away from the radar—within the radar's measurement volume. By taking into account the fact that measurements near the radar are over smaller volumes than those farther away due to the spread of the radar beam, and hence sample less of the wind field's full variability, the SW may be converted into an atmospheric turbulence metric, EDR, independent of range from the radar. The connection between Doppler SW and turbulence intensity has long been recognized, and indeed, the introduction of the U.S. network of Weather System Radar-88 Doppler (WSR-88D, a.k.a. NEXRAD) radars included a turbulence detection algorithm (TDA). However, the SW measurement is exceptionally sensitive to low signal-to-noise ratios, overlaid echoes, and contamination by non-atmospheric scatterers, and the original TDA over-warned so much that it was removed from the NEXRAD algorithm suite. An essential component of the NTDA is a fuzzy-logic procedure for performing comprehensive quality control on each SW measurement and then computing appropriately weighted averages to extract high-quality turbulence measurements. Additionally, each NTDA turbulence measurement is given a quality estimate known as a "confidence," which is useful in weighting turbulence measurements from multiple radars when they are combined into a 3-D mosaic. The NTDA product has proven valuable both as a stand-alone information source and as an input to a product that fuses data from multiple sources to diagnose the turbulence in and around thunderstorms (Williams 2014).

This chapter is organized as follows. Section 2 describes sources of contamination of the SW measurement and the NTDA's fuzzy-logic method for performing quality control. Section 3 explains how SW is related to EDR. Section 4 presents the mosaic algorithm used to merge data from multiple radars into a 3-D grid. Section 5 provides objective and subjective evaluations of the NTDA via research flight data, operational demonstration feedback, and case studies. Finally, Sect. 6 summarizes and concludes the chapter.

7.2 SW Quality Control

7.2.1 The Need for Quality Control

An essential component of the NTDA is careful quality control of the SW field. The SW is the square root of the second central moment of the radial velocity spectrum. Unfortunately, SW measurements are much more easily contaminated than the radar reflectivity (the zeroth moment) or radial velocity (the first moment). Sources of contamination include ground clutter, radiation from the sun, lightning strikes, radio frequency interference, wind farms, birds and insects, errors due to low signal to noise ratio (SNR), and overlaid echoes from other radar signal "trips" (Doviak and Zrnić 1993; Fang et al. 2004). The NEXRAD Open Radar Data Acquisition system includes methods for detecting and censoring ground clutter (Hubbert et al. 2009), censoring overlaid echoes, and removing regions with low SNR. However, the censoring thresholds have been tuned to optimize coverage and accuracy of the radar reflectivity and radial velocity products, with less attention given to the less-used SW. Thus, additional quality control is required for the SW data in order to eliminate or mitigate spurious measurements. Moreover, there have been errors in the NEXRAD signal processing that have caused erroneous data values and limited the usefulness of the SWs. Some of the most serious of these, including a problem with automatic gain control that caused clipping of the signal and erroneously high SW values, were corrected around 2001 (Sirmans et al. 1997; Fang et al. 2004). The operational implementation of the NTDA in the Open Radar Products Generator Build 10 in 2007 brought increased attention to the NEXRAD SW and its value for aviation turbulence detection. Subsequently, the SW estimator was substantially improved through the implementation of a more accurate and less SNR-sensitive "hybrid" SW estimator (Meymaris et al. 2009, 2011). Nevertheless, SW quality control remains essential.

7.2.2 Censoring

Doppler measurements with very low SNR or overlaid echoes (see below for a discussion of overlaid power ratio) are removed prior to transmission of the radar data (Doviak and Zrnić 1993). The NTDA attempts to locate remaining SW data that are deemed so contaminated as to be useless and replace them with "missing" before any calculations are performed.

7.2.2.1 Artifacts

Certain non-physical patterns, or artifacts, sometimes appear in Doppler radar data, and image processing techniques may be used to identify them. One common NEXRAD artifact consists of "ring" structures of repeated values. These rings typically consist of 4 or more range gates containing repeated values in the radial velocity and SW fields, with this pattern being repeated over several adjacent azimuths, though possibly with different repeated values. The fact that the artifact is duplicated in the radial velocity and SW fields makes automated identification relatively straightforward, with little danger of removing good data. Fortunately, these artifacts are rarely, if ever, observed in modern NEXRAD data.

7.2.2.2 Sun Spikes

"Sun spikes" occur when the radar is pointed near the sun, and are typically evidenced by increased reflectivity values along one or two adjacent radials, along with noisy radial velocities and elevated SWs. The NTDA removes sun spikes by computing the position of the sun based on date, time, and radar position and then censoring beams within an azimuthal interval of $\pm 2.0°$ and elevation within $\pm 1.75°$. These intervals were determined empirically by examining several hundred incidents of sun spike contamination and recording the displacements between the azimuth and elevation of the contaminated beams and the sun's computed position. These censoring intervals have proven sufficient to remove essentially all of the sun spike contamination.

7.2.3 Fuzzy Logic Confidence

For those data that aren't censored, the NTDA uses an artificial intelligence approach known as "fuzzy logic" to perform additional "softer" quality control (Williams 2009 and references therein). Fuzzy logic allows information from a number of sources to be combined in a way that allows the preponderance of evidence to be weighed. For the NTDA, the goal is to determine the quality of each SW measurement, assigning it a “confidence” value between 0 and 1 prior to its use in deriving turbulence, and to provide a confidence value for the final EDR measurement. The SW confidence is assembled from several relevant quantities, each of which is passed through an “interest map” to produce a 0 to 1 “proto-confidence” value. These proto-confidences are then combined via a geometric average, as described in greater detail below, to compute the SW confidence. Those SW measurements with higher confidence are given greater weight in the final spatial average of SW-derived EDR values, and they help determine a final EDR measurement confidence that can be used when merging the resulting polar grids into a 3-D Cartesian mosaic. The relevant quantities and their associated interest maps are described below.

7.2.3.1 Signal-to-Noise Ratio

One of the most important quantities for determining the quality of an SW measurement is the signal-to-noise ratio (SNR) of the radar return. For NEXRADs, the NTDA estimates the SNR in dB for every Doppler measurement point using the measured reflectivity, dBZ, and the range r (km) of the measurement from the radar via the following formula:

$$\mathrm{SNR_{dB}} = \mathrm{dBZ} - C - 20\,\log_{10}(r) \tag{7.1}$$

where C is a current radar constant (labeled db*z0* in the NEXRAD Level II data) which relates radar received echo power and radar reflectivity (Doviak and Zrnić 1993). In implementing the NTDA for use with Gematronik radars in Taiwan, the radar constant C was not available in the radar metadata and was estimated from the reflectivity data, which was known to be censored at a specified SNR threshold.

The accuracy of SW measurements obtained from signals with low SNR is generally poor, and accuracy improves as the SNRs grow. Thus, the SW proto-confidence interest map based on SNR must assign 0 confidence at small SNR values before increasing to 1 for higher values. However, the relationship between SNR and the error in the SW measurement is dependent on a number of aspects of the radar’s operational mode and the SW estimator being employed. For instance, the dwell time, pulse repetition frequency (PRF), Nyquist velocity, signal processing, and estimator used in deriving the SW from the raw returned signal all affect the magnitude of the errors. In order to handle the large set of possible operational modes for Doppler weather radars in a consistent way, a methodology was developed for automatically creating the required interest maps based on a database of simulated radar data. The method is called “dynamic SNR” (DSNR) quality control and works as follows.

SW performance statistics (mean and standard deviation) are precomputed for various SNR values, number of pulses per dwell, “true” input SWs, and SW estimators via simulations that include upstream signal processing such as window functions. The true input SWs and performance statistics are all normalized by the

Nyquist velocity, allowing the statistics to be adapted to any Nyquist velocity (which is a function of wavelength and PRF). The performance statistics are computed empirically via a time-series (in-phase and quadrature; see Doviak and Zrnić 1993) simulation adapted from the method described in Frehlich and Yadlowsky (1994), coupled with an implementation of the SW estimator including all relevant upstream signal processing. These statistics are then stored in a database for use in the real-time system. Figure 7.1 shows a depiction of simulation results for an SNR of 10 dBZ given a Nyquist velocity of 30 m s^{-1}.

During NTDA processing of an elevation sweep, the relevant simulation table for the radar's operational mode is retrieved, and the normalized values of input SW and performance statistics are converted to m s^{-1} using the current Nyquist velocity. The performance statistics are then interpolated to estimate performance for a set of input ("true") SW values, ranging from 0.5 to 10 m s^{-1} in steps of 0.25 m s^{-1}, since typical atmospheric SWs fall in this range.

For each SNR and input SW, the probability of a *large* error is computed as one minus the probability that the spatially averaged SW measurement falls within an acceptable interval around the input SW. A simplifying assumption is made that the SW error distribution is Gaussian with mean and standard deviation determined from the simulation performance statistics. The mean is used as is, but the standard deviation is divided by the square root of half of the number of radar measurements in a fixed (approximately) square at that range from the radar; this accounts for the fact that the SW-derived EDRs are averaged spatially in the NTDA's final step, reducing the random measurement error. Thus, the DSNR confidence map also varies with range from the radar, since more azimuths fall within the averaging radius near the radar than farther away. While the Gaussian error assumption is not truly satisfied by SW data, it allows for a simple algorithm and made little difference in practice compared to an algorithm that used a more accurate SW error distribution.

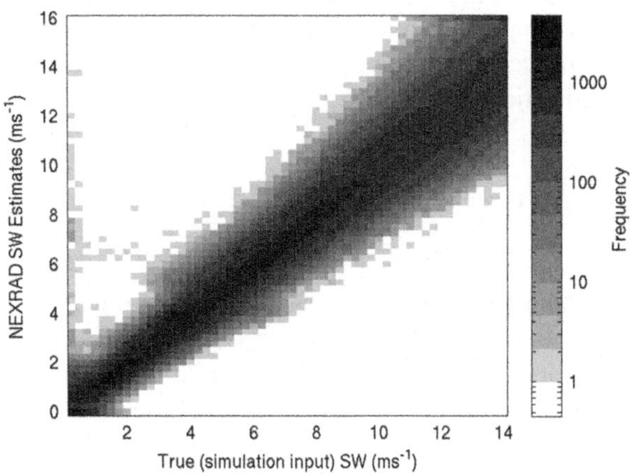

Fig. 7.1 Sample simulation results for an SNR of 10 dB showing true input SW (*x*-axis) with the corresponding distributions of simulated SW measurements, shaded according to the logarithmic scale on the right. A Nyquist velocity of 30 m s^{-1} has been assumed

The lower bound of the allowed SW measurement interval is the smallest of: a constant called *lower bound largest value* (currently, 6.5 m s^{-1}), the floor of the input SW minus 1.0 m s^{-1}, and the floor of 80 % of the input SW. The lower bound is further clipped to be no less than 0 m s^{-1}. The upper bound is the largest of the ceiling of the input SW plus 1.0 m s^{-1} and the ceiling of 120 % of the input SW. (Here the "floor" of a number is the largest integer less than or equal to it, and the "ceiling" is the smallest integer greater than or equal to it.) Thus, for smaller input SWs, the absolute error (about $\pm$1.0 m s^{-1}) is used, and for larger input SWs, the relative error (about $\pm$20 %) is used to bound the allowed interval. The addition of the *lower bound largest value* constant ensures that for large input SWs, the SW estimator need only to register a "largish" value (currently 6.5 m s^{-1}) to be considered good. This allows for some saturation of the SW estimator (see Melnikov and Zrniç 2004) and improves the spatial coverage of the NTDA.

For a particular SNR, range, and SW, the maximum probability of a *large* error is computed using the simulation results and definitions above. If the maximum probability of a large error for all SW values is less than 10 %, then the confidence associated with that SNR and range is set to 1. If the maximum probability is greater than 20 %, then the confidence is set to 0. Linear interpolation is performed in between. These criteria define an SNR confidence interest map adapted to the range from the radar and the radar's current settings. If the radar's SW estimator or relevant upstream processing is changed, only the addition of an appropriate table of simulated performance statistics is required in order for the NTDA to adapt to the change.

In practice, Doppler weather radar reflectivity comes from long-pulse data that is not collected at the same time as the Doppler data. For the lowest "surveillance" tilts, this problem is particularly acute: reflectivity and Doppler data are collected on different revolutions of the radar dish, and the reflectivity is sometimes measured at a coarser resolution. For mid-level "batch" tilts, the radar typically switches back and forth between reflectivity and Doppler moment measurements for each azimuth. Thus, the reflectivity-derived SNR value does not always precisely align with the SW measurement, and any locations assigned low SNR confidence need to be spatially spread via a local minimum filter to ensure that all potentially low-quality data are identified.

7.2.3.2 Overlaid Power Ratio

The time from the emission to receipt of a Doppler radar's energy pulse is an indication of the distance to the targets reflecting the signal. However, when a series of pulses are emitted, the return from the most recent pulse can be coincident with an earlier pulse returning from a more distant target. When pulses are produced at regular intervals, the distance the signal can travel, be reflected back, and arrive at the receiver before the next pulse is emitted is called the "unambiguous range." Signals returning from targets separated by radial distances that are multiples of the unambiguous range arrive at the receiver at the same time and interfere with one another. Sorting out the strongest return from others requires examining the SNR derived from reflectivity data, which is generally collected using a lower pulse repetition frequency (PRF) and thus not subject to the same range ambiguity. The velocity and SW from the returned signal are assigned to the candidate range gate with the highest SNR. However, if the combined SNRs of the other candidate ranges are close to that SNR value, then their contribution to the returned signal may cause significant errors in the velocity and SW measurements. In general, SW is broadened by overlaid echoes because targets at different ranges (and, therefore, heights above the ground) likely have different radial velocities. Velocity is less dramatically affected and may be accurate when the SW is not. The overlaid power ratio (PR) is the ratio of the echo power for a given measurement point to the sum of all overlaid echo powers, i.e., echo powers for ranges differing from the given point by multiples of the unambiguous Doppler range, R_{max}. Let $\mathrm{SNR_{lin}} = 10^{\mathrm{SNR_{dB}}/10}$ represent SNR in linear space (as opposed to dB) and define $\mathrm{SNR_{lin}} = 0$ whenever reflectivity (and hence SNR) are missing. PR in dB at range r from the radar may be expressed via the formula

$$\mathrm{PR_{dB}}(r) = 10\log_{10}\frac{\mathrm{SNR_{lin}}(r)}{\sum_{k\neq 0}\mathrm{SNR_{lin}}(r - kR_{max})} \tag{7.2}$$

where k takes values of all relevant integers except 0; when the denominator in Eq. (7.2) is zero, we define $\mathrm{PR_{dB}}(r) = +\infty$. Doppler moments (radial velocity and SW) are censored by the NEXRAD Radar Data Acquisition (RDA) system when their PR is less than a specified threshold (Doviak and Zrnić 1993). In practice, this threshold is usually chosen to eliminate poor velocity estimates, whereas SW estimates are much more easily contaminated by overlaid echoes and require additional quality control. The interest map used to convert $\mathrm{PR_{dB}}$ to an SW proto-confidence takes a value of 0 for small values of $\mathrm{PR_{dB}}$ and then rises to 1 for larger values.

7.2.3.3 Clutter and Overlaid Clutter

Radar beams spread and bend due to the refractive index gradient in the atmosphere. Sometimes part of the beam, or a sidelobe, can be reflected off ground targets—stationary objects like buildings or mountains, or cars, trees, waves or wind turbines. Such targets can be very effective at reflecting the radar signal and can overwhelm the power of weather returns. Thus, when the radar signal is contaminated with ground target returns, the Doppler moments may not provide useful information about the state of the atmosphere. In particular, the SW is usually quite narrow for a ground clutter return and if used by the NTDA could lead to underestimating atmospheric turbulence. If the power of the clutter return is on par with the weather signal, it will contaminate the velocity spectrum and can cause spuriously large SW estimates. Moreover, “second trip” returns or, more generally, returns from clutter targets that are a multiple of the unambiguous range from a measurement point can similarly contaminate the velocity spectrum. Because a ground clutter return is usually narrow and centered at zero velocity, a "clutter filter" can be applied to the data to remove the velocity values near 0, and the remaining signal may be used to estimate the atmospheric reflectivity, velocity and SW (Doviak and Zrnić 1993). Radar operators and automated algorithms (e.g., Hubbert et al. 2009) identify regions at which the clutter filter should be applied; these are known as "clutter maps." Unfortunately, while reflectivity and radial velocity can often be accurately recovered from clutter-filtered data, SW is easily contaminated by any residual clutter left by the filter or by the filter itself when the weather's radial velocity is near zero. Because the effect on SW is so unpredictable, the clutter interest map assigns an SW proto-confidence value of 0.5 to locations within a clutter map area or a multiple of the ambiguous range away; otherwise, its value is 1. A more sophisticated algorithm might use the radial velocity and SW measurements to estimate the likelihood that the SW measurement was adversely affected by clutter filtering. However, because clutter is most prevalent at measurement locations near the surface and the NTDA was designed primarily to detect en route aviation turbulence, the simple solution was used instead.

7.2.3.4 Anomalous Propagation Clutter

Under certain meteorological conditions, an unusually high refractivity gradient may cause the radar beam from low elevation tilts to bend more than usual, illuminating surface objects and causing more widespread clutter than anticipated by the clutter maps described above. This "anomalous propagation" (AP) clutter may not always be properly identified and filtered. The Radar Echo Classifier (REC; Kessinger et al. 2003; Williams et al. 2009) is a fuzzy-logic algorithm that uses local features of the reflectivity, radial velocity, and SW fields to discriminate weather echoes from measurements contaminated by ground clutter. The NTDA scales the REC's 0-1 clutter likelihood output, assigning an SW proto-confidence of 1 for small REC values (low likelihood of AP clutter) and falling to 0 for large REC values.

7.2.3.5 SW Texture

Despite the NTDA's attempts to account for various known sources of contamination, some corrupted SWs may pass through the previous quality control steps without detection. Thus, additional steps are taken to evaluate whether patterns in the data appear unphysical. If one assumes that the "true" turbulent wind field SW (i.e., the weighted standard deviation of radial velocities over the radar illumination volume) changes only slowly in space, then the standard deviation of the SW measurements over a local spatial area may provide a good approximation to the SW estimation error. Thus, a high local variance in SW values may indicate the presence of poor quality measurements. The SW variance over a range-azimuth "patch" of radius 1.0 km is mapped to an SW proto-confidence by assigning small values a confidence of 1 and very large values a confidence of 0.

The variance of the residuals from a local linear fit of the SW measurements is also computed. The residuals are small if the variability in the SW is due to a linear change, but may be large if considerable nonlinearity or random noise is present, indicating a possible quality problem. The linear fit is computed along radials only, for efficiency, over a radius of 1.0 km. The confidence interest map assigns SW proto-confidence 1 to small residual variance values and 0 to large values. A second quality indicator is derived from dividing the linear fit residual value at the center point by the residuals' standard deviation (i.e., the residual value's "z-score"). A high z-score may indicate an outlier relative to neighboring SW values. However, dividing by small standard deviations can lead to misleading results, so a minimum denominator of 0.5 $m\ s^{-1}$ is enforced. The modified z-score-based confidence is computed by assigning small z-score values an SW proto-confidence of 1 and large values a proto-confidence of 0. For all three SW texture-based proto-confidence fields, interest map thresholds were fine-tuned using case studies and statistics over a large number of cases, ensuring that good data will rarely be eliminated by these filters.

7.2.3.6 Flying Fauna

Summertime Doppler radar data from the Midwest USA often exhibit very high values of SWs in the boundary layer in low-reflectivity or "clear air" conditions. It is hypothesized that these spurious values result from swarms of insects, bats and birds, which do not act as good tracers of the wind field. Hence, a confidence interest function was designed to assign low confidence to measurements near the ground having low reflectivity, with lower reflectivity values being permitted as the height above ground increases. Here the height of a measurement point above the ground, HT (km), is determined using a fast but accurate approximation to the "4/3 earth model" (Doviak and Zrnić 1993) to account for radar beam bending. The reflectivity-height interest maps are based on the quantity dBZ + 3.5 HT, with low values assigned proto-confidence 0 and larger values assigned proto-confidence 1.

7.2.3.7 Measurement Range from Radar

The decrease in the quality of measured SWs due to generally lower values of SNR and PR as the range from the radar increases has already been accounted for. In addition, the broadening radar beam begins to span more altitudes at greater distances, capturing more wind shear and thus increasing values of SW. In principle, it may be possible to compute the vertical shear in the radial velocity field and adjust the SWs to remove its effect, but this is not currently done. On the other hand, wider measurement volumes at larger ranges are more likely to be only partially filled by hydrometeors, or to have inhomogeneous reflectivity, decreasing the SW. Thus, the range-based interest map assigns SW proto-confidence 1 near the radar and then eventually diminishes with increasing range. Spurious values of SW are often observed at ranges slightly more distant than the unambiguous range, possibly due to undetected overlaid echoes, so those ranges are assigned even smaller confidences.

7.2.3.8 Final SW Confidence Calculation

In order to produce a single confidence value for each SW measurement, the values of all the individual SW proto-confidence indicators described above are combined using a method similar to a geometric average or Naïve Bayes approach: they are multiplied together, but with different exponents for the different confidence measures based on their "importance." Denoting the proto-confidences as $C_1, C_2, \ldots, C_9$ and the corresponding exponents as $p_1, p_2, \ldots, p_9$, the final SW confidence at each range and azimuth location is

$$C_{\mathrm{SW}} = C_1^{p_1}\, C_2^{p_2} \ldots C_9^{p_9} \tag{7.3}$$

Thus, if any one of the SW proto-confidences is zero, the final SW confidence also becomes zero; if all the proto-confidences are 1, then $C_{\mathrm{SW}} = 1$; and if all of the values are between 0 and 1, then C_{SW} will also be somewhere in between.

For split cuts—the separate surveillance and Doppler sweeps performed at the smallest elevation angles of most volume coverage patterns (VCPs)—the combined confidence computation shown in Eq. (7.3) is performed in two steps. First, the product of all the confidences derived from radar reflectivity is replaced with the minimum value over a disc having radius 1.5 km. This "smearing" of low confidence accommodates the imperfect alignment between the reflectivity and Doppler azimuths, as well as the advection that may occur between the surveillance and Doppler sweeps. The "smeared" reflectivity-based SW confidence is then combined with the product involving the other proto-confidences.

7.2.3.9 Local Coverage Calculation

Because turbulence is a statistical quantity, it can only be meaningfully computed by averaging at least several measurements having some degree of independence. Averaging is also important for diminishing the effects of random noise in the constituent SW measurements. Furthermore, a region in which many SW values are missing is typically suspect, since the remaining values may be contaminated by the same source that led to their neighbors being censored. Hence, another indicator of confidence for the final averaged EDR, described below, is given by the local "coverage" of the 2.0 km averaging disc by valid (confidence above a specified threshold) SW measurements. The ratio of the number of valid SW measurements to the total number of points in the disc is passed through a piecewise linear interest map to yield the "coverage" confidence, C_{cov}, ranging from 0 for small ratios to 1 for coverage ratios near 1.

7.2.3.10 Contamination Sources Not Addressed

Since the NTDA was designed primarily to detect convective turbulence for en route aircraft, it does not address a number of factors that can compromise SW measurements at low altitudes or outside convective clouds. For example, Melnikov and Doviak (2009) used special vertical radar scans to show that thin shear layers in stratiform clouds cause high SW measurements that do not appear to correspond to severe turbulence; furthermore, the effect of the shear across the measurement volume cannot easily be estimated when a NEXRAD is operating in a normal VCP with sweeps at fixed elevations. Thus, it is likely that the current NTDA overestimates turbulence under these circumstances. Istok and Doviak (1986) showed that coherent shear regions are rare in thunderstorms, and Fang and Doviak (2008) showed that attempting to adjust the SW for the shear contamination using only NEXRAD volume data was difficult, producing unphysical negative values. A future version of NTDA might mitigate these problems by detecting and censoring coherent radial areas of high along-beam shear or attempting to adjust the measured SW for the shear.

Similarly, nonuniform reflectivity over the radar measurement volume (e.g., due to strong gradients at cloud edges) reduces the effective measurement volume, tending to reduce the measured SW. Like shear, this effect is difficult to quantify and thus is another source of error that is not accounted for.

A third issue that affects the interpretation of SW measurements as turbulence is the variation in hydrometeor inertia and fall speeds. For small droplets, the assumption that the hydrometeors track the wind field is well justified; therefore, the SW is indicative of the variability of the wind within the radar measurement volume. For large, heavy raindrops or hail, this assumption may break down, and the radar SW measurement may underestimate the true value of the radial wind's standard deviation. A second effect is caused by hydrometeors falling across the radar measurement volume: those near the top of the beam have greater radial velocity than those lower, causing variability in radial velocity across the radar beam that broadens the SW. Neither hydrometeor inertial effects nor fall speeds are accounted for in the NTDA. The justifications for this omission are (1) that NTDA was designed to work with ground-based radars that typically operate with small elevation angles, and (2) that NTDA was developed for aviation users who would avoid high reflectivity regions indicative of heavy rain or hail, making accurate turbulence detection in those regions operationally insignificant. However, with the advent of dual-polarization data, this issue could be more fully addressed.

Finally, SWs can be significantly contaminated by radio frequency interference (RFI) from other radars or microwave sources. In the USA, NEXRADs are assigned slightly different frequencies from one another to mitigate contaminating each other's measurements, but RFI has recently begun to appear as radial artifacts in data from U.S. NEXRADs and occurs even more frequently in Taiwan radar data. Fortunately, RFI often resembles clutter and is mitigated by clutter filtering or flagged as AP by the REC, while remaining radial artifacts in the SW data are mitigated by the texture-based interest maps. As a result, numerous case studies have shown that RFI contamination in NTDA output is almost completely mitigated. Wind farms can cause similar radial artifacts when the radar signal reflects from blades and other targets and may cause a more significant source of contamination in the future as wind farms proliferate. Thus, it may be useful for a future version of NTDA to use image processing techniques to explicitly identify and mitigate radial artifacts.

7.3 Converting SW to EDR

7.3.1 Adjusting the Measured SW

When the radar beam rotates rapidly, the motion of the dish itself causes broadening of the SW measurement, as one side of the dish moves toward the target and the other side moves away. This positive, additive error is proportional to the radar rotation speed and should be subtracted from the measured SWs prior to their scaling into turbulence values (described below); however, this adjustment will sometimes cause unphysical negative values in the final averaged EDR, which must be replaced with 0. The NTDA allows the user to supply the constant of proportionality for the radar rotation rate adjustment, though this is set to 0 in the examples provided below, meaning that the NTDA EDR may be biased high by slightly different amounts depending on the tilt and VCP.

7.3.2 Scaling SW

For any SW measurement that has not been removed by censoring or assigned zero confidence, a "raw" (as opposed to final) eddy dissipation rate, EDR_{raw}, is estimated by multiplying the adjusted SW by a "scaling" factor dependent on range. The scaling factor, depicted in Fig. 7.2, is largest near the radar, where the radar pulse volume is small and the measured SW is therefore much smaller than the wind field's standard deviation. It then decreases with range from the radar as the wind field (including non-turbulence wind shear) is sampled more and more. The scale factor was computed via a theoretical formula derived under the assumptions that the radar illumination function is a 3-D Gaussian, that radar reflectivity is uniform within the illumination volume, and that the turbulence has a Kolmogorov energy spectrum; see Cornman et al. (1996) and Chap. 5. The use of the infinite outer

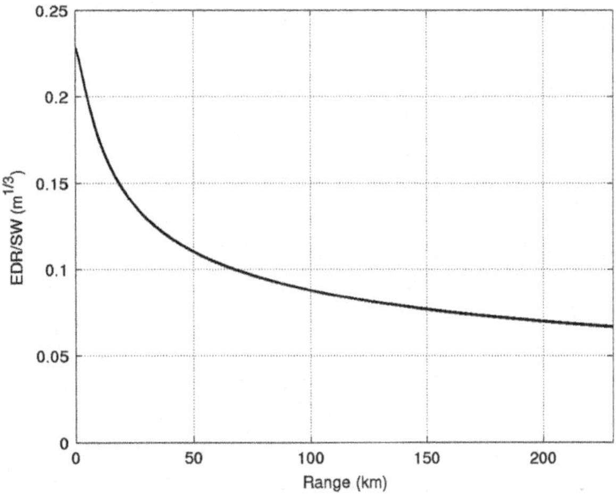

Fig. 7.2 NTDA's scaling factor for mapping SW to "raw" EDR as a function of range from the radar

length scale was chosen by analyzing the relationship between a large set of commercial aircraft in situ EDR measurements (Cornman et al. 1995; Sharman et al. 2014) with collocated radar SW values as a function of range from the radar.

7.3.3 Computing EDR

Turbulence theory relates the ensemble average of the measured second moment of isotropic, homogeneous turbulence to the EDR (see Chap. 6). The real atmosphere, particularly under disturbance by clouds and convection, is neither isotropic, homogeneous, nor stationary. Implicit in the NTDA's approach is the assumption that the turbulence is at least locally homogeneous and approximately stationary over the fraction of a second it takes the radar to scan a region. Thus, a spatial average over a small range and azimuth domain approximates an ensemble average over multiple realizations of the turbulent wind field. The assumption of isotropy is more problematic; measurements with instrumented aircraft show that atmospheric turbulence is often non-isotropic (e.g., Sharman and Frehlich 2003). This is an issue because variability in the vertical component of the wind field causes the most hazardous aircraft response, whereas ground-based radar measures predominantly the horizontal wind variations. The NTDA deals with these issues by (1) using empirical data—aircraft measurements of turbulence—to validate the choice of the SW to EDR scaling function; (2) using local range, azimuth disc averaging of the resulting EDR estimates to create its turbulence measurement; and (3) combining data from multiple radars when possible in the final 3-D mosaic provided to users. However, the non-homogeneity and non-isotropy of convective wind fields likely limit the NTDA's performance in comparison to aircraft-measured turbulence.

Turbulence theory suggests that the final EDR estimate should be the root-mean-squared (RMS) average of the raw EDR estimates, EDR_{raw}, and the original 2007 version of the NTDA was deployed with a confidence-weighted RMS averaging technique. However, since contamination of SW estimates generally leads to enlarged values and the RMS average gives outsized weight to the highest values, it was found that the RMS average values were undesirably sensitive to undetected contamination. As a result, the current version of the NTDA uses a simple confidence-weighted average of the raw EDRs ($\varepsilon^{1/3}$) over a range, azimuth disc surrounding the point of interest. Although the average would be biased low if the measured SW values were unbiased, in practice, this approach counteracts a generally high bias in the SW estimates. As described above, the final NTDA EDR is associated with a confidence value, which is obtained from the disc average of the individual SW confidences, multiplied by the local coverage confidence. Given an output polar grid point (r, az) and a minimum confidence threshold $C_{\min} \geq 0$, we denote the set of all ranges and azimuths within a disk of radius 2 km that have $C_{\rm SW} > C_{\min}$ as $D_{\rm 2km}(r, az)$. Then the final NTDA EDR and confidence calculations are represented by the formulas

$$\mathrm{EDR}(r, az) = \frac{\sum_{(r', az') \in \mathrm{D}_{2\mathrm{km}}(r, az)} C_{\mathrm{SW}}(r', az')\ \mathrm{EDR}_{\mathrm{raw}}(r', az')}{\sum_{(r', az') \in \mathrm{D}_{2\mathrm{km}}(r, az)} \mathrm{C}_{\mathrm{SW}}(r', az')} \tag{7.4}$$

and

$$C(r, az) = C_{\mathrm{cov}}(r, az) \times \frac{\sum_{(r', az') \in \mathrm{D}_{2\mathrm{km}}(r, az)} C_{\mathrm{SW}}(r', az')}{\sum_{(r', az') \in D_{2\mathrm{km}}(r, az)} 1} \tag{7.5}$$

In Eq. (7.5), the denominator represents the number of points in $D_{\rm 2km}(r, az)$. The NTDA typically produces final EDR estimates and associated confidences on a subset of the radar's Doppler moment polar grid, for instance, at intervals of 1° in azimuth and range multiples of 1 km (every fourth NEXRAD Doppler range gate) up to the range at which the radar beam exceeds a specified MSL altitude, e.g., 55,000 ft.

7.4 Real-time Processing and Mosaic

In order to be useful to aviation users, the polar grids of EDR and confidences produced for each radar tilt must be combined into a 3-D, gridded map, or "mosaic." The NTDA's mosaicking algorithm works as follows:

1. At specified time intervals (e.g., every 5 min), a set of 3-D latitude–longitude–altitude grids (denoted A, B, and C) is initialized with zeros. These grids will store the results from different computations. Typically, the NTDA mosaic grid

uses a horizontal resolution of 0.02° (approximately 2 km) and a vertical resolution of 3000 ft up to 45,000 ft, but these are changeable parameters.

2. The latitude, longitude, and MSL altitude of each relevant polar-coordinate radar data point is computed, with altitude adjusted for beam bending using a modified version of the 4/3 earth model. In the future, an estimate of atmospheric conditions based on NWP model output could be used to make more accurate estimates of the beam height.
3. Radar data including reflectivity, velocity, SW, and radar operational mode metadata arrive asynchronously and are immediately parsed and processed by the NTDA EDR computation described above, producing EDR and confidence values for each elevation tilt on a polar grid.
4. As the NTDA EDR and confidence for a radar tilt become available, a loop through ranges and azimuths is performed, and each EDR measurement having confidence over a specified threshold is "pushed" onto the latitude–longitude–altitude grids as follows: a distance weight comprising a horizontal Gaussian range of influence and a vertical weighting function intended to replicate a linear interpolation with radar tilts above and below is added to each nearby point on grid A; this distance weight times the EDR confidence is added to the points on grid B; and the product of the distance weight, the EDR confidence, and the EDR value itself is added to grid C.
5. At a specified completion time (e.g., every 5 min), the mosaic algorithm tabulates the tilts from each radar that have been merged into the grids. If any tilts in a radar's VCP are missing, the most recent one within a specified lookback period is located and step (4) is executed.
6. The confidence-weighted EDR mosaic values are obtained by dividing the values in grid C (total distance weights times confidences times EDRs) by those in grid B (total distance weights times confidences); the associated confidences are obtained by dividing grid B by grid A (total distance weights).
7. For each latitude–longitude location in the grid, the highest altitude with EDR confidence above a given threshold is determined, and these are then 2D filtered with a 90th percentile function over a sliding rectangular window to find a "top" for the NTDA measurements. Columns whose highest altitude value is below this "top" have their topmost value replicated to fill in up to this top. This step avoids the "sawtooth" pattern that otherwise results from the radar's scanning geometry.

A significant limitation of the NTDA is that it is only able to detect turbulence within "in-cloud" regions where radar reflectivity is sufficiently strong to produce reliable Doppler measurements and where signal contamination due to ground clutter, second-trip echoes, or other factors is minimal. It is therefore typical for much of the NTDA EDR mosaic to contain no good data. However, it is important to recognize that "no data" does not mean "no turbulence"; rather, no turbulence information at all is provided by the NTDA in these regions.

7.5 NTDA Evaluation

7.5.1 NASA Turbulence Flight Tests

In the spring of 2002, eleven research flights were performed by the instrumented NASA Langley Boeing 757 (B757) aircraft as part of a successful test of an airborne radar turbulence detection algorithm (Cornman et al. 2003). The high-rate wind data recorded by the aircraft comprise a high-quality in situ dataset that is also ideal for evaluating the performance of the NTDA, which was run on archived Level II data from NEXRAD radars along the flight paths. This dataset is unique in that the B-757 was deliberately maneuvered into developing thunderstorms and their associated turbulence. Commercial pilots typically avoid regions of likely convective turbulence, making commercial aircraft in situ EDR datasets much more difficult to use for NTDA verification.

The B-757's 20 Hz vertical wind data were used to estimate EDR using a single parameter maximum likelihood model that assumes a von Karman energy spectrum form (Sharman et al. 2014). In particular, a sliding window of length 256 points was used, with spectral frequency cutoffs set at 0.5 and 5 Hz. This temporal window corresponds to an along-path distance of about 3 km at the aircraft's average cruising speed, though the windowing function applied before taking the Fourier transform reduces the effective path somewhat. The NTDA was run on Level II data from nearby NEXRADs obtained from the National Climatic Data Center archive. A number of overlay plots such as the one shown in Fig. 7.3 were created. In this

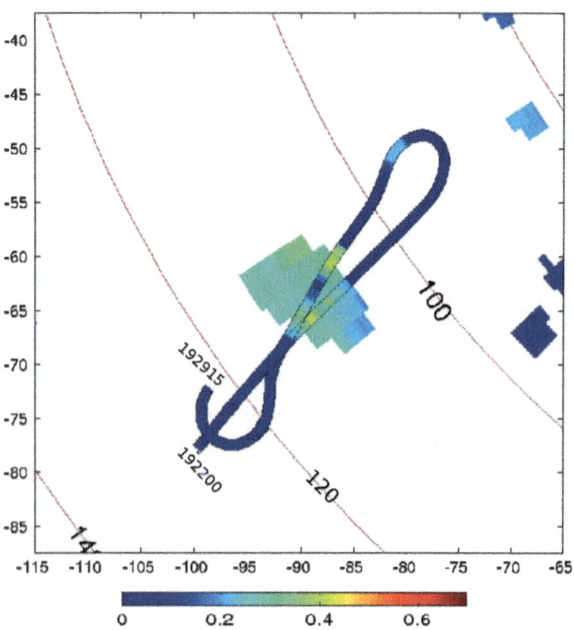

Fig. 7.3 NASA B757 track from 19:22:00-19:29:15 UTC, color shaded by in situ EDR (0–0.7 $m^{2/3}\ s^{-1}$), overlaid on similarly scaled NTDA EDR values from the KLTX (Wilmington, NC) 2.4° elevation sweep beginning at 19:25:26 UTC. The labels on the range rings and the axes represent the distance from KLTX in km

case, which shows good correlation of aircraft and radar turbulence measurements, the aircraft is within about 1 km vertically and within about 4 min of the fixed-elevation radar sweep throughout this flight segment, and the radar reflectivity ranges from about 5-30 dBZ (not shown) within the turbulent region.

The full set of comparison plots was analyzed to subjectively score the ability of the NTDA to detect moderate-or-greater turbulence, defined here as EDR ≥ 0.3 $m^{2/3}$ s^{-1}, encountered by the aircraft from 55 flight segment "events" over the eleven flights of the NASA flight test. A similar scoring exercise performed using the output of the airborne radar turbulence detection algorithm identified 34 correct detections, 8 misses, 4 nuisance alerts, and 9 correct nulls (Cornman et al. 2003), producing a probability of detection (PoD) of 81 % with a false alarm rate (FAR) of 11 %. For the NTDA analysis, 15 events had no available archived NEXRAD data intersecting them. Of the remaining 40, subjective scoring identified 32 correct detections, 2 misses, and 6 false alarms, yielding a PoD of 94 % and a FAR of 16 %. This analysis suggests that the NTDA may have skill comparable to that of the airborne radar algorithm for detecting hazardous turbulence. However, care must be taken in extrapolating these results to commercial aircraft operations.

An objective statistical evaluation was performed by extracting the median NTDA EDR value from a disc of radius 2 km around each aircraft location as projected onto each nearby radar sweep. A scatterplot of the aircraft EDR vs. collocated NTDA EDR is shown in Fig. 7.4, with the EDR confidence represented by color shading of the scatterplot points. Here only comparisons within a spatial distance of 0.5 km to the radar sweep and a time difference of 60 s are included. Showing only comparison points with confidence greater than 0.5 (bottom plot) removes many of the outliers, and the remaining points have higher correlation and better proximity to the 1:1 line. This confirms that the EDR confidence is functioning as designed. Receiver operating characteristic curves produced from these data showed that the NTDA achieved a PoD of 80 % for turbulence exceeding 0.15, 0.35, and 0.55 $m^{2/3}$ s^{-1} (roughly light, moderate, and severe turbulence) thresholds with accompanying false alarm rates of 24 %, 23 %, and 16 %, respectively. Though not quite as good as the subjective scoring, these results are still impressive given the exacting point-to-point nature of the comparison in contrast to the aggregation of turbulence "events." Williams et al. (2005) provide statistical comparisons of NTDA EDR with in situ reports from commercial aircraft. While not as compelling as the research flight results, the statistics nevertheless show that the NTDA EDR has value in regions that pilots evidently feel are safe to penetrate.

7.5.2 Operational Demonstration

Between 2005 and 2007, the FAA Aviation Weather Research Program sponsored a demonstration in which real-time NTDA data were provided to United Airlines dispatchers and en route pilots. The demonstration began with 16 NEXRADs

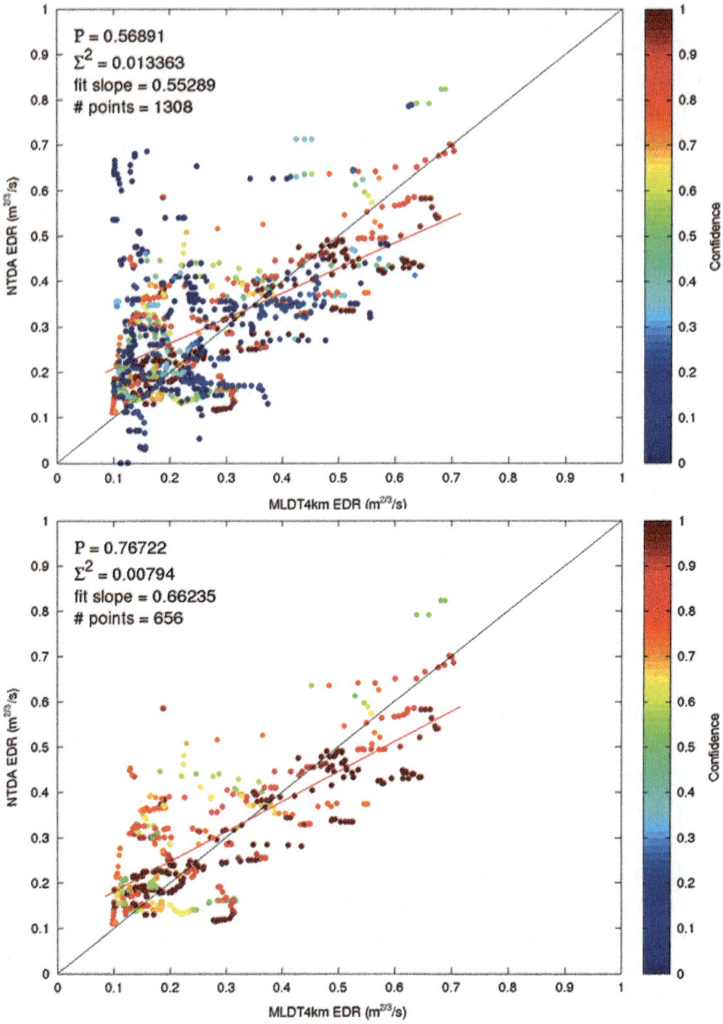

Fig. 7.4 Scatterplot of NTDA EDR vs. in situ EDR computed from the NASA B757 during a 2002 flight test, with associated confidence represented by the colorscale shown at right. Each aircraft EDR is compared to an NTDA EDR polar grid within 0.5 km and 60 s of the aircraft measurement time. (*Top*) All points; (*bottom*) only NTDA EDR points having confidence > 0.5

around Chicago and then grew to cover most of the CONUS east of the Rockies in the final year. NTDA EDR and similarly mosaicked reflectivity data were provided to dispatchers via a Java web-based display. A sample of the display is shown in Figs. 7.5 and 7.6, showing reflectivity and NTDA-derived turbulence categories for a storm complex spanning eastern Colorado and western Kansas at 22:30 UTC on 28 September 2007. The plan views in Fig. 7.5 show that the severe and some of the moderate turbulence at 24,000 ft are near areas of elevated reflectivity. However,

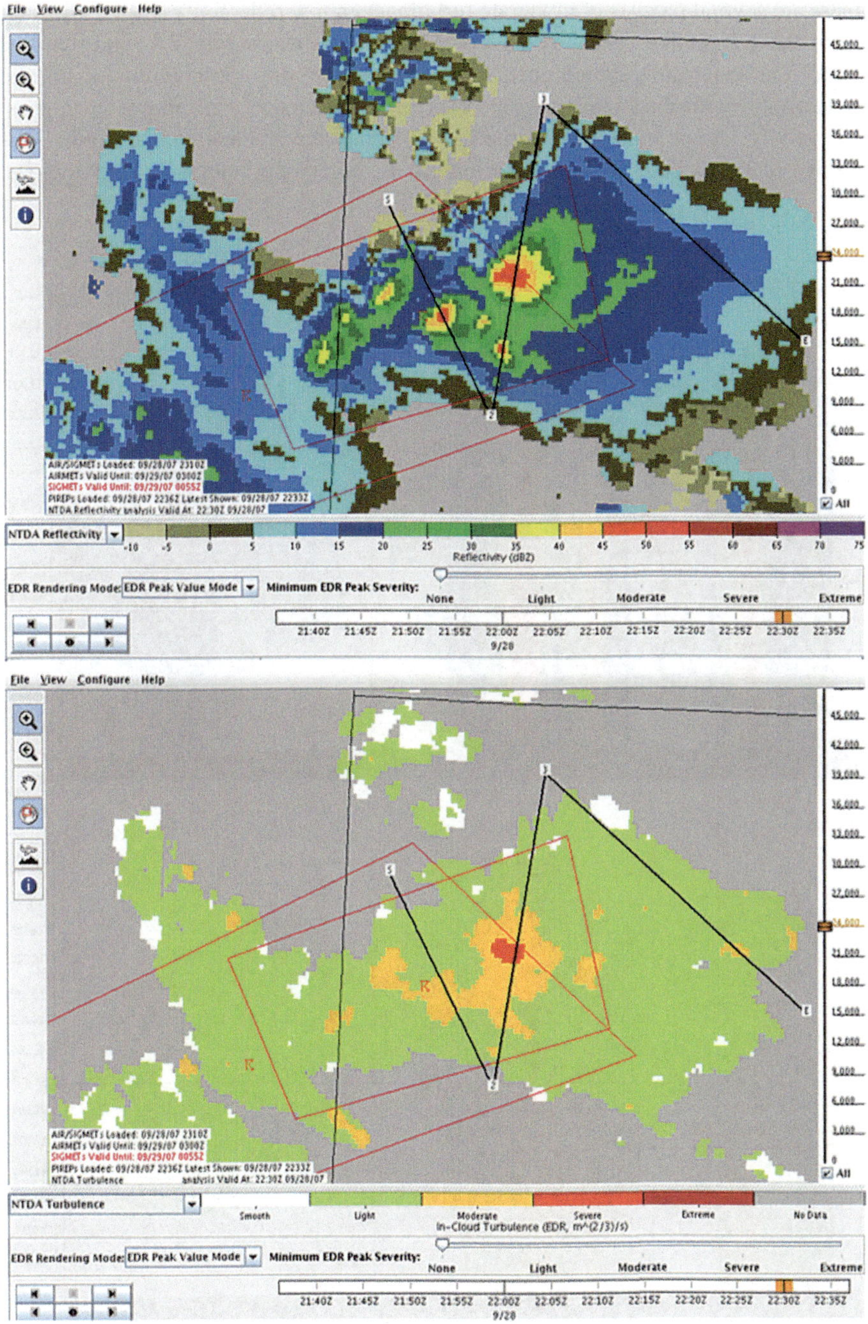

Fig. 7.5 Real-time Java display showing a plan view at 24,000 ft of reflectivity (*top*) and NTDA turbulence (*bottom*) valid at 22:30 UTC on 28 September 2007. Here turbulence has been color shaded as Smooth, Light, Moderate, Severe or Extreme to facilitate easy interpretation by aviation users; grey shading indicates regions where no NTDA data is available. The black line indicates the path for a vertical cross-section view

there are several patches of moderate turbulence in low reflectivity anvil regions far from the convective cores. The vertical cross-sections in Fig. 7.6 reinforce this point: the severe turbulence occurs in a "V" shape atop the columns of highest reflectivity, with the largest regions of severe turbulence well above the highest reflectivity. Again, moderate turbulence patches are evident in the anvil. This ability of the NTDA to detect turbulence in otherwise benign-looking regions,

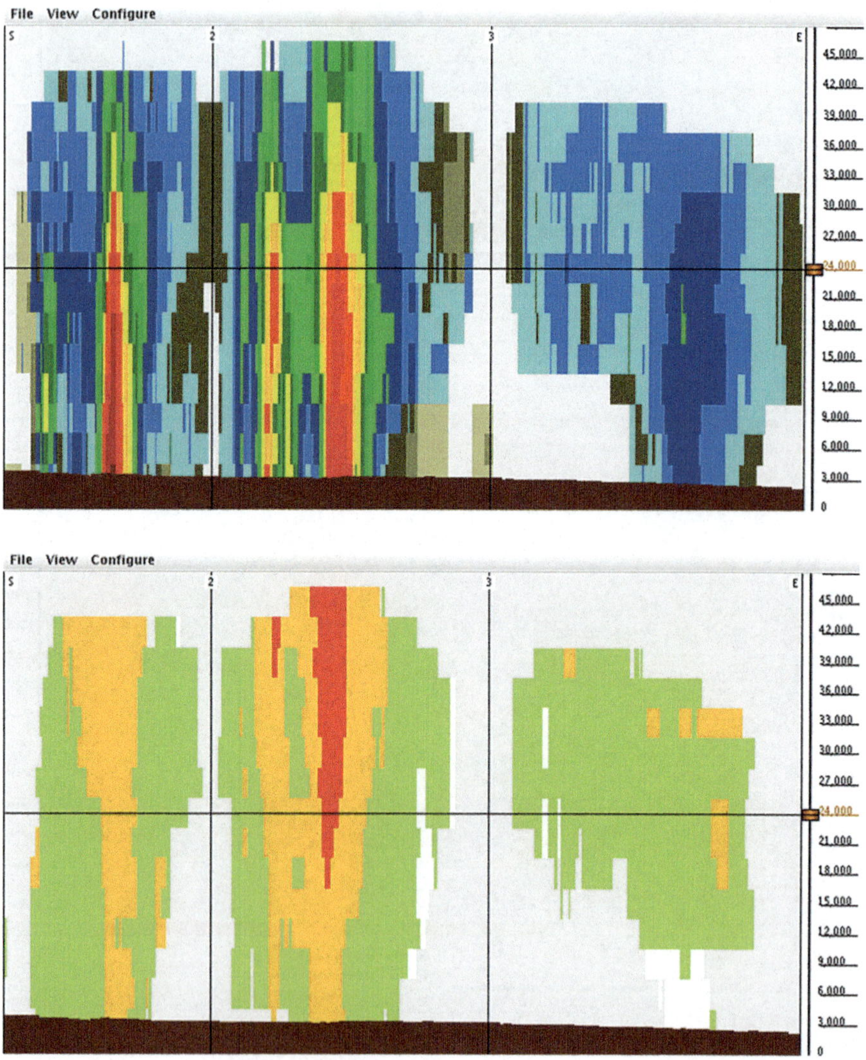

Fig. 7.6 Real-time Java display showing cross-section views of reflectivity (*top*) and NTDA turbulence (*bottom*) valid at 22:30 UTC on 28 September 2007. The cross-section is along the path indicated in Fig. 7.5, and color shading is identical

including those with reflectivity below the threshold commonly displayed on airborne radar cockpit displays, was repeatedly illustrated in the demonstration.

In addition to the Java display, text maps showing horizontal and vertical cross-sections of NTDA turbulence 0 to 100 nautical miles ahead were uplinked to the cockpit Aircraft Communications Addressing and Reporting System (ACARS) printer of participating aircraft piloted by Line Check Airmen. The system generated text maps for every United Airlines aircraft every 5 min, but they were only uplinked if the turbulence proximity, severity, and extent met certain predefined thresholds. After a flight, pilots could visit an NCAR webpage to review all the text maps generated for their flight and provide feedback. Pilots' comments were generally quite positive. One pilot wrote, "The uplink messages I've received in the cockpit gave a very accurate picture of turbulence location and intensity. The turbulence level of detail is beyond what current onboard weather radars can detect." A second pilot commented, "In most of the instances of reports I have received there were radar returns that I was watching in the aircraft and was aware of the probability of turb. In this instance there were NO radar returns.... When the report printed I was rather surprised to see one. But the accuracy was right on for all four reports. At 2345Z we had about 30 s of mod chop. Lt/mod chop started at 2346Z as noted on the 2343 report.... The noted "M" at 2354Z seemed right on. There was lightning well below us but no radar returns...." ("M" was the character used to represent moderate turbulence on the text maps.) Williams et al. (2006) describe the demonstration and provide some early statistical performance results; Craig et al. (2008) provide additional details. The demonstration clearly showed the value of NTDA turbulence information to improve pilot situational awareness. As of this writing in 2015, the FAA is again studying uplinking NTDA tactical turbulence information to pilots via onboard WiFi and iPad-type displays as part of the Weather Technology in the Cockpit Program (Lindholm et al. 2015).

7.5.3 Case Studies

7.5.3.1 United Airlines Flight 1727

On 4 April 2012, en route from Tampa, Florida to Houston, Texas, United Airlines Flight 1727 encountered severe turbulence over the northern Gulf of Mexico south of Louisiana. Aircraft Situation Display to Industry (ASDI) data showed that the altitude of the Boeing 737 decreased from flight level (FL) 380 (about 38,000 ft MSL) to FL 321 in one minute near 11:57 UTC, suggesting either temporary loss of control or a very sudden descent. The pilots declared an emergency, and emergency personnel met the aircraft when it landed in Houston at 7:47 am CDT. A United Airlines spokesman reported that five passengers and two flight attendants were injured and that at least three were transported to a hospital.

The turbulence encounter occurred as the aircraft maneuvered between two cells of a mesoscale convective system (MCS) that was developing southward. While it did so, the aircraft's path penetrated the southern edge of a region of NTDA-detected severe turbulence. As shown by the plan view of reflectivity in Fig. 7.7, the reflectivity between the two cells at 36,000 ft was about 21 dBZ at 11:50 UTC, likely too low to be displayed on the airborne radar display. Figure 7.8 shows a

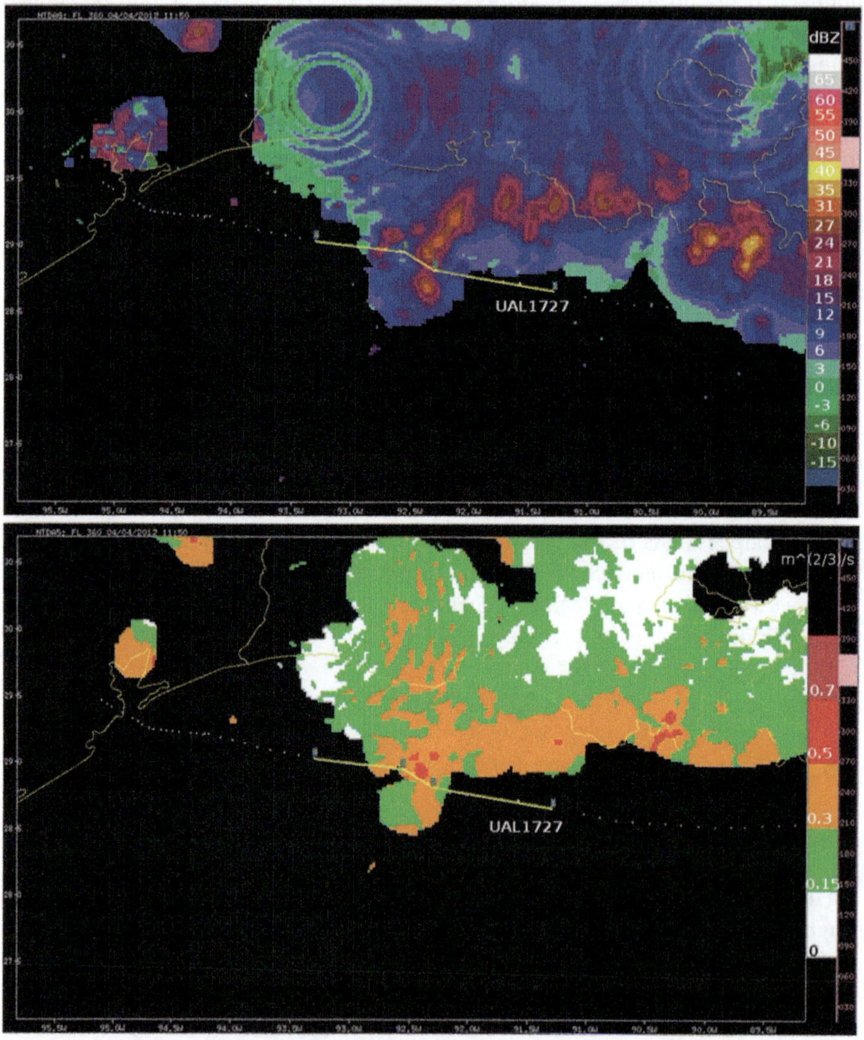

Fig. 7.7 Plan view at 36,000 ft of reflectivity (*top*) and NTDA turbulence (*bottom*) valid at 11:50 UTC on 4 April 2012. The small white dots show ASDI positions of United Airlines Flight 1727, flying east to west, and the yellow line indicates the path for a vertical cross-section. Colors in the bottom panel represent smooth (*white*), light (*green*), moderate (*orange*), severe (*red*), and extreme (*pink*) turbulence.

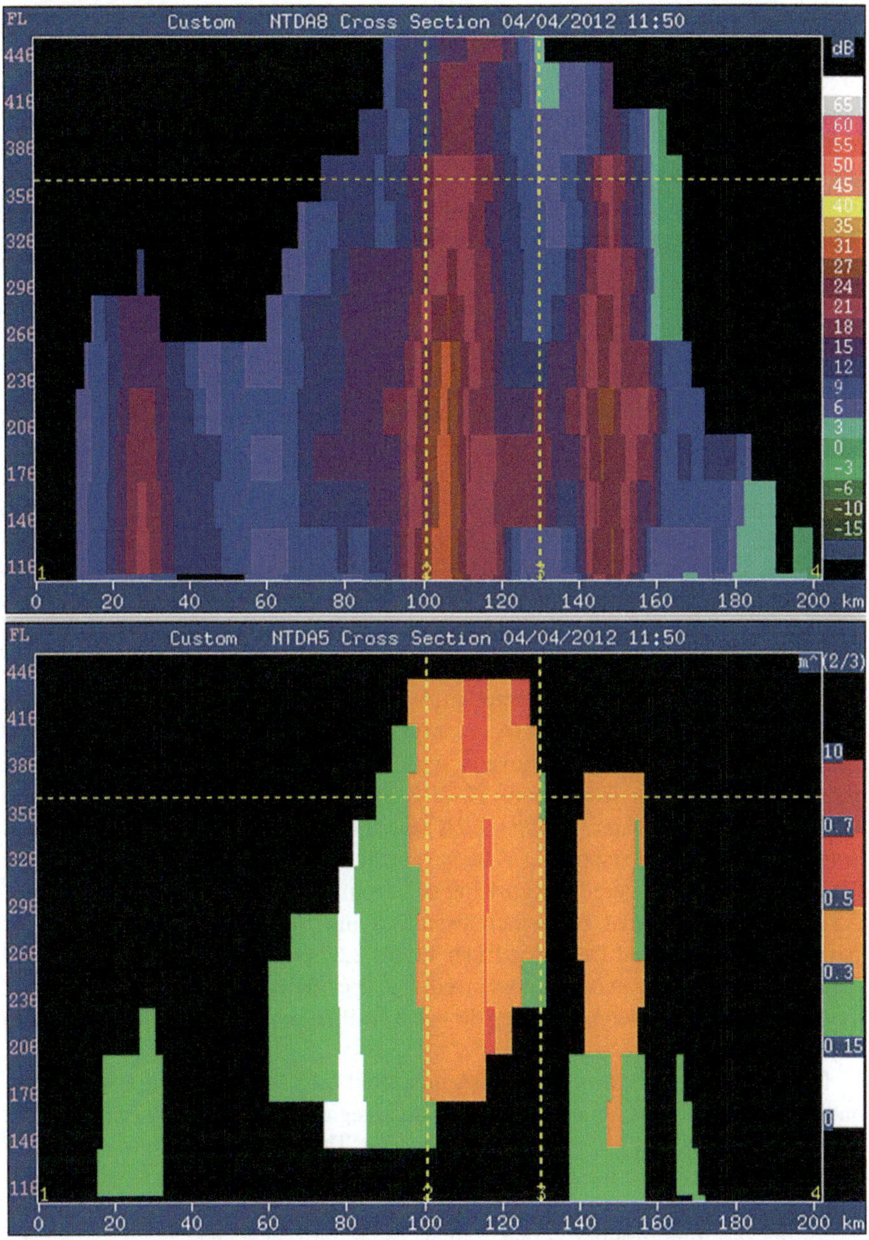

Fig. 7.8 Vertical cross-section views of reflectivity (*top*) and NTDA turbulence (*bottom*) for altitudes of 10,000 ft and above valid at 22:30 UTC on 28 September 2007. The cross-section is along the path indicated in Fig. 7.7, and the color scales are identical. The vertical and horizontal dashed yellow lines indicate the vertices of the cross-section path and the 36,000 ft altitude of the plan view, respectively

cross section along the yellow line in Fig. 7.7, which approximates the flight's east to west path. At an altitude of FL 380, the aircraft flew directly through the column of severe turbulence in the center of the short flight segment. Running on NCAR's real-time prototype system, the NTDA EDR valid at 11:50 UTC, shown here, would be available for transmission by 11:51 UTC, allowing several minutes for the information on the turbulence hazard to be conveyed to the pilot. This case again illustrates that the location of enhanced reflectivity regions may not provide accurate guidance on the location or severity of convective turbulence.

7.5.3.2 United Airlines Flight 967

On July 21, 2010, en route from Washington, D.C. to Los Angeles, California, United Airlines Flight 967 encountered severe turbulence over west central Missouri at about 00:14 UTC. A passenger was quoted by *The Washington Post* as saying that the plane dropped and then "slammed" back up. The flight was diverted to Denver. A United Airlines spokesman said that 25 people, including four crew members, were injured, and a Denver Health spokeswoman reported that 21 people were transported to local area hospitals.

Flight 967 was in cruise at FL 340 south of a large MCS that extended from eastern Kansas to central Missouri when the encounter occurred. The aircraft was flying in the MCS anvil, with reflectivities of about 20 dBZ along the flight track and slightly higher values below. A radar reflectivity cross section revealed that the encounter location was just above a rapidly developing storm cell, with core reflectivity increasing from about 27 dBZ around the time of the encounter to 40 dBZ five minutes later, during which time the top of the cell rose rapidly from within the anvil to overshooting it by several thousand feet. Low visibility in the MCS anvil likely obscured the growing cell from the pilots. However, analysis of ASDI data showed that two other aircraft flying in the same area, Southwest Airlines Flight 481 and Delta Air Lines Flight 2275, changed course and avoided the location where the encounter occurred. It is speculated that the other flights may have detected the growing cell using downward-tilted onboard radar and diverted to avoid it.

In this case (not shown), the NTDA measured only light turbulence at the time and location of the encounter, though a large region of moderate turbulence was detected at the site within the 10 min following it. Thus, this case offers a cautionary tale about the limits of the NTDA's capabilities: it may not detect hazards associated with the organized kinematics of a young, developing cell, even when it does detect the resulting turbulence that evolves several minutes later.

7.6 Conclusion

The NCAR/NEXRAD Turbulence Detection Algorithm is designed to provide a rapid-update, 3-D, wide-area view of in-cloud turbulence that can promote improved, common situational awareness between pilots, dispatchers, and air traffic controllers. An essential component of the NTDA is careful quality control of radar spectrum width (SW), which provides valuable measurements of the wind field's variability but is very sensitive to contamination. NTDA employs a fuzzy logic quality control algorithm, assigning a "confidence" to each SW measurement by evaluating factors that might cause contamination as well as features and patterns in the SW data. SW measurements are scaled to EDR using a range-dependent function that takes into account the radar measurement volume's filtering of the wind field as a function of range from the radar. The scaled values are then combined using a local confidence-weighted average, and a confidence is assigned to the final EDR based on the individual SW confidences and the local coverage of good SW data. The NTDA is accompanied by a mosaic technique that merges EDR and associated confidence data from multiple radars to create 3-D grids of turbulence and confidence. The mosaic procedure provides an additional averaging step that improves the EDR quality and presents the results in a Cartesian format appropriate for aviation users.

The NTDA and mosaic software have been highly optimized; they run in real time on a single multi-processor server at NCAR, ingesting and processing data from over 140 NEXRADs and producing a CONUS mosaic every 5 min with minimal latency. The NTDA was also operationally deployed in Taiwan in 2014 as part of the Taiwan Civil Aeronautics Administration's Advanced Operational Aviation Weather System, providing in-cloud turbulence detection for the island and surrounding oceanic airspace.

The NTDA product has been verified via comparison with research flight data and commercial aircraft turbulence reports and validated in an operational demonstration and numerous turbulence accident case studies. In addition to its value as a standalone product, the NTDA output can be combined with satellite, in situ, and numerical weather prediction model data to identify and forecast regions of hazardous convectively induced turbulence both in and near storms (Williams et al. 2011; Williams 2014). The resulting turbulence "nowcast" capability could significantly improve aviation safety, air traffic flow, and passenger comfort during convective events.

In addition to its value to the aviation community, NTDA EDR data provide valuable information about kinematics that is being used to study convective storms. For instance, Deierling et al. (2011) and Al-Momar et al. (2015) use NTDA data to show a relationship between in-cloud turbulence and lightning. Such a relationship may form a basis for using global lightning data to gauge the likelihood of significant aviation turbulence in areas without Doppler weather radar coverage, as discussed further in Chap. 8. It also suggests that NTDA EDR might be useful in predicting the onset of lightning to enhance warnings to the affected public.

Acknowledgements The authors wish to thank NASA Langley Research Center for providing aircraft data from the spring, 2002 Boeing 757 flight tests. Many colleagues assisted with the NTDA project: Jason Craig, Steve Carson, Jaimi Yee, Gary Blackburn, Seth Linden, Andy Cotter, and Shelly Knight helped design, implement, and optimize the real-time NTDA, 3-D mosaic and operational demonstration software system; Larry Cornman and Kent Goodrich provided the turbulence theory on which the SW to EDR scaling function computation was based; and Frank McDonough, Tina Kalb, and Wiebke Deierling performed numerous case studies. We greatly appreciate their contributions.

References

Al-Momar, S., Deierling, W., Williams, J.K.: Examining in-cloud convective turbulence in relation to total lightning and the 3D wind field of severe thunderstorms. In: AMS 17th Conference on Aviation, Range, and Aerospace Meteorology, Paper 5.3 (2015)

Bedka, K.M., Brunner, J., Dworak, R., Feltz, W., Otkin, J., Greenwald, T.: Objective satellite-based overshooting top detection using infrared window channel brightness temperature gradients. J. Appl. Meteor. Climatol. **49**, 181–202 (2010)

Cornman, L.B., Carmichael, B.: Varied research efforts are underway to find means of avoiding air turbulence. ICAO J. **48**, 10–15 (1993)

Cornman, L.B., Morse, C.S., Cunning, G.: Real-time estimation of atmospheric turbulence severity from in-situ aircraft measurements. J. Aircraft **32**, 171–177 (1995)

Cornman, L.B., Goodrich, R.K.: The detection of atmospheric turbulence using Doppler radars. Preprints, Workshop on Wind Shear and Wind Shear Alert Systems. Oklahoma City, 13-15 November. Am. Meteor. Soc., Boston (1996)

Cornman, L.B., Williams, J., Meymaris, G., Chorbajian, B.: Verification of an airborne radar turbulence detection algorithm. In: 6th International Symposium on Tropospheric Profiling: Needs and Technologies, 9–12 (2003)

Craig, J.A., Williams, J.K., Blackburn, G., Linden, S., Stone, R.: Remote detection and real-time alerting for in-cloud turbulence. In: AMS 13th Conference on Aviation, Range and Aerospace Meteorology, Paper 9.4 (2008)

Deierling, W., Williams, J.K.: The relationship of in-cloud convective turbulence to total lightning. In: AMS 15th Conference on Aviation, Range, and Aerospace Meteorology, Paper 2.3 (2011)

Doviak, R.J., Zrnić, D.S.: Doppler Radar and Weather Observations. Academic, San Diego, CA (1993)

Fang, M., Doviak, R.J.: Coupled contributions in the Doppler radar SW equation. J. Atmos. Oceanic Technol. **25**, 2245–2258 (2008)

Fang, M., Doviak, R.J., Melnikov, V.: SW Measured by WSR-88D: Error Sources and Statistics of Various Weather Phenomena. J. Atmos. Oceanic Technol. **21**, 888–904 (2004)

Frehlich, R.G., Yadlowsky, M.J.: Performance of mean-frequency estimators for Doppler radar and lidar. J. Atmos. Oceanic Technol. **11**, 1217–1230; corrigenda, **12**, 445–446 (1994)

Hubbert, J.C., Dixon, M., Ellis, S.M.: Weather radar ground clutter. Part II: Real-time identification and filtering. J. Atmos. Oceanic Technol. **26**, 1181–1197 (2009)

Istok, M.J., Doviak, R.J.: Analysis of the relation between Doppler spectral width and thunderstorm turbulence. J. Atmos. Sci. **43**, 2199–2214 (1986)

Kaplan, M.L., Huffman, A.W., Lux, K.M., Charney, J., Riordan, A.J., Lin, Y.-L.: Characterizing the severe turbulence environments associated with commercial aviation accidents. Part 1: a 44-case study synoptic observational analysis. Meteor. Atmos. Phys. **88**, 129–153 (2005)

Kessinger, C., Ellis, S., Van Andel, J.: The radar echo classifier: a fuzzy logic algorithm for the WSR-88D. In: 3rd AMS Conference on Artificial Intelligence Applications to Environmental Science, Long Beach, 9–13 Feb (2003)

Lane, T.P., Sharman, R.D., Trier, S.B., Fovell, R.G., Williams, J.K.: Recent advances in the understanding of near-cloud turbulence. Bull. Am. Meteor. Soc. **93**, 499–515 (2012)

Lindholm, T.A., Frazier, E., Barron, B., Blackburn, G., Kessinger, C., Delemarre, M., Williams, J.K.: Demonstrating feasibility of tactical turbulence alerts. In: AMS 17th Conference on Aviation, Range and Aerospace Meteorology, Paper 13.3 (2015)

Melnikov, V.M., Doviak, R.J.: Turbulence and wind shear in layers of large Doppler SW in stratiform precipitation. J. Atmos. Oceanic Technol. **26**, 430–443 (2009)

Melnikov, V.M., Zrniç, D.S.: Estimates of large SW from autocovariances. J. Atmos. Oceanic Technol. **21**, 969–974 (2004)

Meymaris, G., Williams, J., Hubbert, J.: An improved hybrid SW estimator. In: 34th AMS Conference on Radar Meteorology, Paper P5.20 (2009)

Meymaris, G., Williams, J., Hubbert, J.: Hybrid SW estimator. NCAR Report submitted to the NEXRAD Radar Operational Center, 6 pp (2011)

Sharman, R., Frehlich, R.: Aircraft scale turbulence isotropy derived from measurements and simulations. In: Proc. AIAA 41st Aerospace Science Meeting and Exhibit, Paper 2003-194 (2003)

Sharman, R.D., Cornman, L.B., Meymaris, G., Pearson, J., Farrar, T.: Description and derived climatologies of automated in situ eddy-dissipation-rate reports of atmospheric turbulence. J. Appl. Meteor. Climatol. **53**, 1416–1432 (2014)

Sirmans, D.R., Gunther, R., Windes, J.: Engineering study of spectrum width anomaly. Informal Report submitted to the Operational Support Facility of the National Weather Service, Norman, OK, 10 pp (1997)

Williams, J.K.: "Introduction to Fuzzy Logic" (Chapter 6). In: Haupt, S.E., Marzban, C., Pasini, A. (eds.) Artificial Intelligence Methods in the Environmental Sciences, 424 pp. Springer, New York (2009)

Williams, J.K.: Using random forests to diagnose aviation turbulence. Mach. Learn. **95**, 51–70 (2014)

Williams, J.K., Cornman, L., Yee, J., Carson, S.G., Cotter, A.: Real-time remote detection of convectively-induced turbulence. In: AMS 32nd Radar Meteorology Conference, Paper P12R.1 (2005)

Williams, J.K., Cornman, L.B., Yee, J., Carson, S.G., Blackburn, G., Craig, J.: NEXRAD detection of hazardous turbulence. In: AIAA 44th Annual Aerospace Sciences Meeting and Exhibit, Paper AIAA 2006-0076 (2006)

Williams, J.K., Kessinger, C., Abernethy, J., Ellis, S.: "Fuzzy Logic Applications" (Chapter 17). In: Haupt, S.E., Marzban, C., Pasini, A. (eds.) Artificial Intelligence Methods in the Environmental Sciences, 424 pp. Springer, New York (2009)

Williams, J.K., Meymaris, G., Craig, J., Blackburn, G., Deierling, W., McDonough, F.: Measuring in-cloud turbulence: the NEXRAD Turbulence Detection Algorithm. In: AMS 15th Conference on Aviation, Range, and Aerospace Meteorology, Paper 2.1 (2011)

Chapter 8
Relationships Between Lightning and Convective Turbulence

Wiebke Deierling and John K. Williams

Abstract Total lightning has been shown to correlate well with thunderstorm dynamics. Thus, correlations may exist between in-cloud convective turbulence and total lightning activity. Over the continental United States, the NEXRAD Turbulence Detection Algorithm (NTDA) provides three-dimensional (3D) fields of in-cloud Eddy Dissipation Rate (EDR), which is a quantitative measure of turbulence. Comprehensive total lightning measurements are available in some regions of the United States such as in Northern Colorado and New Mexico. Herein, investigations of the temporal and spatial relationships of 3D total lightning and NTDA EDR measurements for non-severe single cell thunderstorms in New Mexico and severe storms in Colorado are described. Results suggest that the occurrence of moderate or greater turbulence at upper levels of storms is related to storm total lightning flash rate, though exact relationships depend on the storm type. Areas of moderate and greater turbulence are also observed to coincide with the footprint of horizontal lightning flash extents. Time series show that lighter turbulence begins to occur prior to lightning initiation in the storms and also extends past the storm's lightning activity. Furthermore, trends of storm total moderate or greater turbulence are correlated to storm total lightning flash rate and extent. These observations are consistent with other recent studies by the authors and others that suggest that total lightning may be useful in diagnosing areas of potentially hazardous atmospheric turbulence. With the expected advent of geostationary satellite based lightning mapping in the next few years, possible relationships between lightning and turbulence characteristics might provide useful information to aviation users, particularly in oceanic or remote areas where ground-based Doppler weather radar observations are unavailable.

W. Deierling (✉)
Research Applications Laboratory, National Center for Atmospheric Research, Boulder, CO, USA
e-mail: deierlin@ucar.edu

J.K. Williams
Research Applications Laboratory, National Center for Atmospheric Research, Boulder, CO, USA

The Weather Company, an IBM Business, Andover, MA, USA
e-mail: john.williams@weather.com

R. Sharman, T. Lane (eds.), *Aviation Turbulence*,
DOI 10.1007/978-3-319-23630-8_8

8.1 Introduction

In-cloud convective turbulence and cloud electrification are both connected to storm dynamics. Observational studies have shown that significant thunderstorm electrification resulting in lightning is related to moderately strong updrafts (e.g. 5–10 m s^{-1}) within the mixed phase region (e.g. Workman and Reynolds 1949; Williams and Lhermitte 1983; Dye et al. 1989; Zipser and Lutz 1994; Deierling et al. 2008; Deierling and Petersen 2008). It is thought that the main mechanism responsible for storm electrification is the noninductive charging mechanism that involves rebounding collisions between riming ice particles and ice crystals when supercooled liquid water (SLW) is present (Takahashi 1978; Saunders 1993, 2008; Takahashi and Miyawaki 2002). Some charging may also occur through rebounding collisions of ice in the absence of SLW (Dye and Willett 2007), and other mechanisms may also contribute to storm electrification such as from inductive charging (Saunders 2008).

Furthermore, several observational and modeling studies have investigated relationships between storm dynamics and lightning characteristics. They found that several parameters such as updraft volume or maximum updraft speed are correlated with total (in-cloud and cloud-to-ground) lightning flash rate. For example, using Doppler radar data, Deierling and Petersen (2008) performed a dual-Doppler synthesis to derive the three-dimensional (3D) wind field for 11 different storms of different types from Northern Alabama and the High Plains of the United States. Estimated storm total updraft volume exceeding vertical velocities (w) of 5, 10, and 20 m s^{-1}, maximum updraft velocities and total lightning flash rates were determined. Deierling and Petersen (2008) found that storm total updraft volumes with vertical velocities $w > 5$–10 m s^{-1} had the highest correlation with storm total lightning flash rates. Similarly, Wiens et al. (2005) and Kuhlman et al. (2006) found good agreement between total flash rate and normalized updraft volume with $w > 10$ m s^{-1} throughout the lifetime of supercell storms based on observations and model simulations. Palucki et al. (2011) compared kinematic and microphysical properties of two weakly electrified convective areas within a larger mesoscale convective system. A comparison of the two convective regions revealed that the convective region with a larger updraft area of $w > 5$ m s^{-1} supported slightly larger amounts of graupel as indicated by dual-polarimetric radar measurements compared to the other convective region. They conclude that this is why the convective region with the larger updraft area produced several lightning flashes, whereas the other region produced no flashes. These observations support the idea that larger areas of higher updraft speeds within a storm produce more and larger ice hydrometeors in the mixed phase region resulting in higher probabilities of ice hydrometeor collisions. This leads to increased charge separation which in turn leads to more lightning production. Recently, Bruning and MacGorman (2013) investigated possible connections between lightning characteristics and turbulence. Taking flash observables as a basis, they derived flash size and energy metrics from idealized electrostatic theory that describes flash initiation, breakdown, and energetics.

Observations of two Oklahoma supercell storms showed that throughout the lifetime of both storms an energetically scaled lightning flash spectrum—represented as flash width versus flash rate dimensionalized energy—exhibited a 5/3 power-law relationship for flash length scales of a few kilometers reminiscent of turbulence. They conclude that the electrical energy spectra are comparable to kinetic energy spectra in a thunderstorm such that flash size and rate may be controlled by kinematic properties of the storm.

Inside thunderstorms, turbulence is generally tied to moist instability within convective updrafts leading to shears that cause mixing within cloudy air, including in downdrafts and anvil regions (Lane et al. 2003, 2012). Generally, larger and more intense updrafts are likely to produce greater shears and more intense and widespread convectively induced turbulence (CIT). Thus, it seems plausible that a strong coupling between storm electrical and kinematic properties (e.g., turbulence) may exist.

This chapter contains a study of in-cloud CIT and total lightning characteristics for several non-severe and severe single cell thunderstorms, making use of 3D turbulence and radar reflectivity measurements together with 3D total lightning observations. Section 8.2 describes the lightning and radar data used; Sect. 8.3 presents analyses of lightning and turbulence in storms in New Mexico and Colorado; and Sect. 8.4 summarizes and concludes the chapter.

8.2 Lightning and Turbulence Data

Lightning emits electromagnetic radiation over a wide range of frequencies from the Very Low Frequency (VLF)/Low Frequency (LF) to the Very High Frequency (VHF)/Ultra High Frequency (UHF) range. Different parts of a lightning discharge radiate at various frequencies. For example, return strokes of cloud-to-ground (CG) flashes radiate strongly at VLF/LF whereas other parts of in-cloud (IC) and CG lightning discharges, such as negative leaders and dart leaders, emit strongly at VHF (Shao et al. 1999; Thomas et al. 2001), allowing for a very comprehensive detection of all lightning discharges at these wavelengths. In this study, 3D total lightning data from the New Mexico Institute of Mining and Technology (NMIMT) Lightning Mapping Arrays (LMA) operating at White Sands Missile Range in New Mexico and in Northern Colorado are used. LMA systems detect the time of arrival (TOA) of VHF sources emitted from lightning at around 60–66 MHz (Thomas et al. 2001), which allows for mapping entire 3D flash structures. The detection efficiency of IC and CG lightning within the range of an LMA network is near 100 % (Lang et al. 2004), and VHF sources are detected with a vertical resolution better than 1 km, making the LMA detections ideal for comparison with the NTDA's high-resolution EDR measurements. Lightning source locations are retrieved using a least squares χ^2 minimization technique (Thomas et al. 2004). To filter out noise, only LMA detected VHF source data that was measured by seven or more stations and had $\chi^2 < 2$ were used in this study. From the quality controlled LMA data, 5 min accumulations of VHF source densities (the number of VHF sources per km^2) were computed. The 5 min accumulation time was chosen to match the temporal resolution of the NTDA turbulence measurement data described below. Furthermore, LMA measured VHF sources were grouped into flashes by applying spatial and temporal constraints following McCaul et al. (2009). Taking the flash location as the position of the first VHF source in a flash, 5 min accumulations of storm total flash rate (flash count per identified storm, where storms were identified based on lightning and radar reflectivity data) were computed. Horizontal flash extent densities (flash count per each 1 km^2 grid box) were also determined by counting the number of flashes that passed over each grid box within the 5 min interval.

In-cloud turbulence measurements were obtained from the NCAR/NEXRAD Turbulence Detection Algorithm (NTDA) 3D mosaic of Eddy Dissipation Rate (EDR, units of $m^{2/3}\ s^{-1}$) produced every 5 min for the continental United States (US). The NTDA product also provides a corresponding 3D NEXRAD radar reflectivity mosaic. To estimate EDR, NTDA uses the Doppler spectrum width measured by the operational NEXRAD radars, after performing extended quality control to eliminate potentially bad radar data. The NTDA has been verified via comparisons to in situ measurements of EDR from both research and commercial aircraft (Williams et al. 2006; Chap. 7). EDR is used as a standard for turbulence reporting in the global airspace. The correspondence of in situ EDR (the peak value over 1-min intervals) to traditional turbulence intensity categories as reported by pilots ("light", "moderate", "severe") is aircraft dependent. Following Sharman et al. (2014), EDR values from 0.15–0.22 $m^{2/3}\ s^{-1}$, 0.22–0.34 $m^{2/3}\ s^{-1}$, and 0.34+ $m^{2/3}\ s^{-1}$ are interpreted as representing light, moderate, and severe turbulence, respectively. These definitions are used below to compute storm total light, moderate, and severe turbulence volumes. Furthermore, the 3D structures of EDR within storms are compared to LMA measured 3D lightning information. These analyses are performed for non-severe single cell thunderstorms as well as severe storms over the US High Plains.

8.3 Lightning–CIT Relationships

8.3.1 Non-severe Storms in New Mexico

Thunderstorms in New Mexico are typically diurnally driven, with storms developing in the afternoon and evening hours. These thunderstorms are often fairly short-lived, stationary storms that are non-severe in nature, and triggered by elevated terrain or outflow boundaries (Saxen et al. 2008).

To investigate the temporal and spatial development of CIT and total lightning characteristics for these non-severe storms, NTDA EDR data were compared to 3D VHF source and 2D flash extent densities as well as radar reflectivity over a

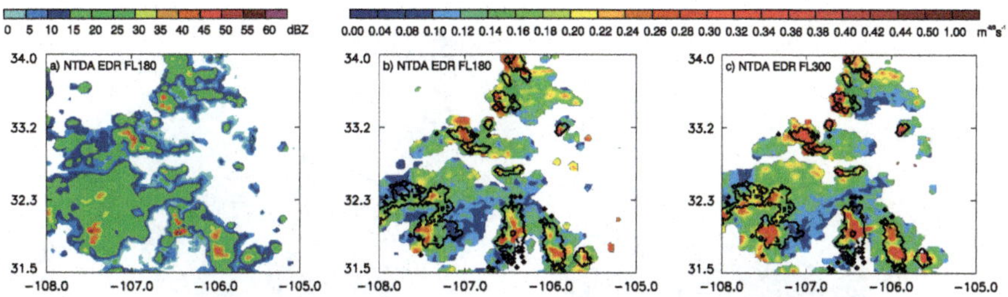

Fig. 8.1 (**a**) Horizontal cross-section of radar reflectivity at flight level 180, (**b**) NTDA EDR at flight level 180, and (**c**) NTDA EDR at flight level 300, all at 21:35 UTC on 31 July 2010 at White Sands Missile Range (WSMR), New Mexico. Areas of 5 min LMA total lightning flash extent densities ≥1 flash/km^2 are overlaid as *black contours*

150×150 km^2 area centered on White Sands Missile Range, New Mexico, a region well covered by an LMA. A number of short-lived storms initiated around 20 UTC on 31 July 2010. Figure 8.1 shows horizontal cross-sections of NTDA radar reflectivity at flight level (FL) 180 (Fig. 8.1a) and NTDA EDR at FL 180 and 300, respectively (Fig. 8.1b, c), for these storms at 21:35 UTC. The freezing level was approximately at FL 180 (5.5 km above sea level). Areas of LMA total lightning flash extent densities ≥1 flash per km^2 per 5 min are overlaid as black contours. As can be seen in Fig. 8.1b, c, areas encompassing lightning flash extents are well aligned with areas of moderate turbulence (orange and warmer colors representing EDR > 0.22 m$^{2/3}$ s^{-1}) at higher altitudes in the storms. Observations of numerous non-severe storms in this area of New Mexico have shown that this is a typical result: 2D total lightning flash extent densities are frequently co-located with areas of moderate or greater convective turbulence at higher altitudes in the storms.

Vertical cross-sections of the EDR, reflectivity, and VHF source densities in two thunderstorm cells during their mature and dissipation phase are shown in Fig. 8.2. They illustrate that maximum EDR values are mostly (but not exclusively) located above the maximum VHF source densities and both, in turn, are mostly located above the storms' reflectivity cores. This "stacked" pattern is found throughout the lifetime of these New Mexico non-severe storms. Furthermore, maxima in the storms' total flash extent densities and flash rates coincide with maxima of storm total moderate and severe turbulence volumes as computed from NTDA EDR (Fig. 8.3). Storm total volumes of light turbulence increase before the onset of lightning activity in the storm and last tens of minutes past the end of the lightning activity. Increasing lightning flash extent densities and storm total lightning flash rates generally coincide with increasing precipitation ice in the cloud as seen by the time series of the total storm volume of radar reflectivity >35 dBZ for temperatures colder than 0 °C in Fig. 8.3. Statistical analyses of lightning extent densities as compared to EDR volumes show a strong correlation on a storm-by-storm basis and also for a 2-week period composed of multiple storms on several days. For the same case shown in Fig. 8.3, Fig. 8.4a shows a scatter plot of storm volume above FL 180 of EDR ≥ 0.22 m$^{2/3}$ s^{-1} (corresponding to moderate and greater turbulence) to

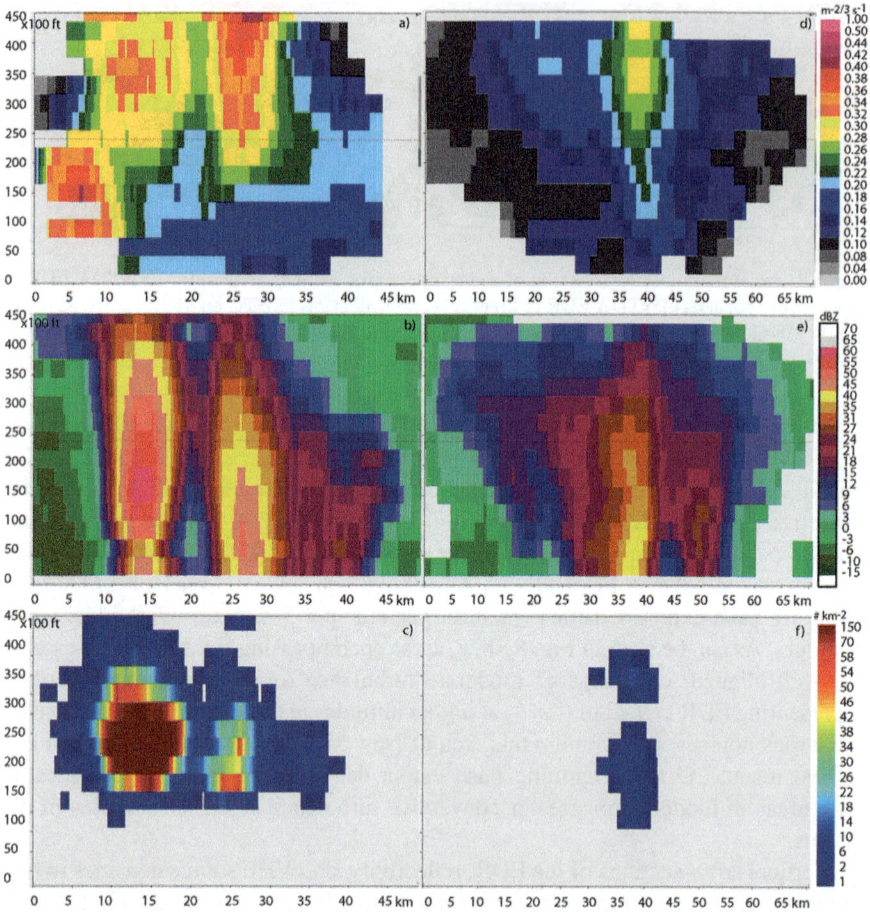

Fig. 8.2 Vertical cross-sections of storms during their mature (*left panels*) and dissipation phase (*right panels*) at WSMR on 31 July 2010 at 22:30 UTC. Plots show EDR with a continuous color scale (**a**, **d**), radar reflectivity (dBZ) (**b**, **e**), and 5 min 3D VHF source densities (km^3) as measured by the LMA (**c**, **f**)

the storm total flash extent densities. Here dots represent the values at 5-min intervals. The two fields have a correlation coefficient of $r = 0.87$, showing good correspondence between the intensity of the lightning and moderate and greater turbulence. Figure 8.4b is similar to Fig. 8.4a but for EDR volumes and lightning data over the 2 week period. The correlation coefficient is slightly higher, at $r = 0.90$. These results suggest that, for these summertime non-severe convective storms in New Mexico, the relationship between lightning and significant in-cloud turbulence is consistent and robust.

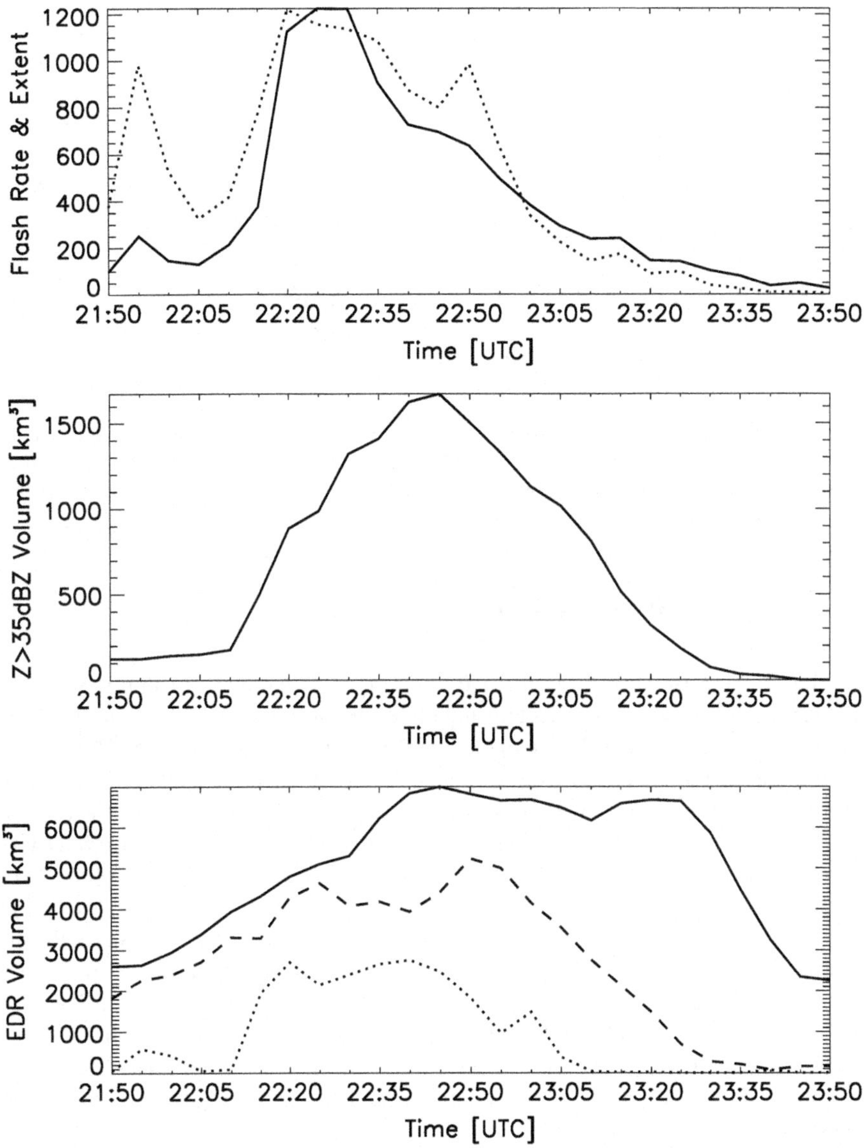

Fig. 8.3 Time series of various storm total lightning, turbulence, and radar reflectivity characteristics for the same New Mexico storm cells shown in Fig. 8.2. *Upper panel*: Time series of 5 min storm total flash extent densities [flash count km^{-2}, *solid line*] and 5 min storm total flash rate [flash count, *dashed line*] multiplied by a factor of 10 for visualization. *Middle panel*: 35 dBZ radar reflectivity volume for temperatures colder than 0 °C as an indicator of storm total precipitation ice. *Lower panel*: Total storm volume of light (*solid line*), moderate (*dashed line*), and severe (multiplied by a factor of 10 for visualization, *dotted line*) turbulence

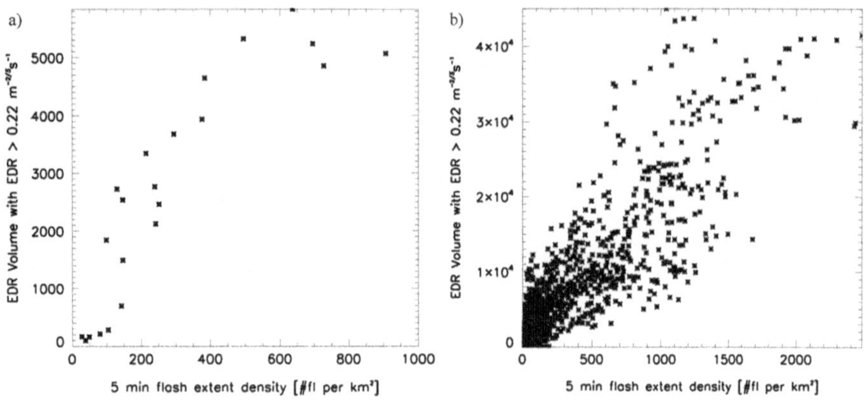

Fig. 8.4 Scatter plot of LMA flash extent and NTDA EDR volume characterized by having EDR values >0.22 (moderate or greater turbulence) for (**a**) storm cells of Figs. 8.2 and 8.3 and (**b**) all storms in the same area over a 2 week period. The correlation coefficients are $r = 0.87$ and $r = 0.90$, respectively

8.3.2 Severe Storms in Colorado

Similar to the New Mexico cases, total lightning information from an LMA system in Northern Colorado was collected, and three-dimensional VHF source densities as well as two-dimensional flash extents and flash rates were computed. These were then compared to NTDA EDR and radar reflectivity characteristics for individual storms and for a period of 4 weeks during June 2012. Figure 8.5 shows vertical cross-sections of NTDA EDR, VHF source densities, and NTDA radar reflectivity for a severe storm in its mature stage that occurred between 21 UTC on 6 June 2012 and 4:40 UTC on 7 June 2012 just south of Denver International Airport in Colorado. The national weather service reported hail and strong winds with this storm. The extent of VHF source densities reaching up to high altitudes in Fig. 8.5 suggests the presence of a strong updraft. In its mature stage, the storm has also developed a large stratiform region. This region exhibits in places high values of EDR associated with less frequent lightning but typically larger spatial lightning extent (resulting in smaller flash extent densities in these regions). Around and within the convective core of the storm, a large area of severe CIT is present. It is located in the upper portions of the storm, and a narrower column extends vertically throughout the storm. This pattern was observed in other severe storms in Colorado as well. In addition, for these storms, maxima in VHF source densities are not necessarily "stacked" between maxima in CIT and radar reflectivity as observed for the New Mexico non-severe single cell storms, though source density maxima are

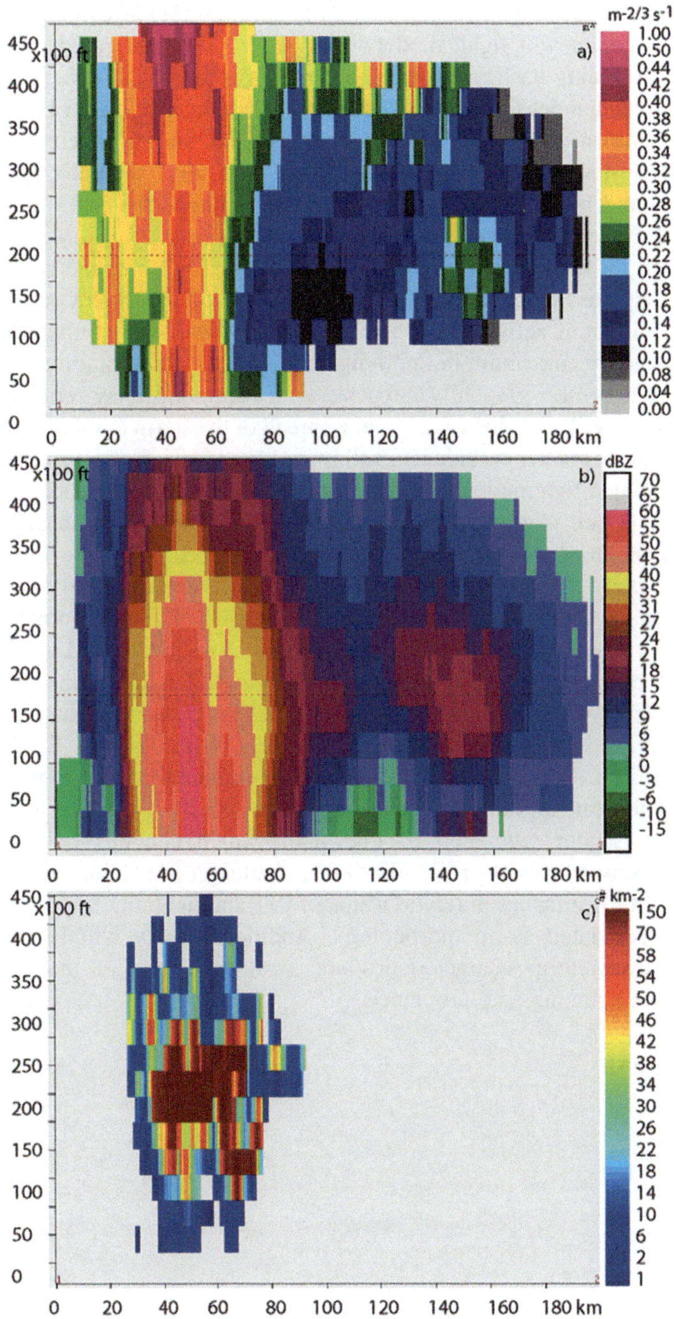

Fig. 8.5 Vertical cross-sections of a severe storm cell in its mature phase at 1:45 UTC on 7 June 2012 that occurred in Colorado. Panel (**a**) shows NTDA EDR data with a continuous color scale, panel (**b**) shows NTDA radar reflectivity (dBZ), and panel (**c**) shows 3D VHF source densities as measured by the Colorado LMA

observed near the base of the larger region of severe turbulence at upper altitudes. Total storm volumes of light, moderate, or severe turbulence were compared to storm total lightning flash rate or flash extents, as shown in Fig. 8.6. Total storm volume flash extent densities, flash rates, and CIT volumes of higher EDR values in this severe storm are much larger than was observed in the non-severe New Mexico storms. While total flash extent densities and flash rates generally increase and decrease in tandem with storm total NTDA EDR volumes, the peaks of total storm volumes of moderate and severe turbulence do not necessarily coincide with each other, unlike for the non-severe New Mexico storms. Still, for the case shown in Fig. 8.6, moderate and greater total storm turbulence volumes increase rapidly with the onset of lightning activity, and the peak of storm total severe turbulence volume coincides with the maximum in lightning activity. The storm total light turbulence volume increases more gradually and peaks with a temporal delay compared to total lightning activity. This is likely due to the spread of the storm anvil, much of which contains light or greater turbulence, following the period of the strongest updraft and its associated lightning generation. Similar to the temporal trends for the New Mexico single cell storms, storm total light turbulence volume begins to grow before the onset of total lightning and remains present after the last lightning flash of the storm. Storm total lightning activity corresponds well with the storm's total precipitation ice volume, as represented by the 35 dBZ total storm volume for temperatures colder than 0 °C. The relationship between CIT and total lightning activity appears more complicated for the severe Colorado storms than for the non-severe New Mexico storms described above. This is also illustrated by Fig. 8.7, which shows a scatter plot of total flash extent versus NTDA EDR volumes exceeding 0.22 $m^{2/3}\ s^{-1}$. Higher flash rates still correlate well with higher amounts of light-to-moderate turbulence (correlation coefficient $r = 0.83$). However, Fig. 8.7 shows more scatter compared to Fig. 8.4b, which contains only data from non-severe New Mexico storms. The larger scatter in the Colorado storms may be due in part to variances in relationships of CIT and lightning for different storm types and associated storm morphology. Additional study will be required to determine what storm characteristics are most predictive of the relationship between total lightning and turbulence.

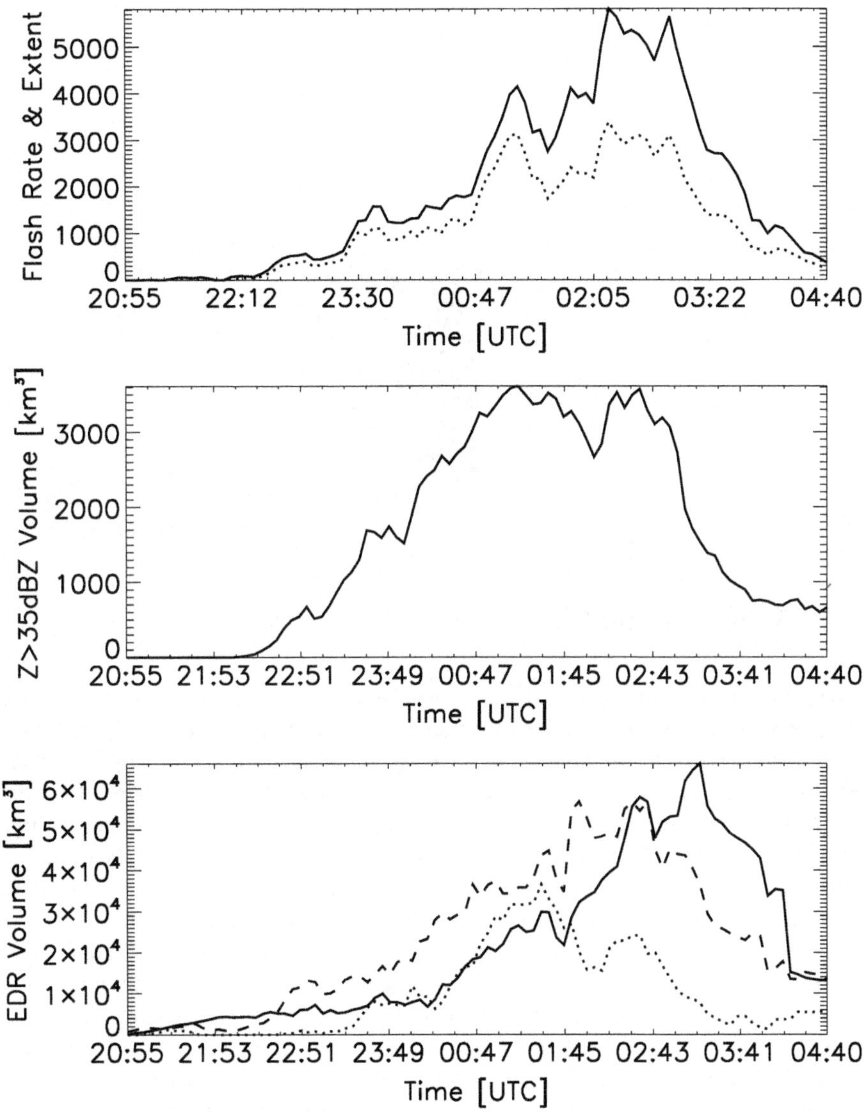

Fig. 8.6 Time series of various storm total lightning, turbulence, and radar reflectivity characteristics for a severe storm that occurred in Colorado. *Upper panel*: Time series of 5 min storm total flash extent densities [flash count km^{-2}, *solid line*] and 5 min storm total flash rate [flash count, *dashed line*] multiplied by a factor of 3 for visualization. *Middle panel*: volume of >35 dBZ radar reflectivity for temperatures colder than 0 °C as an indicator of storm total precipitation ice. *Lower panel*: Total storm volume of light (*solid line*), moderate (multiplied by a factor of 1.5 for visualization, *dashed line*), and severe (multiplied by a factor of 2 for visualization, *dotted line*) turbulence

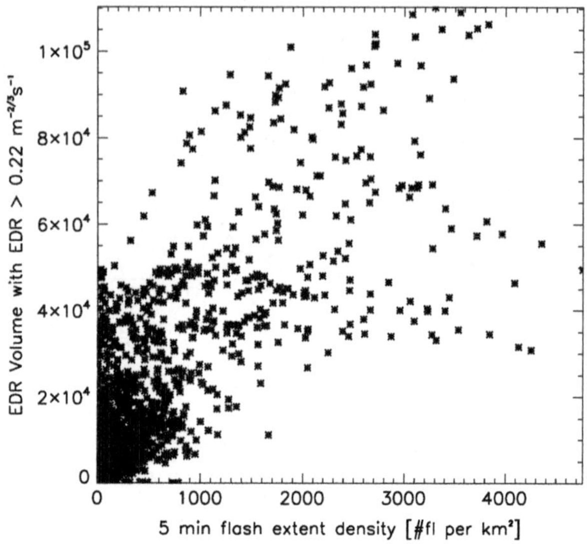

Fig. 8.7 Scatter plot of LMA flash extent and NTDA EDR volume above FL 180 characterized by having EDR values $>0.22\ m^{2/3}\ s^{-1}$, where the correlation coefficient $r = 0.83$

8.4 Summary and Conclusions

The relationships between total lightning characteristics and in-cloud convective turbulence have been investigated for non-severe single cell storms in New Mexico and severe storms in Colorado using regional comprehensive 3D total lightning and 3D EDR and reflectivity data derived from NEXRAD Doppler weather radars. For the New Mexico storms, the occurrence of moderate or greater turbulence at high altitudes appears to be well correlated with the storm total lightning flash rate. In addition, areas of moderate and greater turbulence, as derived from NTDA EDR, coincide spatially with areas of horizontal flash extent, with the most intense turbulence occurring above the lightning maximum, which in turn occurs above the reflectivity core. In the severe storms observed in Colorado, the trend of storm total lightning activity is also correlated with moderate or greater storm total turbulence volumes, though the relationship is less strong. Also, stratiform and convective areas of the storm can both be accompanied by severe turbulence, but lightning flash extents are higher in convective cores than in stratiform regions. This is likely because, whereas turbulence readily spreads into the anvil, lightning initiates mostly in convective storm regions and may or may not extend out into anvil or stratiform storm regions. The latter has been documented by Bruning and MacGorman (2013). Onsets of increasing storm total moderate and greater CIT volumes coincide with onsets of increasing lightning activity. This may be indicative of an intensifying updraft resulting in both higher lightning activity and higher turbulence production. For both severe and non-severe storms, light turbulence occurs before the onset of lightning, peaks after the moderate and greater turbulence maxima, and also decays much more slowly, persisting several tens of minutes or longer after the last lightning flash of a storm. A strong updraft is a prerequisite for lightning production, and the shears and gravity waves produced by the updraft could also create turbulence before the onset of lightning. Decaying storms accompanied with a decrease of lightning would likely also coincide with moderate and greater turbulence decaying into lighter turbulence before dissipating altogether, which can explain the gradual decrease and persistence of storm total light turbulence volumes. Generally, it appears that total lightning activity could be used—at least in a limited way—to indicate the location and to estimate the magnitude of CIT.

When Doppler weather radar observations are available, the NTDA or a similar technique may be used to measure the location and severity of in-cloud convective turbulence more directly. However, relationships between CIT and lightning could be useful for identifying CIT over oceans and remote regions where radars are sparse, blocked by terrain or nonexistent, but where observations of lightning are available. With the advent of lightning sensors on geostationary satellites, such observations may soon be available nearly everywhere, raising the prospect of global lightning-based CIT diagnosis. Additional case studies and statistical evaluations of storms in different locations, seasons, and synoptic situations would be necessary to refine predictive relationships between turbulence and lightning. For instance, oceanic storms generally exhibit much weaker updrafts and less lightning than continental storms, but it is not clear whether they also produce proportionally less turbulence. Numerical weather prediction (NWP) models might be used to help distinguish conditions likely to produce non-severe or severe storms, allowing the appropriate relationship between turbulence and lightning to be selected. Additionally, NWP model data could be used to estimate the freezing level and the tropopause height, which may assist in constraining the likely altitude of CIT. In summary, studies like the ones detailed in this chapter suggest that the use of total lightning to identify areas of potentially hazardous CIT appears promising, but much more work remains to be done to create an operational turbulence diagnostic based on this concept.

References

Bruning, E.C., MacGorman, D.R.: Theory and observations of controls on lightning flash size spectra. J. Atmos. Sci. **70**, 4012–4029 (2013). doi:10.1175/JAS-D-12-0289

Deierling, W., Petersen, W.A.: Total lightning activity as an indicator or updraft characteristics. J. Geophys. Res. **113**, D16210 (2008). doi:10.1029/2007JD009598

Deierling, W., Petersen, W.A., Latham, J., Ellis, S., Christian, H.J.: The relationship between lightning activity and ice fluxes in thunderstorms. J. Geophys. Res. **113**, D15210 (2008). doi:10.1029/2007JD009700

Dye, J.E., Willett, J.C.: Observed enhancement of reflectivity and the electric field in long-lived Florida anvils. Mon. Weather Rev. **135**, 362–3380 (2007)

Dye, J.E., Jones, J.J., Weinheimer, A.J., Winn, W.P.: Observations within two regions of charge during initial thunderstorm electrification. Q. J. Roy. Meteorol. Soc. **114**, 1271–1290 (1989)

Kuhlman, K.M., Ziegler, C.L., Mansell, E.R., MacGorman, D.R., Straka, J.M.: Numerically simulated electrification and lightning of the 29 June 2000 steps supercell storm. Mon. Weather Rev. **134**, 2734–2757 (2006)

Lane, T., Sharman, R., Clark, T., Hsu, H.-M.: An investigation of turbulence generation mechanisms above deep convection. J. Atmos. Sci. **60**, 1297–1321 (2003)

Lane, T.P., Sharman, R.D., Trier, R.D., Fovell, R.G., Williams, J.K.: Recent advances in the understanding of near-cloud turbulence. Bull. Am. Meteorol. Soc. **93**, 499–515 (2012)

Lang, T.J., Miller, L.J., Weisman, M., Rutledge, S.A., Baker III, L.J., Bringi, V.N., Chandrasekar, V., Detwiler, A., Doesken, N., Helsdon, J., Knight, C., Krehbiel, P., Lyons, W.A., MacGorman, D., Rasmussen, E., Rison, W., Rust, W.D., Thomas, R.J.: The severe thunderstorm electrification and precipitation study. Bull. Am. Meteorol. Soc. **85**, 1107–1125 (2004)

McCaul, E.W., Goodman, S.J., LaCasse, K.M., Cecil, D.J.: Forecasting lightning threat using cloud-resolving model simulations. Weather Forecast. **24**, 709–729 (2009)

Palucki, J.L., Biggerstaff, M.I., MacGorman, D.R.: Schuur: comparison between low-flash and non-lightning-producing convective areas within a mature mesoscale convective system. Weather Forecast. **26**, 468–486 (2011). doi:10.1175/WAF-D-10-05012.1

Saunders, C.P.R.: A review of thunderstorm electrification processes. J. Appl. Meteorol. **32**, 542–655 (1993)

Saunders, C.: Charge separation mechanisms in clouds. Space Sci. Rev. **137**, 335–353 (2008). doi:10.1007/s11214-008-9345-0

Saxen, T.R., Mueller, C.K., Warner, T.T., Steiner, M., Ellison, E.E., Hatfield, E.W., Betancourt, T. L., Dettling, S.M., Oien, N.A.: The operational mesogamma-scale analysis and forecast system of the U.S. Army evaluation test and evaluation command. Part IV: The white sands missile range auto-nowcast system. J. Appl. Meteorol. Climatol. **47**, 1123–1139 (2008)

Shao, X.M., Rhodes, C.T., Holden, D.N.: RF radiation observations of positive cloud-to-ground flashes. J. Geophys. Res. **104**, 9601–9608 (1999)

Sharman, R., Cornman, L., Meymaris, G., Pearson, J., Farrar, T.: Description and derived climatologies of automated in situ eddy dissipation rate reports of atmospheric turbulence. J. Appl. Meteorol. Climatol. **53**, 1416–1432 (2014). doi:10.1175/JAMCD-13-0329.1

Takahashi, T.: Riming electrification as a charge generation mechanism in thunderstorms. J. Atmos. Sci. **35**, 1536–1548 (1978)

Takahashi, T., Miyawaki, K.: Reexamination of riming electrification in a wind tunnel. J. Atmos. Sci. **59**, 1018–1025 (2002)

Thomas, R.J., Krehbiel, P.R., Rison, W., Hamlin, T., Harlin, J., Shown, D.: Observations of VHF source powers radiated by lightning. Geophys. Res. Lett. **28**, 143–146 (2001)

Thomas, R., Krehbiel, P.R., Rison, W., Hunyady, S.J., Winn, W.P., Hamlin, T., Harlin, J.: Accuracy of the lightning mapping array. J. Geophys. Res. **109**, D14207 (2004). doi:10.1029/2004JD004549

Wiens, K.C., Rutledge, S.A., Tessendorf, S.A.: The 29 June 2000 supercell observed during STEPS. Part II: Lightning and charge structure. J. Atmos. Sci. **62**, 4151–4177 (2005)

Williams, E.R., Lhermitte, R.M.: Radar tests of the precipitation hypothesis for thunderstorm electrification. J. Geophys. Res. **88**, 10984–10992 (1983)

Williams, J., Cornman, L., Yee, J., Carson, S., Blackburn, G., Craig, J.: NEXRAD detection of hazardous turbulence. AIAA Aerospace Sciences Meeting, Reno, NV (2006)

Workman, E.J., Reynolds, S.E.: Electrical activity as related to thunderstorm cell growth. Bull. Am. Meteorol. Soc. **30**, 142–149 (1949)

Zipser, E.J., Lutz, K.R.: The vertical profile of radar reflectivity of convective cells: a strong indicator of storm intensity and lightning probability? Mon. Weather Rev. **122**, 1751–1759 (1994)

Chapter 9
LIDAR-Based Turbulence Intensity for Aviation Applications

P.W. Chan

Abstract Turbulence could increase the workload of pilots, and timely provision of turbulence information would be beneficial to aircraft operation. This paper summarises the use of a Doppler Light Detection and Ranging (LIDAR) system to measure turbulence intensity at an operating airport. The cube root of eddy dissipation rate (EDR) is used for representing the turbulence intensity, according to the requirement of the International Civil Aviation Organization (ICAO). The quality of EDR so calculated is established by comparing with the aircraft data. Examples of EDR maps based on conical scans of the LIDAR are presented. The EDR maps so obtained should be useful for aviation operations, and the EDRs calculated from the glide-path scans of the LIDAR may be used for turbulence alerting applications.

9.1 Introduction

Low-level turbulence (below 1600 ft from the elevation of the airport) can be hazardous to landing and departing aircraft. Such turbulence may be measured using high-resolution data available from ground-based anemometers. For instance, following the requirements of the International Civil Aviation Organization (ICAO), 10-m-high anemometers may be set up along the runway with high temporal resolution wind data (down to 1 s) available from such equipment for calculating the turbulence intensity, namely, the cube root of eddy dissipation rate (EDR). These calculated EDRs are useful for monitoring the level of turbulence at the airport area.

However, once airborne, the turbulence the aircraft experiences may be different from that measured at the ground. In that case, remote-sensing meteorological equipment would be useful in providing an estimate of EDR along flight paths. One possible source of EDR comes from microwave weather radars, such as the Terminal Doppler Weather Radar (TDWR) installed at some airports. The

P.W. Chan (✉)
Hong Kong Observatory, Hong Kong, China
e-mail: pwchan@hko.gov.hk

R. Sharman, T. Lane (eds.), *Aviation Turbulence*,
DOI 10.1007/978-3-319-23630-8_9

spectral width measured in the return signal of TDWR might be used to provide an estimate of the turbulence intensity, which could be useful for monitoring the level of turbulence in rainy conditions. For non-rainy weather conditions, such radars might not give persistently good quality signals for weather monitoring purpose.

To monitor the turbulence in such conditions, the Doppler Light Detection and Ranging (LIDAR) systems are used increasingly at the airports around the world. In Hong Kong, the complex terrain can on occasion give rise to turbulent airflow in the airport area. The LIDAR introduced to this airport in 2002 provides very useful information for the depiction of the airflow around the airport, particularly when there is disruption of airflow by nearby terrain. The LIDAR data are used to calculate the EDR, and there are interesting observations of the turbulence intensity in different kinds of weather conditions. The locations of meteorological equipment at the airport in Hong Kong are found in Fig. 9.1.

This chapter focuses on the LIDAR-calculated EDR for airport operations at HKIA. The LIDAR has been used to do glide-path and conical scans, and the

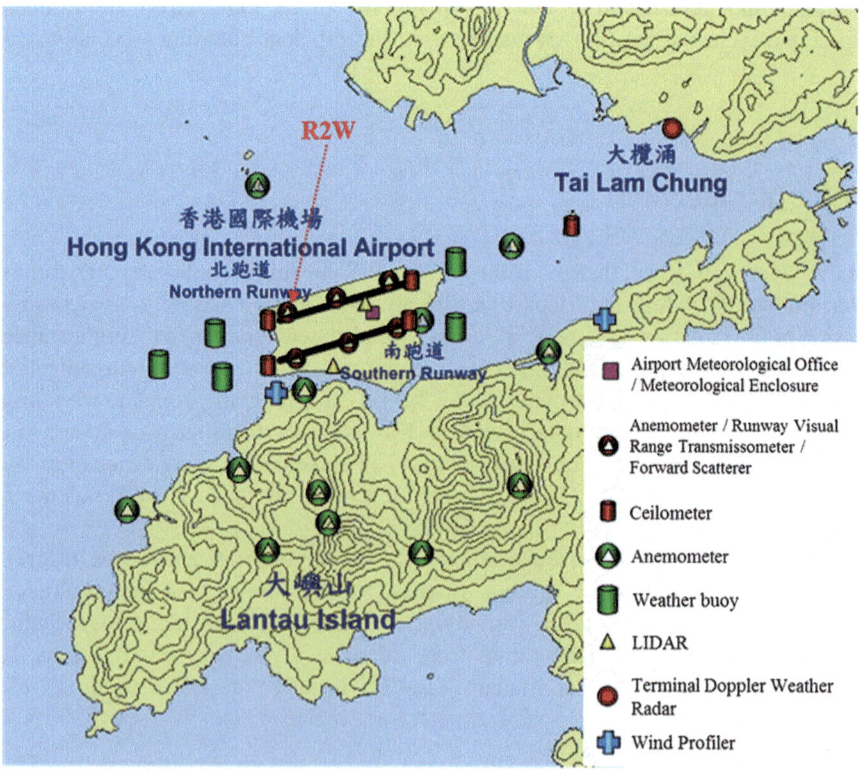

Fig. 9.1 Locations of meteorological equipment at the Hong Kong International Airport. Height contours are in 100 m. The length of the runway is about 3.8 km, which can be used as a length scale of the picture

mathematical formulation of the EDR calculation for such scans is reviewed. It is followed by EDR observations in the airport area for typical cases of terrain-disrupted airflow. The quality of EDR from the LIDAR is then studied by comparison with that obtained from the research aircraft. Application of LIDAR-derived EDR in the alerting of low-level turbulence is discussed, and finally the observations in tropical cyclone (TC) cases are reviewed.

9.2 Calculation of EDR Based on Glide-Path Scans

In the glide-path scan mode, the laser beam of the LIDAR is configured to scan along the oblique line of the glide path in space by orchestrating the azimuthal and elevation motions of the laser scanner. For alerting of wind shear, i.e. significant changes of headwind, only the radial velocity data collected within a confined rectangular block around the glide path are utilised to construct the headwind profile to be encountered by the aircraft (Shun and Chan 2008). In calculating turbulence intensity, all the wind data in the measurement sector of the glide-path scan are considered. The whole measurement sector is divided into a number of overlapping subsectors (each with a size of 10 range gates and 16 azimuth angles, overlapping by 5 range gates and 8 azimuth angles). EDR is calculated in each subsector by adopting a spatial fluctuation method in the structure function approach (Frehlich et al. 2006).

For more accurate determination of the turbulence intensity, azimuthal averaging is minimised by reducing the horizontal rotation speed of the laser beam given the time constraint that the LIDAR is also required to perform other non-glide-path scans for operational purposes. For the current setting of the LIDAR, the gate length Δp is 105 m. Based on results of previous studies (Frehlich 2001; Frehlich and Cornman 2002), $\Delta h/\Delta p$ ($\Delta h = \text{range} \times \text{azimuthal span}$) should be much less than 1 so that azimuthal averaging is small compared with the range scale length (averaging along the range gate). For real application, this ratio could be set at a maximum value of 0.2. Taking the maximum range of 10 km for the LIDAR and the data output frequency of 10 Hz from the LIDAR, the azimuthal rate α should be $\Delta h = 10{,}000\ \text{m} \times \alpha \times 0.1\ \text{s} \leq 0.2 \times 105\ \text{m}$, which implies that $\alpha \leq 0.021\ \text{rad s}^{-1}$ or $1.2°\ \text{s}^{-1}$. This provides a realistic value of the upper bound for the azimuthal rate over a small sector considering operational scanning requirements and laser safety consideration. The azimuthal rate of about $0.8°\ \text{s}^{-1}$ has been implemented for the glide-path scans over the arrival runway corridors of HKIA, including the two most used arrival corridors: 07LA (arrival at the north runway of HKIA from the west) and 25RA (arrival at the north runway of HKIA from the east).

Each subsector in the glide-path scanning area has a size of 10 range gates times 16 radials. For a particular scan *k*, the radial velocity "surface" within this subsector (as a function of range R and azimuth angle θ) is fitted with a plane using the singular value decomposition method. This is essentially the removal of the linear trend. The velocity fluctuation $\hat{v}'$ at each point in the space (R, θ) is taken to be the difference between the measured radial velocity $\hat{v}$ and the fitted velocity $\overline{v}$ on the plane:

$$\hat{v}'(R,\theta,k) = \hat{v}(R,\theta,k) - \overline{v}(R,\theta,k). \tag{9.1}$$

Both longitudinal and azimuthal structure functions are calculated (Frehlich et al. 2006). The longitudinal structure function is given by:

$$\hat{D}_L(R_1,R_2) = N^{-1}\sum_{\theta,k}[\hat{v}'(R_1,\theta,k) - \hat{v}'(R_2,\theta,k)]^2 - E(R_1,R_2) \tag{9.2}$$

where the summation is made over all the possible azimuthal angles and scans over 15 min (about 7–8 glide-path scans, with the scan at each runway corridor updated every 2 min or so), and N refers to the total number of entries in the summation. The choice of 15 min is a balance between the collection of sufficient statistics of turbulent eddies moving across the measurement domain (which requires a longer period of time considering the revisit time of 2 min for glide-path scans) and the time required for capturing the stronger turbulence (which requires a shorter period of time because the use of longer period would tend to smooth out the stronger turbulence in the sampling period). The error term E is calculated using the covariance method on the radial velocity difference (Frehlich 2001; Frehlich et al. 2006) by estimation of the covariance values of different azimuthal changes by linear approximation to lag 0:

$$E = \hat{C}(0) - 2\hat{C}(\Delta\theta) + \hat{C}(2\Delta\theta). \tag{9.3}$$

The azimuthal structure function is calculated in a way similar to Eq. (9.2), but the radial velocity difference of two azimuthal angles is considered:

$$\hat{D}_{AZ}(R\cdot\theta_1,R\cdot\theta_2) = N^{-1}\sum_{R,k}\left[\hat{v}'(R,\theta_1,{}^1k) - \hat{v}'(R,\theta_2,k)\right]^2 - E(R). \tag{9.4}$$

Again, the error term is estimated using the covariance method. The covariance of velocity estimate is:

$$\hat{C}(R,n\Delta\theta) = N^{-1}\sum_{l,k}\hat{v}'(R,l\Delta\theta,k)\times\hat{v}'(R,(l+n)\Delta\theta,k). \tag{9.5}$$

The error term of velocity difference is taken as two times the error of the velocity estimate [hence the factor of 2 at the beginning of the right-hand side of Eq. (9.6)]:

$$E = 2\left(\hat{C}(0) - 2\hat{C}(\Delta\theta) + \hat{C}(2\Delta\theta)\right). \tag{9.6}$$

$EDR^{1/3}$ is determined by fitting the longitudinal or the azimuthal structure function with the theoretical von Kármán model (Frehlich et al. 2006). As an example, the calculation of EDR based on azimuthal structure function is given here.

Let $s = R\ (\theta_1 - \theta_2)$ for two azimuthal angles θ_1 and θ_2. According to Frehlich et al. (2006), for homogeneous von Kármán turbulence over a two-dimensional plane (R, s),

$$D_{AZ}(R,s) = 2\sigma^2\left[\Lambda(q/L_0) + \Lambda_D(q/L_0)\left(1 - R^2/q^2\right)\right], \tag{9.7}$$

where $q = \left(R^2 + s^2\right)^{1/2}$,

$$\Lambda_D(x) = (0.29627426)x^{4/3}K_{2/3}(x), \tag{9.8}$$

σ^2 is the variance of the radial velocity, L_0 is the outer scale of turbulence, $\Lambda(x)$ is a universal function and $K_{2/3}(x)$ is the modified Bessel function of order 2/3. Since the LIDAR is configured to scan in the azimuthal direction very slowly such that the transverse dimension of the LIDAR sensing volume for each radial velocity estimate is much less than the range resolution of about 105 m, Eq. (9.7) could be simplified to be:

$$D_{AZ}(s,\sigma,L_0) = 2\sigma^2 G_\theta(s/\Delta p,\mu,\chi), \tag{9.9}$$

where $\mu = (2\ln 2)^{1/2}\Delta p/\Delta r, \chi = \Delta p/L_0$, Δr is the full width at half maximum of the LIDAR sensing volume in range that defines the extent of the illuminated aerosol targets and G_θ is given as Eq. (46) in Frehlich et al. (2006) and repeated below:

$$G_\theta(m,\mu,\chi) = 2\int_0^\infty F(x,\mu)\left[\Lambda\left(\chi\sqrt{m^2+x^2}\right) - \Lambda(\chi x) + \Lambda_D\left(\chi\sqrt{m^2+x^2}\right)m^2/\left(m^2 + x^2\right)\right]\mathrm{d}x,$$

where $F(x,\mu) = \frac{1}{2\sqrt{\pi}\mu}\left\{\exp\left[-\mu^2(x+1)^2\right] + \exp\left[-\mu^2(x-1)^2\right] - 2\exp(-\mu^2x^2)\right\} + \frac{1}{2}\{(x+1)\mathrm{Erf}[\mu(x+1)] + (x-1)\mathrm{Erf}[\mu(x-1)] - 2\cdot\mathrm{Erf}(\mu x)\}$ and $\mathrm{Erf}(x)$ is the standard error function.

The azimuthal structure function is calculated from the LIDAR's radial velocity and then fitted with the theoretical von Kármán model to give σ^2 and L_0. EDR (also denoted as ε) is given by:

$$\mathrm{EDR} = (0.933668)\frac{\sigma^3}{L_0}. \tag{9.10}$$

The fitting involves the minimization of a cost function to obtain two unknown parameters, namely, the variance of radial velocity and the outer scale of turbulence.

The values calculated from both azimuthal and longitudinal methods are studied in a number of cases of turbulent flow with different intensities, and they are found to be comparable with each other with a correlation coefficient (R^2) reaching 0.9. In the following discussions, only the $EDR^{1/3}$ determined from azimuthal structure function will be used.

The turbulence intensity distribution based on the two-dimensional plan position indicator (PPI) scans is calculated using a similar method.

9.3 Examples of LIDAR-Based Turbulence Intensity

To illustrate the observation of EDR in different kinds of turbulent flow at HKIA, two examples of PPI-based maps are presented in this section.

9.3.1 Springtime Case: 23–25 March 2004

During the morning of 24 March 2004, synoptic-scale high pressure over the south-eastern coast of China brought an easterly airstream to Hong Kong. As shown in the LIDAR's velocity data in the 0° elevation PPI scans (Fig. 9.2a), the wind was mainly from the east, with a slight southerly component. The magnitude of the maximum radial velocity was about 10 m s^{-1}. The $EDR^{1/3}$ calculated from these scans (Fig. 9.2b) was generally small in the vicinity of HKIA at that time, in the order of 0.15 $m^{2/3}$ s^{-1} or less (light and dark blue). There were a few localised regions with higher $EDR^{1/3}$ values. For instance, $EDR^{1/3}$ reached about 0.35 $m^{2/3}$ s^{-1} over a continuous region to the east-northeast of the LIDAR just beyond its blind zone.

Later that day, the wind increased and veered gradually, which resulted in a more southerly component. Accelerated flows emanating from the valleys of Lantau Island became more apparent (Fig. 9.2c). Compared to the situation in the morning, the $EDR^{1/3}$ was generally higher in the vicinity of HKIA (Fig. 9.2d) as a result of higher wind speeds and greater disruption of the southerly airflow by the Lantau terrain. In the evening $EDR^{1/3}$ was on the order of 0.3 $m^{2/3}$ s^{-1} or above within the first 2 km downstream of Lantau Island and gradually decreased northward over the sea. Moreover, the $EDR^{1/3}$ was generally higher (exceeding 0.5 $m^{2/3}$ s^{-1}) just downstream of the Lantau terrain to the west of the LIDAR compared with similar locations to the east of the LIDAR. These regions of higher $EDR^{1/3}$ appear as "flares of red" emanating from the terrain in Fig. 9.2d. The more turbulent airflow in that location may be related to the convergence between the prevailing

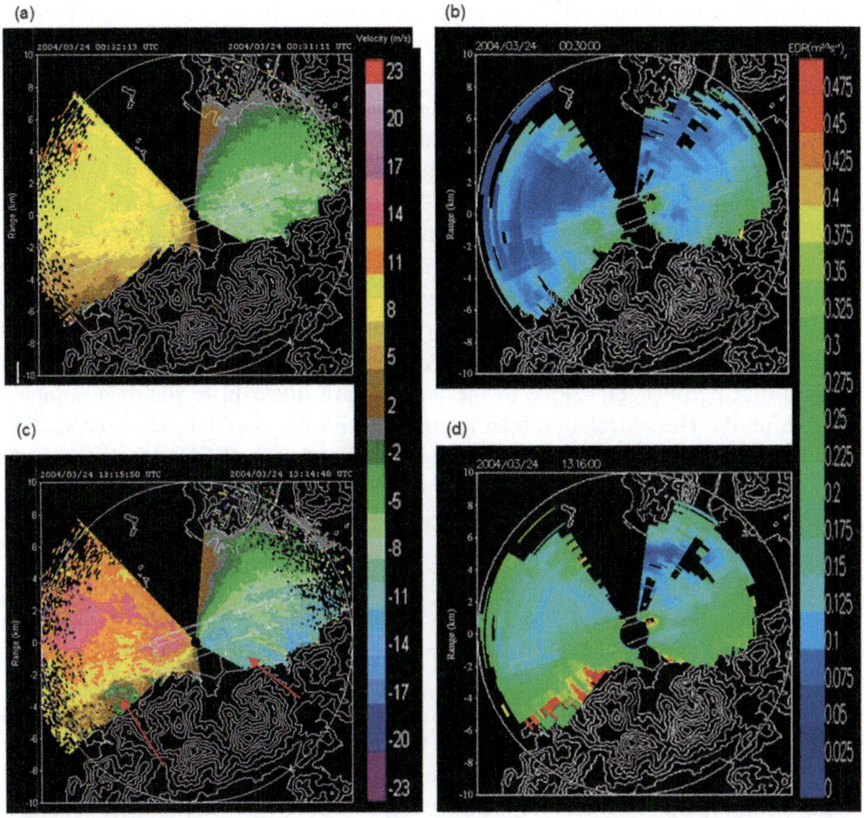

Fig. 9.2 Radial velocity imagery (*left-hand side*) and EDR map (*right-hand side*) obtained from 0° PPI scans of the LIDAR at about 00:30 UTC [(**a**) and (**b**)] and 13:16 UTC [(**c**) and (**d**)] of 24 March 2004. The major valley flows from Lantau Island are indicated by *red arrows* in (**c**). The colour scales of the figures are given on the *right-hand side* of each figure

easterly over HKIA and the southerly valley flow from Lantau Island and the jump/mountain wake in the valley flow (Szeto and Chan 2006).

To examine the quality of the LIDAR-based $EDR^{1/3}$ estimates, we compare them with the $EDR^{1/3}$ values calculated from the wind velocity measurements (at 20 Hz output) of a 3-D sonic anemometer on Lantau Island (see Fig. 9.1 for location). The sonic anemometer was set up at the top of a wind mast of 10 m above the ground. It was located at the top of a small hill having a height of about 166 m above mean sea level.

The $EDR^{1/3}$ calculation method is similar to that used in a past experiment performed on a rather flat area (Chan 2004). However, since the anemometer in this case is mounted on top of a conical hill, the vertical velocity is not always negligible in comparison to the horizontal wind speed. In fact, it is not uncommon to have an upward velocity of several metres per second in strong east to southeasterly winds. Following the common practice, a co-ordinate transformation is applied to the three components of the wind to align the longitudinal axis with the mean wind direction. The $EDR^{1/3}$ is then calculated using maximum likelihood estimator method (Smalikho 1997). This involves the fast Fourier transform of the time series of the horizontal wind velocity (normally the longitudinal component of the wind is used). Technical details are found in Smalikho (1997).

The scatter plot of the $EDR^{1/3}$ calculated from the vertical velocity data of the sonic anemometer and that from the 1° elevation PPI scans of the LIDAR at the location nearest to the anemometer has been prepared (not shown), with the y-intercept set to be zero. The elevation of these PPI scans is chosen because the height of the laser beam at the anemometer's location (137 m AMSL) is close to that of the anemometer itself (176 m AMSL). The plot covers the period 23–25 March 2004. Both LIDAR and anemometer $EDR^{1/3}$ values are determined over 30-min periods. The slope of the least-square linear fit to the data points is close to unity. The correlation coefficient between the two datasets exceeds 0.9, which is considered very high, judging from the fact that the $EDR^{1/3}$ from the sonic anemometer is essentially a point measurement, whereas the corresponding value from the LIDAR represents the turbulence over a much larger area (1 km in range and 20° in azimuth).

Table 9.1 summarises the slopes and correlation coefficients of the least-square linear fits (with y-intercepts set to be zero) between the $EDR^{1/3}$ determined from each of the three components of the wind data of the sonic anemometer and that from the 1° elevation PPI scans of the LIDAR. The slopes are close to unity and the correlation coefficients are about 0.9 in all cases. The sonic anemometer is situated at about 5 km from the LIDAR, which is half of the LIDAR's maximum measurement range. From the present results, the effect of velocity averaging of the LIDAR in the longitudinal and transverse dimensions on $EDR^{1/3}$ estimation (Hannon et al. 2005) is not significant, at least at this range from the LIDAR.

The vertical velocity spectrum is obtained from the sonic anemometer, and structure functions are calculated from the LIDAR data for cases of both weak (04:00 local time) and strong turbulence (21:30 local time). For the vertical velocity spectrum, in general the inertial subrange (with a slope of $-5/3$) is well captured. For the structure function, the calculated values generally fit well with the theoretical model. It is noted that the sampling volume of the sonic anemometer and that of the LIDAR is very different. For the sonic anemometer, it is essentially a point in space. For the LIDAR, it has a range span of about 1 km and an azimuthal span of 20 beams. Moreover, the sonic anemometer is close to the ground (10 m above the hilltop), and the LIDAR's laser beam is much higher above ground (at least 50 m above ground for horizontal scans; even higher for 1° elevation scan). The sampling periods for both instruments are 30 min. The $EDR^{1/3}$ value based on sonic anemometer data and that based on LIDAR data could be quite different due to the differences in the sampling volume and the height above ground.

Table 9.1 Slope (m) and correlation coefficient of the least-square linear fit ($y = m \times x$) between the $EDR^{1/3}$ derived from different wind components of the sonic anemometer (y) and that calculated from 1° PPI scan of the LIDAR (x) at the anemometer's location

Wind component of the sonic anemometer to calculate $EDR^{1/3}$	Slope	Correlation coefficient (R)
Vertical	0.9981	0.911
Longitudinal	1.1274	0.903
Transverse	0.9177	0.887

9.3.2 Typhoon Case: 23–27 August 2003

Typhoon Krovanh tracked west-northwestward across the northern part of the South China Sea in late August 2003. It brought east to southeasterly storm-force winds to Hong Kong. For instance, at about 1:30 a.m. of 25 August 2003 (17:30 UTC of 24 August), the 1° elevation PPI scans of the LIDAR show radial velocities close to 20 m s^{-1} in the vicinity of HKIA (Fig. 9.3a). Because of the high humidity, the measurement range of the LIDAR was reduced. There were not much velocity data beyond 8 km or so to the west of the LIDAR.

Similar to the above springtime case, the airflow is more turbulent in the areas just downstream of the Lantau terrain to the west of the LIDAR compared with the corresponding locations east of the LIDAR, as revealed by the presence of many small-scale features (with horizontal dimensions of several hundred metres) in the radial velocity imagery (Fig. 9.3a). This pattern is also confirmed from the EDR map (Fig. 9.3b). Once again, "flares" of high $EDR^{1/3}$ values (around 0.5 $m^{2/3}$ s^{-1}) emanate from the Lantau terrain to the southwest of LIDAR. They are more extensive in size compared to the springtime case and affect almost the whole western approach corridor of the south runway.

The $EDR^{1/3}$ determined from 1° elevation PPI scans of the LIDAR is compared in Table 9.2 with that derived from sonic anemometer winds. Compared to the springtime case, the slopes deviate slightly more from unity and the correlation coefficients are generally smaller, but still reach at least 0.92 and 0.8, respectively. The calculation of $EDR^{1/3}$ in the typhoon case is more challenging because of the limited measurement range of the LIDAR and the missing/erroneous wind data from the sonic anemometer in episodes of heavy rain. Given that, the present comparison results are considered to be satisfactory.

9.4 Comparison with Flight Data

The Observatory regularly obtains flight data from commercial jets and a fixed-wing aircraft operated by the Government Flying Service (GFS). Such data would be useful in the development of turbulence detection algorithms and verification of

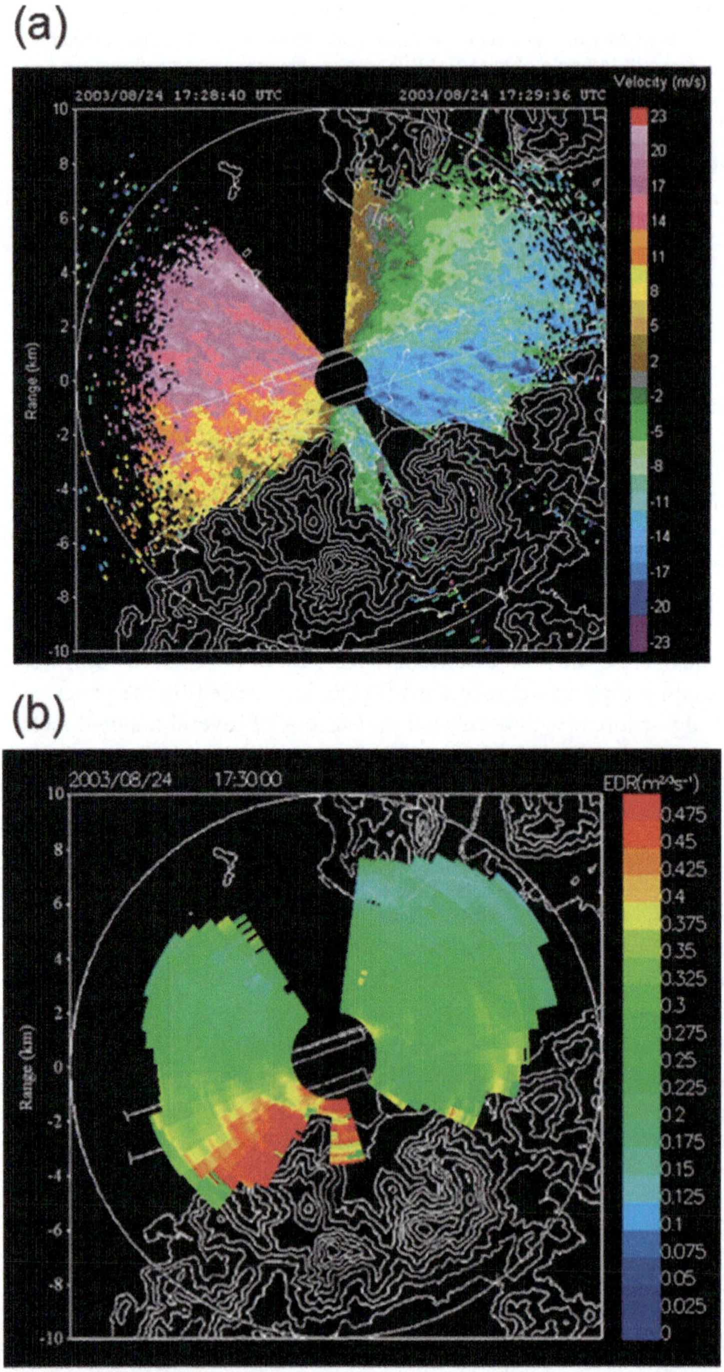

Fig. 9.3 Radial velocity imagery from 1° PPI scan of the LIDAR (**a**) and the EDR map (**b**) at around 17:30 UTC, 24 August 2003. The colour scales of the figures are given on the *right-hand side* of each figure

Table 9.2 Same as Table 9.1, but for the period 23–27 August 2003

Wind component of the sonic anemometer to calculate $EDR^{1/3}$	Slope	Correlation coefficient (*R*)
Vertical	0.9283	0.851
Longitudinal	0.9881	0.799
Transverse	0.9247	0.839

the turbulence intensity as calculated from different remote-sensing instruments, including the LIDARs. A comparison between headwind data as estimated by LIDAR glide-path scans, and as measured on board an instrumented aircraft of GFS over corridors 07LA and 25RA, is first presented here.

For the GFS fixed-wing aircraft, a high-resolution data probe, capable of measuring horizontal wind components, pressure and humidity at a frequency up to 20 Hz, is installed. Technical specifications of the data probe (AIMMS-20) can be found in Beswick et al. (2008). Data collection flights are regularly carried out (about three times every week) during which the aircraft performs 2–3 take-off/touchdown cycles. With reference to the LIDAR glide-path scanning strategy, elevation angles of 3 and 6° are observed, respectively, for touchdown and take-off motions. Between the months of January and September from 2010 to 2012, a total of 333 (270) measured headwind profiles have been available over 07LA (25RA).

Given the high temporal resolution (20 Hz) of wind data collected by AIMMS-20, some form of data thinning is required for a fair comparison with LIDAR glide-path scans, which come with a line-of-sight resolution of about 100 m. In the current study, the distance *d* is the distance along the extended centre line of the runway away from the runway threshold of each LIDAR datapoint is first identified, and a corresponding datapoint is extracted from aircraft measurements at the same position *d*. Spatial location of the two data points would coincide perfectly if they follow the same elevation angle (3° for arrival) as well as touchdown location, here assumed at the runway threshold. In reality this is not always the case. Additional quality control (QC) is therefore implemented by filtering pairs of data points with horizontal distance over 0.25 nautical miles (about 460 m) away from each other.

Figure 9.4 shows the scatter plot of LIDAR-estimated and measured headwinds over corridors 07LA and 25RA. In both cases strong positive correlations can be observed between the two data sources over a wide range of velocities from −10 to +30 knots (about −5 to +15 m s^{-1}). Linear regression gives a slope of 0.88 (0.90) and intercept of 0.29 (0.04) knot with a correlation coefficient R^2 of 0.80 (0.77) over 07LA (25RA), indicating that on average LIDAR values are about 10 % higher than GFS measurements. For both corridors over 90 % of data pairs (histogram not shown) are within 5 knot (2.5 m s^{-1}) from each other.

An algorithm for calculating EDR from AIMMS-20 measurements based on direct fitting of the inertial subrange of the turbulence spectrum, developed by the

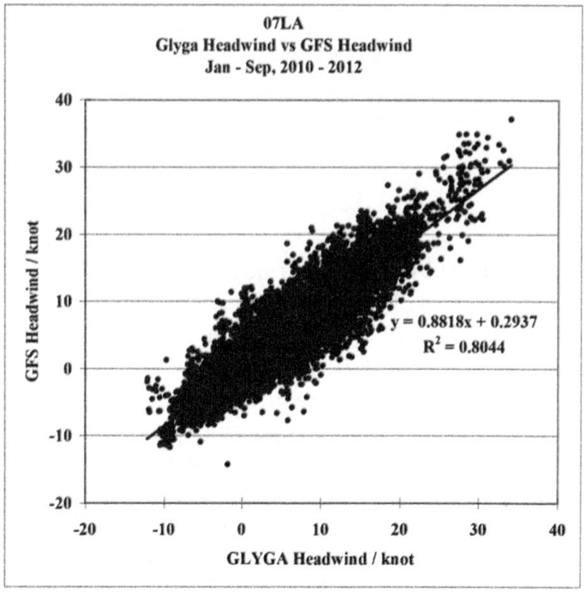

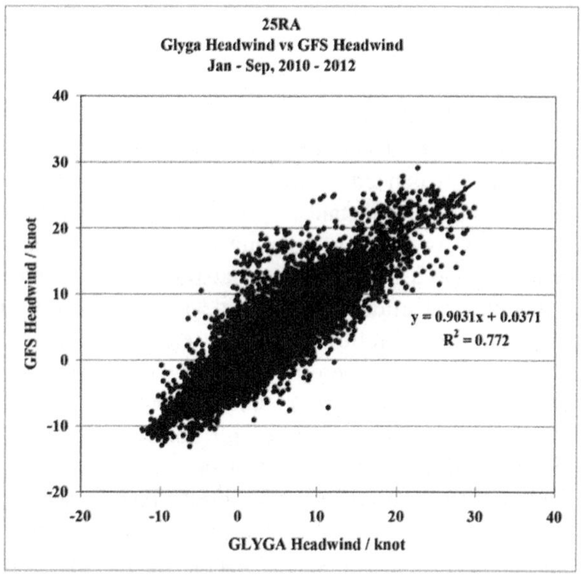

Fig. 9.4 Comparison of headwind measured on board a GFS fixed-wing aircraft during touchdown/take off at HKIA and that from the corresponding LIDAR glide-path scan for corridors 07LA (*top*) and 25RA (*bottom*)

National Aerospace Laboratory and similar to De Bruin and Haverdings (2007), has been available to process data collected during the above GFS flights. Comparison of EDR values derived from LIDAR and flight data are shown in Fig. 9.5, where a positive linear correlation can be observed for runway corridor 07LA.

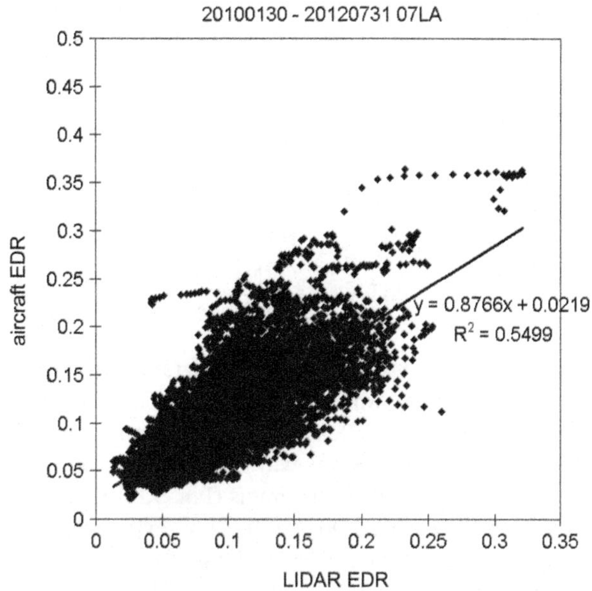

Fig. 9.5 Comparison of EDR between research aircraft and LIDAR for runway corridor 07LA

9.5 Performance on Alerting of Low-Level Turbulence

Having established the correlation between in situ measurements and LIDAR-derived analogues, for both headwinds, and EDR profiles, we now explore the application of EDR profiles for diagnosing low-level turbulence at HKIA.

Following the method described in Sect. 9.2, EDR profiles along aircraft glide paths are calculated at 15-min intervals (available at 00, 15, 30 and 45 min of each hour) for the months of January to September 2008–2012 over arrival corridors 07LA and 25RA of HKIA. These profiles are available at a line-of-sight resolution of about 100 m and cover a distance up to 6–8 km (about 3–4 nautical miles) along the extended centre line of the runway away from the respective runway threshold, depending on meteorological conditions. A threshold-based approach is then applied such that an alert will be triggered whenever a representative value of EDR from the LIDAR-derived profile exceeds a predetermined threshold. Each alert is valid for 15 min and is based on the EDR profile calculated from LIDAR sector/glide-path scans over the previous 15 min (e.g. an alert issued at 11:15 UTC is based on the EDR profile calculated using data between 11:00 and 11:15 UTC and will be valid until 11:30 UTC). In what follows, the use of both the maximum and the median EDR values from each profile as the representative EDR value will be explored.

These alerts are then validated against pilot reports of significant low-level turbulence (intensity of moderate or above) received during the study period. A total of 89 (48) turbulence events have been reported over 07LA (25RA). An event of turbulence is taken as a "hit" if its time of occurrence falls within a 15-min interval where an EDR-based alert is valid. By applying a range of alert thresholds, overall skill level of the EDR-based alerting method can be assessed using a modified relative operating characteristics (ROC) diagram. The vertical and horizontal axes are, respectively, the percentage of detection (PoD) and the percentage of time on alert (PoTA), defined as:

$$\text{PoD} = \frac{\text{No. of "Hits"}}{\text{Total No. of Pilot Reports}} \times 100\,\% \tag{9.11}$$

$$\text{PoTA} = \frac{\text{No. of Alerts Issued}}{\text{Total No. of 15-min Intervals}} \times 100\,\% \tag{9.12}$$

As with the conventional ROC diagram, high skill level is indicated by proximity to the top-left corner, i.e. high hit rate at low alert duration. Note that by using PoTA as the x-axis instead of the false alarm ratio, a perfect forecast with PoD of 100 % and no false alarms will not lie exactly on the top-left corner.

In the resulting ROC diagrams (Fig. 9.6), EDR-based alerts show considerable skill with PoD around 90 % (80 %) at PoTA of 10 % for 07LA (25RA). For both corridors, a slight gain in PoD is obtained by issuing alerts using the median EDR value (2–3 % for 07LA, 5–10 % for 25RA) while keeping PoTA constant. Also included for comparison is the performance of the HKO winds hear and turbulence warning system (WTWS), currently in operation at HKIA, which is characterised by a moderate PoD of 50.6 % (44.4 %) at a very low alert duration of 0.24 % (0.64 %) for 07LA (25RA) over the period 2008–2012 (October to December included). Comparable performance can be obtained from the EDR-based approach using a higher alert threshold, with PoD and PoTA of 47.2 and 0.56 % for 07LA and 54.2 and 1.0 % for 25RA.

The application of LIDAR-derived EDR profiles can be illustrated in the following case. On 22 June 2011, passage of Tropical Storm Haima across the northern part of the South China Sea was responsible for gale force winds both offshore and in Hong Kong. Interactions of low-level southeasterly flows with the Lantau Mountains led to disturbed winds around HKIA. At 22 UTC, a pilot on board an A330 aircraft landing from the west on corridor 07LA reported an encounter of "moderate turbulence". Despite a positive offset as well as a lack of some fine-scale features, the LIDAR profile corresponds well to the EDR values calculated using the quick access recorder (QAR) data measured on board, which gave a maximum of 0.39 $m^{2/3}\ s^{-1}$ at a similar location.

9.6 Summary and Conclusions

In this chapter, the formulation of LIDAR-derived EDR has been reviewed. It is based on the structure function approach using two calculation methods, namely, azimuthal structure function and longitudinal structure function, and is found to

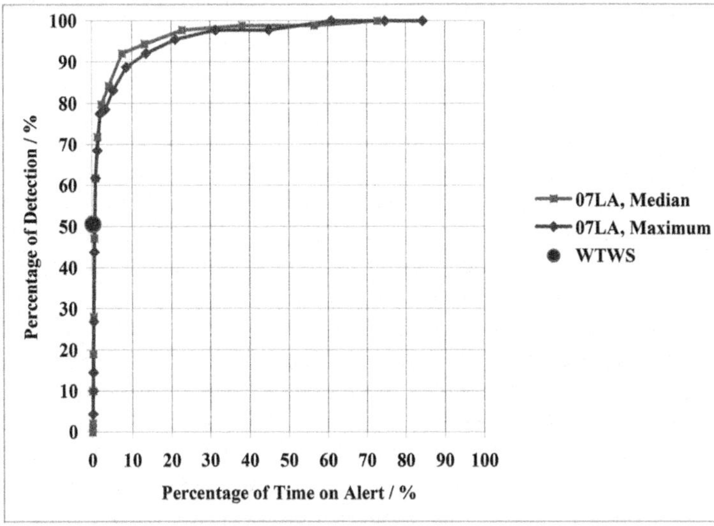

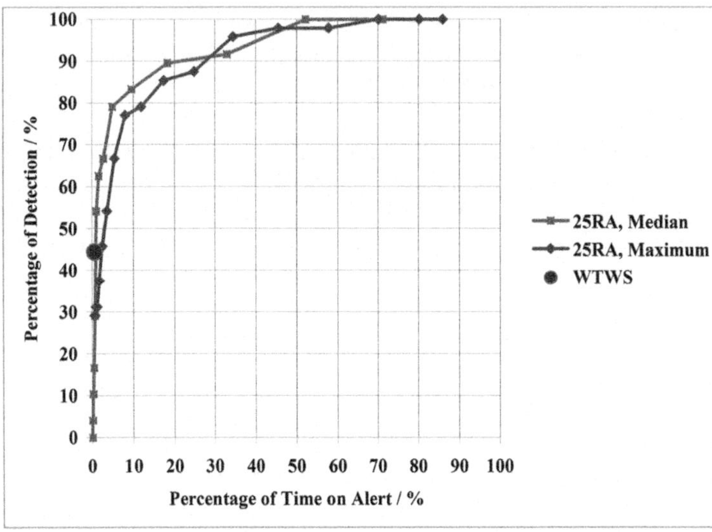

Fig. 9.6 Threshold-based performance of low-level turbulence alerting on corridors 07LA (*top*) and 25RA (*bottom*) at HKIA using LIDAR-derived EDR along glide paths over January to September 2008–2012. Alerts are issued in 15-min intervals either based on the median (*median, grey*) or maximum (*maximum, black*) values from each EDR profile. The performance of the operational wind shear and turbulence warning system of HKO ("WTWS", *dot*) is included for comparison

give comparable results based on the LIDAR measurements at HKIA. An alternative approach is to use the spectrum width of the LIDAR signal return, but there may be background noise in the equipment (e.g. electronic noise) that needs to be removed, and the calculation of EDR from this approach is not so trivial.

The structure function approach may be applied to two kinds of scanning data of the LIDAR, namely, the glide-path scans and the PPI scans. They have been implemented in real time at HKIA and serve different purposes. The glide-path-scan-based EDR can be used to construct the EDR profile, similar to the headwind profile obtained from this kind of scan, and such profiles may be used to alert the aircraft on the possibility of encountering significant turbulence. On the other hand, the PPI-scan-based EDR gives an overview of the distribution of turbulence intensity over the airport area and could be useful in the understanding and monitoring of turbulence associated with terrain-disrupted airflow, sea breeze fronts and gust fronts associated with thunderstorms.

These two kinds of EDR information are studied in detail later in the chapter. The PPI scans are used to review the overall pattern of EDR in two examples of terrain-disrupted airflow, namely, in springtime easterly to southeasterly flow in stable boundary layers and in turbulent flow associated with TCs. In contrast, glide-path-scan EDRs are used to illustrate the accuracy of the LIDAR-based EDR by comparison with the EDR obtained from a research aircraft. The LIDAR-based EDRs and aircraft-based EDRs are found to have good correlations, illustrating the usefulness of the former in the monitoring of turbulence in the airport area.

The performance of LIDAR-based EDR in the alerting of low-level turbulence is then studied by comparisons with pilot reports of turbulence encounters. Though the sample size of pilot reports is relatively small, the skill of the LIDAR-based EDR is demonstrated using ROC curves.

Looking into the future, the EDR formulation may be applied to the short-range LIDAR data, which are used for monitoring wind shear and turbulence associated with buildings, which are much smaller in spatial scale. The matter is still being actively pursued in Hong Kong and would be reported in future publications.

References

Beswick, K.M., Gallagher, M.W., Webb, A.R., Norton, E.G., Perry, F.: Application of the Aventech AIMMS20AQ airborne probe for turbulence measurements during the convective storm initiation project. Atmos. Chem. Phys. **8**, 5449–5463 (2008)

Chan, P.W.: Measurement of eddy dissipation rate by a mini-sodar for aviation application: comparison with tower measurement. In: Proceedings of the 11th Conference on Aviation, Range, and Aerospace Meteorology, American Meteorological Society, Massachusetts (on-line proceedings) (2004)

De Bruin, A.C., Haverdings, H.: Validation of an eddy dissipation rate calculation method, based on flight data recording data. NLR-CR-2007-540, National Aerospace Laboratory, Amsterdam (2007)

Frehlich, R.G.: Estimation of velocity error for Doppler lidar measurements. J. Atmos. Oceanic Technol. **18**, 1628–1639 (2001)

Frehlich, R.G., Cornman, L.: Estimating spatial velocity statistics with coherent Doppler Lidar. J. Atmos. Oceanic Technol. **19**, 355–366 (2002)

Frehlich, R.G., et al.: Measurements of boundary layer profiles in an urban environment. J. Appl. Meteorol. Climatol. **45**, 821–837 (2006)

Hannon, S.M., Pelk, J.V., Benda, P.: Autonomous Doppler Lidar wind and aerosol measurements for Pentagon Shield. In: Proceedings of the 2nd Symposium on Lidar Atmospheric Applications, American Meteorological Society, California (on-line proceedings) (2005)

Shun, C.M., Chan, P.W.: Applications of an infrared Doppler Lidar in detection of wind shear. J. Atmos. Oceanic Technol. **25**, 637–655 (2008)

Smalikho, I.N.: Accuracy of turbulent energy dissipation rate estimation from wind velocity temporal spectrum. Atmos. Oceanic Opt. **10**, 559–563 (1997)

Szeto, K.C., Chan, P.W.: High resolution numerical modelling of windshear episodes at the Hong Kong International Airport. In: Proceedings of the 12th Conference on Aviation, Range, and Aerospace Meteorology, American Meteorological Society, Georgia (on-line proceedings) (2006)

Part III
Nowcasting, Forecasting, and Verification

Chapter 10
A Summary of Turbulence Forecasting Techniques Used by the National Weather Service

David R. Bright, Steven A. Lack, and Jesse A. Sparks

Abstract A summary of turbulence forecasting by the US National Weather Service (NWS) is provided. This chapter begins with a historical perspective, beginning with the Wright brothers and ending with NextGen. Following the historical perspective, present-day aviation-based meteorological operations are described. This discussion covers the services provided by the NWS and its partners and includes a foundational summary of present-day (2015) forecasting techniques. This chapter closes with a narrative on current and future prediction, built around numerical models and ensemble systems. These systems will drive intelligent information extraction and decision support services provided by operational meteorologists, leading to a safer and weather-ready airspace.

10.1 A Short History of National, Government-Based Aviation Meteorological Services

Beginning with the first flight of the Wright brothers in the early 1900s, the US government via the Weather Bureau, now the National Oceanic and Atmospheric Administration (NOAA) National Weather Service (NWS), has supported the safety and efficiency of aviation operations (Whitnah 1961). As aviation operations continued to expand through the early 1920s, the Air Commerce Act of 1926 was established (Congressional Record 1926). This act set the groundwork for the first official requirements for providing aviation weather services. The act stated (section 5, part e):

D.R. Bright (✉)
NOAA/National Weather Service, 5241 NE 122nd Avenue, Portland, OR 97230-1089, USA
e-mail: david.bright@noaa.gov

S.A. Lack • J.A. Sparks
NOAA/National Weather Service, Kansas City, MO, USA
e-mail: steven.a.lack@noaa.gov; jesse.sparks@noaa.gov

R. Sharman, T. Lane (eds.), *Aviation Turbulence*,
DOI 10.1007/978-3-319-23630-8_10

> It shall be the duty of the Secretary of Commerce ... (a) to furnish such weather reports, forecasts, warnings, and advices as may be required to promote the safety and efficiency of air navigation in the United States and above the high seas, particularly upon the civil airways designated by the Secretary of Commerce under authority of law as routes suitable for air commerce, and (b) for such purposes to observe, measure, and investigate atmospheric phenomena and establish meteorological offices and stations.

Air travel boomed in the years following World War II, and the dawn of the jet age in the early 1950s saw the aviation industry on the brink of rapid expansion. With the expansion came accidents and a need to further manage the airspace. The 1958 Federal Aviation Act (Congressional Record 1958) established the modern-day Federal Aviation Administration (FAA), then known as the Federal Aviation Agency. The Federal Aviation Act set an important precedent requiring the Weather Bureau to work with the FAA to provide the necessary aviation weather products to govern the safety and efficiency of flight. Public Law 85-726 Section 310 states:

> The Administrator is empowered and directed to make recommendations to the Secretary of Commerce for providing meteorological service necessary for the safe and efficient movement of aircraft in air commerce. In providing meteorological services, the Secretary of Commerce shall cooperate with the Administrator and give full consideration to such recommendations.

Over the years, the Weather Bureau and later the NWS established specialty forecast products and services to specifically handle aviation elements. Today, the NWS Weather Forecast Offices (WFOs) provide the Terminal Aerodrome Forecast (TAF) at over 600 airports across the USA and its territories. The Center Weather Service Unit (CWSU) provides Center Weather Advisories (CWAs) and Meteorological Impact Statements (MISs) for the 21 air traffic control centers in the USA. The CWSUs are colocated with FAA air traffic operations to provide continuous support for weather impacting the various FAA centers. The National Aviation Weather Advisory Unit (NAWAU), which is now the Aviation Weather Center (AWC), was established in Kansas City, MO in 1982 to provide national aviation support for the continental USA (CONUS). The initial requirements of the center were to issue routinely available area forecasts (FAs), Airman's Meteorological Information statements (AIRMETs), Significant Meteorological Information statements (SIGMETs), and Convective SIGMETs (Fig. 10.1). In addition to the AWC, the Alaska Aviation Weather Unit (AAWU) and the WFO in Honolulu, HI (HFO) provide AIRMET and SIGMET issuances. This information is shared and distributed publicly to pilots via the web and by FAA Flight Service Station specialists. The Aeronautical Information Manual (AIM), specifically Chap. 7, lays out the framework for responsibilities for the creation and dissemination of weather information to pilots (FAA 2014).

AIRMETS are issued for low ceiling and visibility, moderate turbulence, moderate icing, mountain obscuration, and strong surface winds. SIGMETS cover convection, severe turbulence, severe icing, widespread dust or sandstorms, volcanic ash, and tropical cyclones. Specific to turbulence, the AIRMET Tango package forecasts areas of moderate or greater turbulence in two layers of the atmosphere:

Fig. 10.1 AWC forecaster producing convective SIGMETs over the CONUS

below FL180 and at or above FL180 (FL180 refers to Flight Level 18,000 feet). Also included in the Tango AIRMET suite are sustained surface winds over 30 kt and low-level wind shear. Currently, AIRMETs are derived from the Graphical Airmen's Meteorological Information statements (G-AIRMET). The first 6 h of the G-AIRMET correspond to the 6-h period of the AIRMET. The SIGMET is an unscheduled product that is valid for a maximum duration of 4 h. The turbulence SIGMET warns of severe or extreme turbulence or clear-air turbulence (CAT) not associated with thunderstorms. Both AIRMETs and SIGMETs are designed to capture widespread events or, more precisely, conditions affecting areas of 3000 square miles or larger at any moment in time. For smaller areas, the CWSU will often issue unscheduled CWAs as a complement to the AWC products. AIRMETs and SIGMETs will be discussed in more detail shortly.

Going forward, the FAA Modernization and Reform Act of 2012 (Congressional Record 2012) further defines roles and responsibilities for aviation meteorological services in the era of the Next Generation Air Transportation System (NextGen). The enhancement of turbulence forecasts and warnings remains an important milestone for both the FAA and NWS and an ongoing topic of research. Additionally, the National Transportation Safety Board (NTSB) has recently advocated for further NWS operational products related to mountain wave activity (MWA) and turbulence (NTSB 2014).

10.2 Meteorological Operations in the National Weather Service

The NWS has meteorological watch responsibility for the USA and adjacent coastal waters, a large area of the central and eastern Pacific Ocean north of 5°N latitude, and a large portion of the western Atlantic Ocean, Gulf of Mexico, and Caribbean Sea (Fig. 10.2). A meteorological watch office (MWO) is designated by regional air navigation agreement with the International Civil Aviation Organization (ICAO) to provide information concerning the occurrence or expected occurrence of specified en route weather phenomena and natural and other hazards which may affect the safety of aircraft operations within a specified area of responsibility (ICAO Annex 3 2013). This responsibility is shared by three MWOs: the AWC in Kansas City, MO; the AAWU in Anchorage, AK; and the Weather Forecast Office in Honolulu, HI (HFO). These offices provide constant monitoring and notification of turbulence that impacts flight safety.

The SIGMET is the written mechanism for warning aviators of adverse turbulent conditions (or other aviation hazards) and ensuring safety of aircraft operations. SIGMETs are concise, brief descriptions of the occurrence in time and space of specified en route weather phenomena which may affect the safety of aircraft operations. SIGMETs are issued by any of the US MWOs on an event-driven basis when severe turbulence is forecast or observed within the US area of responsibility.

The AIRMET is issued from the MWO as a planning and decision-making tool to help aviators maintain high safety margins away from bad weather while

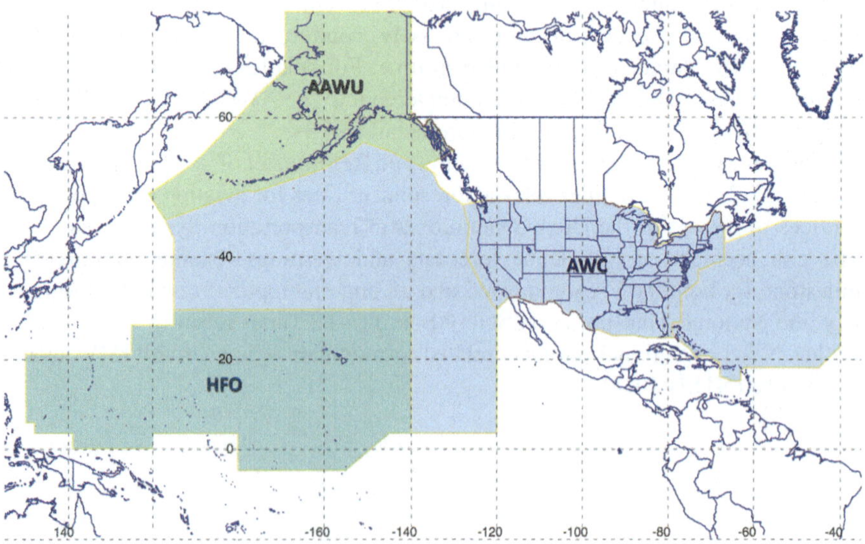

Fig. 10.2 US geographic areas of meteorological watch responsibility

planning more efficient routes. AIRMETs are broad-scale, text-only descriptions of hazardous weather across a region and are valid for a 6-h period with a 6-h outlook. The G-AIRMET is issued by the AWC over the CONUS and produces snapshot graphical forecasts of hazardous weather valid at discrete 3-hour time frames out to 12 hours. The G-AIRMET identifies hazardous weather in space and time more precisely than the AIRMET text. AIRMETs/G-AIRMETs are issued by the US MWOs on a routine-scheduled basis (4 times per day) when moderate or greater turbulence is forecast or observed within the U.S. area of responsibility; note though, only AWC issues the G-AIRMET over the contiguous U.S. and adjacent waters.

Through partnerships with the FAA and aviation industry, NWS meteorologists in Center Weather Service Units (CWSUs) are embedded within the FAA's 21 Air Route Traffic Control Centers (ARTCCs). The NWS also has National Aviation Meteorologists (NAMs) stationed at the FAA's Air Traffic Control System Command Center (ATCSCC), providing real-time direct weather support to air traffic control and management all over the country. Traffic managers are advised of adverse turbulent conditions by meteorologists on site and through collaboration from US MWOs on a 24-h basis to better plan large-scale traffic management initiatives, smaller-scale air traffic deviations, and airspace allocation. When conditions warrant, a Center Weather Advisory (CWA) is issued by the CWSU. The CWA is an aviation warning that is issued when conditions are approaching SIGMET/AIRMET criteria. They are issued on an event-driven basis regardless of areal extent of the hazard.

Outside the USA, the World Area Forecast System (WAFS) was established by ICAO and the World Meteorological Organization (WMO) to improve the quality and consistency of en route guidance provided for international aircraft operations. Two World Area Forecast Centers (WAFCs) were established: WAFC-Washington (USA) and WAFC-London (UK). Both WAFCs are simultaneously generating graphical representations of significant weather (SIGWX) forecasts between FL250 and FL630 across the globe four times daily. Forecasts are provided over the North Atlantic domain for significant weather between FL100 and FL450. These SIGWX forecasts indicate the 24-h future position of jet streams, tropopause heights, convection, and moderate or greater turbulence. They provide global airline operations with all the weather hazard information needed to plan medium and long-haul flights most efficiently. The SIGWX forecast provides both clear-air turbulence (CAT; shear-based) and in-cloud turbulence (ICTURB; instability-based) forecasts. The latter is limited to the North Atlantic domain.

The WAFCs are also generating gridded WAFS forecasts from a blend of numerical model output between the US Global Forecast System (GFS) and the UK Meteorological Office model (UKMET). The final forecast is provided in two sets, a mean grid output and a maximum grid output, that give the user the average between and the maximum of the GFS and UKMET grids. Both the US and UK

CAT algorithms are based heavily on the Ellrod Index (Ellrod and Knapp 1992) while the ICTURB algorithms are based on the model indicating the presence of a cloud, and the change in equivalent potential temperature with height within the cloud, which is a measure of instability. These data are used as objective, numerical input into user decision models.

10.3 Operational Forecasting Techniques

Meteorologists in the NWS employ turbulence forecasting techniques, based on recognizing synoptic-scale patterns conducive to turbulence-generating mechanisms, for analyses and numerical weather prediction output. The latter includes the application of mesoscale and synoptic-scale diagnostics and algorithms to diagnose the ingredients necessary for turbulence generation. Observations of turbulence from pilots and aircraft sensors are the ultimate "truth" used to verify and improve turbulence forecasts.

10.3.1 Turbulence-Generating Mechanisms

Forecasting turbulence requires an understating of the mechanisms that generate turbulent eddies in the atmosphere. These mechanisms are described in more detail in other chapters of this book. NWS forecasters are constantly assessing the state of the atmosphere to diagnose areas where these mechanisms may occur. Large-scale conditions that are conducive to the formation of turbulence are briefly described as follows:

(a) Wind Shear Along a Density Interface Wind shear is the most important ingredient for generating turbulence. The wind shear in the atmosphere can be vertical or horizontal and speed or directional. A density discontinuity occurs on the boundary of dissimilar air masses and is characterized meteorologically by any of the following:

- Deformation
- Frontogenesis
- Temperature gradients/thermal advections aloft
- Moisture gradients/moisture advections aloft
- Lapse rate discontinuities

When shear is sufficiently large, turbulent waves can form along the density interface (Kelvin-Helmholtz waves).

(b) Breakdown of Anticyclonic Flow Anticyclonic flow (clockwise in the northern hemisphere; counterclockwise in the southern hemisphere) can only build to a certain limit until the absolute rotation opposes the planetary rotation (i.e., the absolute vorticity becomes the opposite sign of the planetary vorticity). This manifests as inertial instability, which causes a horizontal acceleration of air parcels and increased wind shear. In addition, the subsequent spontaneous breakdown of balance results in gravity wave formation which may also lead to turbulence.

(c) Unbalanced (Ageostrophic) Flow Perfectly geostrophically balanced flow is rare in the atmosphere. Geostrophic adjustment, the process which brings unbalanced flow back toward geostrophic balance, results in gravity wave formation. Ageostrophic flow also enhances vertical wind shear. Significantly unbalanced flow results from the following regimes:

- Convergence/divergence
- Jet entrance/exit regions
- Translating jet streaks
- Intensifying jet streaks
- Curvature in the flow (troughs and ridges)

(d) Mountain Waves/Terrain Enhancement Waves in the lee of mountains and low-level rotors just downstream of the ridge crest can occur when there is appreciable cross-mountain flow. These waves can become turbulent if (1) an inversion exists just over the ridge top with slightly less stable conditions above the inversion and strong wind shear below the inversion. This would result in low- to mid-level turbulence. (2) Strong wind shear above the ridge top with stable conditions throughout the atmosphere. This would result in turbulence near the tropopause level.

These processes result in generation of atmospheric gravity waves. If the atmosphere is in a stable state and static stability is low enough, these waves continue to grow in amplitude and can begin to overturn or break resulting in turbulence.

10.3.2 Pattern Recognition of Meteorological Features

The following synoptic-scale patterns are commonly associated with moderate or greater turbulence and are identified frequently by meteorologists as an aid in forecasting turbulence. This list is by no means an all-inclusive list of patterns conducive to turbulence:

(a) Amplifying Ridge Turbulence is most likely where the jet core experiences the greatest latitudinal displacement in the amplifying ridge. In this pattern, the breakdown of anticyclonic flow releases inertial instability, while the translating jet streak contributes to subsidence on the periphery of the building ridge. Furthermore, the curvature of the flow increases geostrophic imbalance and strong wind shear is present with the jet streak.

(b) Strong Wind Max Upstream of Trough Turbulence is most likely where the jet streak approaches the rear of an upper trough. In this pattern, convergence upstream of the trough axis contributes to subsidence and gravity wave generation. Also, curvature of the flow associated with the trough axis increases geostrophic imbalance. The ageostrophic response enhances vertical wind shear while a region of cold-air advection aloft behind the trough axis enhances the air mass density discontinuity. The resulting wind shear/density discontinuity generates KH waves and turbulence.

(c) Surface Cyclogenesis Turbulence is most likely on the cold side of the surface low pressure. In this region, a strong wind shear/density discontinuity exists in regions north/northwest (south/southwest in the southern hemisphere) of the surface low. Baroclinic waves tilt poleward with height, resulting in a sloped upper front toward the west/northwest (west/southwest in the southern hemisphere) of the surface low. A wind and temperature cross section from the surface to the jet axis reveals isotachs that are sloped from the jet axis to the surface cold front, with strong temperature gradients and wind shear along the frontal interface. The poleward tilt of the upper front eventually intersects a relatively low tropopause, where jet streak-induced tropopause folding is occurring and turbulence is enhanced.

(d) Upper Cutoff Low Turbulence is most likely in the "throat" region of an upper level cutoff low. Within this region is a strong shear zone which separates the two opposing flow regimes between the cutoff low and the main westerly flow. As a rule of thumb, sufficient shear exists for moderate or greater turbulence when the strength of the two opposing flows is at least 50 kt. Strong deformation and subsidence in the shear zone enhances the horizontal density gradient providing the necessary density discontinuity.

(e) Positively Tilted Trough Turbulence is most likely near the trough axis, particularly when there is stronger flow on the downstream side. Very strong directional shear exists as weak anticyclonic flow west of trough impinges on cyclonic flow east of the trough. Divergence downstream of the trough axis contributes to ascent and gravity wave generation. Curvature associated with the upstream trough increases geostrophic imbalance. The embedded downstream jet streak induces tropopause folding and mid-level drying, therefore enhancing the density discontinuity.

(f) Strong Low-Level Flow Low-level turbulence is most likely where (1) steep low-level lapse rates exist under strong lower tropospheric flow (convective boundary layer turbulence—see below), and (2) the flow is normal to the orientation of terrain features (low-level terrain-induced turbulence—see below).

(g) Convective Boundary Layer Turbulence Forecasters are generally looking for steep low-level lapse rates where buoyant mixing will transfer stronger winds and momentum from the top of the boundary layer. Strong low-level flow with cold-air advection and strong surface heating is a particularly favorable setup for turbulent mixing.

(h) Low-Level Terrain-Induced Turbulence The stronger the flow and the more normal the directional component is to the terrain, the more likely it is that low-level turbulence will be generated.

10.3.3 Numerical Diagnostics and Turbulence Algorithms

Many NWP-based algorithms exist in an effort to diagnose sufficient atmospheric ingredients for turbulence generation. A few of the more common algorithms used by NWS forecasters are listed below.

(a) Absolute Vorticity Absolute Vorticity <0 (i.e. inertial instability) in the northern hemisphere (and >0 in the southern hemisphere) indicates potential for turbulence (Knox 1997). As a result of this instability, horizontal acceleration of air parcels occurs and a spontaneous breakdown of balance may result in gravity wave formation.

(b) Richardson Number (Ri) Static stability/vertical wind shear (VWS) (Endlich 1964). This is a measure of dynamic (Kelvin-Helmholtz) instability caused by wind shear. A value <1 indicates that VWS dominates the stability and allows KH waves to grow in amplitude and result in turbulence. A value >1 means that it is too stable for turbulence and KH waves will dampen out.

(c) Divergence Tendency Total divergence tendency causes gravity waves in a stable environment due to unbalanced flow and geostrophic adjustment (McCann 2001). Divergence tendency also reduces the Ri which is more favorable for turbulence. At AWC, divergence tendency is computed from the primary terms in the nonlinear balance equation.

(d) Ellrod (Ellrod and Knapp 1992) This index is computed over specified composite layers. It accounts for the dynamical generation of turbulent waves and eddies from wind shear and stretching. It is very skillful in determining the *existence* of turbulence but has limited skill in determining the *intensity* of turbulence. It does not account for static stability present in the atmosphere which could play an important role in determining intensity.

(e) Ellrod-Knox (Ellrod and Knox 2010) This index combines Ellrod with divergence tendency to pinpoint areas where shear and unbalanced flow are contributing to turbulence.

10.3.4 Turbulence Detection and Incorporation into Forecasts

Observations of turbulence are crucial to creating better forecasts and more timely warnings of turbulence. There are two primary sources of turbulence observations available to NWS forecasters:

(a) Pilot Reports (PIREPs) PIREPs are relayed manually from pilots and indicate turbulence conditions encountered by an aircraft while in flight. The approximate location of the turbulence is based on distance and direction to a known navigation aid. PIREPs are used continuously by forecasters to substantiate turbulence forecasts and refine when needed. PIREPs indicating severe turbulence, when occurring in a region favorable for turbulence, are usually the primary trigger for issuing SIGMETs. Guidelines on turbulence intensity used in PIREPs are outlined in the AIM, Chap. 7, pp. 47–48 (FAA 2014).

While PIREPs are the most abundant en route observations received by forecasters, they are very qualitative and subjective in nature since they depend upon the pilot's sensitivity and perception of the intensity of turbulence, which in turn depends upon the size, speed, angle of attack, and aerodynamic characteristics of the aircraft being flown. In addition, significant inaccuracies can exist in the reported time and location of the PIREP (Schwartz 1996; Sharman et al. 2014).

(b) In Situ EDR Software to estimate energy (or eddy) dissipation rate (EDR) have been installed on several commercial air carriers (e.g., Sharman et al. 2014) and provide quantitative, objective, and aircraft-independent measurements of atmospheric turbulence intensity. Mean and peak turbulence intensity (EDR) is automatically computed every minute with accuracy of the occurrence of turbulence within 1 min in time and ~10 km in space. Therefore, these observations are very reliable and tremendously useful to helping forecasters initiate and amend forecasts of turbulence.

10.4 The Future of Aviation Meteorology

The role of the meteorologist in the modern forecast process continues to evolve (e.g., Novak et al. 2008). As models increase in resolution and sophistication, the role of the operational meteorologist shifts from pattern recognition and conceptual downscaling to data mining and information extraction for decision support. Historically, forecasters bias corrected and downscaled synoptic output from deterministic NWP models, based on conceptual models and experience, to predict almost all weather phenomena (e.g., heavy precipitation, thunderstorms, clouds, visibility, icing, turbulence). Beginning in the early 2000s, forecasters began to interrogate directly mesoscale-gamma (horizontal grid scales ~4 km; Orlanski 1975) NWP output for explicit hazards and the magnitude of potential impacts

(Kain et al. 2006). Additionally, the emergence of ensemble systems at the global, regional, and convection-allowing scales has provided forecasters with predicted atmospheric data that directly lends itself to probabilistic forecasting, risk assessment, and customer-based decision support (e.g., Sobash et al. 2011).

Turbulence and turbulent eddies, however, with space and time scales on the order of 10–100 m and 10–100 s, respectively (Fujita 1986), remain well below the resolvable grid scale of today's operational models. The explicit prediction of turbulence for operational forecasting must await the computing power necessary to run large-eddy simulation (LES) models in real time (i.e., grid scales on the order of 100 m and 10 m horizontally and vertically, respectively) in order to capture explicitly the larger, turbulent-sized eddies affecting aircraft. Until then, conceptual models and considerable statistical post-processing must be utilized to quantify the hazards associated with turbulence. And even with the utilization of operational LES models, a completely accurate deterministic forecast of the thermals and turbulent eddies at any given moment in time will be impossible; however, statistical assessments over space and time may indeed provide accurate turbulence forecasts out several days (Stull 1988). Similarly, the extension of these forecasts to ensemble systems will collectively capture the phenomena across multiple simulations, such that individual eddies may have an extremely short-lived (and perhaps operationally useless) predictability, but capturing their existence across larger spatial, temporal, and numerical domains may very well yield skillful probabilistic forecasts.

In the USA, the NWS is the federal agency responsible for weather services to protect life and property and enhance the commerce of the nation. To that end, in 2013 the NWS transitioned its entire NWP modeling suite to the new Weather and Climate Operational Supercomputing System (WCOSS; NOAA 2013). WCOSS is a fully redundant operational system; it is in fact a pair of identical supercomputers located in Reston, VA and Orlando, FL. The pair allows one system to carry the entire operational inventory and the second system to carry the development and testing load while also providing an immediate and full backup to ensure an uninterrupted continuity of operations.

The NWS has a strategic goal to become a Weather-Ready Nation (WRN) by 2020 (NOAA 2012), and WCOSS is the supercomputing platform to ensure the numerical capability is sufficient. A WRN is a society prepared to predict and respond to weather-related events in the most appropriate and effective manner possible (NOAA 2010). NWP is integral to building a WRN, with NOAA's operational NWP suite moving toward ensemble-based prediction systems (including post-processing) across all scales from global to convection-allowing (NOAA 2015). This is a paradigm shift for the operational meteorologist, as forecast production becomes less human and more computer intensive (often called Meteorologist Over the Loop or MOTL) and operational meteorologists build a better understanding of human, societal, and economic sciences and impacts and subsequent impact-based decision support services.

The research side of NOAA looks to increase the lead time of high-impact warnings through a research program called Warn-on-Forecast (WoF; Stensrud et al. 2009). In NWS aviation, the comparable service to WoF is a SIGMET-on-Forecast (SoF). WoF serves as the cornerstone to a larger forecast paradigm called

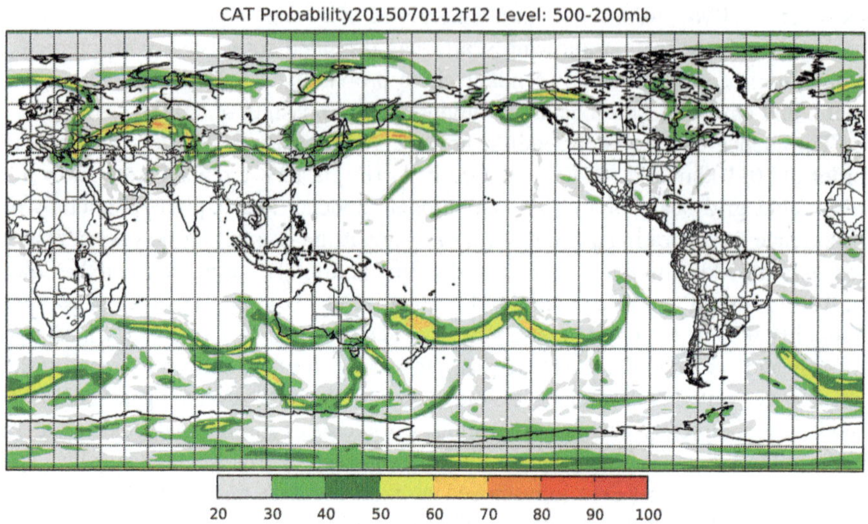

Fig. 10.3 Example of NOAA's Global Ensemble Forecast System (GEFS) post-processed guidance, showing the percentage of ensemble members indicating moderate or greater turbulence (12-h forecast valid at 00 UTC 02 July 2015)

Fig. 10.4 Operational and research meteorologists discuss turbulence forecasting during a NOAA Aviation Weather Testbed (AWT) real-time experiment (February 2012)

Forecasting a Continuum of Environmental Threats (FACETs; NOAA 2014). FACETS is a next-generation, probabilistic-based, high-impact weather forecasting concept that provides user-adaptable threat assessments, which allows various classes of users to base decisions on their specific situations and vulnerabilities. In aviation turbulence forecasting, rapidly updating probabilistic forecasts of turbulence (e.g., Fig. 10.3) based on post-processed WCOSS ensemble system output evolves into NWS meteorologist-approved SIGMETs (i.e., warnings) at certain space, time, and probabilistic thresholds (which may or may not be user specific).

The NWS research-to-operations linkage between present aviation operations and future aviation operations occurs through the NOAA Aviation Weather Testbed (AWT; Ralph et al. 2013). The AWT is located within the AWC, side by side the AWC operations center. The AWT as a stand-alone NOAA Testbed was established in 2009 to provide a physical and technical environment conducive to the efficient transfer of research into meteorological operations (or R2O). Since its inception, the AWT has hosted numerous experiments (Fig. 10.4) and migrated well over 20 products from research into NWS operations. Its goal as a NOAA Testbed is to ensure a rapid and efficient transfer of research into NWS operations. More information on the AWT, its charter and standards of execution, may be found at the AWT website, http://testbed.aviationweather.gov.

References

Congressional Record: 69th Congress, 1st session, 20 May 1926. 44 Stat. 568

Congressional Record: 85th Congress, 2nd Session, 23 Aug 1958, 72 Stat. 731, Public Law 85-726

Congressional Record: 112th Congress, 2nd Session, 14 Feb 2012. 126 Stat. 11, Public Law 112-95

Ellrod, G.P., Knapp, D.L.: An objective clear-air turbulence forecasting technique: Verification and operational use. Weather Forecast. **7**, 150–165 (1992)

Endlich, R.M.: The mesoscale structure of some regions of clear-air turbulence. J. Appl. Meteorol. **3**, 261–276 (1964)

Ellrod, G.P., Knox, J.A.: Improvements to an operational clear-air turbulence diagnostic index by addition of a divergence trend term. Weather Forecast. **25**, 789–798 (2010)

FAA: Aeronautical Information Manual: Official Guide to Basic Flight Information and ATC Procedures. http://www.faa.gov/atpubs (2014)

Fujita, T.T.: Mesoscale classifications: their history and their application to forecasting. In: Ray, P.S. (ed.) Mesoscale Meteorology and Forecasting, pp. 18–35. American Meteorological Society, Boston (1986)

ICAO: Annex 3 Meteorological Service for International Air Navigation, 18th edn, Chapter 3, pp. 2–3 (2013)

Kain, J.S., Weiss, S.J., Levit, J.J., Baldwin, M.E., Bright, D.R.: Examination of convection-allowing configurations of the WRF model for the prediction of severe convective weather: the SPC/NSSL spring program 2004. Weather Forecast. **21**, 167–181 (2006)

Knox, J.A.: Possible mechanisms of clear-air turbulence in strongly anticyclonic flows. Mon. Weather Rev. **125**, 1251–1259 (1997)

McCann, D.: Gravity waves, unbalanced flows, and aircraft clear air turbulence. Nat. Weather Dig. **25**(1, 2), 3–14 (2001)

NOAA: http://www.ppi.noaa.gov/wp-content/uploads/NOAA_NGSP.pdf (2010)

NOAA: http://www.nws.noaa.gov/com/weatherreadynation/#.VXYONkZqM1V (2012)

NOAA: http://www.noaanews.noaa.gov/stories2013/2013029_supercomputers.html (2013)

NOAA: http://www.nssl.noaa.gov/projects/facets/ (2014)

NOAA: http://www.ncep.noaa.gov/director/strategic_plan/strategic_plan.pdf (2015)

NTSB: http://www.ntsb.gov/safety/mwl/Pages/mwl7_2014.aspx (2014)

Novak, D.R., Bright, D.R., Brennan, M.: Operational forecaster uncertainty needs and future roles. Weather Forecast. **23**, 1069–1084 (2008)

Orlanski, I.: A rational subdivision of scales for atmospheric processes. Bull. Am. Meteorol. Soc. **56**, 527–530 (1975)

Ralph, F.M., Intrieri, J., Andra Jr., D., Atlas, R., Boukabara, S., Bright, D., Davidson, P., Entwistle, B., Gaynor, J., Goodman, S., Jiing, J.-G., Harless, A., Huang, J., Jedlovec, G., Kain, J., Koch, S., Kuo, B., Levit, J., Murillo, S., Riishojgaard, L.P., Schneider, T., Schneider, R., Smith, T., Weiss, S.: The emergence of weather-related test beds linking research and forecasting operations. Bull. Am. Meteorol. Soc. **94**, 1187–1211 (2013)

Schwartz, B.: The quantitative use of PIREPs in developing aviation weather guidance products. Weather Forecast. **11**, 372–384 (1996)

Sharman, R.D., Cornman, L.B., Meymaris, G., Pearson, J., Farrar, T.: Description and derived climatologies of automated *in situ* eddy dissipation rate reports of atmospheric turbulence. J. Appl. Meteor. Climatol. **53**(6), 1416–1432 (2014). doi:10.1175/JAMC-D-13-0329.1

Sobash, R.A., Kain, J.S., Bright, D.R., Dean, A.R., Coniglio, M.C., Weiss, S.J.: Probabilistic forecast guidance for severe thunderstorms based on the identification of extreme phenomena in convection-allowing model forecasts. Weather Forecast. **26**, 714–728 (2011)

Stensrud, D.J., Wicker, L.J., Kelleher, K.E., Xue, M., Foster, M.P., Schaefer, J.T., Schneider, R.S., Benjamin, S.G., Weygandt, S.S., Ferree, J.T., Tuell, J.P.: Convective-scale warn-on-forecast system. Bull. Am. Meteorol. Soc. **90**, 1487–1499 (2009)

Stull, R.B.: An Introduction to Boundary Layer Meteorology. Kluwer, Dordrecht (1988)

Whitnah, D.R.: A History of the United States Weather Bureau, 267 pp. University of Illinois Press, Urbana (1961)

Chapter 11
An Airline Perspective: Current and Future Vision for Turbulence Forecasting and Reporting

Melissa Thomas, Stephanie Klipfel, Emily N. Wilson, and Tom Fahey

Abstract The flight planning process is the most critical time for the turbulence forecaster. It is the time that routes and altitudes are selected, and fuel is added for any necessary deviations. Aviation meteorologists can have a significant impact at this stage of flight planning. Internationally, flights are given preference for route and altitude selection based on the time the flight plan is filed. It is difficult for flights to change altitude once the flight plan has been accepted, since flights over the oceans cannot be seen on radar and have to be sequenced in space and time. If turbulence is not considered at the planning stage, altitude deviations for turbulence may not be allowed if there are other aircraft above and below the flight. Therefore, advanced knowledge of turbulence gives the advantage to the airlines that forecast turbulence accurately.

The turbulence forecaster can also provide significant value during the flight following process. Monitoring real-time satellite and weather radar observations, as well as automated aircraft and manual pilot reports, can help verify model data and identify areas of un-forecasted turbulence. This enables Human-in-the-Loop adjustments to the Numerical Weather Prediction (NWP) models and ability to provide updates to flights while en route. A case study from 30 June 2014 over the upper Midwest USA demonstrates this process in detail.

As NWP models improve, the role of the operational aviation meteorologist will evolve. In tandem, additional insight by the research community into the current operational forecast processes is important to successfully transition to automated turbulence forecasting. Improved methods of observing and reporting turbulence, and more efficient ways of distributing and displaying the turbulence forecasts for avoidance, will be needed.

M. Thomas (✉) • S. Klipfel • E.N. Wilson • T. Fahey
Meteorology Department #091, Delta Air Lines, Inc., P.O. Box 20706, Atlanta, Ga 30320-6001, USA
e-mail: melissa.thomas@delta.com; Stephanie.Klipfel@delta.com; emily.n.wilson@delta.com; tom.fahey@delta.com

R. Sharman, T. Lane (eds.), *Aviation Turbulence*,
DOI 10.1007/978-3-319-23630-8_11

11.1 An Airline Perspective: The Current State of Turbulence Forecasting

Delta Air Lines (DAL) is one of the few remaining airlines that retain a company-employed staff of aviation meteorologists. A significant portion of their responsibilities involve producing forecasts and reports of aviation weather hazards as part of the Turbulence Plot (TP) system. When this system originated in the 1960s, turbulence was the primary focus (see Chap. 2). Since that time, a number of hazards in addition to turbulence have been included in the system, but the name has remained unchanged.

11.1.1 The Current Turbulence Plot Program

Currently, DAL meteorologists produce two products that include turbulence information. Turbulence Plot (TP) messages are used to report and provide short-term forecasts of turbulence for en route avoidance. Upper Air Depictions provide longer-term forecasts of turbulence for preflight avoidance when the route to be flown is selected. TP messages and Upper Air Depiction charts are similar to the ICAO SIGMET product and Significant Weather (SIGWX) Chart, respectively.

DAL meteorologists manually create Upper Air Depiction charts and TPs for every region Delta operates in, and both are used to help minimize or avoid hazardous weather conditions. Once an Upper Air Depiction is created for preflight planning use, Forecast Turbulence Plot (TP) messages are then issued for the turbulence areas, as well as for areas with high ozone concentrations, mountain wave activity, space weather, and/or volcanic ash clouds. Observed TPs are issued as needed for thunderstorms, volcanic eruption, and tropical cyclone activity, as well as for turbulence once it has been confirmed by reports. Both Forecast and Observed TP messages are continually monitored for accuracy by DAL meteorologists and updated as needed with new information. Upper Air Depictions are a scheduled product and are only rarely updated when conditions change significantly between scheduled issue times. They are manually drawn using an in-house tailored, computer-aided design software product. This extensively customized, commercial off-the-shelf software includes automated overlays of forecasted winds as well as longer-term forecasts of the same computer model parameters used for TP production. (See section "Graphical and Text Communications" for examples of Delta's Upper Air Depiction charts and TPs.)

When the meteorologist prepares a TP message, the information is sent to a number of databases. The TP product can then be displayed as text, graphics, or a combination of both. TP messages are displayed graphically at DAL flight dispatcher workstations on the graphical flight following applications. Another display method is on the Delta Meteorology weather website for access by flight crews at gates and crew lounges and by station personnel for international preflight weather package preparation.

Text distribution is done worldwide for both the preflight and en route phases of Delta flights. All applicable TP messages are automatically printed on the flight crew's preflight paperwork. For en route phases of flight, dispatchers evaluate how new or updated TPs apply to each flight and then send applicable TPs to aircraft en route. A flight crew can also initiate the uplink of a current TP to their flight deck via Aircraft Communications, Addressing and Reporting System (ACARS).

11.1.1.1 Mountain Wave Avoidance

The TP system is much more than just products. Distribution and display of the products are important. Avoidance procedures are another critical aspect of the TP system. For example, avoidance of a layer of altitudes near the tropopause and use of mountain wave avoidance routes are two procedures in use at Delta today to reduce the exposure to wave action and the associated turbulence. A number of options are available for flight planning during mountain wave conditions. If moderate wave conditions are forecasted or observed, one can operate through the active mountain wave area but avoid either the tropopause layer using "altitude avoidance" or avoid the highest, steepest dropping terrain using a "deviation route." In strong wave conditions, if operating through the active area, use of a combination of *both* a deviation route, if available, and avoiding the listed altitudes is recommended. In both moderate and strong wave conditions, a "bypass route" can be used that avoids the active mountain wave area entirely.

11.1.2 Current Delta Meteorology Turbulence Forecasting Techniques

Currently, Delta Meteorology uses many techniques to forecast turbulence. They all rely on analyzing the synoptic scale, upper air flow, as well as pattern recognition to identify features that have long been known to cause turbulence. One of the first features that meteorologists look for is a significant area of wind shear, especially around jet streams. The speed and slope of trough lines are investigated, since fast moving, deep, and sharp vertically sloped troughs can often cause turbulence. Vertical motions in the atmosphere can cause turbulence for aircraft, so meteorologists investigate any large-scale feature that may create rising and sinking motions, including mountain waves. Monitoring all available turbulence reports, including manual pilot reports (PIREPs) and automated reports, and monitoring satellite imagery and radar for developing turbulence are all important in a real-time mode. Meteorologists watch for thunderstorm development and identify cloud features, like banded cirrus, that may cause bumpiness in flight for DAL aircraft.

11.1.3 GTG: Strengths and Weaknesses

The Graphical Turbulence Guidance (GTG) product takes a different approach to forecasting turbulence. It produces a forecast by using a weighted combination of several different turbulence diagnostics (Sharman et al. 2006). There were originally two versions of this forecast product, the GTG and the GTG2. The GTG2 expanded on the original GTG by computing turbulence in the mid-levels of the atmosphere (down to FL100) and by adding diagnostics to forecast turbulence related to mountain waves (Sharman et al. 2006). To test the effectiveness of the GTG, turbulence forecasts were produced from the GTG combination and from each one of the GTG's components, and each forecast was compared to PIREPs for verification. It was determined that the GTG combination performed better than any single turbulence diagnostic (Sharman et al. 2006). Out of all of the different turbulence diagnostics that are combined into the GTG, frontogenesis has been shown to be the single best predictor of turbulence (Sharman et al. 2006).

In recent years, there has been an update to the GTG2, creating an algorithm known as the GTG2.5. This update replaced the RUC model, the model that the GTG2 ingests, with the newer Rapid Refresh model, creating a higher-resolution forecast product (Benjamin et al. 2006; Wandishin et al. 2011). Another change to the new GTG2.5 product is that it predicts eddy dissipation rate (EDR), rather than predicting turbulence potential like the GTG2 (Wandishin et al. 2011). In an analysis comparing the two GTG forecast products, it was found that, when verified by PIREPs, the GTG2 outperforms the GTG2.5 at forecasting moderate-or-greater (MOG) turbulence events (Wandishin et al. 2011). However, the superior performance of the GTG2 over the GTG2.5 in the analysis may be entirely dependent on the selection of turbulence thresholds for the two products. The GTG2.5 threshold for moderate turbulence is set to yield fewer forecasts of MOG turbulence than that produced by GTG2, and as a result, GTG2.5 captures fewer turbulence events than GTG2 (Wandishin et al. 2011). Using the current GTG2.5 forecast thresholds involves an implicit choice to have fewer false alarms at the cost of an increase in missed turbulence events (Wandishin et al. 2011).

While the combination of several distinct turbulence diagnostics is what makes the GTG a useful forecast tool, it can also be a weakness for the program. After all, the forecasts produced by the GTG can only be as good as the individual diagnostics that are ingested into the program. An improvement to any single turbulence diagnostic should lead to an overall improvement in the GTG. Finally, while today's turbulence forecast models have greatly improved the ability to detect areas in which turbulence may occur, all of these models still have one similar weakness: the numerical weather prediction (NWP) model data that is used to make the forecasts. Even today's advanced NWP models cannot resolve features on the much smaller scale at which turbulence occurs. The current grid spacing of these models is about two orders of magnitude too large to resolve features at the scale of an aircraft (Sharman et al. 2006).

11.1.4 Turbulence Forecasting and Human-in-the-Loop

Throughout the decades, meteorologists have been applying new theories and technologies to turbulence forecasting. Application of concepts in turbulence forecasting was forever changed with the advent of computer models for NWP and computer processing to handle the high volume of computations necessary to generate a forecast and display the graphical results that meteorologists use today. NWP has become more sophisticated over the years, and with the leap that was made in supercomputing, more complicated algorithms and finer resolution in space and time have become a reality. NWP has improved in accuracy and has been developed for situation-specific weather (i.e., convection, tropical activity, turbulence, etc.). Yet, meteorologists have remained operationally necessary.

The assumption that meteorologists will be completely replaced by computer models has not materialized. What computers have done, instead, is make them more efficient. It has allowed meteorologists to access more and better information faster. Our theories and understanding of meteorological processes have continually been refined. Computers, researchers, modelers, and operational forecast meteorologists were necessary to advance the science to this point, and all are needed to continue to improve.

It is up to the trained operational forecast meteorologist to recognize when and which NWPs are doing well and when they are lacking, requiring manual forecast adjustments. This is what will be referred to as "human-in-the-loop" (see Table 11.1). A meteorologist can be educated to understand and apply current conceptual theories, model algorithms, and biases (strengths and weaknesses) and can be taught pattern recognition. Computers can also be programmed to do these things, but pattern recognition remains a difficult task to program. Both processes: NWPs and human-in-the-loop, have their own strengths and weaknesses that, if managed well, can be complimentary.

Table 11.1 Forecast process terms defined. Comparison of wind forecasting processes used from 1980s to 2000s to turbulence forecasting processes used currently and planned for the future by DAL

Turbulence fcst process	NWA wind fcst process time period	Time period/DAL turbulence product	Process description
Manual	Forecasting in early 1980s	Today/TP messages	All hand-entered text
Human-in-the-loop	Forecasting in late 1980s	Today/upper air depictions	Use of some model data and some hand-entered text
Human-over-the-loop	Forecasting in 1990s	Future/upper air depictions and TPs	Use of all model data and by exception manual adjustment
Automated	Forecasting in 2000s	Future/upper air depictions and TPs	Use of all model data

11.1.4.1 Value Added: Case Study, 30 June 2014

On 30 June 2014, a large area of convection was present across most of Iowa prior to 1800 UTC (see Fig. 11.1). The convection was a remnant of the previous nights' mesoscale convective system (MCS). These thunderstorms were intensifying rapidly between 1700 UTC and 1800 UTC due to diurnal heating. Air traffic control (ATC) was making an evening (1900 UTC–0300 UTC) routing plan for the air traffic crossing through the upper Midwest USA. On this day, the Minneapolis (ZMP) Air Route Traffic Control Center (ARTCC) chose to route many coast-to-coast flights over central and southern Minnesota (identified by solid blue lines in Fig. 11.1) to bypass the thunderstorms across Iowa.

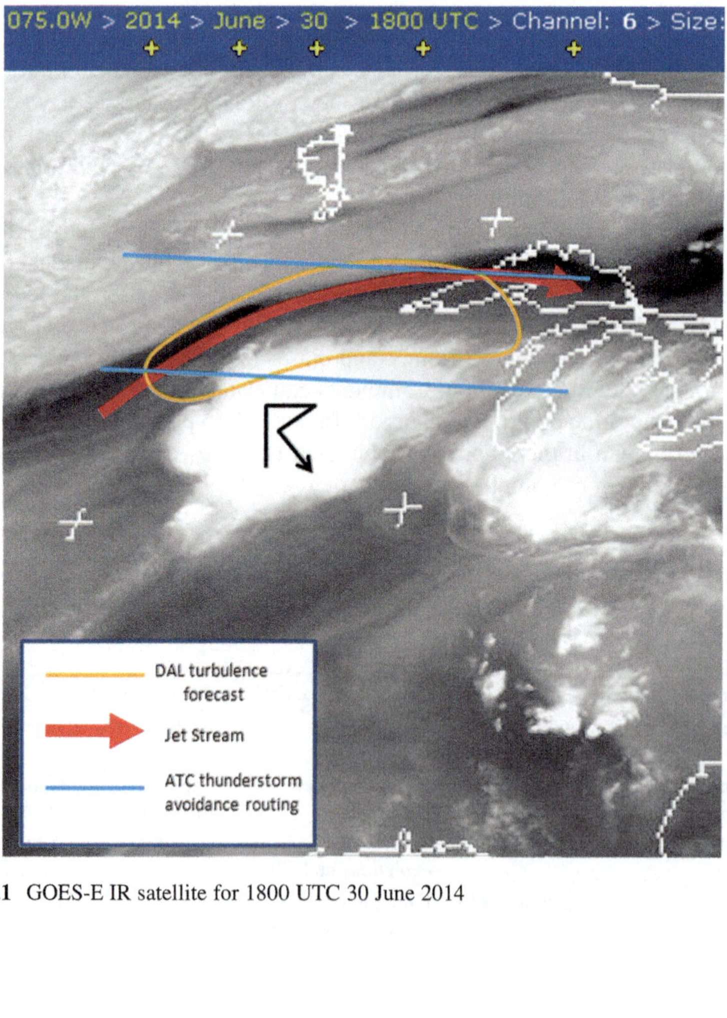

Fig. 11.1 GOES-E IR satellite for 1800 UTC 30 June 2014

Verbal Communications

The meteorologists on duty at DAL recognized a significant potential for turbulence across southern MN due to an intensifying jet stream and wind shear related to the radial cloud bands (identified by the orange line in Fig. 11.1). At 1742 UTC on 30 June 2014, a new turbulence forecast was issued to alert pilots and dispatchers that routes across Minnesota would be expected to have significant turbulence and that significant spatial and altitude deviations may be required when using the routes provided by ZMP. Delta's dispatchers then began planning flights to go around the identified area to the north and/or avoiding altitudes affected on ATC selected routes. The dispatchers filed flight plans and were able to get altitudes approved before other airlines, thereby getting preference for those routes and flight levels due to their early requests. The on-duty meteorologists also contacted ZMP to alert them to the potential for significant turbulence, given that many extra aircraft would be rerouted across their airspace.

Short- and long-term computer models generally do not forecast this type of turbulence well. This day proved to be no different. Research on the cause of radial bands by Trier and Sharman (2009) and Trier et al. (2010) identified that intense convection can cause significant turbulence away from the main precipitation area, and this research was used as training by DAL to make this type of turbulence forecast. Meteorologists can and need to be taught to forecast this turbulence accurately and, in doing so, can add value to computer models that do not yet recognize it.

Reaction Time

We know that NWPs and meteorologists are not always right. Situational awareness is an advantage for the human; the meteorologist can change faster than a computer model in some cases. The most useful product to the customer is the one that is most timely, pertinent and accurate, and in that order of priority. While a NWP may be able to weight its algorithms to pilot reports or automated turbulence reports, it will not be able to quickly ignore areas of turbulence that are over-forecast and eliminate them from the end product. Rather, the NWP may phase the feature out with time, leading to a less accurate forecast in the short term. Currently, there is no easy way for the end user to know the level of confidence or accuracy of a computer-produced turbulence forecast, if that is the only product available.

Graphical and Text Communications

DAL uses multiple methods of conveying turbulence forecast information. One method is with a turbulence depiction chart, shown in Fig. 11.2.

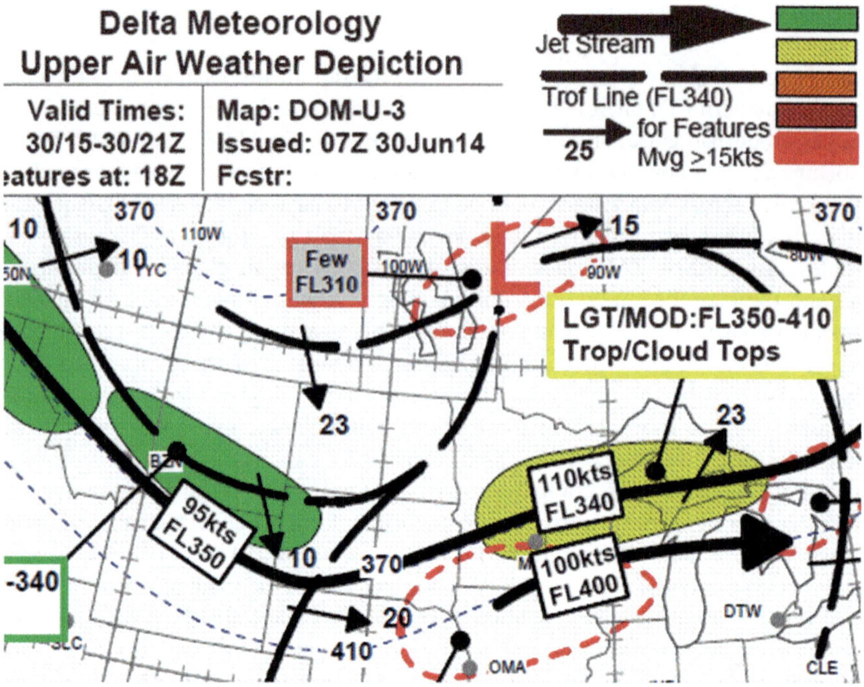

Fig. 11.2 Example of the Delta Air Lines Upper Air (turbulence) Weather Depiction chart valid for 18z, 30 June 2014

The Upper Air Depiction chart can be issued many hours before a flight leaves the ground, when the flight is in the planning stages. It is necessary to plan turbulence avoidance early, since fuel can be added to avoid areas or altitudes of expected turbulence. This is the first stage of turbulence avoidance. Naturally, unlike some military aircraft, it is impossible for commercial aircraft to add fuel once the aircraft is en route. During this second stage of turbulence avoidance, while aircraft are en route, DAL's TP messages are used. An example is shown in Fig. 11.3.

TPs can be issued to alert pilots and dispatchers of changes in the forecast. Any area of forecast turbulence can be updated to change the position, intensity, altitude, or level of confidence. DAL meteorologists can upgrade an area of *forecasted* turbulence to *actual* turbulence based on pilot reports or EDR data. This conveys to pilots and dispatchers a heightened level of risk and increased confidence that turbulence is likely. Information can also be included about best and/or worst altitudes and be shared in real time between the meteorologists and the customers using these turbulence plots. None of this is possible with current computer model forecasts.

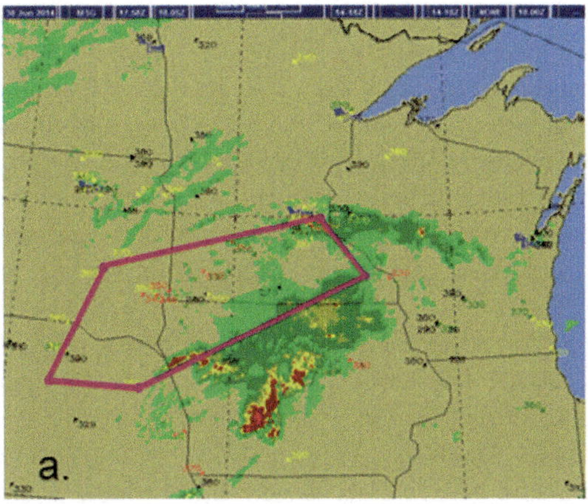


```
TP  CN26  301742Z
1. NE SD IA MN
2. * ALERT *
VOR: OBH ONL FSD MSP RST
HAZ: TURBC-OTHER
SOURCE: PIREPS
AREA: 45N093W 44N098W
      42N099W 42N097W
      44N092W
TIME:POSN AT 30/1741Z
ALTS:FL310-390 MOD
INFO:TSTM BLOWOFF.
  SPORADIC REPTS FL310-320.
  BEST AOB FL300
3. VALID 301742Z/010000Z
4. CANCEL CN25
```

b.

Fig. 11.3 (**a**) NEXRAD radar reflectivity display with overlay of TP points (*purple lines*) and pilot reports labeled by altitude, color-coded by intensity (*black* occasional light or smooth, *green* light, *yellow* light occasional moderate, *orange* moderate). (**b**) Example of the TP text identified by the points in (**a**)

11.1.4.2 Turbulence and Flight Planning

Aviation meteorology departments, such as the one at DAL, produce many different types of forecasts that are specifically tailored to commercial, large aircraft operations. Due to the long duration of overseas flights, and the geographical area these flights cover, turbulence forecasting is an important consideration for flight planning. International flight planning has a turbulence forecast time frame of up to 18 h in advance. Domestic US flight planning usually occurs less than 12 h prior to flight completion. These time frames require a longer-term forecast than is currently available from some turbulence forecasting products that extend to only +12 h. Considering there is approximately a 2 h time lag between the current model runtime and when the data is available for viewing, turbulence models must be available out to 24 h for most international flight planning purposes. Once the flight plan has been generated, the turbulence forecast products can be continually updated for tactical use by the pilots, but major route changes become more difficult after the flight plan has been submitted to ATC.

11.2 An Airline Perspective: The Current State of Turbulence Reporting

The Meteorological Data Collection and Reporting System (MDCRS), which has been in use in North America for over 25 years (Taylor et al. 1990), was developed by Aeronautical Radio Inc. (ARINC) under contract with the Federal Aviation

Administration (FAA). The system accepts meteorological reports from commercial aircraft downlinked through ACARS in a variety of company-specific formats and processes the reports into a common format for transmission to the NWS and/or FAA. The FAA and ARINC have renewed the MDCRS contract numerous times over the years, and the FAA currently has a contract with ARINC (now owned by Rockwell Collins) to handle some of the airlines' aircraft-based observations. Originally, in the 1990s the contract included reports of winds and temperatures only. Over the years it has been expanded to include ARINC's forwarding of automated turbulence reports.

In 2015 NOAA/NWS awarded a contract to ARINC to handle the same type of data as the FAA MDCRS contract. NOAA also has two separate contracts to obtain automated reports of water vapor. One contract is with United Parcel Service (UPS), and the other contract is with Southwest Airlines (SWA). Lastly, there currently is no contract or clear structure between US-based airlines and the FAA nor with NOAA/NWS on how automated turbulence data are handled.

Airlines For America (A4A) is a trade organization representing approximately ten commercial US airlines. The A4A Meteorology Work Group issued an industry position paper in 2014 supporting automated reporting and proposing initiatives to help advance the effort. The recommendations covered four areas: (1) data; (2) cost and benefit balancing; (3) turbulence measurement methods, standardization of calculation, and quality control; and (4) expanding the number of participants reporting automated aircraft weather information.

As of date of writing, June 2015, NOAA/NWS has agreed to act as the responsible organization for data issues, including storage, quality control, access, and security. In addition it has been agreed that the Meteorological Assimilation Data Ingest System (MADIS) will be the designated database.

11.3 An Airline Perspective: The Future of Turbulence Reporting and Forecasting

Automated reporting of turbulence addresses to a large degree the subjectivity of manual reports. But there is still a fair amount of work to be done to standardize turbulence calculations, such as EDR. In addition, there are needs for both the turbulence state of the atmosphere which EDR provides and measurements of intensity of turbulence encountered by individual aircraft.

From a turbulence forecasting perspective, until objective computer-derived forecasts are more accurate, humans will be required to use their strengths to improve upon the computer models' weaknesses. Operational meteorologists must work together with the research community to help identify the weaknesses in the models so that those events can be researched and understood, theory adjusted, and algorithms then written and applied, to continually improve the forecasts. This feedback loop is necessary but currently has a long time frame for implementation. In the meantime, meteorologists can still use current research and apply it with the available computer data, until such time as it can be successfully incorporated into the objective turbulence forecast products. There are many roles the meteorologists must have as human-in-the-loop evolves to human-over-the-loop in the foreseeable future.

11.3.1 Evolution from Manual Forecasting to Automated

11.3.1.1 Past Wind Forecasting Evolution

The current state of turbulence forecasting for commercial aviation and resultant avoidance efforts can be compared to the state of the science for wind velocity forecasting for flight planning in the 1980s. The following description and comparison will be used to describe the turbulence forecasting processes: manual human-in-the-loop, human-over-the-loop, and automated.

In the 1950s and 1960s, wind speeds and directions expected to be encountered at cruise level were manually forecast by aviation meteorologists. On longer flights, the route to be flown was then manually selected to take advantage of tailwinds and avoid headwinds whenever possible. By the 1980s some airlines had discontinued this practice and were using the automated winds produced by government-funded national weather forecast models instead. In the 1980s, some airlines continued to manually forecast the winds and manually select the route to be flown based on the most favorable winds and to avoid significant areas of potential turbulence. Most airlines had adopted automated flight planning systems, using computer model-generated wind forecasts, but some utilized in-house meteorologists to manually adjust the winds in selected areas where the models were known to be in error.

Northwest Airlines (NWA) and DAL are examples of airlines with in-house meteorology staff who continued the manual wind forecasting process into the 1990s. In the early 1980s, NWA meteorologists were referencing the latest government model forecast of winds (global spectral, baroclinic, and barotropic models), but they were manually entering all weather forecast data into a mainframe computer system for flight plan calculation. This included a forecast of wind velocity and temperature at FL300 (300 mb). The direction and magnitude of the vertical wind shear was manually entered, as well as the tropopause height and temperature lapse rate above in the stratosphere. A lapse rate below the tropopause was always assumed to be 2.4 C/1000 ft. Wind forecasts for regions of the USA for domestic flights and all international flights were manually produced by the in-house meteorology team. Flight dispatchers manually adjusted the route of long-haul domestic flights when necessary for clear air turbulence or mountain wave avoidance. The NWA meteorologist was responsible for selecting the exact route to be flown by every international flight and for manually producing a forecast for winds, temperatures, and tropopause height along the entire route of flight, including climb, cruise, and descent, until the early 1990s.

In the mid- to late 1980s, NWA transitioned from an in-house, legacy system that used the completely manually produced wind forecasts to the use of government-produced wind models for forecasting winds. During this transition, NWA temporarily contracted with an external flight planning service provider, while NWA bought and adapted for internal use an existing flight planning system that had been developed by another airline. Domestic flight planning was cutover to the new in-house system in the late 1980s, and international cutover was done in stages, with all international flight planning cutover by June of 1990. During the decade of the 1990s, NWA meteorologist manually reviewed and adjusted the government forecast winds when needed. By the 1990s, route selection was being done by the computerized flight planning system with the flight dispatcher now responsible for manually adjusting the route when needed.

11.3.1.2 Future Turbulence Forecasting Evolution

The past evolution of wind forecasting for commercial aircraft can be used as a model for future evolution of turbulence forecasting. Table 11.1 displays the comparison of wind forecasting processes used in the 1980s and 1990s by NWA to the turbulence forecasting processes used currently by DAL meteorologists for producing TP messages and Upper Air Depictions. The examples and descriptions of manual, human assisted (with distinctions between human-in-the-loop and human-over-the-loop), and automated are included.

11.3.2 Manual to Automated Reporting: Transition Issues

The volume of data that operational forecast meteorologists analyze on any given day is often overwhelming. New data has to prove itself useful for meteorologists to change their processes to incorporate the data. There are certainly benefits of automated turbulence reporting versus the manual reporting process. Aviation has spent decades with subjective pilot reports. Errors in altitude, location, and time are common. With automated reports, knowing the exact time, location, and altitude is ideal. Once meteorologists learn how to incorporate the new data into their daily processes, a transition will begin with more reliance on the automated reports and less on the manual reports.

Integration and proper training must also be had so meteorologists understand the strengths and weaknesses of the automated turbulence data. Simple questions around location accuracy (x, y, and z axis), intensity accuracy, and equipment calibration will need to be understood. There is also the current pending Radio Technical Commission for Aeronautics (RTCA) Special Committee 206 Aeronautical Services Data Link, subgroup 4, task of developing a Minimum Operating Performance Standard (MOPS) for EDR calculations, which currently are not standardized. In addition, there needs to be a process to address perception issues.

For example, if meteorologists believe that B757 aircraft seem to report higher values then other aircraft types, is it accurate to conclude that B757 aircraft are more susceptible to turbulence and the calculated EDR is higher than other aircraft? Is there an error in the algorithm that calculates the EDR value? Also important is the need to understand how a given EDR value, say 0.25 ($m^{2/3}$ s^{-1}), affects different-sized aircraft. Can we conclude that turbulence forecasting with EDR will become more like temperature forecasting? Fifty degrees can feel warm to someone who has been in subfreezing temperatures but feel cold to someone who has been in the desert, but it is still the same temperature value. Turbulence forecasts will eventually just be a number at a point in the atmosphere that may feel very turbulent to some aircraft and just light turbulence for others.

Another perspective may be that both a standardized state of the atmosphere value for turbulence and aircraft g-load are needed; using calculated EDR values and the actual measured value from the aircraft accelerometer, provided as a g-load value, is also needed from every automated turbulence reporting aircraft. Computer models, modelers, researchers, airlines, and aviation meteorologists will all need access to readily understood and usable reports of turbulence for the foreseeable future, and this later solution could address all needs.

11.3.3 Manual to Automated Turbulence Forecasting: Transition Plan

Automation can help meteorologists actively manage the data to obtain needed information. For example, rather than just receiving a turbulence report, an alert can be sent to the meteorologist when a higher turbulence value is reported in an unexpected area or when a low turbulence value is received in a forecast turbulent area. Meteorologists are tasked to evaluate the atmospheric conditions that cause the discomfort of turbulence for passengers on aircraft. The general philosophy is the more dynamic/complex the atmosphere, the higher our confidence grows of turbulent areas developing and the increased likeliness of strong turbulence.

As researchers continue to refine computer models to gain a better understanding of dynamic/complex atmospheric conditions, we will be able to better rely on the computed turbulence values. At first, we could display the model-based forecast to the users and allow the meteorologist to issue products that indicate areas where they don't agree with the model data. Similar to forecasting other parameters, such as temperature, a meteorologist will be able to view different models and adjust their forecasts based on additional real-time analysis and trends. For example, if the model data was indicating low values of turbulence, but we were receiving reports of higher values, the meteorologist could issue a TP-like product to indicate a manual forecast with updated and more accurate information than the model provides. But over time, software could be developed and integrated to allow the meteorologist to change the model-based turbulence values, and the separate product could be eliminated. Some days may require a lot of adjustments, but other days may not. As the algorithms continue to evolve, required modifications will occur less and less.

11.3.3.1 Text to Graphics Distribution: Transition Plan

Although graphical weather technology has been available for years to private pilots, tablet technology is still being integrated into commercial airline cockpits. This will allow graphical weather distribution both preflight and en route! Once we are able to provide accurate turbulence values in a gridded format, we will be able to integrate turbulence values in the flight planning system. An airline can determine which turbulence values to avoid by aircraft size. Then when the dispatcher generates a flight plan, the route will be the most economical route available while avoiding the maximum turbulence values. When a pilot does the preflight, the route overlaid with turbulence in both horizontal and vertical views and eventually even 3-D will be available! If the forecast values change while a flight is en route, automation can compare the route to the new forecast. An alert can be sent to the pilot and the dispatcher so adjustments can be made. The gridded data can be displayed with contours and meaningful colors on flight following tools. Real-time and forecast data can also be displayed on tablets in the cockpit along with the current position and the planned route. This allows dispatcher and pilots to maintain a common awareness of the turbulence risk and to develop plans to mitigate significant turbulence. It can be used by the flight attendants to plan when to provide service to the customers.

With the introduction of electronic flight bag (EFB) capability, the use of portable devices such as tablet computers in the cockpit will change the roles for the pilot, dispatcher, and aviation meteorologist. Technology is now allowing pilots and dispatchers to view turbulence forecasts as a vertical cross section along the planned route of flight. This allows a pilot to see where forecasted turbulence will lie above or below their flight level, as well as a plan view around them. This should help pilots make faster tactical decisions to avoid prolonged turbulence exposure, which is expected to keep flight attendants and passengers more comfortable, less fatigued, and reduce the risk of injuries. This is assuming, of course, that the product provided is accurate. In order to keep up with more advanced technology available in the cockpit, meteorologists, dispatchers, and pilots will need to transition from the current plan view, 2-D view, to a gridded format, 3-D way of thinking. Conceptual turbulence theory and visualization will become exceedingly important to make this transition to 3-D gridded forecasts.

11.4 Conclusions

Meteorology, as a maturing science, needs more initiatives to help guide the research community's efforts, based on a better understanding of the operational meteorologist role and needs. It was with this collaborative perspective that the above information has been shared.

11.4.1 Observations of Turbulence

Automated measurements and calculations of turbulence must and will eventually replace subjective, human-produced reports. But this will not occur until work is completed on standardization of turbulence calculations, such as EDR. In addition, there will always be needs for information about both the turbulent state of the atmosphere as well as the intensity of turbulence encountered by specific aircraft.

Regardless of whether we are using manually produced or automatically calculated or automatically measured turbulence values, data issues need to be addressed, including storage, quality control, public versus restricted access, and security.

Finally, NWS and/or FAA contracts in the future may be more likely to foster an increase in airline contributions of automated reports of turbulence if the contract model used for water vapor was adopted: Individual airlines signing automated aircraft weather report contracts directly with either the NWS or the FAA.

11.4.2 Product Distribution and Display

From a general perspective, the advancement of aviation meteorology must continue to be a global and coordinated effort. Commercial airlines, flying to all corners of the world, need consistent, reliable, and accurate weather information (observation as well as forecasts, turbulence as well as other hazards) regardless of their departure airport, route, or destination. Pilots flying in the USA may have the luxury of having weather hazard information at their fingertips, but pilots flying to or from remote areas of the world should have access to the same, including turbulence observations and forecasts or the processes that enable turbulence hazard risk mitigation.

11.4.3 Turbulence Forecasting and Avoidance Processes

From an operational meteorologist perspective, automated turbulence forecasts such as the GTG are not yet ready to be used as the sole source for preflight or en route turbulence avoidance decisions. In time, it is anticipated that the GTG and other automated turbulence forecasts will increase in accuracy, cover more than just the domestic USA, and extend more than 12 h in the future.

It is certain that as technology advances, the advancement of meteorology as a science will continue, and airlines will continue to pursue the safest and most economical routes. As computer modeling for turbulence improves, the aviation meteorology industry will take the opportunity to automate the turbulence forecast and avoidance processes. This will allow operational meteorologists to focus on other aviation hazards. There is no lack of other challenges. Safe and economical routing decisions based on identifying space weather impacts on aircraft routed over polar regions and/or operating at high altitudes, observed and forecasted volcanic ash concentrations, or high ice water content concentrations that both can cause aircraft engine problems are just a few examples. Even reliable long-range forecasts for airline flight and flight crew schedule development, as well as for fuel planning, will hopefully someday be feasible.

References

Benjamin, S.G., Devenyi, D., Smirnova, R., Weygandt, S., Brown, J.M., Peckham, S., Brundage, K., Smith, R.L., Grell, G., Schlatter, T.: From the 13km RUC to the rapid refresh. In: 12th Conference on Aviation, Range, and Aerospace Meteorology, Atlanta, GA, American Meteorological Society, CD-ROM. 9.1 (2006)

Sharman, R., Tebaldi, C., Wiener, G., Wolff, J.: An integrated approach to mid and upper-level turbulence forecasting. Weather Forecast. **21**, 268–287 (2006)

Taylor, D.L., Landot, D., Ligler, G.T.: Automated meteorological reports. In: Proceedings of the Aeronautical Telecommunications Symposium on Data Link Integration, pp. 141–144, Annapolis, MD, May 1990

Trier, S.B., Sharman, R.D.: Convection-permitting simulations of the environment supporting widespread turbulence within the upper-level outflow of a mesoscale convective system. Mon. Weather Rev. **137**(6), 1972–1990 (2009)

Trier, S.B., Sharman, R.D., Fovell, R.G., Frehlich, R.G.: Numerical simulation of radial cloud bands within the upper-level outflow of an observed mesoscale convective system. J. Atmos. Sci. **67**(9), 2990–2999 (2010)

Wandishin, M.S., Pettegrew, B.P., Petty, M.A., Mahoney, J.L.: Quality assessment report: graphical turbulence guidance, version 2.5. United States National Oceanic and Atmospheric Administration, Earth System Research Laboratory, Global Systems Division (2011). Available online at http://docs.lib.noaa.gov/noaa_documents/OAR/GSD/NOAA_TM_GSD-39.pdf

Chapter 12
Automated Turbulence Forecasting Strategies

John A. Knox, Alan W. Black, Jared A. Rackley, Emily N. Wilson, Jeremiah S. Grant, Stephanie P. Phelps, David S. Nevius, and Corey B. Dunn

Abstract Forecasting aviation turbulence remains a challenge in the twenty-first century because of the small temporal and spatial scales of the phenomenon and the multiplicity of causes of the turbulence. In this chapter we discuss both the theoretically derived methods and the empirical techniques used to forecast aviation turbulence. The focus is mainly on clear-air turbulence (CAT), but forecasting techniques for low-level, convectively induced, and mountain wave turbulence (MWT) are also surveyed. We conclude with a brief glimpse into the future of aviation turbulence forecasting.

12.1 Introduction

Forecasting strategies for aviation turbulence vary both with the type of turbulence and with the evolution of aviation meteorology from a largely empirical discipline emerging from pilot observations to a discipline increasingly informed by theory, scientific observations, and numerical modeling.

J.A. Knox (✉) • J.S. Grant • D.S. Nevius • C.B. Dunn
Department of Geography, University of Georgia, 210 Field Street, Room 204, Athens, GA, USA
e-mail: johnknox@uga.edu

A.W. Black
IIHR-Hydroscience and Engineering, 100 C. Maxwell Stanley Hydraulics Laboratory, Iowa City, IA 52242-1585, USA
e-mail: alan-black@uiowa.edu

J.A. Rackley
Oak Ridge National Laboratory, P.O. Box 2008 MS6315, Oak Ridge, TN 37831-6315, USA
e-mail: rackleyja@ornl.gov

E.N. Wilson
Meteorology Department #091, Delta Air Lines, Atlanta, GA 30320-6001, USA
e-mail: emily.n.wilson@delta.com

S.P. Phelps
APEX Clean Energy, Charlottesville, VA, USA

R. Sharman, T. Lane (eds.), *Aviation Turbulence*,
DOI 10.1007/978-3-319-23630-8_12

The forecasting of turbulence for aviation purposes is vexed by the reality that, even in the twenty-first century, operational numerical models cannot simulate the phenomena that are directly responsible for the turbulent motions that affect the aircraft. As a result, forecasting strategies must rely on identifying larger-scale phenomena or patterns that can be correlated with the occurrence of turbulence. Research and operations have focused more on nowcasts and avoidance based on observations than on actual forecasting techniques for some types of turbulence.

For this reason, we devote the bulk of this chapter to techniques for forecasting clear-air turbulence (CAT), which has received much attention in both research and operations during the past several decades. Then we address more briefly the forecasting of turbulence at other altitudes and in other contexts. The reader is also directed to the excellent and comprehensive review of upper-level turbulence by Sharman et al. (2012b).

12.2 Clear-Air Turbulence Forecasting

12.2.1 Definition

CAT represents disorganized fluid motions in the form of microscale eddies that can take place within cloud-free (clear air) or limited cloud patches in the free atmosphere, which causes in-flight bumpiness of aircraft (Chambers 1955; Lester 1993; Knox 1997). These turbulent eddies are most effective at causing noticeable bumpiness when their dimensions are about the size of the aircraft (~100 m) flying in the free atmosphere (Sharman et al. 2006; McCann et al. 2012). The free atmosphere is above the planetary boundary layer (PBL), where surface effects on temperature, humidity, wind speed, and wind direction are negligible. Therefore, CAT is not affected by surface conditions and low terrain, as opposed to low-level turbulence (LLT). CAT also occurs away from mountains and thunderstorms, distinguishing it from mountain wave turbulence (MWT) and convectively induced turbulence (CIT), respectively. (As we shall see later in the chapter, the distinctions between CAT, MWT, and CIT can be unclear in some situations.)

CAT is generally but imprecisely defined as occurring above about 15,000–18,000 ft (roughly 5–5.6 km) AGL; it can occur at lower altitudes above the PBL (Lester 1993; Ellrod et al. 2015). CAT tends to occur in regions of strong vertical and/or horizontal wind shears and strong static stability and therefore can occur within troughs/ridges of planetary waves, at the tropopause (i.e., the layer of air between the troposphere and stratosphere) where upper-level jet streams tend to be located, along upper-level fronts, and in the statically stable stratosphere where jet streams penetrate (Colson 1963; Dutton and Panofsky 1970).

In the subsections below, we discuss the early work of pattern recognition in CAT forecasting based on empirical evidence and then turn to the dynamics that can induce CAT, including how the Richardson number (Ri) has been employed in CAT forecasting since the 1960s. Next, kinematic-dynamical approaches to forecasting CAT that became widely used in the 1990s will be surveyed, with emphasis on the Ellrod–Knapp index. Improved and alternative forecasting diagnostics that recognize the importance of waves and instabilities in CAT production are then addressed. The discussion culminates in an overview of the advances in merging multiple diagnostics into a statistically optimized index for CAT forecasting, the Graphical Turbulence Guidance (GTG) diagnostic.

12.2.2 Empirical Strategies

Prior to the development of diagnostic indices and statistical methods for forecasting CAT, the primary means for minimizing CAT encounters relied on empirical recognition of meteorological patterns and features. Because of its microscale nature, CAT cannot be resolved in NWP models and cannot be directly analyzed using upper air analysis charts. Instead, features of the larger-scale environment that are favorable to the development of shearing gravity waves and turbulence must be analyzed, and the presence and intensity of CAT must be estimated from these analyses. Multiple studies using pilot reports, field analyses, observations, and dynamical theory have determined that CAT is most likely to occur in and around jet streams, jet stream fronts, and the tropopause (Ellrod and Knapp 1992). These regions are particularly conducive to the strongly sheared, stable layers required for CAT formation. Indeed, wind shear has long been known to be a useful indicator of CAT probability, both horizontal and vertical shear (e.g., Chambers 1955), although sheared layers may exist without the presence of turbulence.

Two-thirds of all CAT-related aviation incidents occur near the jet stream, and thus pattern recognition of the jet and its features is critical to CAT forecasting (Lester 1993). Jet stream analysis most often involves the use of the 300-, 250-, or 200-hPa pressure charts. Jet streams with greater than 110 kt (55 m s^{-1}) cores are likely to have regions of significant turbulence near them in both the sloping tropopause above the core and in the jet stream front below the core. CAT is most likely to occur in the narrow band of high wind speeds on the cyclonic side of the jet stream axis (the jet stream frontal zone). In particular, significant CAT is more frequent in this area along the ridges of Rossby waves. Further, the probability of CAT occurrence increases as the curvature of the jet axis and jet speed increase. For example, a sharply curved upper-level ridge or trough that is rapidly growing has much greater CAT probability than the surrounding areas due to the large horizontal wind shear across the area. However, the probability of CAT is at its minimum near the maximum winds in the jet core, because vertical shear is also at a minimum.

Greater probabilities also exist in areas with jet confluence or diffluence. Confluent areas downstream of a cutoff low or diffluent areas upstream of the cutoff low are particularly favorable regions for CAT. Regions to the north or northeast of a developing cyclone also exhibit increased CAT frequency (Serebreny et al. 1962; Chandler 1986; FAA 1988; Stack 1991). These preferred locations for CAT are summarized in plan view in Fig. 12.1 (from Ellrod et al. 2015).

A US climatology of upper-level turbulence via pilot reports (Fig. 12.2, from Wolff and Sharman 2008) bears out to some extent the relationship between pilots'

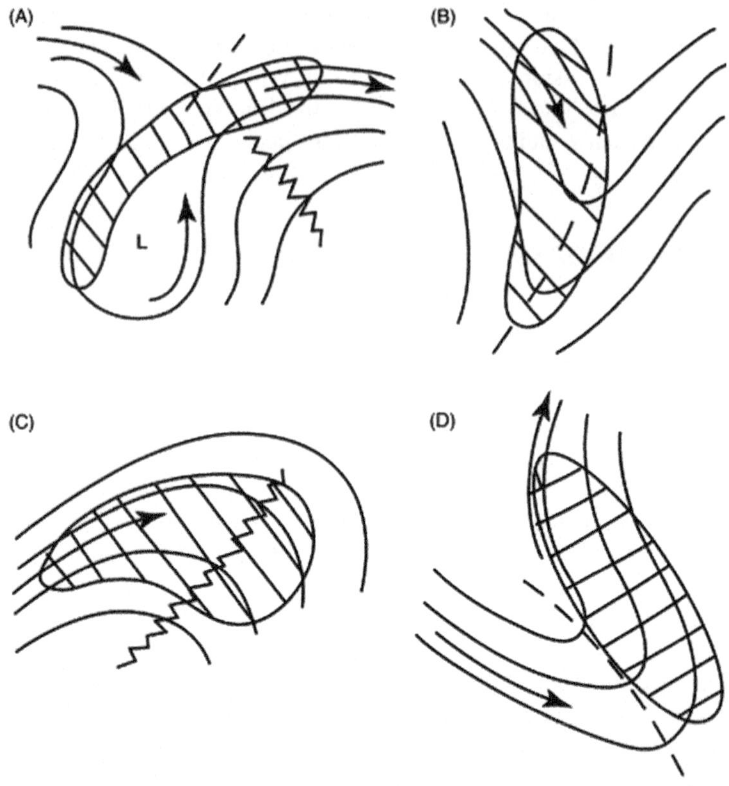

Fig. 12.1 Idealized synoptic-scale flow patterns conducive to CAT: (**a**) deformation zone, (**b**) sharp trough, (**c**) ridge, (**d**) negatively tilted trough. From Ellrod et al. (2015, Fig. 1)

experiences and the pattern recognition techniques. Reports in Eastern USA tend to move north and south with the seasons, tracking the jet stream. However, contributions from mountain waves in the West, and convection in the Deep South, are also present—highlighting the complicated nature of differentiating between separate mechanisms of turbulence in the upper troposphere.

12.2.3 Energetics Approaches

The Richardson number (Ri) is a dynamical diagnostic for forecasting CAT. It is defined in derivative form (often called the "gradient Richardson number") as

$$\mathrm{Ri} = \frac{(g/\theta)\partial\theta/\partial z}{(\partial u/\partial z)^2 + (\partial v/\partial z)^2}, \tag{12.1}$$

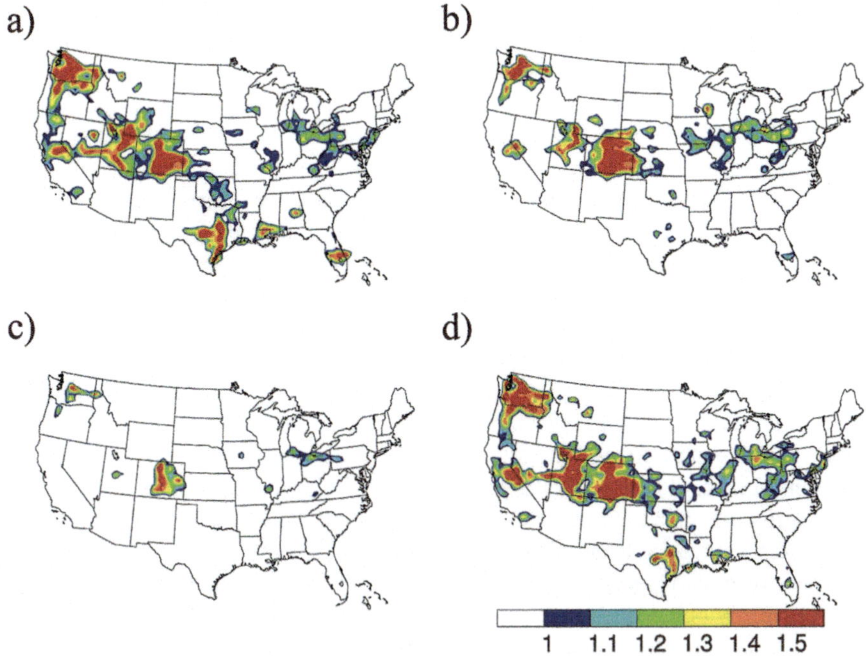

Fig. 12.2 The ratio of moderate-or-greater turbulence versus total pilot reports for (**a**) January–March, (**b**) April–June, (**c**) July–September, and (**d**) October–December for the years 1994 through 1995. From Wolff and Sharman (2008, Fig. 5)

in which the numerator is the static stability and the denominator is the vertical wind shear; often the denominator is calculated only for the zonal wind. For simplicity of notation, the numerator will be referred to as N^2, the Brunt–Väisälä frequency, and denominator as *VWS*.

The origins of Ri trace back to Richardson's (1920) groundbreaking work on the energetics of atmospheric eddies. US Air Force meteorologist George Kronebach conducted one of the earliest systematic investigations of the utility of Ri for aviation forecasting, inspired by empirical rules of thumb for both vertical and horizontal shear pioneered by aviation meteorologists at the United Airlines and Eastern Airlines (Kronebach 1964). He found that "horizontal shear was much inferior to vertical shear in identifying CAT." Seventy-eight percent of 593 cases of CAT was found collocated with vertical shear of at least 6 kt/1000 ft during a 2-week period in March 1962, as determined from synoptic analyses from the National Meteorological Center (later NCEP). Furthermore, "(a)reas where Ri was less than one were found to contain approximately 40 % of all the CAT which occurred during the next 12 h." These results focused future research on vertical, rather than horizontal, shear.

The utility of Ri for forecasting CAT was given considerable theoretical justification at around the same time that aviation meteorologists began focusing on vertical shear as a forecast diagnostic. The fluid dynamicists John Miles and Louis Howard (Miles 1961; Howard 1961; Miles and Howard 1964) proved the conjecture of another, even more celebrated fluid dynamicist, G. I. Taylor (Taylor 1931), that the criterion for stability of stratified flow is $\mathrm{Ri} > ¼$. [Richardson (1920), based on physical insight, had speculated that the criterion for stability would be $\mathrm{Ri} = 1$. This was confirmed later by Miles (1986) based on energy considerations.] According to this theory, values of $\mathrm{Ri} \leq ¼$ constitute a necessary, but not sufficient, condition for instability. The two variables in Ri, static stability and vertical shear, are related to the existence of turbulence because as Ri lowers to its critical value (e.g., via increasing shear), the flow will become dynamically unstable and waves will begin to form as shear generation of turbulence overcomes buoyant suppression of turbulence. As these waves grow larger, they break due to static instability (Stull 1988), generating turbulent flows. This dynamical instability, Kelvin–Helmholtz instability, fits the general spatial and temporal requirements for CAT generation, growing on the timescales of a few minutes and wavelengths of a few kilometers for typical tropospheric conditions.

These discoveries led Dutton and Panofsky (1970) to speculate that "a mystery may be unfolding" regarding CAT, stating in a much-cited *Science* article that "(i)t appears today that the secret of clear-air turbulence may have been discovered—in what seems to be a sudden convergence of almost classical theory and new empirical data from the laboratory, the ocean, and the atmosphere… an apparently complex phenomenon has been explained in quite simple terms."

Despite the initial excitement regarding an explanation of CAT, the results of studies conducted to investigate the accuracy of Ri as a CAT predictor were mixed. Although some studies found favorable correlations between Ri and the presence of CAT (e.g., McCann 1993), many flaws in this method for CAT prediction have been found. Measurements of vertical shear are difficult to obtain over large spatial areas because such measurements can only be made using radiosondes or vertical profilers. Therefore, it is difficult to evaluate stability and wind shear estimates for areas away from where profiles are made. Also, because of the discrete nature of both observational data and model output, the bulk Richardson number

$$\mathrm{Ri}_B = \frac{g/\theta(\Delta\theta/\Delta z)}{(\Delta u/\Delta z)^2 + (\Delta v/\Delta z)^2} \tag{12.2}$$

must be used. Values of Ri_B derived from radiosondes under turbulent conditions often are larger than the ¼ threshold for the theoretically derived Ri, but are usually smaller than 1, making it difficult to deduce the presence of CAT (Ellrod and Knapp 1992). It was also found that low Ri can be possible during both turbulent and laminar conditions (Endlich 1964). Some simple statistical comparisons revealed disappointing overall correlations between CAT and Ri (e.g., Dutton 1980). McCann (1993) pointed out the fact that the Ri method could produce erroneous

CAT forecasts since Ri is a ratio. This means that small values of static stability and vertical wind shear would result in the same Ri as conditions with large values of static stability and vertical wind shear which would be more likely to cause robust turbulence. McCann also noted the fundamental weakness of Ri in predicting the intensity of turbulence, as distinct from its presence, a weakness that was evident as early as the mid-1960s (e.g., Colson and Panofsky 1965).

Inspired by the work of Roach (1970), Keller (1990) created a method that could circumvent some of the problems with the *value* of Ri_B by deriving a CAT forecasting technique based on its *tendency*, or more specifically the tendency of ln(Ri). While it continues to be used in multi-diagnostic forecasting approaches (e.g., Sharman et al. 2006), the Richardson number has not proven to be a "silver bullet" forecasting diagnostic for CAT.

An alternate line of thinking regarding an energetics criterion for CAT forecasting was pursued by Colson and Panofsky (1965). The authors developed a turbulent kinetic energy (TKE) equation for CAT forecasting based on dimensional arguments, yielding

$$\mathrm{CP} = \lambda^2 \times \mathrm{VWS} \times \left[1 - \frac{\mathrm{Ri}}{\mathrm{Ri_{crit}}}\right], \tag{12.3}$$

in which λ is a length scale equal to the vertical grid spacing and $\mathrm{Ri_{crit}}$ was chosen to be somewhere between the theoretical value of ¼ and a more empirically useful value of ¾. A more advanced diagnostic using a one-and-a-half turbulence closure scheme (see Stull 1988, Ch. 6.5) was created by Marroquin (1998):

$$\varepsilon = K_M\left(\frac{c_1}{c_3}\mathrm{VWS} - \frac{c_2}{c_3}\frac{N^2}{\mathrm{Pr}}\right), \tag{12.4}$$

in which the constants $c_1 = 1.44$, $c_2 = 1.0$, and $c_3 = 1.92$ (Stull 1988, p. 219) and the eddy diffusivity of momentum K_M and the Prandtl number Pr are tunable constants. This diagnostic, known as DTF3 (see Appendix A of Sharman et al. 2006), has, like the Colson–Panofsky index, enjoyed continued use.

As with the condition for instability itself, Richardson number and other energetics approaches for forecasting CAT have proven to be necessary, but not sufficient. Other techniques that incorporate a combination of kinematic and dynamical insights have proven to be equally effective and to some extent complementary rather than duplicative.

12.2.4 Kinematic Approaches

Another way to tackle the problem of forecasting CAT is to focus on the kinematics of weather patterns that create strong vertical shears, such as fronts (via the thermal wind law; see Holton and Hakim 2013, Ch. 3.4). This approach, encouraged by the long-standing empirical evidence that CAT is collocated with fronts, has spawned several forecasting indices.

Continuing the theory proposed by Mancuso and Endlich (1966) and Roach (1970), Ellrod and Knapp (1992) identified deformation as an important factor in turbulence production. Several previous studies comparing satellite imagery and pilot reports of turbulence (PIREPs, e.g., Schwartz 1996) illustrated that turbulence frequently occurred along deformation zones (Anderson et al. 1982; Ellrod 1985). Ellrod and Knapp then proposed two diagnostics, the most popular of which simply multiplies VWS with deformation DEF:

$$\mathrm{TI} = \mathrm{VWS} \times \mathrm{DEF}, \tag{12.5}$$

in which

$$\mathrm{DEF} = \left(\mathrm{DST}^2 + \mathrm{DSH}^2\right)^{1/2}, \tag{12.6}$$

with the stretching deformation DST defined as

$$\mathrm{DST} = \frac{\partial u}{\partial x} - \frac{\partial v}{\partial y} \tag{12.7}$$

and the shearing deformation DSH defined as

$$\mathrm{DSH} = \frac{\partial v}{\partial x} + \frac{\partial u}{\partial y}. \tag{12.8}$$

The vertical wind shear term in TI, also known as the "Ellrod-Knapp index" or "Ellrod1," has its roots in Petterssen (1956)'s equation which relates deformation to increased frontogenetic intensity. Ellrod and Knapp connect this to turbulence through this chain of reasoning: "Frontogenesis will therefore result in an increase in vertical wind shear, and an increase in the likelihood of CAT occurrence."

Earlier studies supported Ellrod and Knapp's claim that the product of deformation and vertical wind shear appeared to be a good indicator of turbulence potential, as it was noted that this combination of parameters resulted in the highest correlation with observed moderate or severe turbulence (Mancuso and Endlich 1966).

Since its creation, the Ellrod–Knapp index has been widely used internationally because of its combination of accuracy and ease of computation. It seems to capture turbulence in and near the jet streams as well as near fronts while not simply replicating the results obtained from using Ri as a diagnostic (Fig. 12.3, from Jaeger and Sprenger 2007). It is not the only way to incorporate frontogenesis into a CAT diagnostic, however. The magnitude of the horizontal temperature gradient is a straightforward measure of a front:

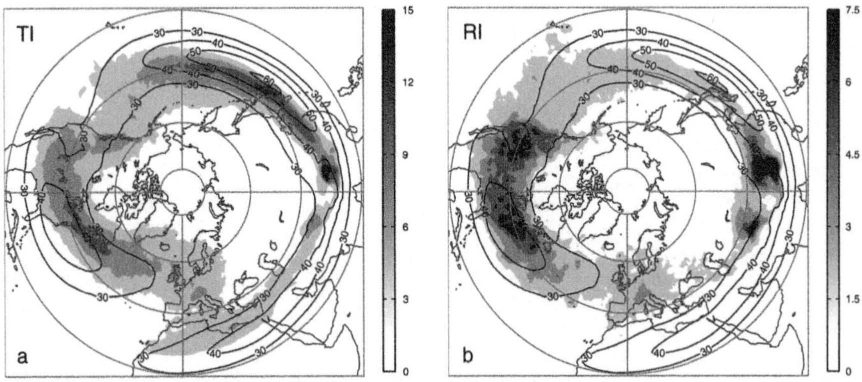

Fig. 12.3 December–February average frequencies of (**a**) high values of the Ellrod–Knapp turbulence index (TI) and (**b**) low values of the Richardson number (RI) for the period 1958–2001, as determined from reanalyzed data. From Jaeger and Sprenger (2007, Fig. 2)

$$|\nabla_H T| = \left[\left(\frac{\partial T}{\partial x}\right)^2 + \left(\frac{\partial T}{\partial y}\right)^2\right]^{1/2}. \quad (12.9)$$

Alternatively, a simple rewriting of the frontogenesis function results, by way of the thermal wind relation, an expression relating frontogenesis F to vertical shear (here shown in pressure coordinates for simplicity):

$$F \propto \frac{\partial u}{\partial p}\frac{D}{Dt}\left(\frac{\partial u}{\partial p}\right) + \frac{\partial v}{\partial p}\frac{D}{Dt}\left(\frac{\partial v}{\partial p}\right). \quad (12.10)$$

This index also provides accuracy similar or slightly superior to Ri at upper levels as well as in the mid-troposphere (Sharman et al. 2006).

12.2.5 Dynamical Approaches

A weakness of energetics and kinematic methods is that they do not incorporate an understanding of the full time-dependent equations of motion. The full primitive equations, and even balance approximations to the primitive equations that include nonlinear terms, retain crucial dynamical asymmetries between high- and low-pressure systems that include divergent motions on a variety of scales that can, in turn, generate turbulence.

One concern is that frontogenetical diagnostics, if applied to regions that are not frontogenetical, may give erroneous results. Knox (1997) critiqued several existing CAT diagnostics and determined that CAT in strongly anticyclonic flows cannot be explained by frontogenetical CAT indices and that strong shears in such flows are

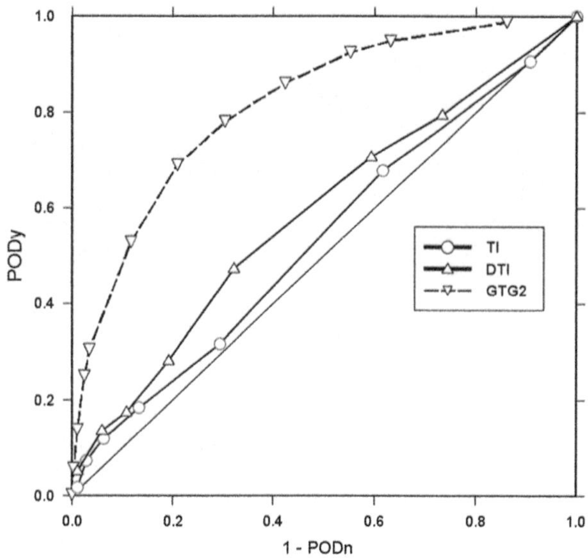

Fig. 12.4 Relative operating characteristic (ROC) curve comparing the Ellrod–Knapp turbulence index (TI) with the divergence trend-added Ellrod–Knox index (DTI) 6-h forecasts, with the graphical turbulence guidance multi-diagnostic index (GTG2) for reference, for December 2007. From Ellrod and Knox (2010, Fig. 8)

not necessarily linked to either deformation or frontogenesis. This work stimulated an extension of the Ellrod–Knapp index to incorporate a divergence trend term to capture CAT in both strong ridges and sharp troughs (Ellrod and Knox 2010). In case studies (Wilson 2012) as well as climatological (Fig. 12.4, from Ellrod and Knox 2010; also Knox et al. 2011) and operational use at the Aviation Weather Center (Holicky and Silberberg 2015), the Ellrod–Knox index has proven superior to the Ellrod–Knapp index by a variety of metrics of forecast accuracy. In practice, the Ellrod–Knox method captures more than enough additional CAT occurrences in troughs and amplifying ridges to offset a somewhat higher false alarm rate versus the Ellrod–Knapp index. Some drawbacks for implementation are that it requires time differencing and that it must be tuned to each model to optimize the benefit from the divergence trend term.

One mechanism for CAT not explicitly included in any of the diagnostics discussed so far is internal gravity waves with wavelengths of a few kilometers to a few hundred kilometers (Sutherland 2010). Turbulence has been correlated with gravity waves since at least Reiter's early work on the subject in the 1960s (e.g., Reiter and Nania 1964). If an aircraft were to fly into a large-amplitude internal gravity wave, the aircraft would likely encounter strong gusts and vertical motions, which can be interpreted as CAT. Gravity waves also can reduce the local Ri, triggering Kelvin–Helmholtz instabilities (e.g., Koch et al. 2005). Koch et al. concluded that "The findings here do suggest that automated turbulence forecasting algorithms should include some reliable measure of gravity wave activity."

One approach to including gravity waves in a CAT forecasting diagnostic is to estimate the unbalanced nature of the flow. The residual of the nonlinear balance equation

$$\text{UBF} = -\nabla^2\Phi + 2J(u,v) + f\zeta - \beta u \tag{12.11}$$

has been used by several researchers as a diagnostic related to gravity wave activity and/or turbulence (e.g., McCann 2001; Koch and Caracena 2002; Zhang et al. 2000).

Another approach to gravity wave-based CAT forecasting was developed by Knox et al. (2008), who developed a forecasting method based on the Lighthill (1952) and Ford (1994) shallow-water theory for generation of sound waves (Lighthill) and inertia-gravity waves (Ford) via spontaneous emission. This work was stimulated by fluid dynamics tank experiments (Williams et al. 2005). In this method, the amplitude of the gravity waves was set proportional to the forcing terms derived from the theory, namely:

$$\text{LHFK} = f\mathbf{V}\cdot\nabla\zeta + 2Df\zeta - f\mathbf{k}\cdot\mathbf{V}\times\nabla D - 2\frac{\partial}{\partial t}J(u,v), \tag{12.12}$$

in which D is the horizontal divergence. This amplitude was then employed in the McCann (2001) ULTURB algorithm to compute TKE production. Theoretically, the terms on the right-hand side of Eq. (12.12) were shown to share commonalities with both inertial-advective diagnostics and with the Ellrod–Knox index and also coincide in large part with the forcing terms found by Medvedev and Gavrilov (1995) in their study of nonlinear generation of gravity wave generation.

Operationally, the Lighthill–Ford method has shown promise in some applications (Knox et al. 2008; Steel 2011; McCann et al. 2012) and not in others (e.g., Wilson 2012), the latter in which the gravity wave physics has been replaced with simple division of the forcing term by Ri. There is also debate about the applicability of the shallow-water theory to the observed atmosphere and the interpretation of the mechanism of spontaneous emission of gravity waves as the cause of CAT (Plougonven et al. 2009; Knox et al. 2009). Prediction of gravity wave-caused CAT remains a challenging subject.

12.2.6 Multi-diagnostic Statistical Approaches

Given that there is no "silver bullet" diagnostic that can reliably predict CAT because of its varied causal mechanisms, a reasonable response is to create an algorithm which combines different diagnostics in a statistically weighted and optimized manner. This is the approach of Sharman and collaborators in the development of the GTG system. The diagnostics used for upper levels (above 20,000 ft) in GTG2 are Eqs. (12.2)–(12.5), (12.9), and (12.10) (in isentropic coordinates), and (12.11) in this chapter, as well as three other diagnostics (see Sharman et al. 2006 for more details). The GTG approach has been repeatedly shown to outperform any one diagnostic in a wide range of applications (e.g.,

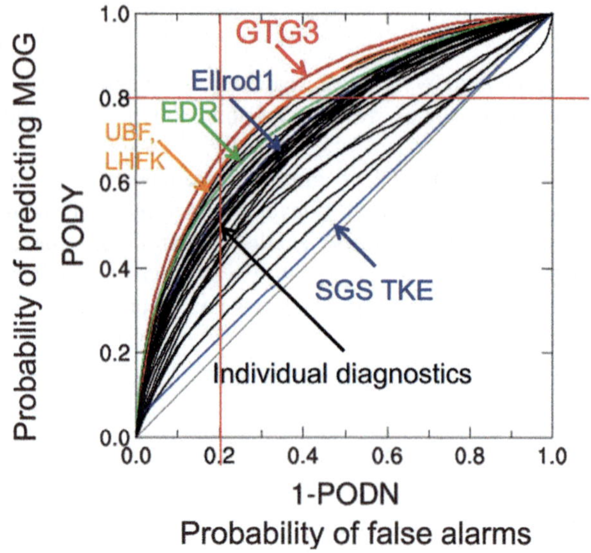

Fig. 12.5 Relative operating characteristic (ROC) curves for the third generation of graphical turbulence guidance (GTG3) and individual diagnostics that are used in it, for 6-h forecasts over 12 months. From Slide 9 of http://www.rmets.org/sites/rmets.org/files/presentations/15012014-sharman.pdf

Sharman et al. 2006; Kim et al. 2011), as seen in both Figs. 12.4 and 12.5. One drawback from the perspective of aviation meteorologists is the inability to link GTG output to specific causal mechanisms, since different indices representing different mechanisms are combined to form GTG. This complaint notwithstanding, GTG is currently viewed as the state-of-the-art approach for CAT forecasting and is discussed extensively elsewhere in this book.

12.3 Low-Level and Convectively Induced Turbulence

LLT and CIT present major challenges to forecasting because of the small spatial and temporal scales of the triggering phenomena involved (e.g., thunderstorms, eddies associated with obstacles, or terrain). Lead times have been very short historically, focusing attention on nowcasting versus forecasting. The serious risk of accidents posed by LLT and CIT has also placed a high priority on avoidance based on real-time detection by observational systems, for example, Guidelines 5 and 6 of the US Federal Aviation Administration (FAA 2012) state:

Guideline 5. Do avoid by at least 20 miles (laterally; 36 km) any thunderstorm identified as severe or giving an intense radar echo. This is especially true under the anvil of a large cumulonimbus.

Guideline 6. Do clear the top of a known or suspected severe thunderstorm by at least 1000-ft altitude for each 10 knots (5 m s^{-1}) of wind speed at the cloud top. This should exceed the altitude capability of most aircraft.

One older empirically based technique for forecasting LLT involves a nomogram which relates horizontal temperature gradients and frontal speed to the presence of wind shear and turbulence (see Lester 1993, pp. 2–37). According to this nomogram, a combination of larger temperature gradients and more rapid frontal speed is correlated with a greater likelihood of turbulence and/or wind shear. This correlation makes sense in the context of the previous discussion of frontogenesis, Kelvin–Helmholtz instability, and CAT.

Mesoscale convection can create turbulence both within the storm's updrafts and downdrafts and also beyond the periphery of the convection. For example, convection can also create internal gravity waves, thus producing turbulence. This convectively initiated CAT is created when strong updrafts penetrate the tropopause or some capping layer, resulting in internal gravity waves propagating away from the thunderstorm in stable layers in the troposphere and stratosphere. Therefore, an aircraft flying near a thunderstorm is susceptible to CAT from these propagating waves. Turbulence is also associated with transverse bands in the cirrus clouds associated with thunderstorm anvils, with its causes a topic of current research (Lenz et al. 2009; Knox et al. 2010; Trier et al. 2010; Kim et al. 2014). Underneath thunderstorm anvils and other mid-level cloud decks, turbulence can also arise via cooling due to sublimation of snow that leads to absolute instability (Kudo 2013).

A growing body of research indicates that CIT can extend well beyond the convective core of thunderstorms and is called "near-cloud turbulence" (NCT). While NCT has been known since the 1960s (e.g., Burns et al. 1966), the recent review article by Lane et al. (2012) has expanded our understanding that convection can cause turbulence at least 70 km away and over 3 km above cloud-top level, probably as a result of gravity waves triggered by the convection—a much wider sphere of influence than the FAA guidelines reflect. The distinction between NCT and CAT is thus an evolving boundary, with potential forecasting implications.

12.4 Mountain Wave Turbulence

Mountain waves are a type of internal gravity wave. They are created when stable airflow passes over significant topographical barriers (usually not just one mountain peak). In contrast to CAT, the triggering mechanism for MWT is straightforward and the theory for how these waves are formed and interact with their environment is better established, from Scorer (1949) and Eliassen and Palm (1961) forward. Vertically propagating mountain waves amplify through both the decrease of atmospheric density with height and also when they encounter "critical layers" in which the wind reverses direction with height (see Sharman et al. 2012b for additional details and references). In this linear theory, wave instability occurs near when the inverse Froude number

$$\mathrm{Fr}^{-1} = \frac{NH}{U} \tag{12.13}$$

exceeds 1. In Eq. (12.13), H is the topographic obstacle height and U is the low-level velocity. Mountain waves can also propagate horizontally through stable layers and are known as trapped lee waves.

According to McCann (2006), "Numerous investigators have published research articles furthering the understanding of mountain waves, but few have focused on their turbulence-producing potential. As a result practical turbulence forecast methods are rare." One such method stems from gravity wave drag (GWD) parameterizations used in numerical weather prediction and global climate models. Bacmeister (1993) and Bacmeister et al. (1994) introduced an algorithm based on McFarlane's (1987) orographic GWD scheme that improved upon use of Ri to forecast MWT in the lower stratosphere. This line of approach does not incorporate non-hydrostatic and nonlinear effects. Direct numerical simulations "provide the most robust method for reconstructing or predicting mountain waves and MWT" (Sharman et al. 2012a). However, the requirement of 3D, non-hydrostatic modeling runs with realistic terrain, and very high horizontal (on the order of 1 km) and vertical (a few hundred meters or less) resolution has slowed operational implementation.

As in the case of CAT forecasting, therefore, MWT forecasting has historically relied on empirical rules of thumb (e.g., Lee et al. 1984), most prominently the requirement of strong winds blowing nearly perpendicularly to sharp terrain. Researchers and operational forecasters also rely on analysis of sounding data from the larger-scale flow rather than direct numerical simulation. Conditions conducive to both vertically propagating waves and trapped lee waves can be identified via decreases in the Scorer parameter (Scorer 1949):

$$l^2(z) = \frac{N^2}{U^2} - \frac{1}{U}\frac{\partial^2 U}{\partial z^2}. \tag{12.14}$$

Constant l^2 conditions with height promote vertical propagation; strongly decreasing l^2 with height promotes trapping of lee waves. Operational forecasters look for strong winds and strong vertical wind curvature within these regions for the potential for wave generation and horizontal wave propagation downstream, respectively. In extreme cases of wave trapping, horizontally rotating flows downstream of the terrain known as atmospheric rotors can develop. MWT, including rotors, has been the subject of a recent observational and modeling campaign (Doyle et al. 2011a) which found "relatively low predictability of key characteristics of topographically forced flows" (Doyle et al. 2011b).

12.5 Future Strategies

As noted by Sharman et al. (2012b), "turbulence forecasting remains one of the last great challenges of numerical weather prediction." An even greater challenge, then, is predicting the future of turbulence forecasting strategies. As discussed in detail elsewhere in this book, deterministic forecast approaches are yielding to probabilistic techniques, as they are throughout numerical weather prediction. However, as Sharman et al. (2006) discussed, "better diagnostics... is a continued research area at major laboratories and universities." Fundamental improvements in our understanding of turbulence that lead to new and/or improved diagnostics will benefit multi-diagnostic approaches such as GTG in addition to being useful diagnostics in their own right. In particular, improvements or syntheses in our understanding of gravity wave generation, propagation, and breaking seem to be at the heart of any significant advances in forecasting. As has been the case repeatedly in the history of turbulence forecasting, a rich interplay between observations, theory, and modeling is most likely to provide the fertile ground for these advances.

Acknowledgments Thanks to Erik Galicki, Maria Augutis, and Patrick Malone for their efforts in the research seminar at the University of Georgia that culminated in this chapter. We are very grateful to James McCormick for suggestions regarding the MWT portion of this chapter, and to the editors and reviewers who made this chapter possible and whose suggestions improved its content.

References

Anderson, R.K., Gurka, J.J., Steinmetz, S.J.: Application of VAS multispectral imagery to aviation forecasting. In: Proceedings of the 9th Conference on Weather Forecasting and Analysis, pp. 227–234. American Meteorological Society, Seattle (1982) (Preprints)

Bacmeister, J.T.: Mountain wave drag in the stratosphere and mesosphere inferred from observed winds and a simple mountain wave parameterization scheme. J. Atmos. Sci. **50**, 377–399 (1993)

Bacmeister, J.T., Newman, P.A., Gary, B.L., Chan, K.R.: An algorithm for forecasting mountain wave-related turbulence in the stratosphere. Weather Forecast. **9**, 241–253 (1994)

Burns, A., Harrold, T.W., Burnham, J., Spavins, C.S.: Turbulence in clear air near thunderstorms. Tech. Memo. 30, National Severe Storms Laboratory, Norman, 20 pp. (1966)

Chambers, E.: Clear air turbulence and civil jet operation. J. R. Aeronaut. Soc. **59**, 613–628 (1955)

Chandler, C.L.: Turbulence forecasting. In: Camp D.W., Frost W. (eds.) Atmospheric Turbulence Relative to Aviation, Missile, and Space Programs, pp. 137–154. NASA Conference Publication 2468 (1986)

Colson, D.: Summary of high level turbulence over United States. Mon. Weather Rev. **91**, 605–609 (1963)

Colson, D., Panofsky, H.A.: An index of clear-air turbulence. Q. J. R. Meteorol. Soc. **91**, 507–513 (1965)

Doyle, J.D., Jiang, Q., Smith, R.B., Grubišić, V.: Three-dimensional characteristics of stratospheric mountain waves during T-REX. Mon. Weather Rev. **139**, 3–23 (2011a). doi:10.1175/2010MWR3466.1

Doyle, J.D., Gaberšek, S., Jiang, Q., Bernardet, L., Brown, J.M., Dörnbrack, A., Filaus, E., Grubišić, V., Kirshbaum, D.J., Knoth, O., Koch, S., Schmidli, J., Stiperski, I., Vosper, S., Zhong, S.: An intercomparison of T-REX mountain-wave simulations and implications for mesoscale predictability. Mon. Weather Rev. **139**, 2811–2831 (2011b)

Dutton, M.J.O.: Probability forecasts of clear air turbulence based on numerical model output. Meteorol. Mag. **109**, 293–308 (1980)

Dutton, J., Panofsky, H.A.: Clear air turbulence: a mystery may be unfolding. Science **167**, 937–944 (1970)

Eliassen, A., Palm, E.: On the transfer of energy in stationary mountain wave. Geof. Publ. **22**, 1–23 (1961)

Ellrod, G.P., Knapp, D.L.: An objective clear-air turbulence forecasting technique: verification and operational use. Weather Forecast. **7**, 150–165 (1992)

Ellrod, G.P., Knox, J.A.: Improvements to an operational clear-air turbulence diagnostic index by addition of a divergence trend term. Weather Forecast. **25**, 789–798 (2010)

Ellrod, G.P.: Detection of High Level Turbulence Using Satellite Imagery and Upper Air Data, 30 pp. NOAA Tech. Memo. NESDIS 10, U.S. Department of Commerce, Washington, DC (1985)

Ellrod, G.P., Knox, J.A., Lester, P.F., Ehernberger, L.J.: Clear air turbulence. In: North, G.R., Pyle, J., Zhang, F. (eds.) Encyclopedia of Atmospheric Science, 2nd edn, pp. 177–186. Academic Press, Elsevier (2015)

Endlich, R.: The mesoscale structure of some regions of clear-air turbulence. J. Appl. Meteorol. **3**, 261–276 (1964)

FAA: FAA Aeronautical Information Manual. Available online at www.faa.gov/air_traffic/publications/atpubs/aim (2012)

FAA: Rules of thumb for avoiding or minimizing clear air turbulence, 56 pp. Federal Aviation Administration, Washington, DC (1988)

Ford, R.: Gravity wave radiation from vortex trains in rotating shallow water. J. Fluid Mech. **281**, 81–118 (1994)

Holicky, E., Silberberg, S.: Improving turbulence SIGMET decision support with Ellrod-Knox guidance. In: 17th American Meteorological Society Aviation, Range and Aerospace Meteorology Conference, Phoenix, AZ. https://ams.confex.com/ams/95Annual/webprogram/Paper268888.html

Holton, J.R., Hakim, G.J.: An Introduction to Dynamic Meteorology, 5th edn., 532 pp. Academic Press, New York (2013)

Howard, L.N.: Note on a paper of John W. Miles. J. Fluid Mech. **10**, 509–512 (1961)

Jaeger, E.B., Sprenger, M.: A northern-hemispheric climatology of indices for clear air turbulence in the tropopause region derived from ERA40 re-analysis data. J. Geophys. Res. **112**, D20106 (2007). doi:10.1029/2006JD008189

Keller, J.L.: Clear air turbulence as a response to meso- and synoptic-scale dynamic processes. Mon. Weather Rev. **118**, 2228–2243 (1990)

Kim, J.-H., Chun, H.-Y., Sharman, R.D., Keller, T.L.: Evaluations of upper-level turbulence diagnostics performance using the graphical turbulence guidance (GTG) System and pilot reports (PIREPs) over East Asia. J. Appl. Meteorol. Climatol. **50**, 1936–1951 (2011)

Kim, J.-H., Chun, H.-Y., Sharman, R.D., Trier, S.B.: The role of vertical shear on aviation turbulence within cirrus bands of a simulated western Pacific cyclone. Mon. Weather Rev. **142**, 2794–2813 (2014)

Knox, J.A.: Possible mechanisms of clear-air turbulence in strongly anticyclonic flows. Mon. Weather Rev. **125**, 1251–1259 (1997)

Knox, J.A., McCann, D.W., Williams, P.D.: Application of the Lighthill-Ford theory of spontaneous imbalance to clear-air turbulence forecasting. J. Atmos. Sci. **65**, 3292–3304 (2008)

Knox, J.A., McCann, D.W., Williams, P.D.: Application of the Lighthill-Ford theory of spontaneous imbalance to clear-air turbulence forecasting: reply. J. Atmos. Sci. **66**, 2511–2515 (2009)

Knox, J.A., Bachmeier, A.S., Carter, W.M., Tarantino, J.E., Paulik, L.C., Wilson, E.N., Bechdol, G.S., Mays, M.J.: Transverse cirrus bands in weather systems: a grand tour of an enduring enigma. Weather **5**, 35–41 (2010)

Knox, J.A., Ellrod, G.P., Silberberg, S.R., Wilson, E., Black, A., Galicki, E., Rackley, J., Dunn, C., Malone, P., Phelps, S., Augutis, M., Grant, J.: Verification of clear air turbulence (CAT) forecast indices during two winters. In: National Weather Association Annual Meeting, Birmingham. http://www.nwas.org/meetings/abstracts/display.php?id=1258 (2011)

Koch, S.E., Caracena, F.: Predicting clear-air turbulence from diagnosis of unbalanced flow. In: Proceedings of 10th Conference on Aviation, Range, and Aerospace Meteorology, pp. 359-363. Portland, OR (Preprints) (2002)

Koch, S.E., Jamison, B.D., Lu, C., Smith, T.L., Tollerud, E.L., Girz, C., Wang, N., Lane, T.P., Shapiro, M.A., Parrish, D.D., Cooper, O.R.: Turbulence and gravity waves within an upper-level front. J. Atmos. Sci. **62**(11), 3885–3908 (2005)

Kronebach, G.: An automated procedure for forecasting clear-air turbulence. J. Appl. Meteorol. **3**, 119–125 (1964)

Kudo, A.: The generation of turbulence below midlevel cloud bases: the effect of cooling due to sublimation of snow. J. Appl. Meteorol. Climatol. **52**, 819–833 (2013)

Lane, T.P., Sharman, R.D., Trier, S.B., Fovell, R.G., Williams, J.K.: Recent advances in the understanding of near-cloud turbulence. Bull. Am. Meteorol. Soc. **93**(4), 499–515 (2012). doi:10.1175/BAMS-D-11-00062.1

Lee, D.R., Stull, R.B., Irvine, W.S.: Clear air turbulence forecasting techniques. Air Force Weather Service Report AFGWC/TN-79-001 (REV) (1984)

Lenz, A., Bedka, K., Feltz, W., Ackerman, S.: Convectively-induced transverse band signatures in satellite imagery. Weather Forecast. **24**(5), 1362–1373 (2009)

Lester, P.F.: Turbulence: A New Perspective for Pilots. Jeppesen Sanderson, Englewood, CO (1993)

Lighthill, M.J.: On sound generated aerodynamically. I. General theory. Proc. R. Soc. Lond. **211A**, 564–587 (1952)

Mancuso, R.L., Endlich, R.M.: Clear air turbulence frequency as a function of wind shear and deformation. Mon. Weather Rev. **94**, 581–585 (1966)

Marroquin, A.: An advanced algorithm to diagnose atmospheric turbulence using numerical model output. In: Proceedings of 16th Conference on Weather Analysis and Forecasting, pp. 79–81. American Meteorological Society, Phoenix, AZ (Preprint) (1998)

McCann, D.: Gravity waves, unbalanced flows, and aircraft clear air turbulence. Nat. Weather Dig. **25**(1, 2), 3–14 (2001)

McCann, D.W.: Diagnosing and forecasting aircraft turbulence with steepening mountain waves. Nat. Weather Dig. **30**, 77–92 (2006)

McCann, D.: An evaluation of clear-air turbulence indices. In: Proceedings of the 5th International Conference on Aviation Weather Systems, pp. 449–453. American Meteorological Society (1993)

McCann, D.W., Knox, J.A., Williams, P.D.: An improvement in clear-air turbulence forecasting based on spontaneous imbalance theory: the ULTURB algorithm. Meteorol. Appl. **19**, 71–78 (2012). doi:10.1002/met.260

McFarlane, N.A.: The effect of orographically excited gravity wave drag on the general circulation of the lower stratosphere and troposphere. J. Atmos. Sci. **44**, 1775–1800 (1987)

Medvedev, A.S., Gavrilov, N.M.: The nonlinear mechanism of gravity wave generation by meteorological motions in the atmosphere. J. Atmos. Terr. Phys. **57**, 1221–1231 (1995)

Miles, J.W.: On the stability of heterogeneous shear flows. J. Fluid Mech. **10**, 496–508 (1961)

Miles, J.: Richardson's criterion for the stability of stratified shear flow. Phys. Fluids **29**, 3470–3471 (1986). doi:10.1063/1.865812

Miles, J.W., Howard, L.N.: Note on a heterogeneous shear flow. J. Fluid Mech. **20**, 331–336 (1964)

Petterssen, S.: Weather Analysis and Forecasting, vol. 1, 428 pp. McGraw-Hill Book, New York (1956)
Plougonven, R., Snyder, C., Zhang, F.: Comments on "Application of the Lighthill-Ford theory of spontaneous imbalance to clear-air turbulence forecasting.". J. Atmos. Sci. **66**, 2506–2510 (2009)
Reiter, E.R., Nania, A.: Jet-stream structure and clear-air turbulence. J. Appl. Meteorol. **3**, 247–260 (1964)
Richardson, L.F.: The supply of energy from and to atmospheric eddies. Proc. R. Soc. Lond. Ser. A **97**, 354–373 (1920)
Roach, W.T.: On the influence of synoptic development on the production of high level turbulence. Q. J. R. Meteorol. Soc. **96**, 413–429 (1970)
Schwartz, B.: The quantitative use of PIREPs in developing aviation weather guidance products. Weather Forecast. **11**, 372–384 (1996)
Scorer, R.S.: Theory of waves in the lee of mountains. Q. J. R. Meteorol. Soc. **75**, 41–56 (1949)
Serebreny, S.M., Wiegman, E.J., Hadfield, R.G.: Some characteristic fields of the jet stream complex during selected synoptic conditions. J. Appl. Meteorol. **2**, 137–153 (1962)
Sharman, R., Tebaldi, C., Wiener, G., Wolff, J.: An integrated approach to mid- and upper-level turbulence forecasting. Weather Forecast. **21**(3), 268–287 (2006). doi:10.1175/WAF924.1
Sharman, R.D., Doyle, J.D., Shapiro, M.A.: An investigation of a commercial aircraft encounter with severe clear-air turbulence over western Greenland. J. Appl. Meteorol. Clim. **51**(1), 42–53 (2012a)
Sharman, R.D., Trier, S.B., Lane, T.P., Doyle, J.D.: Sources and dynamics of turbulence in the upper troposphere and lower stratosphere: a review. Geophys. Res. Lett. **39**, L12803 (2012b). doi:10.1029/2012GL051996
Stack, D.T.: Turbulence avoidance. In: Proceedings of the 4th International Conference on Aviation Weather Systems. American Meteorological Society, pp. 283–286, Paris, France (Preprints) (1991)
Steel, M.: Improved CAT algorithms. UK Met Office, 42 pp. (2011)
Stull, R.: An Introduction to Boundary Layer Meteorology. Springer, Dordrecht (1988)
Sutherland, B.R.: Internal Gravity Waves. Cambridge University Press, Cambridge (2010)
Taylor, G.I.: Effect of variation in density on the stability of superposed streams of fluid. Proc. R. Soc. Lond. A **132**, 449–523 (1931)
Trier, S.B., Sharman, R.D., Fovell, R.G., Frehlich, R.G.: Numerical simulation of radial cloud bands within the upper-level outflow of an observed mesoscale convective system. J. Atmos. Sci. **67**(9), 2990–2999 (2010)
Williams, P.D., Haine, T.W.N., Read, P.L.: On the generation mechanisms of short-scale unbalanced modes in rotating two-layer flows with vertical shear. J. Fluid Mech. **528**, 1–22 (2005)
Wilson, E.N.: Case studies of clear-air turbulence: evaluation and verification of new forecasting techniques, 127 pp. M.S. thesis, University of Georgia (2012)
Wolff, J.K., Sharman, R.D.: Climatology of upper-level turbulence over the contiguous United States. J. Appl. Meteorol. Climatol. **47**, 2198–2214 (2008)
Zhang, F., Koch, S.E., Davis, C.A., Kaplan, M.L.: A survey of unbalanced flow diagnostics and their application. Adv. Atmos. Sci. **17**, 165–183 (2000)

Chapter 13
Aviation Turbulence Forecast Verification

Philip G. Gill

Abstract Forecast verification is an essential part of any forecast system. Verification results can be used in many ways: in the development process to ensure changes are leading to the desired improvements and to assure users and relevant authorities the forecasts have the required level of skill and can help to maximise the value a user gains from the forecast. In this section, we look at the challenges of verifying both deterministic and probabilistic turbulence forecasts. We investigate the different potential sources of truth data for use in verification with some discussion of the advantages and disadvantages of each. We also look at the possible methodologies and metrics that can be applied with some suggestions on the most appropriate ways to verify turbulence forecasts.

13.1 Introduction

Verification should be an integral part of any forecast service or development project. It is important to be able to assess any changes to forecast production to ensure that they are leading to overall improvements or at the very least not leading to a deterioration of the forecast service. If implemented at the start of a project, then an objective verification scheme can facilitate more rapid developments later on.

Forecast verification is simply the process of comparing a forecast value with the truth value that occurred. However, it is not without significant challenges, and aviation turbulence is no exception. The common challenges of availability of truth data are fundamental issues to ensuring a robust verification system. Representativeness of the truth data is also an important issue that needs to be considered carefully. Moderate or severe aviation turbulence that is of interest to the aviation community is a rare event and as a result leads to issues with quality control of the truth data to ensure that extreme values are not thrown out when they are real. Care

P.G. Gill (✉)
Met Office, FitzRoy Road, Exeter, Devon EX1 3PB, UK
e-mail: philip.gill@metoffice.gov.uk

Published by Springer International Publishing AG Switzerland 2016.
R. Sharman, T. Lane (eds.), *Aviation Turbulence*, DOI 10.1007/978-3-319-23630-8_13

should be taken in the use of many metrics which tend to trivial scores as the event gets rarer.

If in situ observations are used, then clearly verification is limited entirely to where these aircraft fly. However as turbulence forecasts improve, then if these are used effectively to influence flight planning the encounters of turbulence will reduce even if the frequency of turbulence does not, making the event rarer and more difficult to verify.

Due to the small scale at which turbulence occurs, it is not possible, particularly on a global scale, to resolve turbulence itself in forecast models. This means that in many cases we may be forecasting one quantity that is believed to be related to turbulence and observing another which complicates the verification process further.

In addition to leading to improvements to the forecast service, verification can also be used by the user to make more informed decisions in managing risk. Verification statistics on past performance can help a user decide where to set operating thresholds to maximise the value they obtain from the forecast service.

13.2 Use of Truth Data for Verification

In order to verify a forecast, we need some source of observational truth to enable us to compare a forecast with what actually occurred. For in situ data, reports of turbulence are often obtained from Pilot Reports (PIREPS) which are issued by aircraft pilots describing the turbulence that they have encountered. Automated reports are routinely available by Aircraft Meteorological Data Relay (AMDAR) and also within the flight data recorder itself.

At short range it is possible to use aircraft mounted Lidar to detect turbulence. There are also some possibilities for using satellite data to identify conditions that may relate to turbulence. We look at each of these possible sources of truth data highlighting the advantages and disadvantages of both.

13.2.1 Pilot Reports

PIREPS are written by the pilot when turbulence is encountered (FAA 2014). They are useful observations of what was actually experienced within the aircraft and come with a level of severity which is the subjective view of the pilot. The severity is categorised as null/light/mod/severe/extreme. The reports are in the form of abbreviated text and include a time of issue and location with a latitude and longitude, flight level and aircraft type. These reports occasionally have additional information attached to them which can be particularly useful when looking at case studies of the more severe events. For example, the type of turbulence may be categorised and additional location information given.

PIREPS are available in real time which makes them useful for nowcast applications (Sharman et al. 2006), although this is not generally necessary for routine verification.

Whilst PIREPS are an invaluable source for case studies, they have significant drawbacks for use in an automated objective verification system (Kane et al. 1998). Firstly, they are subjective reports that are aircraft specific. The timing and location of the event are not necessarily exactly where the event occurred, as control of the aircraft must always be the priority and producing the report comes later. Null reports of turbulence are produced but are not routine, meaning that only a partial picture of turbulence encounters is produced. To provide meaningful objective verification, it is just as important to know where turbulence has not occurred as to know where it has occurred.

There is also some evidence that reporting of turbulence does not always take place as this can lead to additional administration after the flight and in the case of severe turbulence taking the aircraft out of service whilst maintenance checks are carried out.

Example of a PIREP containing turbulence information:

UA/OV EXT/TM 2334/FL310/TP DC10/TB MDT

This can be decoded by breaking the message down into sections delimited by the "/" as follows:

UA	Routine report as opposed to UUA for an urgent report
OV EXT	Location is close to Exeter Airport (EXT)
TM 2334	Time of report is 23:34 UTC
FL310	Flight level of report is FL310 (31,000 ft)
TP DC10	Aircraft type is a DC10
TB MDT	Turbulence severity is moderate

13.2.2 AMDAR

AMDAR reports are a World Meteorological Organisation (WMO) standard code that are produced by commercial aircraft and can generally be obtained and made available to researchers at meteorological centres (WMO 2003). The data are available on the WMO Global Telecommunications System (GTS). AMDAR are routine reports that often include turbulence information. The advantages of AMDAR reports are they are all produced to a WMO standard and they are widely available, although there are costs involved in receiving them. Where aircraft are equipped to report turbulence information, then these reports do contain null reports of turbulence as well as actual reports which is very important for verification. They are also an objective measurement of turbulence for the specific aircraft involved. AMDAR reports are available in real time.

However, AMDAR frequency is variable depending on the phase of flight. The highest reporting frequencies are usually for the ascent and descent rather than at cruise level which is generally of interest for verification studies. Some AMDAR report instantaneous turbulence measurements which has the advantage of a precise location and time to compare to the corresponding forecast. However, it is not clear if some AMDAR may be reporting a maximum turbulence since the previous report. Reporting frequency at cruise level is generally between 7 and 21 min which may not be sufficient for accurate verification of even the coarser global model grids in the case of a maximum value. In the case of instantaneous reports, this low temporal frequency will severely limit the sample size for these relatively rare events, and therefore longer sample periods and/or more aircraft will be required to produce meaningful verification.

Most AMDAR turbulence reports are categorical light/mod/severe determined by vertical acceleration measurements. However, there are some aircraft that are equipped to produce turbulence reports from calculations of Derived Equivalent Vertical Gust (DEVG) (e.g. Australian AMDAR reports) used by Overeem (2002) and Eddy Dissipation Rate (EDR) (e.g. American AMDAR reports). These two are much more desirable as they give aircraft independent measures of turbulence.

13.2.3 Flight Recorder Data

All aircraft store highly detailed records of each flight. The main advantages of the flight recorder data are they are high frequency objective measurements, and algorithms can be used to output the turbulence measurements in different ways to give aircraft independent measurements using EDR or DEVG. This enables full and detailed verification to be carried out and is currently the best source of truth data for turbulence verification.

Flight recorder data has the benefit of being a continuous value that can then be categorised into severity.

Due to the amount of data involved, these are not generally available in real time and sometimes not for several days after the flight. For verification purposes though this is not usually a problem.

The only real drawback to flight recorder data is availability. Currently, availability is subject to individual contracts between airlines and research institutions. Usage is often restricted so that individual cases cannot be used in publications, and the source of data cannot be revealed. The reasons for this appear to be related to pilot concerns that the information may be used against them. Airlines are understandably sensitive to any adverse publicity regarding turbulence that could result. For verification it does not matter where the data comes from. There are indications that in time collaboration between airlines could lead to data being more widely available in an anonymous way, to enable research institutions to carry out turbulence research which will ultimately benefit all airlines.

13.2.4 Lidar Data

Aircraft can be equipped with lidar instruments that are able to detect short range turbulence. Currently, this is limited to a few aircraft, and the data are not widely available. Nevertheless, lidar data could be used in future to verify turbulence in an objective way. The European DELICAT (Demonstration of Lidar based clear air turbulence) project to investigate this has recently been completed (Barné 2014).

13.2.5 Satellite Data

There are some ways that satellite data can be used to identify regions where turbulence may occur. This form of data has the advantage that it is likely to be at high spatial resolution and cover wide areas, not being restricted to aircraft tracks. The use of several satellites spanning the globe could be used to give near global coverage. Satellite products have been produced to identify regions of tropopause penetrating convection which is likely to be associated with strong vertical and horizontal wind shear turbulence in the vicinity but may also generate gravity waves that can produce strong turbulence at significant distances (Bedka et al. 2010). This could be particularly useful for improving convective turbulence predictors.

Satellite products have also been used to detect areas of tropopause folding where turbulence may occur (Wimmers and Feltz 2007, 2010).

Satellite data are of use in categorising turbulence events as clear air turbulence and with some processing can be used to detect deep convection which could contribute towards turbulence in the surrounding area.

The significant drawbacks at present are that satellite data do not give us a definitive observation of turbulence, only a partial one and therefore full verification is not possible although it may be able to give us important additional information particularly in areas where turbulence observations are extremely limited using other data sources.

13.3 Quality Control of Observations

Each of the data sources needs to have adequate quality control to ensure the observations give as reliable a picture of the truth as possible. This will ensure that the verification that uses these observations is as accurate as possible.

Syntax errors are a common form of error that may be present both in human sources of observations and corrupted machine produced observations. Checks to ensure that data are within range can help to remove these. Simple checks for consistent times and positions through a flight can be made to remove some of these errors (Jerrett and Turp 2005). Positional errors are also possible and in the case of PIREPS have already been discussed.

Instrument errors are always a possibility, and methods need to be developed to attempt to remove these although this is not straight forward as turbulence is sporadic and noisy in its nature.

Manoeuvring aircraft may give false readings of turbulence, and these too need to be filtered out. One simple way of doing this is to just restrict verification to cruise levels where significant manoeuvres and related errors are less likely to occur.

Extreme values can also be removed where extreme values are continually reported although care needs to be taken where to set the threshold to avoid removing any real severe turbulence cases. There are also examples where the values have been observed to slowly drift across a large range of values which are clearly not realistic observations (Tenenbaum 2014).

Consistency of other meteorological parameters such as wind and temperature can be a useful additional check as these might reveal gross instrument errors that lead to erroneous turbulence observations.

Consistency of subjective reports in PIREPS is difficult to ensure making them unsuitable for use in an objective verification scheme (Schwartz 1996).

For the very rare severe turbulence reports, manual quality control may be necessary to ensure that these are genuine.

13.4 Turbulence Observation Types

Each of the observation types may be reported in different ways. Currently, the ICAO standard for reporting turbulence is EDR, and this should be the truth type used where possible in the verification.

13.4.1 Pilot Assessment

Pilot assessment of turbulence encounters is available via PIREPS. As already noted, these are aircraft dependent subjective reports that are of great use for case studies to accompany objective verification but are unlikely to be sufficient on their own.

13.4.2 Eddy Dissipation Rate

EDR has the benefits of being an objective, automated, aircraft independent measure that is potentially available from flight recorder data. There is a further benefit in that both the numerical predictors and the observations can be produced in terms of EDR, enabling a full comparison between the two as both continuous and categorical variables. We need to be sure that the link between EDR and turbulence experienced on the aircraft is strong to ensure that not only are the forecasts good at predicting EDR but also at predicting turbulence. Currently, availability of EDR data for verification is limited mainly to the USA although this should improve now that it is the ICAO standard. It should also be noted that even EDR is reported in more than one way at present and this needs to be standardised for global use. Currently, the best coverage of EDR data is over the USA and has been used successfully for verification (Takacs et al. 2005).

13.4.3 Derived Equivalent Vertical Gust

Another turbulence measure that is in use is DEVG (Sherman 1985). This also has the benefit of being an automated, objective, aircraft independent measure. As with EDR reports, DEVG reports are not available everywhere. DEVG reports are sometimes available in AMDAR data, and the new Tropospheric Airborne Meteorological Data Relay (TAMDAR) reports are also capable of reporting DEVG. Using aircraft flight recorder data, the DEVG can be calculated (Truscott 2000).

At the time of writing, DEVG reports are available from fleets of aircraft operating out of Europe and Australia from both flight recorder data and, for some airlines, AMDAR data.

13.4.4 Vertical Acceleration

Vertical acceleration is perhaps the simplest measurement used for turbulence reporting. Most AMDAR reports that include turbulence use vertical acceleration to measure severity, and this is categorised in the report as null, light, moderate and severe. AMDAR coverage is global which makes this an appealing choice for global verification of turbulence. However, the aircraft dependence and low reporting frequency outweigh this benefit.

13.5 Turbulence Categorisation

Clear air turbulence or in-cloud turbulence can be caused by one or more of the following: wind shear, convection, mountain wave. Categorisation of turbulence events is difficult, and therefore assessing turbulence forecasts as a whole against all available observations is a good way to get around the issue.

However, different predictors generally attempt to predict turbulence from only one of these sources. Therefore, it is useful to be able to categorise the turbulence types in order to assess improvements to individual predictors. This also has the benefit of identifying the areas of turbulence research, where there is the most potential to improve predictors further (Gill and Stirling 2013).

The user community has also expressed the preference to separate turbulence forecasts by source, and so it would be useful for the users to know the performance of the predictors for each turbulence type.

For case studies then PIREPS are an extremely valuable source of observations that may well have a category already assigned by the pilot. For routine verification some examples of how this has been tackled are given below.

13.5.1 Convective Turbulence

- Pilot reports can be used if available.
- Lightning detection reports can be used to locate areas close to convective activity.
- Satellite data can be used to find areas with severe convection by locating areas with overshooting tops. Satellites due for launch soon such as the Geostationary Operational Environmental Satellite R-series (GOES-R) and Meteosat Third Generation (MTG) will be equipped with lighting sensors.

13.5.2 Clear Air and In-cloud Turbulence

- Pilot reports can be used where they are available
- Satellite data can be used to determine areas that are in or out of cloud

13.5.3 Wind-Shear Turbulence

- Pilot reports if available
- Model analyses can be used to identify possible wind shear areas

13.5.4 Mountain Wave Turbulence

- Pilot reports can be used where they are available
- Satellite data can be used to identify interference patterns where turbulence may occur (Feltz et al. 2009)
- Orography fields can be used to identify mountainous areas, although mountain waves may propagate significant distances downstream from mountains before breaking.

13.6 Forecast Data

Forecast turbulence data is generally in gridded form from numerical models and is routinely produced by World Area Forecast Centres (WAFC) (ICAO 2012) and other meteorological organisations. Alternatively, the forecast data may be generated by forecasters using model guidance but presented in a chart form as used for WAFC forecasts presented on a significant weather chart (SIGWX). For an explanation of the interpretation of SIGWX charts, see Lankford (2000). In the case of the chart form, the forecasts are usually binary yes/no forecasts of the risk of a particular severity encompassed by a polygonal area. These can be simply transformed to gridded forecasts at the desired resolution to allow standard verification to take place (Gill 2014).

13.7 Verification Methodology

Case studies have been used in the past to produce verification statistics for turbulence forecasts. This is particularly suited to the more extreme cases where due to the rarity of the event routine verification is unlikely to give robust results unless a long time period of data is available. Case studies can be extremely useful in supplementing long-term objective verification.

Verification generally can be split into area-based verification and station-based verification. An area-based verification system assesses forecast grid points in a user-defined area against some form of truth that may be a gridded truth such as model analyses or gridded observations for example from satellite data. Station-based verification is performed at a series of points where observations are available. To do this gridded model data is interpolated to the observation point, and then a point-wise comparison is carried out.

For aviation turbulence, either approach could be taken for an objective verification scheme. Area-based verification would be particularly suited to verification against a gridded truth dataset if a suitable product could be derived from satellite data. Alternatively, aircraft observations could be gridded up over a certain time period and verified against the corresponding forecast grid.

A station-based approach can be used by counting the instantaneous reports as points (AMDAR) or for high resolution data (flight recorder) by looking at each flight and breaking it down into small segments to compare with the corresponding forecast field. This approach is particularly suited to flight data recorder information where the data are already broken down into individual flights. This can be done with AMDAR data, but care needs to be taken to match up observations to form a single flight. The flights can either be broken down into segments using time or distance. Distance was used in one detailed study of turbulence over the North Atlantic (Dutton 1980), where the idea of the probability of encountering

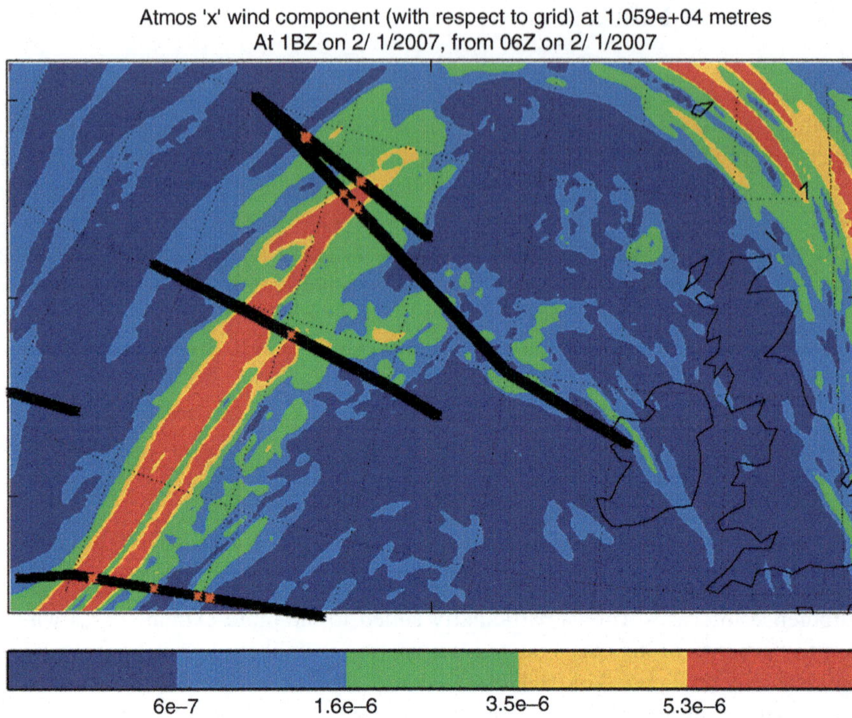

Fig. 13.1 Verification methodology showing a turbulence forecast field together with verifying observations. This figure shows the Ellrod TI1 turbulence predictor (Ellrod and Knapp 1992) in the background with units s^{-2}. The aircraft tracks are shown as *black lines* and turbulent encounters as orange points on the tracks

turbulence in a 100 km stretch was used and pilots filled in forms to assess conditions during each flight.

A similar automated process has been followed using 10 min time intervals to divide forecast tracks into segments to assess WAFC forecasts (Gill 2014). An example can be seen in Fig. 13.1. The background is the turbulence forecast to be verified which is overlaid with any aircraft tracks available in the area during a time window around the validity time of the forecast. In this case, the time window is ±1.5 h from the validity time of the turbulence forecast. Any forecast events are shown on each aircraft track as orange points.

13.8 Verification Metrics

Generally, verification is broken down into continuous verification, where forecast and observation are both the same continuous quantity, or categorical verification. In its simplest form categorical verification compares, binary turbulence/no turbulence forecasts with corresponding observations. Detailed guides to meteorological verification covering the measures referred to in this section can be found in Jolliffe and Stephenson (2012), Wilks (2011) and online reference hosted by the Australian Bureau of Meteorology (Ebert).

The choice of verification metrics will depend upon the measure of turbulence forecast being verified. In the case of EDR forecasts and EDR observations then continuous verification is possible. In the cases where the forecast and observations are different quantities, then categorical verification is more suitable. To enable this, both the forecast and observation severity need to be categorised in a meaningful and standardised way. With turbulence forecasts, this is typically using the categories null, light, moderate and severe.

13.8.1 Continuous Verification

Common metrics used for continuous verification are the mean error which gives us information about any bias in the forecast. A perfect forecast would score 0; however, it doesn't give any information on the magnitude of any errors. It would be possible for large positive and negative errors to cancel each other out resulting in a perfect or near perfect score. This information could be used to minimise these errors through calibration.

$$\text{Mean Error} = \frac{1}{N}\sum_{i=1}^{N}(F_i - O_i)$$

Root Mean Square Errors (RMSE) give us information about the accuracy of a forecast, penalising larger errors more than smaller errors.

$$\text{RMSE} = \frac{1}{N}\sum_{i=1}^{N}(F_i - O_i)^2$$

The Mean Absolute Error is also often used as an accuracy measure that treats all errors equally.

$$\text{MAE} = \frac{1}{N}\sum_{i=1}^{N}|F_i - O_i|$$

Neither the RMSE nor the MAE tells us anything about the sign of the errors, so it is useful to use these measures together with the mean error.

13.8.2 Categorical Verification

Categorical verification can be used to verify different severities of turbulence. As light turbulence is not generally of interest to airlines, it is usually the moderate or greater category that is verified. To do this, 2×2 contingency tables can be constructed with the four possible outcomes (Table 13.1). A forecast event that is observed is known as a hit (a), and a forecast event that is not observed is known as a false alarm (b). An observed event that is not forecast is known as a miss (c) and an event that is not forecast to occur and is not observed to occur is knows as a correct rejection (d). From these tables, a range of performance measures can be calculated.

The hit rate (H) is often quoted and measures the fraction of correctly forecast events given that an event has been observed. It should always be quoted together with the corresponding false alarm rate (F) which measures the fraction of events that were forecast to occur given that the event was not observed. H is also known as the probability of detection (PODy), and F is also known as the probability of false detection (PODn). A perfect score for H would be 1 and the worst score would be 0. A perfect score for F would be 0 and the worst score would be 1.

$$H = \frac{a}{a + c}$$

$$F = \frac{b}{b + d}$$

Many other measures can be calculated from the entries in the contingency table, and it is desirable to look at multiple measures to form a better picture of forecast performance than using one measure alone.

The frequency bias is the ratio of the number of forecast events to observed events and is useful in determining whether an event is under-forecast (score <1) or over-forecast (score >1). A perfect forecast would score 1.

$$\text{Frequency bias} = \frac{a + b}{a + c}$$

The equitable threat score (ETS) (also known as the Gilbert skill score) measures the fraction of correctly predicted events out of all observed and forecast events

Table 13.1 2×2 Contingency table

	Turbulence observed	No turbulence observed
Turbulence forecast	a (Hit)	b (False alarm)
No turbulence forecast	c (Miss)	d (Correct rejection)

after allowing for any hits that may have occurred by chance. A perfect score would be 1 and a no skill forecast would score 0. The lowest score is −1/3.

$$\text{ETS} = \frac{a - a_R}{a + b + c - a_R}$$

Where random hits a_R is defined by

$$a_R = \frac{(a + c)^*(a + b)}{a + b + c + d}$$

Peirce's Skill Score (also known as the True Skill Statistic or the Hanssen and Kuipers' score) is a useful score that benefits from using all the elements in the 2 × 2 contingency table and being independent of the base rate and can simply be written as *H–F*. A perfect score would be 1 and 0 indicates a forecast with no skill. The lowest score is −1.

$$\text{Peirce's Skill Score} = \frac{a}{a + c} - \frac{b}{b + d}$$

For the more severe types of turbulence where the events are rare scores should be viewed with some caution as many scores will be dominated by the non-events and give misleading information. One score that has recently been developed to get around this problem is the Symmetric Extremal Dependency Index (SEDI) score (Ferro and Stephenson 2011). A perfect forecast would score 1 and a forecast with no skill would score 0. The lowest score is −1. The forecast should be calibrated before using this score.

$$\text{SEDI} = \frac{\text{Log}\,F - \text{Log}\,H - \text{Log}(1 - F) + \text{Log}(1 - H)}{\text{Log}\,F + \text{Log}\,H + \text{Log}(1 - F) + \text{Log}(1 - H)}$$

It is advisable to use several metrics to assess forecast performance as each metric gives different information and no one score can measure the performance overall. For example, the hit rate and false alarm rate are both simple to understand and useful measures but should not be used in isolation. Each of these scores uses information from different halves of the contingency table. The sample size and observed frequency should also be used in conjunction with any scores to properly understand forecast performance.

When reporting verification statistics, it is useful to be able to stratify the verification in different ways. Typically, this might be by forecast range, region or season. Reporting results using a rolling number of years is desirable to remove any seasonal effects. Retaining a database of paired forecasts and observations together with other related data makes stratification of verification a simple filtering and aggregation process. This can also be helpful when optimising the blend of different forecast predictors and models.

Finally, it is important to retain all forecasts and observations to demonstrate that any verification results are reproducible.

13.8.3 Probabilistic Forecasts

Probabilistic forecasts also need a different set of metrics. One of the desirable aspects of a probabilistic forecast is that it produces reliable probabilities. In addition, it is still important that the forecast is skilful in discriminating between events and non-events, and that this skill is actually useful to the user to make decisions in an operational setting.

13.8.4 Forecast Skill

The Brier Score (BS) is often used in the verification of probabilistic forecasts. A perfect score would be zero attained when all non-event forecasts are forecast with certainty (probability 0 of forecasting the event), and all event forecasts are forecast with certainty (probability 1 of forecasting the event). Conversely, the worst possible score is 1.

$$\mathrm{BS} = \frac{1}{N}\sum_{i=1}^{N}(F_i - O_i)^2$$

where F_i is the forecast probability and the observation is converted to a binary depending on whether the event is observed or not.

$$\boldsymbol{O_i} = \begin{cases} 0 & \text{if no turbulence observed} \\ 1 & \text{if turbulence observed} \end{cases}$$

The score is quite often positively oriented by presenting it as 1-BS so that a perfect score scores 1 and the worst score scores 0. The Brier Score can give misleading skill for rare events and should be interpreted with caution in these cases. In general, care also needs to be taken when verification metrics are produced to assess skill over areas of varying climatologies (Hamill and Juras 2006). If possible quoting results over areas of similar climatological frequency is desirable, however, the reduction in sample size is likely to cause further issues.

A sequence of contingency tables can be created for a range of probability thresholds. A good way of visualising each of these contingency tables is to plot the hit rate against the false alarm rate to produce a Relative Operating Characteristic (ROC) curve (Fig. 13.2). Skill can be visualised again by the use of a ROC plot. Calculating the area under the ROC curve gives a single measure of skill; however,

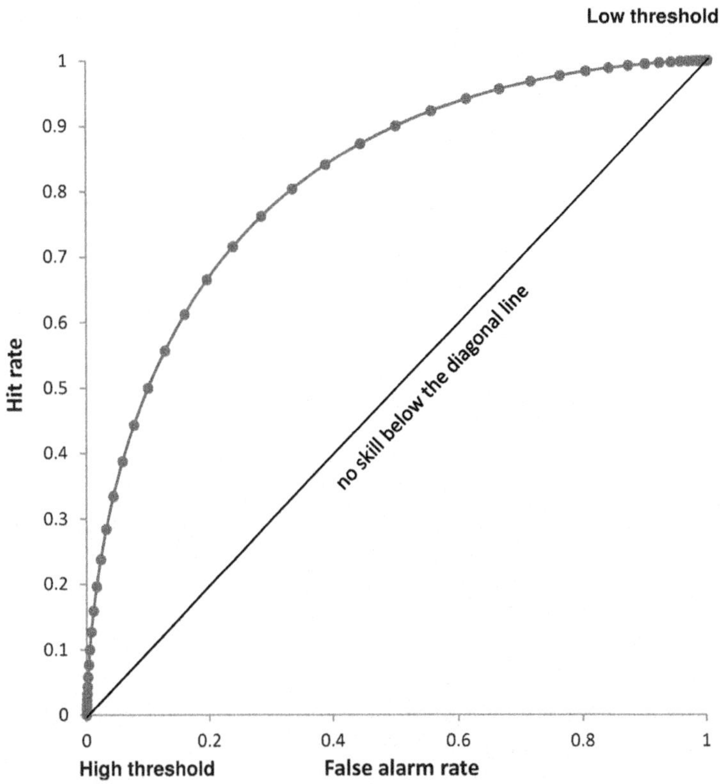

Fig. 13.2 ROC curve

care should be taken in interpreting these values as they can be misleading as not all the skill will be useful in an operational setting. Typically with rare events, the points will cluster towards the lower left corner making it hard to determine the ROC area. A ROC area of 1 indicates a forecast with perfect skill whilst a ROC area of 0.5 represents a forecast with no skill.

The partial ROC area can be calculated by integrating under a portion of the curve. This measure has been used in medical statistics (MCClish 1989) to assess the area of the portion of the curve that is most relevant. A refinement of this would be to determine a partial ROC area by agreeing upper bounds on the false alarm rate and lower bounds on the hit rate with the user. The area under the curve between these points is then calculated and used as a measure of skill that is relevant to the user. The user will want the hit rate to be as high as possible and the false alarm rate to be as low as possible. For example with turbulence forecasts, this could be conditions such as hit rate ≥ 0.4 and false alarm rate <0.3 (Fig. 13.3).

Deterministic forecasts can also be plotted on a ROC plot with a single hit rate and false alarm rate giving one point on the plot. This can then be compared to the curve of a probabilistic forecast with a point below the curve having less skill and a point above the curve having more skill.

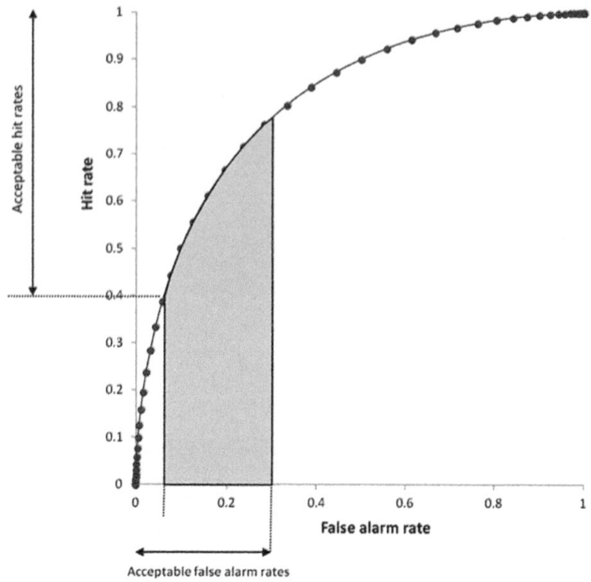

Fig. 13.3 Partial ROC

13.8.5 Forecast Reliability

The reliability performance can be assessed by plotting a reliability diagram (Fig. 13.4). The probabilities are binned, and in each bin the frequency of the event being observed is calculated. Ideally, the frequency with which an event is observed should be equal to the probability of the forecast, resulting in a straight line on the diagonal indicating a perfectly reliable forecast. In practice, some deviation from the line is likely, with points below the line corresponding to over-forecasting (probability forecast higher than the frequency of observations). Points above the line correspond to under-forecasting (probability forecast lower than the frequency of observations). The reliability diagram is more useful if accompanied by a likelihood diagram showing the number of forecasts in each bin.

The discriminatory skill and reliability are independent which means that any bias shown up by the reliability diagram can be corrected through calibration without affecting the skill as long as the calibration is order-preserving.

13.8.6 Relative Economic Value

Another measure that can be used to assess the useful skill or value of a forecast is the relative economic value (Richardson 2000). Values for costs and losses are assigned to the elements of a contingency table (Table 13.2). This is a way of presenting the value of a forecast in economic terms dependent on the cost-loss ratio of the user (Fig. 13.5). The relative economic value is a number between 0 and 1 with a perfect forecast scoring 1 and a climatology forecast scoring 0. The relative

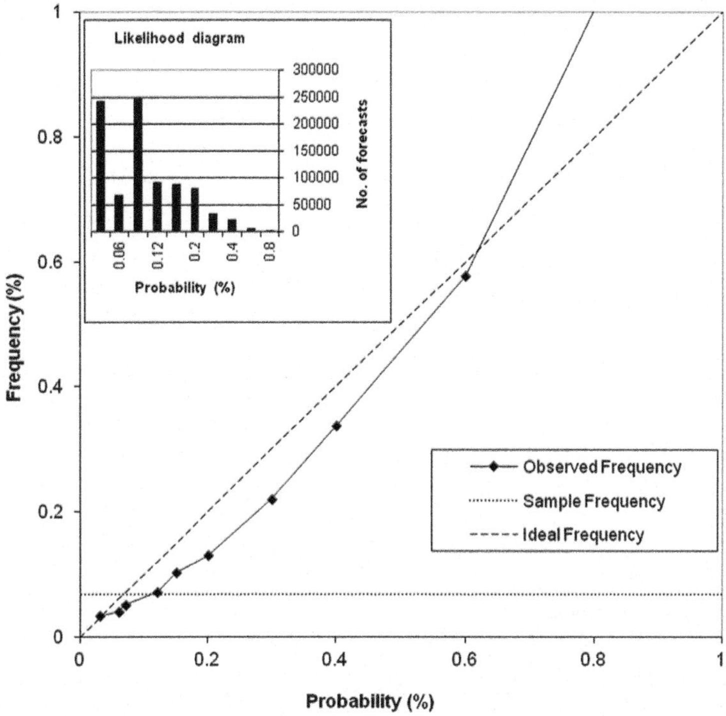

Fig. 13.4 Reliability diagram

Table 13.2 Values assigned to contingency table

	Turbulence observed	No turbulence observed
Turbulence forecast Action taken	Hit COST + REDUCED LOSS	False alarm COST
No turbulence forecast No action taken	Miss LOSS	Correct rejection

economic value can also be produced for deterministic forecasts enabling a useful way of comparing the value of the two.

These three measures together provide a useful way for assessing the performance of a probabilistic turbulence forecast system (Gill and Buchanan 2014).

13.8.7 Confidence

When quoting verification statistics, it is good practice to also quote the confidence in these results. Depending on the metric used, there are statistical methods to estimate the confidence in the results. One of the simpler methods is the use of

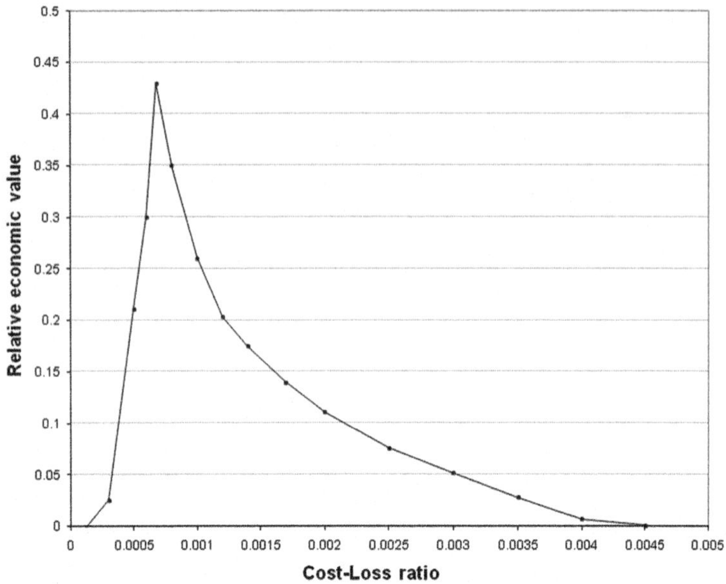

Fig. 13.5 Relative economic value plot

re-sampling techniques with replacement to produce a confidence interval as described in Gilleland (2008).

13.9 Sources of Errors

Errors identified by the verification may come from a variety of sources. Errors may be in the model itself which can be estimated from routine model verification. They may come from errors in the predictors, particularly in the case of turbulence where assumptions are made that link predictors to turbulence. Errors may also occur in the observations and again with turbulence this is clearly an important issue to address through quality control to remove as many erroneous observations as possible. The availability of observations is also a limiting factor that may be influenced by avoidance of forecast turbulence areas. Finally, errors may come from bugs in the code itself. Good code development and management processes including robust testing and peer reviewing will help to minimise this source of errors. Quantifying which of these errors is the most dominant is complicated.

Turbulence measurement errors have been estimated to be around 3 % (Sherman 1985) for vertical acceleration. For DEVG, these are typically likely to be around 3–4 % although these could be as high as 10–12 % in some cases. EDR measurements have been estimated to have 5–10 % errors related to them (WMO 2003). Manoeuvring aircraft can give erroneous high values, and it is possible that active control techniques could dampen the turbulence leading to underestimation of turbulence.

Global NWP models currently have errors of around 10 % for wind speed at cruise level. Many predictors are based on wind components, often involving first derivatives in which case the resulting error would be further increased in the final predictor.

13.10 Applications of Verification

Verification is useful in a number of ways, commonly divided into administrative, scientific and economic. These areas are discussed in the three related sections below relating to decision making, research and development and assurance with examples.

13.10.1 Decision Making

Verification should be an important part of any decision making process, by allowing more informed decisions to be made to increase the chances of success.

For turbulence forecasts, the way a forecast service is provided may change over time, and verification provides useful information to enable decision makers to make timely choices. For example, changes to the way forecasts are produced and the level of automation that is required.

- Decision on implementing changes to a forecast service
- Decisions on providing the best forecaster guidance
- Decisions on introducing probabilistic forecasts

Verification also assists forecast users to make decisions more appropriately to maximise the value they obtain from the forecast service. The use of the reliability diagram can inform the user how likely turbulence is to occur at the forecast probability. Using this information, and possibly comparing the frequency to climatology, the user can make a more informed decision on taking any action.

The ROC diagram can be used to help consider the importance of a hit and false alarm. Past performance will show a user where the trade off between hits and false alarms lies so they can decide how risk averse they want to be. For example, if a user is very risk averse they will want to minimise misses (maximising hit rate) which will be at the expense of incurring more false alarms. This may lead to a point on the ROC curve closer to the upper right corner, representing lower probability thresholds to take action at. Conversely, a user that is not very risk averse may minimise false alarms at the expense incurred from increasing misses (reducing the hit rate). This may lead to a point on the ROC curve closer to the lower left corner, representing a higher probability threshold to take action at.

The use of scores such as the relative economic value score in particular enables users to decide how best to set their own operating thresholds to maximise value and minimise risk. An example of this approach is given in Mylne (2002). For example, if a user can accurately determine their cost/loss ratio, then the probability threshold that generates the maximum value for that cost/loss ratio can be obtained from the value plots.

The user of turbulence forecasts could use the information to help make a decision on taking some form of action to mitigate the risks of flying through a turbulent area. This decision is likely to change depending on the type of turbulence and severity of turbulence expected to be encountered. For light turbulence, simply switching on the seatbelt sign may reduce the risks of any injuries. For more severe turbulence, a change in course to avoid the area may be advisable. For shear turbulence, a change of altitude could achieve this. For convective turbulence, a route around the area may be necessary.

Figure 13.6 (Lunnon 2009) shows an example of two different routes through an area of forecast turbulence. The light grey corresponds to low probability of turbulence, mid-grey to moderate probability of turbulence and black to high probability of turbulence. The graphs below show the probability of turbulence at various points along the two routes. In this case, the cost of flying through the light grey area is high, but the cost of encountering turbulence is lower. For the direct flight, the cost of the flight is lower but the cost of encountering turbulence higher. If the information on the distance flown and flight costs and the costs of encountering turbulence can be obtained together with information on the frequency of turbulence and verification information, then it would be possible to select the best avoidance strategy to minimise costs. An example of how wind and turbulence prediction data could be combined is available in Kim et al. (2015).

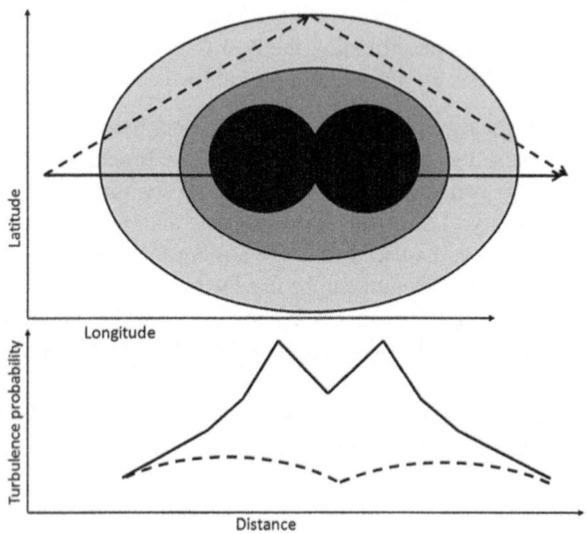

Fig. 13.6 Turbulence avoidance

13.10.2 Research and Development

Verification is a vital part of any new development. This enables researchers to compare the skill, reliability and value of different forecast models or forecast providers to ensure that new developments lead to measurable improvements for the user.

- Optimisation of predictor weightings when combining turbulence predictors and/or forecast models.
- Calibration of turbulence forecasts, both deterministic and probabilistic.
- Confirmation that new turbulence forecast developments are going to improve the forecast service before implementation.

13.10.3 Assurance

Aviation is regulated by national and international organisations that set levels of desired accuracy for the forecast services that are provided. Verification is a way of demonstrating that these levels have been attained.

- Assurance to ICAO and WMO that the WAFCs and NMSs are providing the appropriate levels of accuracy in their forecasts to comply with regulations.
- Assurance to users that they are getting a useful forecast service.
- Forecast monitoring to identify system and model errors at an early stage.

13.11 Summary

The use of forecast verification is an essential part of providing a forecast service. Turbulence forecasting systems are increasingly using multiple predictors which need to be combined in an optimal way, and the resulting combined predictor needs to be calibrated before being used. Verification is a fundamental part of this process.

Verification also enables the performance of the forecast to be monitored to assure users it is fit for purpose and can help users to make the most of the forecasts by maximising the value they are able to obtain from them.

At present, the main challenges are very much in obtaining sufficient high quality turbulence observations to enable robust objective verification. Rigorous quality control is needed to maintain a good data set for verification.

With future turbulence predictions, more likely to be presented in terms of probabilities there will be a need to communicate not just how to use these probabilities but also how to interpret the verification of them.

Verification remains an active area of research, and new metrics and techniques are constantly being developed. Many of these may well bring improved ways of assessing turbulence forecasts in the future.

References

Barné, H.: DELICAT Demonstration of Lidar based clear air turbulence detection, Final report. Thales avionics. http://delicat.inoe.ro/files/DELICAT_Final_report_PUBLIC.pdf (2014)

Bedka, K.M., Brunner, J., Dworak, R., Feltz, W., Otkin, J., Greenwald, T.: Objective satellite-based overshooting top detection using infrared window channel brightness temperature gradients. J. Appl. Meteorol. Climatol. **49**(2), 181–202 (2010). doi:10.1175/2009JAMC2286.1

Dutton, M.J.O.: Probability forecasts of clear-air turbulence based on numerical output. Meteorol. Mag. **109**, 293–310 (1980)

Ebert, E.: Forecast verification issues methods and FAQ. Bureau of Meteorology website: http://www.cawcr.gov.au/projects/verification/

Ellrod, G.P., Knapp, D.L.: An objective clear-air turbulence forecasting technique: verification and operational use. Wea. Forecasting **7**, 150–165 (1992)

FAA: Aeronautical Information Manual 7-1-20. https://www.faa.gov/air_traffic/publications/ATpubs/AIM/ (2014)

Feltz, W.F., Bedka, K.M., Otkin, J.A., Greenwald, T., Ackerman, S.A.: Understanding satellite-observed mountain-wave signatures using high-resolution numerical model data. Wea. Forecasting **24**, 76–86 (2009)

Ferro, C.A.T., Stephenson, D.B.: Extremal Dependence Indices: improved verification measures for deterministic forecasts of rare binary events. Wea. Forecasting **26**, 699–713 (2011)

Gilleland, E.: Confidence intervals for verification. Research Applications Laboratory, National Centre for Atmospheric Research: Boulder, CO, USA (2008)

Gill, P.G.: Objective verification of World Area Forecast Centre clear air turbulence forecasts. Meteorol. Appl. **21**, 3–11 (2014)

Gill, P.G., Stirling, A.J.: Including convection in global turbulence forecasts. Meteorol. Appl. **20**, 107–114 (2013)

Gill, P.G., Buchanan, P.: An ensemble base turbulence forecasting system. Meteorol. Appl. **21**, 12–19 (2014)

Hamill, T.M., Juras, J.: Measuring forecast skill: is it real skill or is it the varying climatology? Q. J. Roy. Meteorol. Soc. **132**, 2905–2923 (2006)

ICAO: Guidance on the Harmonized WAFS Grids for Cumulonimbus Cloud, Icing and Turbulence Forecasts. Version 2.5. (2012)

Jerrett, D., Turp, D.J.: Characterisation of events identified in the GADS aircraft dataset. Met Office Forecasting Research Technical Report. Met Office: Exeter, UK (2005)

Jolliffe, I.T., Stephenson, D.B.: Forecast verification A practitioner's guide in atmospheric science, 2nd edn. Wiley, West Sussex, England (2012)

Kim, J.-H., Chan, W.N., Sridhar, B., Sharman, R.D.: Combined winds and turbulence prediction system for automated air-traffic management application. J. Appl. Meteorol. Clim. **54**(4), 766–784 (2015)

Kane, T.L., Brown, B.G., Bruintjes, R.T.: Characteristics of pilot reports of icing. Preprints of 14th Conf on Probability and Statistics, 11–16 January 1998, Phoenix, AZ. American Meteorological Society: Boston, MA (1998)

Lankford, T.T.: Aviation weather handbook. McGraw Hill, New York, NY (2000)

Lunnon, R.W.: Use of GRIB Hazard Forecasts in Flight Planning. WAFC Science meeting, Washington. DC, USA. ICAO website: http://www.icao.int/safety/meteorology/WAFSOPSG/

Seminars%20and%20Workshops/WAFC%20Science%20Coordination%20Meeting/Presentations/Use%20of%20GRIB%20hazard%20forecasts%20in%20flight%20planning.ppt (2009)
McClish, D.K.: Analyzing a portion of the ROC curve. Med. Decis. Making **9**, 190–195 (1989)
Mylne, K.R.: Decision-making from probability forecasts based on forecast value. Meteorol. Appl. **9**, 307–315 (2002)
Overeem, A.: Verification of clear-air turbulence forecasts. Technical Report TR 244, Koninklijk Nederlands Meteorologisch Instituut: De Bilt; 76 pp. (2002)
Richardson, D.: Skill and relative economic value of the ECMWF ensemble prediction system. Q. J. Roy. Meteorol. Soc. **126**, 649–667 (2000)
Schwartz, B.: The quantitative use of PIREPs in developing aviation weather guidance products. Wea. Forecasting **11**, 372–384 (1996)
Sharman, R., Tebaldi, C., Wiener, G., Wolff, J.: An integrated approach to mid- and upper-level turbulence forecasting. Wea. Forecasting **21**(3), 268–287 (2006)
Sherman, D.J.: The Australian implementation of AMDAR/ACARS and the use of derived equivalent gust velocity as a turbulence indicator. Structures Report No 418, Department of Defence, Defence Science and Technology Organisation, Aeronautical Research Laboratories, Melbourne, Victoria (1985)
Takacs, A., Holland, L., Hueftle, R., Brown, B., Holmes, A.: Using in situ eddy dissipation rate (EDR) observations for turbulence forecast verification. Report to the FAA Aviation Weather Research Program (October 2005). Research Applications Laboratory, National Center for Atmospheric Research: Boulder, CO; 28 pp. (2005)
Tenenbaum, J.: GADS Quality Control and turbulence. Personal communication. 27 April 2014 (2014)
Truscott, B.S.: EUMETNET AMDAR AAA AMDAR Software Developments – Technical Specification. Doc. Ref. E_AMDAR/TSC/003. Met Office. Exeter. UK (2000)
Wilks, D.S.: Statistical Methods in the Atmospheric Sciences. International Geophysics Series, 3rd edn. Elsevier Academic Press, Oxford, UK (2011)
Wimmers, A., Feltz, W.: A prediction scheme for aircraft turbulence at tropopause folds using satellite imagery and EDR data. Joint 2007 EUMETSAT Meteorological Satellite Conference and the 15th Satellite Meteorology & Oceanography Conference of the American Meteorological Society. Amsterdam, The Netherlands, 24–28 September 2007. https://www.eumetsat.int/cs/idcplg?IdcService=GET_FILE&dDocName=pdf_conf_p50_s3_21_wimmers_p&allowInterrupt=1&noSaveAs=1&RevisionSelectionMethod=LatestReleased (2007)
Wimmers, A., Feltz, W.: Tropopause folding turbulence product. Center for satellite applications and algorithm theoretical basis document. http://www.goesr.gov/products/ATBDs/option2/Aviation_Turbulence_v1.0_no_color.pdf (2010)
WMO: AMDAR Reference Manual. WMO-No. 958 (2003)

Chapter 14
Aviation Turbulence Ensemble Techniques

Piers Buchanan

Abstract Global ensemble weather model forecasts have been available over several decades now. Ensemble forecasts allow uncertainties in even short range forecasts to be quantified. This is achieved through the principle of running several slightly different but equally feasible forecasts. These forecasts can be used to generate probabilities of turbulence indicators exceeding certain key thresholds. Through calibration, these probability forecasts can be used to generate the likelihood of turbulence occurring. It is even possible to combine these probabilities with non-meteorological information (e.g. the purpose of the aircraft flight) to produce impact forecasts.

14.1 Introduction

Global ensemble numerical weather prediction (NWP) model forecasts have been available over several decades now. The European Centre for Medium Range Forecasts (ECMWF) has been running ensemble forecasts from its Ensemble Prediction System (or EPS) model since 1992. Similarly, the NCEP GFS model has had ensemble members from 1992. The Met Office has been running ensemble forecasts from MOGREPS (Met Office Global and Regional Ensemble Prediction System) since 2005. Ensemble forecasts allow uncertainties in even short range (approximately 1 day) forecasts to be quantified (Buizza 2008). There is also evidence (Bougeault et al. 2010) that ensembles can give indications of low probability/high impact weather events (e.g. heavy rainfall). Turbulence is, by its nature, a difficult physical phenomenon to predict, and ensemble weather models can assist with the prediction of its occurrence. For example, most shear induced turbulence is associated with the position of the jet stream Ensemble models that can give a range of forecasts of the position of the jet stream and which can then be

P. Buchanan (✉)
Met Office, Fitzroy Road, Exeter, Devon EX1 3PB, UK
e-mail: piers.buchanan@metoffice.gov.uk

Published by Springer International Publishing AG Switzerland 2016.
R. Sharman, T. Lane (eds.), *Aviation Turbulence*, DOI 10.1007/978-3-319-23630-8_14

combined to give its most likely location and therefore the area of highest probability of wind shear turbulence.

14.2 Principles of Ensemble Forecasting

Weather models are fundamentally uncertain due to issues such as the uncertainty in the observations used to form their initial conditions and the admittedly incomplete understanding of the physical processes that occur in the atmosphere (Steiner et al. 2010). Ensemble models aim to provide a representative sample of the possible future states of the atmosphere. These models work on the principle of running several slightly different forecasts or ensemble members instead of just one single forecast. Figure 14.1 shows a schematic of how they work using several alternative forecasts of the Ellrod TI1 (Ellrod and Knapp 1992) turbulence indicator. The initial conditions have a set of perturbations added to them to represent the initial condition uncertainty represented by the circle around the analysis. Additionally, uncertainties in the model physics are generated using stochastic model physics (represented by the wavy lines in Fig. 14.1). This enables several (typically 20–50) alternative but equally feasible ensemble members to be produced. These members enable a range of forecasts to be generated, represented by the solid enclosed shape on the right of Fig. 14.1. This range of forecasts is typically

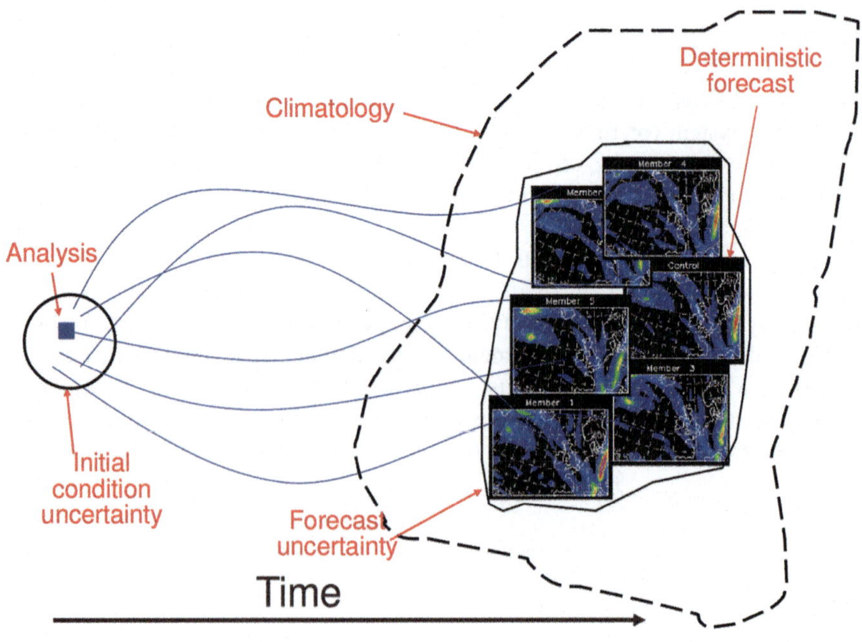

Fig. 14.1 Ensemble schematic

narrower than the underlying model climatology represented by the dashed enclosed shape. For more information see Bowler et al. (2008).

14.3 Ensemble Models Available

There are several global NWP ensembles currently available. These include the Global Ensemble Forecast System (GEFS) provided by NCEP in the USA (Buizza et al. 2005), the MOGREPS-G model provided by the Met Office in the UK (Bowler et al. 2008) and the Ensemble Prediction System (EPS) provided by the European Centre for Medium Range Weather Forecasts (ECMWF) (Buizza et al. 2007). Table 14.1 summarises the key characteristics of these three models. Tables like this can be useful because, other than the objective and subjective verification of the models, users like to know:

1. What forecast lead time ranges models provide forecasts for
2. What range of possible solutions (ensemble members are available).
3. The vertical and horizontal resolution of the models. It can be argued that the vertical resolution is particularly important for turbulence forecasts (although for upper-level turbulence prediction, this will depend on the number of model levels that are located in the cruise altitude range. For example, the MOGREPS-G and ECMWF-EPS models have similar cruise altitude model level separations of 600 m and 500 m respectively. This is despite the ECMWF EPS having significantly more model levels overall (91 model levels compared to the 70 model levels of MOGREPS-G)).

Once one has the ensemble members, it is possible to produce forecast products that represent the uncertainty of certain meteorological events happening. These so-called postage stamp plots show small images of all of the individual member forecasts. Figure 14.2 shows the T+27 hour forecast from 24 MOGREPS-G members for the single Ellrod TI1 turbulence indicator. At first glance, the individual

Table 14.1 Summary of the characteristics of three of the main global ensemble models

Model	MOGREPS-G	GEFS	ECMWF EPS
Horizontal resolution	N400 (33 km) at mid latitudes	T254 (55 km) to T+8 days and then T190 (73 km) T+8 to 16 days	T639 (32 km) to T+10 days then T319 (63 km) T+10 to T+15 days
No of vertical levels	70	42	91
Refresh rate	6 hourly	6 hourly	12 hourly
Output time steps	3 hourly	3 hourly	3 hourly
No. of members (including control	12	21	51
Maximum lead time	7 days	16 days	15 days

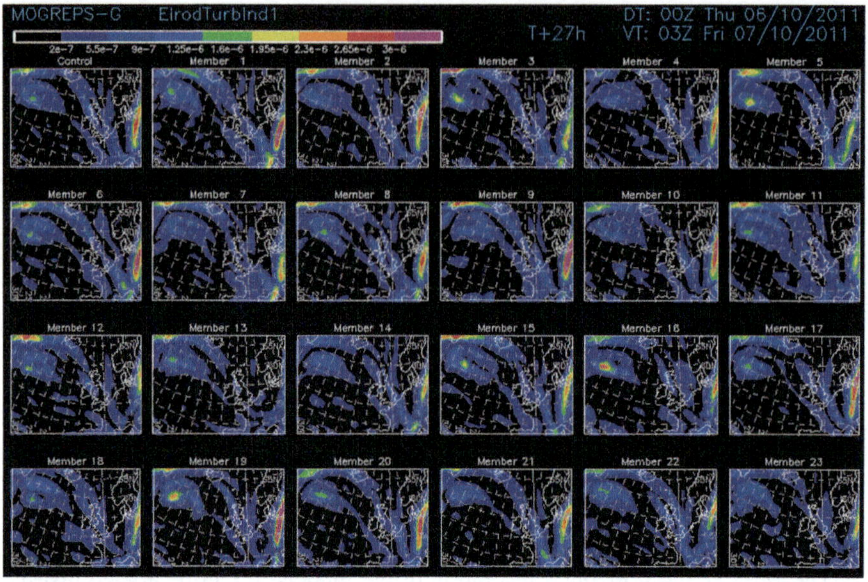

Fig. 14.2 Postage stamps for the T+27 hour MOGREPS-G forecasts for the Ellrod TI1 turbulence indicator

members look quite similar but there are some differences, e.g. member 16 has a significant feature in the North Atlantic. As stated earlier, each of these outcomes is equally likely—there is no reason to trust one particular outcome more than another.

Another way to represent uncertainty is to combine these individual forecasts to calculate the probability of a particular event happening. A particular threshold is chosen, and then the number of ensemble members that are greater than (or less than) the threshold are calculated for each grid square. This value is then divided by the total number of ensemble members. This generates the fractional model probability. It is a "model" probability because it is uncalibrated at this stage. Figure 14.3 shows an example of this type of plot. The MOGREPS-G model probability of the Ellrod TI1 turbulence indicator exceeding 0.0000008 s^{-2} has been shown for Europe and a large section of the Atlantic Ocean. This threshold has been chosen to represent moderate or greater turbulence for medium and heavy weight class aircraft. This model probability also has the potential to be calibrated using previous forecasts and actual forecast outcomes so that the probability is a closer approximation of the actual probability of a given event occurring. This is a key advantage of ensemble models over deterministic models in that the probabilities that they generate for a given event occurring (e.g. turbulence) can be compared to the observed frequency of that particular event, i.e. the user knows what "success" looks like.

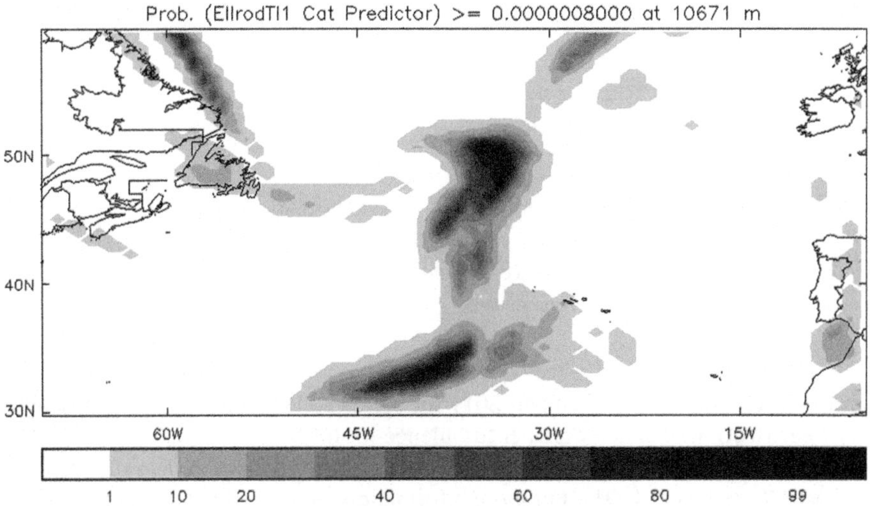

Fig. 14.3 Model probability that the Ellrod TI1 turbulence indicator will exceed 0.0000008 s^{-2} for the T+24 hour forecast for MOGREPS-G. This is the forecast from 00Z on 20th March 2012

There is a potential issue because the probability of turbulence at any given place and at any given time is extremely small. A nonexpert user may have difficulty understanding whether a very low probability value is considerably higher than normal or whether that is just a typical probability value in that area. One way that this could be overcome would be by having a turbulence forecast product which compared the current probability of turbulence with the climatological average probability for each grid point. This would give the user an idea if a given probability was significantly higher than average for the particular location in the world. However, it would still be necessary to supply the absolute probability as well, as the user is still likely to want to know the likelihood of turbulence in a given location. The information that turbulence is more likely there is not likely to help mitigate against its impact.

14.4 Ensemble Turbulence Forecasts

Gill and Buchanan (2014) have developed and trialled an ensemble model-based global turbulence forecasting system at the Met Office. It is based on the MOGREPS-G global ensemble model. It uses 24 ensemble members from its latest two model runs and the principles discussed in Sect. 14.3 to produce probability forecasts for eight different turbulence indicators. Each of the turbulence indicators has five thresholds aiming to cover light to moderate to severe turbulence. The turbulence predictors used are detailed in Table 14.2. For more details on these indicators, see Gill and Buchanan (2014). A trial was run for a 12 month period

Table 14.2 Turbulence indicators for the Gill and Buchanan (2014) ensemble turbulence trial

Predictor
Dutton (s^{-1}) (Dutton 1980)
Brown ($cm^{-2}\ s^{-3}$) (Brown 1973)
Ellrod TI1 (s^{-2}) (Ellrod and Knapp 1992)
Ellrod TI2 (s^{-2}) (Ellrod and Knapp 1992)
Convective precipitation accumulation ($kg\ m^{-2}$) (see Gill and Stirling 2013)
Convective precipitation rate ($kg\ m^{-2}\ s^{-1}$) (see Gill and Stirling 2013)
−Ri (negative Richardson number) (see Gill and Buchanan 2014)
Lunnon (s^{-1}) (see Gill and Buchanan 2014)

from November, 2010 to October, 2011 for the T+24 hour forecasts. The probability forecasts of moderate or greater turbulence for the various turbulence indicators are verified against automated reports from the flight data recorders of a fleet of aircraft (Tenenbaum 1991) (For more information on the verification technique see Gill 2014). For each individual turbulence indicator, the Receiver Operating Characteristic (ROC) curves are plotted and the area under the curve (AUC) is determined. Potentially, the partial ROC area can be calculated by integrating under a portion of this curve. This measure has been used in medical statistics (McClish 1989) to assess the area of the portion of the curve that is most relevant to the user. Previous studies (Sharman et al. 2006) have shown that combining deterministic indicators can lead to significantly greater forecast skill. The eight turbulence indicators are combined with one another optimally using a linear combination to produce a combined indicator. This linear combination is produced by finding the weighting of turbulence predictors that maximises the ROC area from any pair of the eight turbulence predictors (Gill and Stirling 2013). Furthermore, a turbulence climatology has also been produced from aircraft data for light and moderate turbulence (Gill and Stirling 2013). This turbulence climatology has been optimally combined with the eight turbulence indicators linearly as before by maximising the ROC area by varying the weightings to create a combined turbulence predictor (with climatology).

Figure 14.4 shows the ROC plot for the MOGREPS-G forecast combined predictors (the unbroken line) and the Ellrod TI1 deterministic model predictor (the dashed line). The greater area under the combined probabilistic forecast indicates greater skill than the deterministic single indicator forecast. This figure illustrates the difference between the (single, centre) deterministic turbulence forecast—Ellrod TI1—and the potential of using a multi-predictor probabilistic forecast. The former is not unheard of in current operational forecasting systems. Figure 14.5 summarises the ROC areas for the deterministic forecasts as well as the probabilistic forecasts with their confidence intervals. These 95 % confidence intervals were generated using a re-sampling technique as described in Gilleland (2008). From this figure, it is possible to compare the ensemble combined predictor forecast with the deterministic combined predictor forecast (both with and without forecast

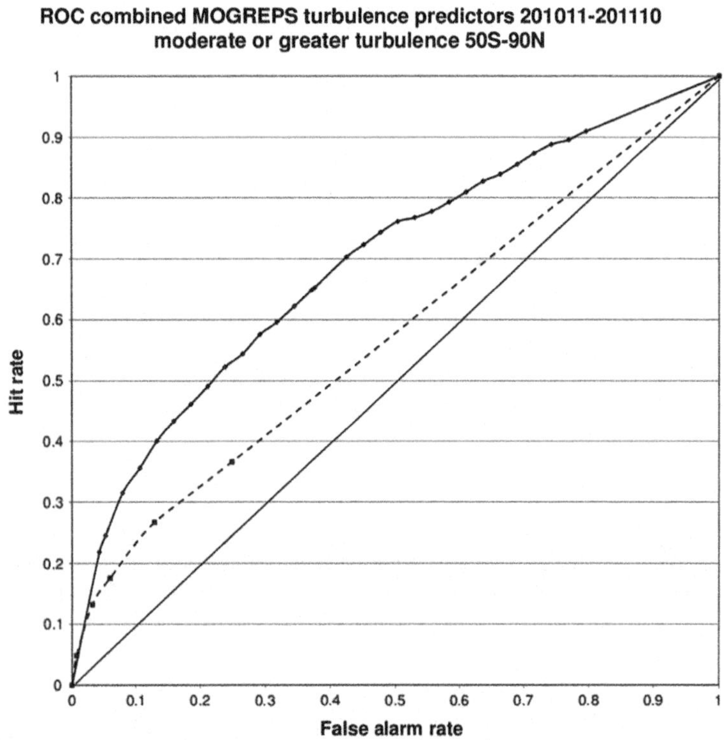

Fig. 14.4 ROC curve comparing probabilistic and deterministic CAT predictors. The results for the combined MOGREPS predictors are shown by the *blue line* and the results for the deterministic Ellrod TI1 predictor is shown with the *red line*

climatology). There is a clear benefit (in terms of ROC score) in using a probabilistic approach. Comparing probabilistic and deterministic ROC scores for the individual indicators indicates that only Dutton (1980) and convective precipitation accumulation indicators show no improvement in using the probabilistic approach. The key conclusions are that:

1. Ensemble forecasts are, on the whole, higher in skill than deterministic forecasts (using this verification metric)—particularly for the combined predictor cases.
2. There is an increase in skill from combining predictors for both deterministic and probabilistic forecasts.
3. There is a further increase in skill by including a turbulence climatology for both deterministic and probabilistic forecasts.

Figure 14.6 shows the reliability plot for the combined MOGREPS-G predictor of moderate or greater turbulence following a simple linear calibration. This calibration is achieved by multiplying the forecast probabilities by scale factor so that the reliability curve (dashed line) lies as close as possible to the ideal line. The resulting probabilities are very low (around 1 %) but show reasonable

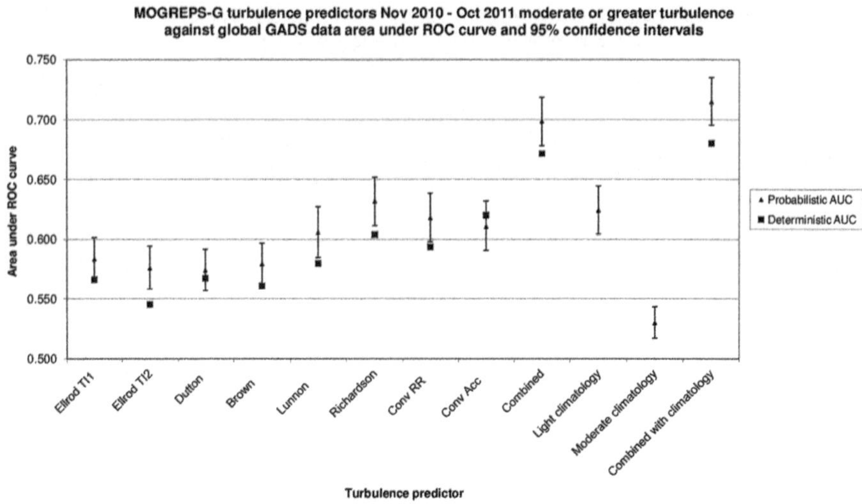

Fig. 14.5 AUC and 95 % confidence intervals for probabilistic and deterministic turbulence predictors. *Black diamond* represents probabilistic AUC, *black square* represents deterministic AUC and bar represents upper and lower bound of 95 % confidence interval on the probabilistic AUC

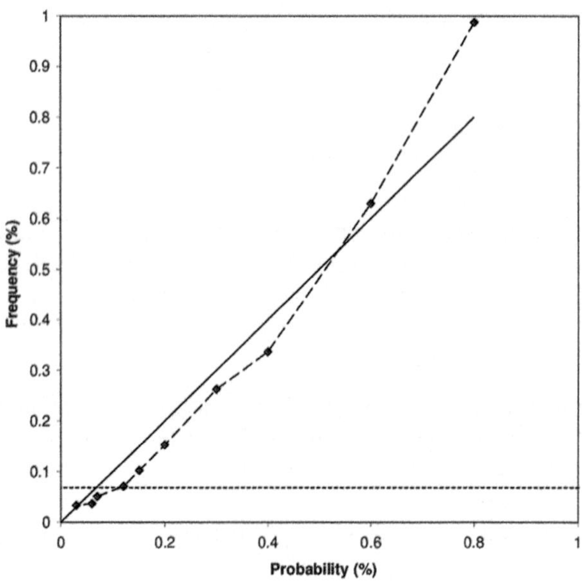

Fig. 14.6 Reliability diagram showing observed frequency (*dashed line*), ideal frequency (*solid line*) and sample frequency (*dotted line*)

discrimination between lower probability and higher probability of turbulence (i.e. lower forecast probabilities are correlated with lower observed frequencies). It is also possible to estimate the economic value of the forecasts. Values for costs and losses are assigned to the elements of the contingency table. This is a way of presenting the value of a forecast in economic terms dependent on the cost–loss ratio of the user (Richardson 2000). Figure 14.7 shows the relative economic value plotted for a range of cost-loss ratios. The combined probabilistic predictor shows the highest value and is at least 25 % greater than the combined deterministic predictor, and the deterministic Ellrod TI1 indicator shows significantly less value than this.

In summary, so far it was shown that a probabilistic ensemble model-based turbulence system offers improvement in overall skill and forecast value compared to equivalent deterministic model-based forecasts. Probabilistic forecasts also allow for confidence in the forecast to be communicated. Moreover, it is still possible to add new turbulence predictors to the probabilistic framework. For example, it would be perfectly feasible to run a probabilistic version of the GTG turbulence prediction system.

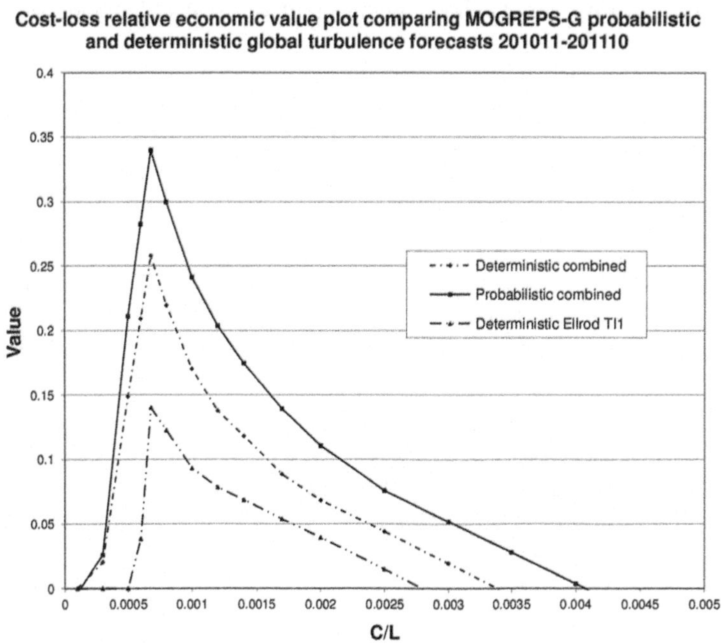

Fig. 14.7 Relative economic value cost–loss plot for combined global probabilistic turbulence predictors (*solid line*), combined deterministic turbulence predictors (*dashed line*) and deterministic Ellrod TI1 (*dotted line*) between November 2010 and October 2012

14.5 Moving Towards Impact-Based Probability Forecasting

As discussed previously in this chapter, ensemble forecasts have been used in more general weather forecasting for single meteorological parameters such as temperature, wind and rain for many years. In more recent years, more tailored forecast systems have been set up for forecasting hazards. An example of this at the Met Office is the Ensemble Prediction System first-guess Warnings (EPS-W) tool, referred to as "MOGREPS-W" in Neal et al. (2014).

EPS-W is an ensemble-based first-guess support tool for severe weather, which was developed to assist Met Office meteorologists in issuing severe weather warnings for the UK as part of the impact-based National Severe Weather Warning Service (NSWWS). EPS-W post-processes ensemble data into a format which mimics the NSWWS impact matrix (Fig. 14.8e). The first-guess warnings use low, medium and high impact thresholds for each parameter, such as wind. Likelihoods (derived from the ensemble frequency) of the three impact levels are combined to derive an overall warning status at each grid point (yellow, amber or red)—these are displayed as overall warning colour maps (Fig. 14.8a–d). Impact colour takes priority over likelihood, giving emphasis to low likelihood high impact

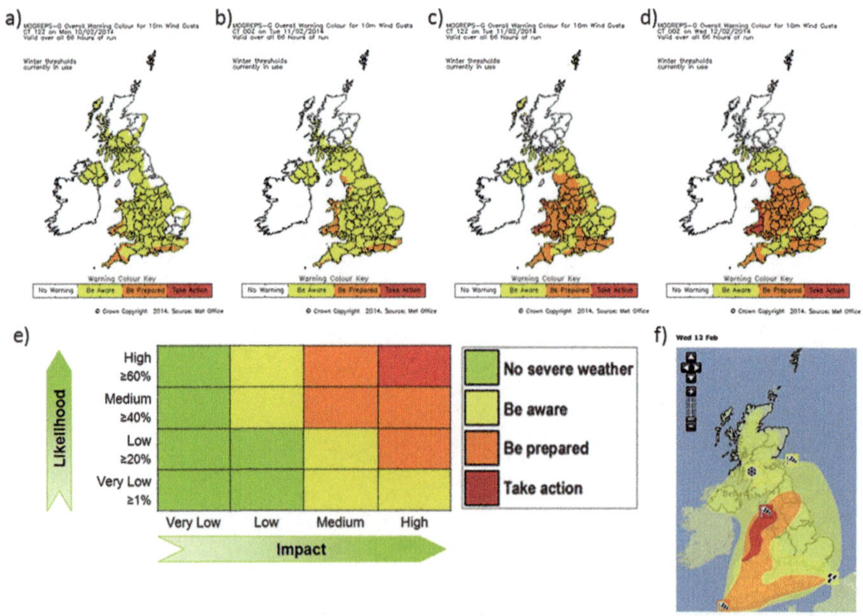

Fig. 14.8 (**a**)–(**d**) show consecutive forecast updates from the MOGREPS-G version of EPS-W (left image is the oldest and right image is the most recent) showing the overall warning colours all valid on 12 February 2014; (**e**) shows the NSWWS weather impact matrix used in EPS-W (*left*) and colour key (*right*); (**f**) shows the final forecaster issued warning for 12 February 2014. Note: Snow and rain warnings were also issued on this day

events. For example using the impact matrix (Fig. 14.8e), a 25 % high impact warning (amber) would take priority over a 75 % low impact warning (yellow), even though the yellow warning has a higher likelihood.

Figure 14.8 shows an example of how EPS-W was used to issue a very rare "red warning" for parts of Wales and north-west England on 12 February 2014. Here, output is displayed from the MOGREPS-G version of EPS-W which generates daily warnings out to 5 days ahead. The proxy impact thresholds used in EPS-W for Wales and north-west England at the time of this event were 65 mph for low impact wind gusts, 70 mph for medium impact wind gusts and 80 mph for high impact wind gusts.

This impact-based approach has potential to be applied to the forecasting of turbulence. The three impact thresholds required for turbulence indicators could be varied according to which part of the world the forecast is for or whether it is a passenger or cargo flight for example. The main benefit of this approach is to turn a series of probabilistic forecasts for a number of thresholds into a simple and usable "yellow, amber and red" overall warning colour map with additional information available when required from the low, medium and high impact breakdowns.

14.6 Summary

There is strong evidence that the introduction of global NWP model ensemble turbulence forecasts increases the skill of turbulence forecasts. Furthermore because they produce probabilities, it is possible to calibrate the probabilities so that they are the probability of the turbulent event happening rather than the probability of a particular turbulence indicator threshold being exceeded. This has the consequence that probabilities produced from different forecasting centres (e.g. USA and UK) can be combined in a reliable way. Given that there is the potential to tailor probability thresholds according to customer needs (or the needs of differing areas of the world), a probabilistic approach to turbulence forecasting could be used to give impact forecasts.

A major focus area over the next few years would seem likely to be calibration of forecasts and combination of forecasts from different centres. The calibration applied by Gill and Buchanan (2014) is a simple linear one, but more sophisticated calibrations are possible and could be investigated. Another logical step would be to increase the number of turbulence predictors in the ensemble system from 8 to the number used in the GTG turbulence prediction system. Furthermore, an impact-based forecasting approach may make these probabilistic hazard forecasts easier for customers to use who are used to deterministic forecasts.

References

Bougeault, P., Toth, Z., Bishop, C., Brown, B., Burridge, D., Chen, D.H., Ebert, B., Fuentes, M., Hamill, T.M., Mylne, K., Nicolau, J., Paccagnella, T., Park, Y.-Y., Parsons, D., Raoult, B., Schuster, D., Silva Dias, P., Swinbank, R., Takeuchi, Y., Tennant, W., Wilson, L., Worley, S.: The THORPEX interactive grand global ensemble. Bull. Am. Meteorol. Soc. **91**, 1059–1072 (2010)

Bowler, N.E., Arribas, A., Mylne, K.R., Robertson, K.B., Beare, S.E.: The MOGREPS short-range ensemble prediction system. Q. J. Roy. Meteorol. Soc. **134**, 703–722 (2008)

Brown, R.: New indices to locate clear air turbulence. Meteorol. Mag. **102**, 347–361 (1973)

Buizza, R.: The value of probabilistic prediction. Atmos. Sci. Lett. **9**, 36–42 (2008)

Buizza, R., Bidlot, J., Wedi, N., Fuentes, M., Hamrud, M., Holt, G., Vitart, F.: The new ECMWF VAREPS (Variable Resolution Ensemble Prediction System). Q. J. Roy. Meteorol. Soc. **133**, 681–695 (2007)

Buizza, R., Houtekamer, P.L., Toth, Z., Pellerin, G., Wei, M., Zhu, Y.: A comparison of the ECMWF, MSC, and NCEP global ensemble prediction systems. Mon. Weather Rev. **133**, 1076–1097 (2005)

Dutton, M.J.O.: Probability forecasts of clear-air turbulence based on numerical output. Meteorol. Mag. **109**, 293–310 (1980)

Ellrod, G.P., Knapp, D.L.: An objective clear-air turbulence forecasting technique: verification and operational use. Wea. Forecasting **7**, 150–165 (1992)

Gill, P.G.: Objective verification of World Area Forecast Center clear air turbulence forecasts. Meteorol. Appl. **21**, 3–11 (2014)

Gill, P.G., Buchanan, P.: An ensemble base turbulence forecasting system. Meteorol. Appl. **21**, 12–19 (2014)

Gill, P.G., Stirling, A.J.: Including convection in global turbulence forecasts. Meteorol. Appl. **20**, 107–114 (2013)

Gilleland, E.: Confidence intervals for verification. Research Applications Laboratory, National Centre for Atmospheric Research, Boulder, CO; 62 pp. (2008)

McClish, D.K.: Analyzing a portion of the ROC curve. Med. Decis. Making **9**, 190–195 (1989)

Neal, R.A., Boyle, P., Grahame, N., Mylne, K.: Ensemble based first guess support towards a risk-based severe weather warning service. Meteorol. Appl. **21**, 563–577 (2014)

Richardson, D.: Skill and relative economic value of the ECMWF ensemble prediction system. Q. J. Roy. Meteorol. Soc. **126**, 649–667 (2000)

Sharman, R., Tebaldi, C., Wiener, G., Wolff, J.: An integrated approach to mid- and upper-level turbulence forecasting. Wea. Forecasting **21**, 268–287 (2006)

Steiner, M., Bateman, R., Megenhardt, D., Liu, Y., Xu, M., Pocernich, M., Krozel, J.: Translation of ensemble weather forecasts into probabilistic air traffic capacity impact. Air Traffic Control Quarterly **18**(3), 229–254 (2010)

Tenenbaum, J.: Jet stream winds: comparisons of analyses with independent aircraft data over southwest Asia. Wea. Forecasting **6**, 320–336 (1991)

Part IV
Observational and Modeling Studies

Chapter 15
Multi-scale Observational and Numerical Modeling Studies of the Turbulence Environment

Michael L. Kaplan

Abstract In this chapter, the focus will be on diagnosing multi-scale dynamical and thermodynamical processes that produce a favorable environment for aviation turbulence. In particular, the chapter will describe possible links among circulations at multiple scales of atmospheric motion that likely are favorable for the downscale generation of severe aviation turbulence. The chapter will be composed of three sections: (1) a synoptic/meso-α scale observational analyses of a broad cross section of turbulence accidents and the larger scale flow patterns that likely create a favorable mesoscale environment for the development of the turbulence event, (2) numerical simulations of the downscale dynamic/thermodynamic focusing of a favorable turbulence environment including the interaction of features at the synoptic, meso-α ($L = 200$–2000 km), and meso-β ($L = 20$–200 km) scales of motion, and (3) an example of an operational mesoscale numerical model-based turbulence forecasting system and its implications for how it can be employed to determine the processes resulting in an environment which organizes severe aviation turbulence. These three sections are based upon research in Kaplan et al. (Characterizing the severe turbulence environments associated with commercial aviation accidents, 2004; Meteorol. Atmos. Phys. 88:129–152, 2005a; Meteorol. Atmos. Phys. 88:153–175, 2005b; Meteorol. Atmos. Phys. 94:235–270, 2006). We conclude with a summary and synthesis of a multi-scale paradigm for the downscale organization of a favorable aviation turbulence environment.

M.L. Kaplan (✉)
Division of Atmospheric Sciences, Desert Research Institute, Reno, NV, USA
e-mail: michael.kaplan@dri.edu

R. Sharman, T. Lane (eds.), *Aviation Turbulence*,
DOI 10.1007/978-3-319-23630-8_15

15.1 Characterizing the Severe Turbulence Environments with Synoptic Scale Datasets

The database for this analysis is predicated on National Transportation Safety Board (NTSB) accident summaries of severe commercial aviation turbulence accidents or incidents over the USA and adjacent oceanic regions. An accident or incident involved injury to at least one flight crew member or passenger during minimally severe aviation turbulence. Forty-four case studies of severe aviation turbulence accidents were compiled by the NTSB between 1990 and 1996 over these locations. These accident reports were made available to the research community by the NASA Ames Research Center.

Meteorological data employed for the categorization of favorable synoptic scale environments in this study included: (1) NCEP-NCAR Global Renanalysis Datasets (Kalnay et al. 1996) with 2.5-degree horizontal resolution for all mandatory pressure levels, (2) NOAA NESDIS high resolution 1-km visible and 4-km infrared satellite imagery, and (3) surface observations of cloud and precipitation where available.

Originally, the NTSB accident report data were classified into four categories of turbulence case studies: (1) (CAT) clear air turbulence widely separated from mountains and moist convection, (2) (MTN) turbulence near mountains in conditions typically favorable for mountain-wave turbulence, (3) (TRW) turbulence occurring within convective storms with strong up and downdrafts, and (4) (CLD) turbulence occurring in cloud covered regions without the requirement of mountain waves, deep convection, or moisture striking the ground. Interestingly, it became apparent that many of the CAT, TRW, and CLD accidents were actually near but not in deep convective up or downdrafts. The pilot reports typically mentioned the proximity to convection but that the aircraft was not directly within the convective cell. Often the satellite data and NTSB report implied that the aircraft was in and out of convection, and the turbulence occurred during both scenarios. Thus, a new category, TRW*, was formulated and after another attempt at categorizing the case studies the breakdown for categories was modified to: (1) CAT-16, (2) TRW*-13, (3) TRW-8, (4) CLD-4, and (5) MTN-3. Thus, 21 of the 44 case studies were unambiguously near or within moist convection. This general category of turbulence involving some degree of moist convective forcing we will refer to as convective-induced turbulence (CIT).

The 44 case study synoptic scale analysis was divided into two general parts, first, an evaluation of the time of year, time of day, location, and elevation was performed, followed by an analysis of key synoptic hydrodynamic and thermodynamic large scale "predictors." "Predictors" were defined as synoptic fields most closely ***associated*** in space and time with turbulence both: (1) overall across all turbulence categories and (2) as a function of individual turbulence category. By far, turbulence accidents were most likely to occur across all categories under general conditions favorable for moist convection, i.e., during summer in general and July in particular, the southeastern USA, during 2100–0000 UTC, and at altitudes ranging between 9001 and 12,000 m MSL, which comprise typical commercial aviation cruising altitudes. The least likely scenario involving turbulence accidents included winter, February, over the southwestern USA at 0900–1200 UTC. In spite of this statistical finding, as will be shown, many of the most severe and widespread turbulence events involve wintertime conditions.

The results of the analyses based on the predictors (Table 15.1) indicate that there is a fundamental consistency when one intercompares the best predictors across all turbulence categories and within each turbulence category. The best predictors across all turbulence categories based on the % of turbulence accident case studies associated with a given predictor are: (1) immediate upstream cyclonic or anticyclonic radius of curvature < than a horizontal wavelength of the closest trough or ridge (98 %), (2) convective clouds <100 km away (86 %), (3) upward vertical motion (82 %), (4) layer averaged absolute vorticity $\leq 10^{-4}\ s^{-1}$ (80 %), and (5) jet stream entrance region location (77 %). Increasing vertical wind shear in time, lapse rate ≥ moist adiabatic, low absolute vorticity at flight level, convective clouds within 30 km, horizontal cold air advection, flight level relative vorticity <0, and leftward-directed ageostrophic flow occurred in at least 64 % of the events with an average % for all 12 of these predictors being ~78 %. While the small sample

Table 15.1 Large Scale "Predictor" Fields developed using output derived from NCEP-NCAR Reanalysis datasets

Immediate Upstream Curvature (Streamline ≤ 100 km from Accident)
Entrance/Exit Region of Jet Stream
Sign of OMEGA
Lapse Rate ≥ Moist Adiabatic
Direction of Ageostrophic Wind Shear Vector
Sign of Horizontal Temperature Advection
Sign of Horizontal Advection of Total Wind Velocity Shear
Magnitude of Vertical Variation of the Brunt–Vaisala Frequency
Flight Level Vertical Absolute Vorticity $\leq 10^{-4}\ s^{-1}$
Flight Level Vertical Relative Vorticity $\leq 0\ s^{-1}$
Absolute Vertical Vorticity Averaged Over 2 Levels $\leq 10^{-4}\ s^{-1}$
Magnitude of Vertical Total Wind Velocity Shear
Magnitude of Isentropic Potential Vorticity
Relative Humidity ≥ 50 %
Sign of Horizontal Advection of the Total Lapse Rate
Magnitude of Ageostrophic Wind Velocity
Magnitude of Vertical Variation of Richardson Number
Magnitude of Vertical Variation of Total Wind Velocity Shear
Magnitude of Richardson Number
Convective Clouds (All Bases) <100 km from Accident
Convective Clouds (All bases) <30 km from Accident
Ellrod Index (Ellrod and Knapp 1992) (EI)
NCSU Modification = EI/IPV

size limited the detail that could be expected in any statistical analysis, this combination of predictors is clearly closely associated with turbulence in general at the larger/hydrostatic scales of atmospheric motion. Within the CAT, TRW, and TRW* categories, the closest association is with curvature, low relative vorticity, jet stream entrance region, low static stability, and nearby convection. Apparently, even clear-air turbulence does not occur much of the time far removed, i.e., >100 km, from convective cloud systems. The MTN category did exhibit some difference from the overall CAT, TRW, and TRW* categories as would be anticipated with vertical wind shear playing a much more dominant role. These results point towards the right entrance region of a jet streak with strong cold air advection producing low static stability, nearby moist convection, reduced vorticity (therefore, low inertial stability), and frontogenesis as a favored area for all but mountain wave-induced turbulence.

15.2 Numerical Simulations of Meso-α/β Scale Processes That Characterize Severe Turbulence Environments

In this section, the focus will be on the meso-β scale dynamical and thermodynamical processes that serve as a bridge between the synoptic scale and turbulent scale adjustments. This will involve both observational analyses employing the North American Regional Reanalyses (NARR; Mesinger et al. 2006) as well as meso-α/β scale numerical simulations. The goal of this section is to arrive at a synthesis of the NCEP-NCAR, NARR, and numerical model-derived fields into a multi-scale paradigm for how the hydrostatic scales of motion create atmospheric circulations, perhaps vortex tubes within microfronts, favorable for organizing the turbulent breakdown of the airflow.

Two severe turbulence case studies are presented in this section, including a CIT and CAT case study not included in the aforementioned 44-case study analyses in Sect. 15.1. These two cases represent two of four case studies described in Kaplan et al. (2005b), which therein were simulated and analyzed using the original NCEP-NCAR 2.5° dataset instead of with the higher-resolution NARR dataset presented herein. The CAT case study, which was reported as a commercial accident to the NTSB, occurred over Cape Girardeau, MO (CGI), on 28 January, 1997 at 1453 UTC and ~400 hPa/7 km. The CIT case study, to be described here, occurred over Wilmington, DE (ILG), on 13 January, 2000 at 1931 UTC and ~775 hPa/2.4 km. This CIT case study represented an FAA Flight Operations Quality Assurance (FOQA) case study wherein equipment capable of recording severe turbulence was on board a commercial aircraft. The magnitude of recorded turbulence qualified this case study as an aviation accident which was reported to the NTSB as well.

The numerical model employed in the hydrostatic real data simulation experiments is the Mesoscale Atmospheric Simulation System (MASS version 5.13) (e.g., Kaplan et al. 2004, 2005b, 2006). The simulations to be described in this section are the 30 km (coarse) and 6 km (fine mesh) resolutions. Initial and time-dependent lateral boundary conditions are derived from the National Weather Service (NWS) Eta analyses for the coarse mesh simulations. All subsequent finer mesh simulations, which are nested, derive their initial and time-dependent lateral boundary conditions from the previous coarser mesh simulation. Climatological soil moisture and sea-surface temperatures as well as an average of both silhouette and envelope terrain, i.e., an average of: (1) smooth terrain that follows a coarse contour of the terrain geometry and (2) highly detailed terrain are utilized in both case studies. Forty-five vertical levels were employed for the CGI case study which was initialized at 0000 UTC 1/28/97 and nested at 1000 UTC 1/28/97 over a 130 × 100 grid matrix. Fifty vertical levels were employed for the ILG case study which was initialized at 0000 UTC 1/13/00 and nested at 1200 UTC 1/13/00 over a similar matrix of grid points.

Figures 15.1 and 15.2 describe the NARR observed features while Figs. 15.3 and 15.4 describe 6 km nested grid and 30 km coarse mesh simulation features for both case studies, i.e., CGI and ILG. Apparent in Figs. 15.1 and 15.2 is the confluence of two different momentum plumes (black arrows) ahead of the cyclonic height perturbation (black dash) and cold/warm front (cold front/warm front symbol) near the accident location and time. In both case studies the curvature and location of the height ridge/trough system (core of geostrophic relative vorticity) lag the downstream confluence of ageostrophic wind flow originating from different directions. For CGI this means that there is the confluence of winds from the southwest and northwest and for ILG the confluence of northwest and westerly flow. However, the confluence of winds is downstream from the height features, i.e., the upstream ridge to the northwest and upstream trough to the southwest, which are also on a collision course in both case studies. In both cases this confluent geometry is reflected in ageostrophic relative vorticity plumes (Figs. 15.1c and 15.2c) that elongate away from the geostrophic vorticity downstream and merge in a region of strong shearing and stretching deformation where the inertial stability is low and there is also a growing mesoscale "bulge" in the preexisting larger scale cold front (Figs. 15.1b and 15.2b). This bulge (Figs. 15.1b and 15.2b) is aligned with the along-stream variation of airflow (Figs. 15.1a and 15.2a) evident near the accident locations, where upstream geostrophic wind maxima are separated from downstream ageostrophic wind maxima.

The confluence of features which NARR depicts in these two cases (Figs. 15.1 and 15.2) represents atmospheric forcing transitioning from the synoptic to meso-α to meso-β scales of motion. To better understand the flow imbalances established by these features in NARR at the meso-β and ultimately finer scales of motion one must rely on the numerical model simulations that use 6 km horizontal grid spacing. Figures 15.3, 15.4, 15.5, and 15.6 illustrate a simulated multi-scale hydrodynamic and thermodynamic set of adjustments for both CGI and ILG that may very well indicate key links between the synoptic/meso-α scale environment discerned from NCEP-NCAR and NARR analyses and the meso-β scale preturbulent environment that the higher resolution model can resolve. The sequence of events can be partitioned into four key phases consistent with the simulated fields in Figs. 15.3,

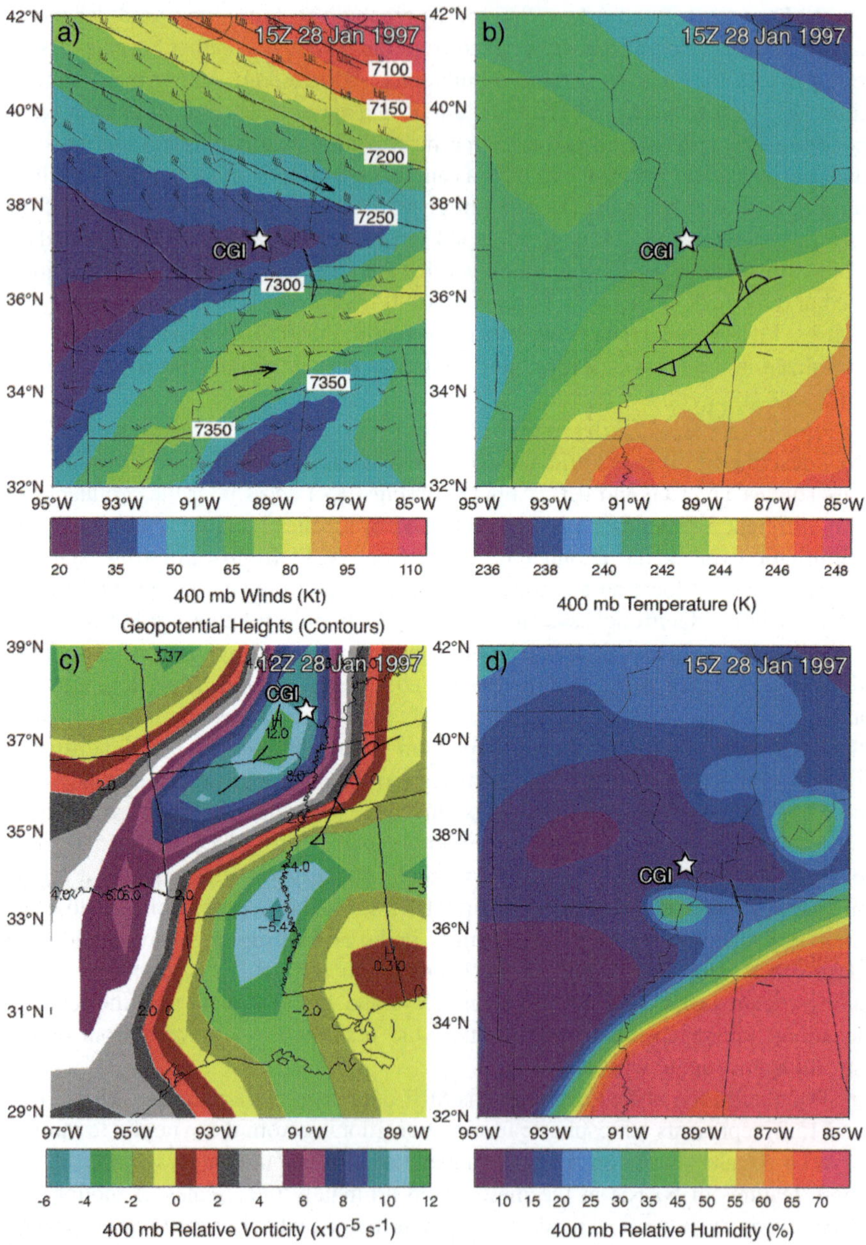

Fig. 15.1 1500 UTC 28 January 1997 NARR 400 hPa (**a**) height (solid, 50 m contour intervals), wind barbs (in knots), and isotachs (color shading in knots), (**b**) temperature (color shading in K), (**c**) relative vorticity at 1200 UTC 28 January 1997 at 400 hPa (color shading in units of $10^{-5}\ s^{-1}$), and (**d**) relative humidity (shaded in %). Relative vorticity from Plymouth State Weather Center archive. The star symbol indicates the location of the turbulence incident near CGI and black arrows/dash represent leading wind maxima, trailing height trough and cold/warm front symbol the front leading the trough (see text)

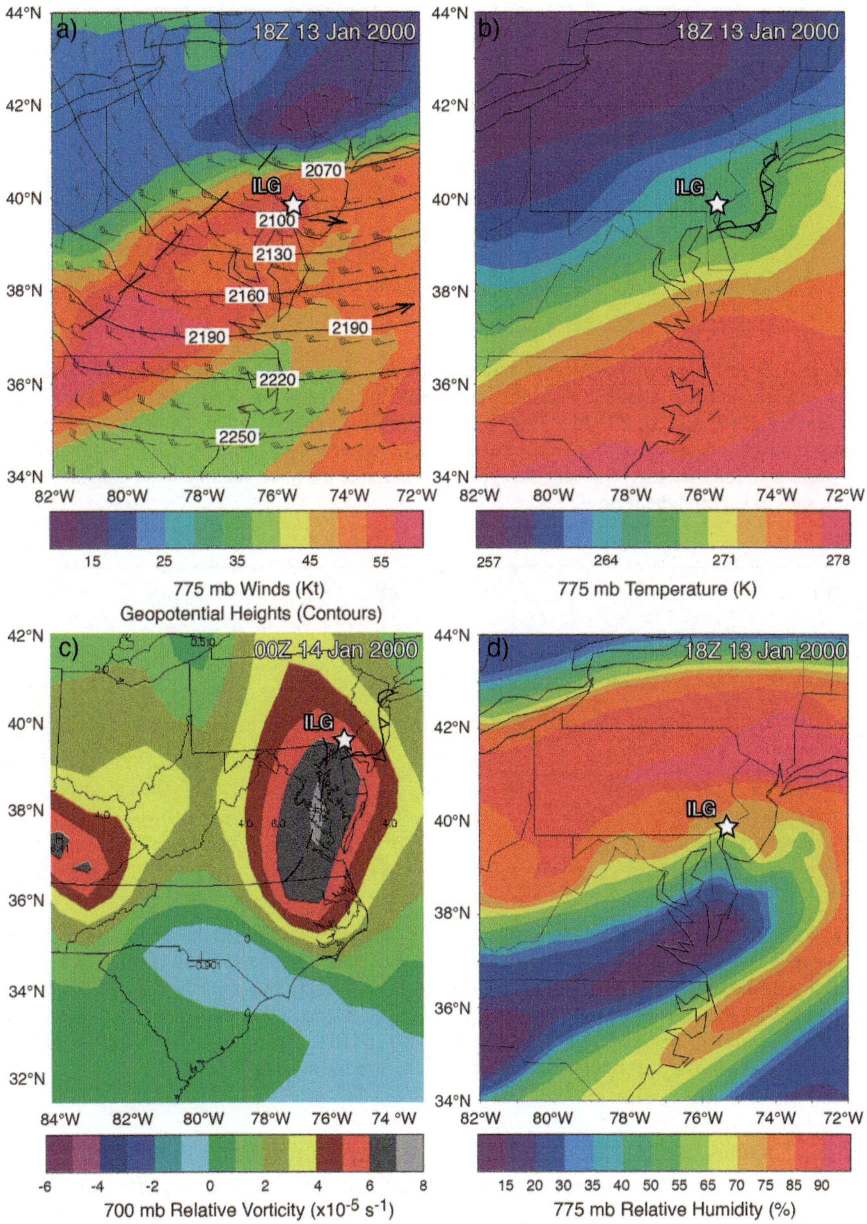

Fig. 15.2 As in Fig. 15.1 but for 1800 UTC 13 January 2000 at 775 hPa (30 m contour intervals) and relative vorticity at 0000 UTC 14 January 2000 at 700 hPa. The star symbol indicates the location of the turbulence incident near ILG

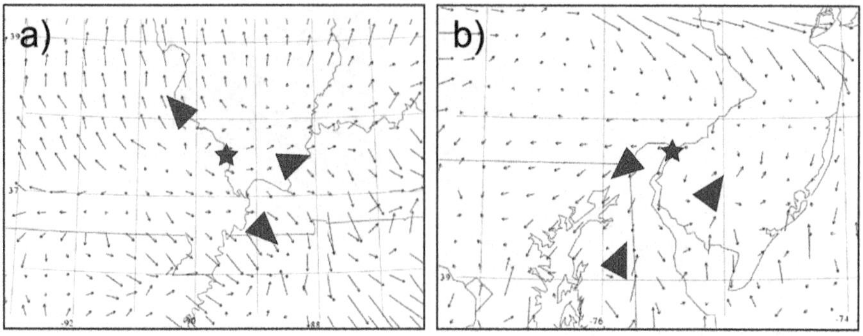

Fig. 15.3 MASS 6 km simulated resultant of four force vectors (along-stream pressure gradient, cross-stream pressure gradient, Coriolis, and Centrifugal) valid on/at (**a**) 400 hPa 1330 UTC 28 January 1997, and (**b**) 775 hPa 1900 UTC 13 January 2000 (from Kaplan et al. 2005b). The star symbols indicate the location of the turbulence incidents near CGI and ILG in parts (**a**) and (**b**), respectively. The triangles represent the cyclonic rotation of the force vectors in the transition zone from sub- to supergradient flow

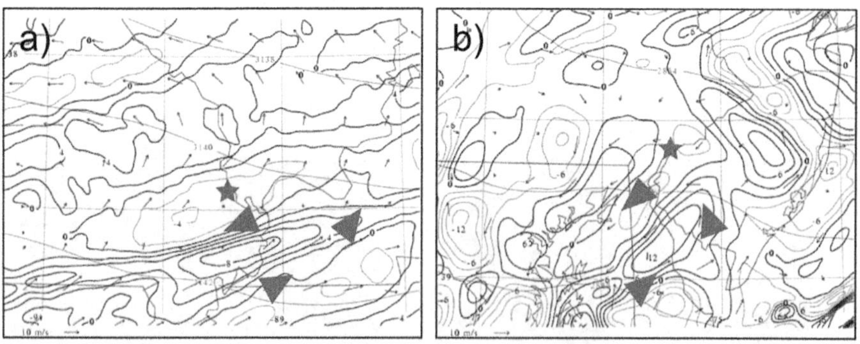

Fig. 15.4 MASS 6 km simulated Montgomery stream function (light solid in $m^2\ s^{-2}$), ageostrophic wind vectors, and ageostrophic Z-space relative vorticity (dark positive solid and negative dashed in $s^{-1} \times 10^{-4}$) valid on/at (**a**) 314 K 1330 UTC 28 January 1997 and (**b**) 287 K 1900 UTC 13 January 2000 (from Kaplan et al. 2005b). The star symbols indicate the location of the turbulence incidents near CGI and ILG in parts (**a**) and (**b**), respectively. The triangles indicate the ageostrophic wind's horizontal shear direction across the micro-cold front

15.4, 15.5, and 15.6: (1) the breakdown of gradient wind balance where the merger of the height and wind features occurs, (2) the development of significant along-stream variation in primarily meridionally forced divergence and ageostrophic vorticity in this location, (3) ageostrophic frontogenesis due primarily to ageostrophic meridional stretching and shearing deformations acting on the meridional temperature gradient, and (4) the concentration of a secondary isentropic vorticity maximum near the region of strongest ageostrophic frontal forcing.

Figure 15.3 depicts the force resultant among: (1) along-stream pressure gradient force, (2) cross-stream pressure gradient force, (3) Coriolis force, and (4) centrifugal

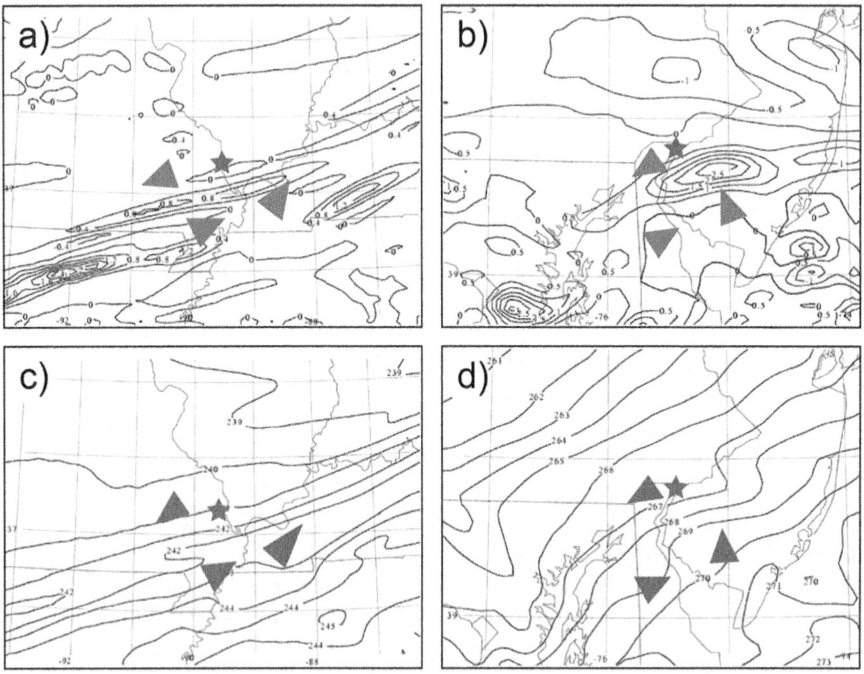

Fig. 15.5 MASS 6 km simulated v wind component divergence forcing function term in Miller's frontogenesis equation ($Km^{-1}\,s^{-1} \times 10^{-8}$) valid on/at (**a**) 400 hPa 1330 UTC 28 January 1997 and (**b**) 775 hPa 1900 UTC 13 January 2000. MASS 6 km simulated temperature (K) valid on/at (**c**) 400 hPa 1330 UTC 28 January 1997 and (**d**) 775 hPa 1900 UTC 13 January 2000 (from Kaplan et al. 2005b). The star symbols indicate the location of the turbulence incidents near CGI and ILG in parts (**a**, **c**) and (**b**, **d**), respectively. Triangles convey the variation in ageostrophic v wind component flow controlling micro-cold front structure

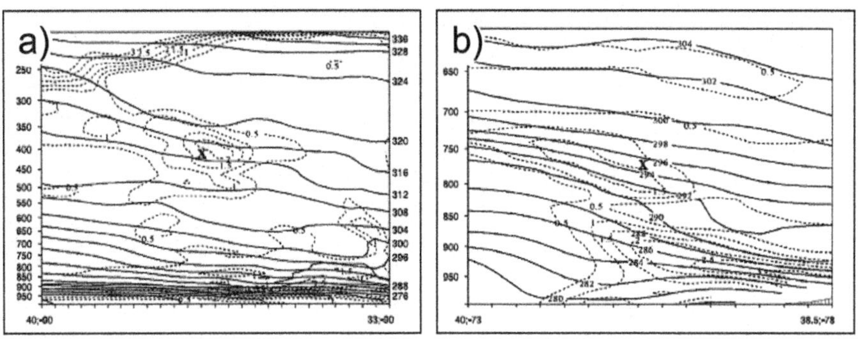

Fig. 15.6 MASS 30 km simulated jet normal vertical cross sections of potential temperature (*solid* in K) and isentropic potential vorticity (*dashed* in $Kmb^{-1}\,s^{-1} \times 10^{-6}$) valid at (**a**) 1200 UTC 28 January 1997, (**b**) 1800 UTC 13 January 2000 ("X" marks approximate accident location) (from Kaplan et al. 2005b)

force for both case studies at the level and the closest available (biased early) time to the severe aviation turbulence. Evident is a cyclonic turning of the resultant vector and a general reduction in vector magnitude along the generally southwest–northeast axis of the cold fronts in both case studies. This geometry reflects the transition along and across the cold front from subgradient to supergradient flow predominantly over southern Missouri and Illinois for CGI and north central Delaware and southwestern New Jersey for ILG. There are weakly varying along-stream pressure gradients accompanying strongly varying centrifugal force and cross-stream height gradients as the along-stream gradient of isoheights is weaker in magnitude than the confluence of streamlines and increasing inertial-advective flows. Hence, the along-stream confluence of meridional winds is large in a region of relatively widely spread isoheights, or the radius of curvature of the isoheights is larger in scale than the radius of curvature of the streamlines.

This inconsistency between radii of curvature of the mass and momentum fields enables curved accelerative flow with significant along-stream variation of force imbalance resulting in ageostrophic vorticity and velocity divergence, i.e., *variations of the v wind component at the mesoscale along the stream.* Figure 15.4 depicts the net effect of the force imbalance in Fig. 15.3 as simulated Z-space ageostrophic relative vorticity, which is displayed with the Montgomery stream function (ψ) on a nearby isentropic surface along with the ageostrophic wind vectors. Narrow corridors or "streamers" of ageostrophic relative vorticity over southeastern Missouri for CGI and Delaware/Maryland/New Jersey for ILG reflect strong gradients of ageostrophic v wind component variation on ψ in the regions of vorticity gradients. Maxima can be seen for CGI nearly directly aligned with the turbulence location and for ILG between central Delaware and southwestern New Jersey or just southeast of the turbulence location. In both locations, the ageostrophic flow in Fig. 15.4, like the acceleration vectors in Fig. 15.3, exhibits a cyclonic gyre which extends the "bulge" in cyclonic ageostrophic relative vorticity downstream along the micro-cold front ahead of the height-driven quasi-geostrophic relative vorticity (note also Figs. 15.1 and 15.2).

The consequences of the kinematics for the mass/temperature field at the meso-β scale can be in part explained by viewing Fig. 15.5 and diagnosing its meaning. Miller's (1948) 3-dimensional frontogenesis Eq. (15.1) is depicted below and contains three terms for hydrostatic adiabatic motions, which are (A) x-space deformation, (B) y-space deformation, and (C) the tilting term:

$$F = -\frac{1}{|\nabla\theta|}\left[\overset{\text{A}}{\frac{\partial\theta}{\partial x}\left(\frac{\partial u}{\partial x}\frac{\partial\theta}{\partial x}+\frac{\partial u}{\partial y}\frac{\partial\theta}{\partial y}\right)}\right.$$
$$\overset{\text{B}}{+\frac{\partial\theta}{\partial y}\left(\frac{\partial v}{\partial x}\frac{\partial\theta}{\partial x}+\frac{\partial v}{\partial y}\frac{\partial\theta}{\partial y}\right)} \quad (15.1)$$
$$\left.\overset{\text{C}}{+\frac{\partial\theta}{\partial p}\left(\frac{\partial\omega}{\partial x}\frac{\partial\theta}{\partial x}+\frac{\partial\omega}{\partial y}\frac{\partial\theta}{\partial y}\right)}\right]$$

For meridional flow that is varying substantially at the meso-β scale in proximity to a preexisting frontal zone with significant meridional temperature gradients, additional scale contraction of the v wind component will enhance the frontal structure locally. Shearing deformation caused by along-stream varying v wind flow in x-space and any additional stretching deformation caused by the varying v wind flow in y-space will enhance the structure of a downscale contracting "microfront" locally. This simulated (B) term in Miller's equation as well as the temperature fields are depicted in Fig. 15.5. One can see that maxima of this term are stretched along the cold front near the time and altitude of both turbulence incidents over southern Missouri and northern Arkansas for CGI (Fig. 15.5a) and northern Delaware and southwestern New Jersey for ILG (Fig. 15.5b). These terms dominate the total frontogenesis function and in time produce meso-β scale bulges along the cold front roughly oriented from west–east nearly directly over CGI and from west–northwest–east–southeast ~50 km east–northeast of ILG. These frontal bulges have a width of ≪100 km and represent the scale contracted and strengthened features very weakly evident in the NARR in Figs. 15.1b and 15.2b. Hence, they represent microfronts reflecting the complex along-stream variation of the primarily meridional wind component's deformation field and hence ageostrophic advection of v momentum.

Finally, Fig. 15.6 depicts a simulated vertical cross section of potential temperature and isentropic potential vorticity (IPV) that bisects both of the microfronts described above. One can see that just upstream from the accidents and their local microfronts, subsynoptic IPV maxima approaching or exceeding 2 IPV units develop. These fronts are still ~ meso-β scale in structure as limited by the relatively coarse 6–30 km horizontal and ~500 m vertical grid spacings. One could speculate that within these IPV maxima, vortex tube formation and overturning could be favored by virtue of the variation in vorticity and static stability, i.e., the slope of potential temperature relative to the orthogonal slope of vorticity magnitude (note Eq. 15.5 in Table 15.2). Such regions may therefore be favored for a continued cascade of energy into turbulence at much finer scales of motion?

Table 15.2 RTTM turbulence parameters/forecast indices

$\frac{(V-V_t)}{\left(0.1+1000\left(\frac{d\theta}{dz}\right)\right)^2},$	(15.2)
$\text{KTI} = f\left(f\left(1-\frac{1}{Ri}\right)+\zeta\right),$	(15.3)
$\text{NCSUI} = (U\cdot\nabla(U))\frac{\lvert\nabla(\zeta)\rvert}{\lvert Ri\rvert},$	(15.4)
$\text{NCSU2} = \lvert\nabla_{\psi}X\nabla\zeta\rvert,$	(15.5)

For more details see Kaplan et al. (2004)

15.3 An Example of the Development and Performance of an Operational Mesoscale Numerical Model-Based Turbulence Forecasting System

In this section, we will briefly address the topic of designing turbulence environment forecast products that could be generated in real time from an operational mesoscale numerical modeling system as well as how they synthesize the analyses presented thus far. The primary goal of this section is to integrate what was learned concerning the synoptic dynamics of the environment which organizes aviation turbulence in Sect. 15.2 into a real-time stand-alone predictive system. As noted in Kaplan et al. (2004, 2006), the numerical model employed for research in Sect. 15.2 or for that matter any numerical model could be modified for use as an operational turbulence potential simulation system by designing a suite of products that are sufficiently flexible for use in forecasting various types of turbulence, i.e., CAT, CIT, and MTN in particular. NASA supported the development of such a modeling system, the Real-Time Turbulence Model (RTTM), which was run operationally at the Langley Research Center to exclusively support real-time forecasts of turbulence potential for use in directing NASA B-757 research aircraft to locations of primarily but not exclusively convective turbulence.

MASS version 5.13 (Kaplan et al. 2000) was employed as the numerical model for the RTTM.

Table 15.2 depicts the forecast indices that are derived from the model output. Index #1 represents the NASA turbulence index developed by Proctor (2000) that is primarily designed to determine layers of neutral static stability during significant velocity flows in which V_t represents 35 m/s. As the relative humidity increases to a value of 95 %, potential temperature θ is replaced by equivalent potential temperature θ_e. This index should be large where lapse rates approach moist neutrality in the presence of strong jet streams. Index #2 is the inertial stability parameter from Knox (1997) based on Stone's (1966) inertial instability criterion. It should be large in cases of strong anticyclonic shear juxtaposed with low Richardson numbers. Indices #3 and #4 will be applied to a case study of widespread aviation turbulence. Both are based on the research reported in this chapter. They are both designed to maximize during highly curved and accelerative flows near significant baroclinic zones in which downscale processes organize mesoscale frontal zones such as the vortex tubes and microfronts described earlier. Additional products are generated by this system to aid forecasters in determining where significant convection will occur and where mountain wave genesis is likely. These products in conjunction with the turbulence indices are designed to be useful tools for the forecasting of CAT, CIT, and MTN turbulence.

Figures 15.7 and 15.8 will now be employed to demonstrate the products and forecast utility of the RTTM for a case of widespread and extreme aviation

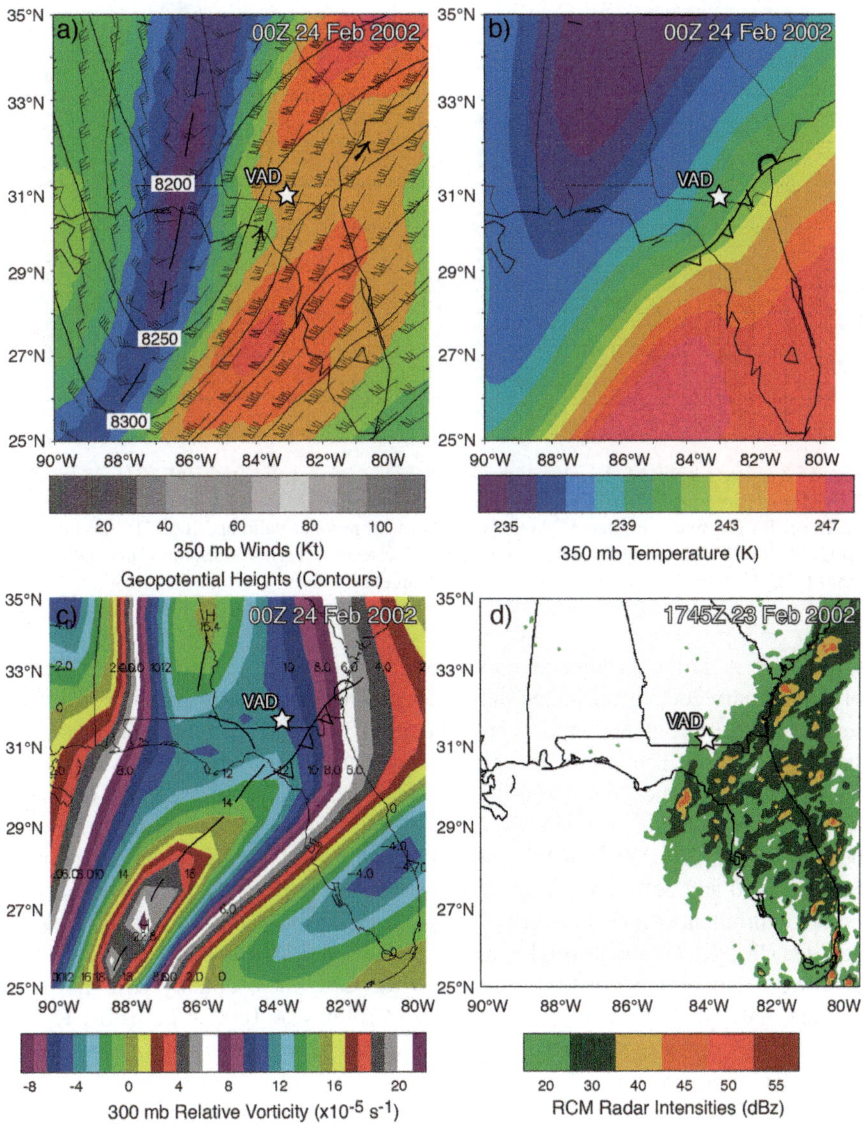

Fig. 15.7 Same as Fig. 15.1, but for 350 hPa and 1800 UTC 23 February 2002 in (**a**), (**b**), and (**c**) relative vorticity at 0000 UTC 24 February 2002 at 300 hPa. (**d**) Radar intensity summary (shaded in dBz) for 1745 UTC 23 February 2002. Radar plot from Plymouth State Weather Center archive

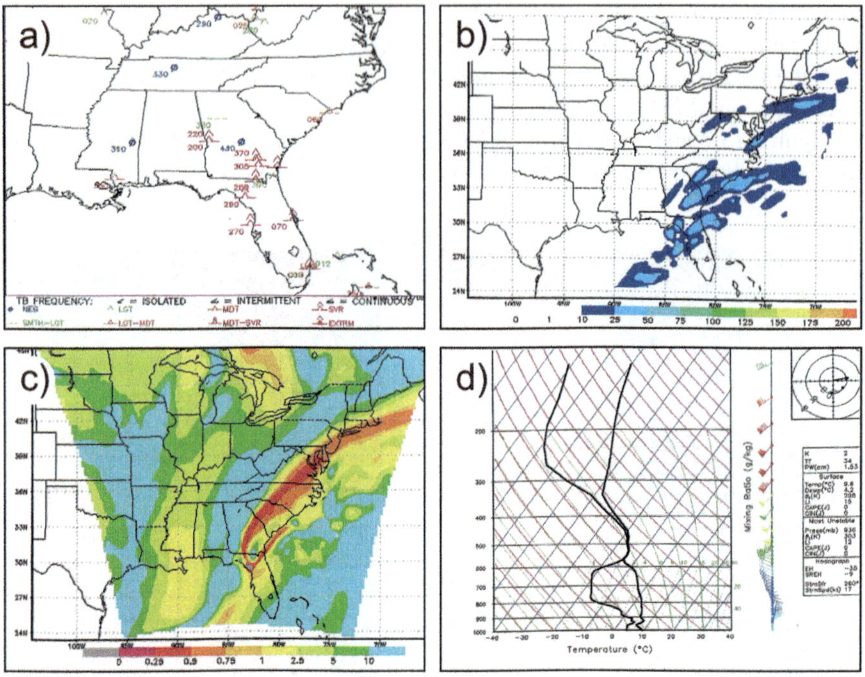

Fig. 15.8 Pireps from NOAA Aviation Digital Data source valid from (**a**) 1641–1836 UTC. RTTM fine1 simulated NCSU2 index (shaded in s^{-3}) for (**b**) 22,000 ft on 1800 UTC 23 February 2002. (**c**) RTTM fine1 simulated 22,000 ft. Richardson number valid at 2100 UTC 23 February 2002. (**d**) RTTM fine1 simulated skew-t sounding located at Charleston, South Carolina, valid at 0000 UTC 24 February 2002 (Adapted from Kaplan et al. 2004)

turbulence. Aviation turbulence reports (pireps) during the 1641–1836 UTC period of 23 February 2002 are depicted in Fig. 15.8a. During the early part of this period, several severe turbulence reports are evident primarily over northern Florida and southern Georgia. The region undergoing severe turbulence then spreads northeastwards up the Atlantic Coast to North Carolina, Virginia, New Jersey, and Connecticut after 1836 UTC (not shown). The elevations of the reports are primarily between 20 and 30,000 ft or between ~450 and 300 hPa with the largest number between 400 and 350 hPa. All of these reports are located in the aforementioned favored turbulence zone between the trough/ridge feature in the height field at 350 hPa (Fig. 15.7a) and the highly ageostrophic jets from the southwest and south–southwest converging downstream and to the east, most notably over northern Florida, eastern Georgia, and the eastern Carolinas at 1800 UTC (Fig. 15.7a). This favored turbulence region propagates downstream to the Mid-Atlantic after 2100 UTC (not shown).

In Fig. 15.7c, a remarkable bulge in the 300 hPa relative vorticity analysis occurs where plumes of positive relative vorticity from western Georgia and the Florida Panhandle to the Georgia Atlantic coast can be seen. Hence, the ageostrophic relative vorticity varies substantially across this region with highly ageostrophic winds resulting in gradients of ageostrophic relative vorticity both along and across the stream of $>10^{-5}$ s^{-1} in less than 100 km in Fig. 15.7c. This strong along-stream variation in relative vorticity is consistent with analyses in Sect. 15.2. Such variation is consistently located within the 350 hPa baroclinic zone over northeastern Florida and southeastern Georgia in the NARR in Fig. 15.7b. Note also in this figure that there are undulations in the 350 hPa baroclinic zone consistent with mesoscale frontogenesis (developing microfronts) simulated and described in the analyses in Sect. 15.2. Furthermore, this strongly perturbed wind and mass field is just northwest of the widespread precipitation along the Atlantic Coast with >40 dBz returns extending from south of the North Carolina coast to northeastern Florida in Fig. 15.7d indicating a likely contribution from upstream convective outflow (e.g., Wolf and Johnson 1995; Hamilton et al. 1998; Kaplan et al. 1998).

Figures 15.8a–d can be employed to inter-compare the RTTM simulated NCSU2 turbulence index (Table 15.2, Eq. 15.5) to the severe turbulence pireps as well as the RTTM Richardson number and vertical sounding fields over the Atlantic Coast at 1800 UTC. The 22,000 ft NCSU2 index exhibits maxima in excess of 75 units at 1800 UTC over southeastern Georgia and along the northern Gulf of Mexico coast. The patterns of values exceeding 50 units are elongated primarily from the lower North Carolina Coast southwestward to the west central part of Florida. This is roughly centered on the early region of severe pilot reports of turbulence (PIREPS) at and above 22,000 ft. Three hours later, the NCSU2 index maximizes at >75 units between central South Carolina to northeastern North Carolina with another maximum along the southeastern New England coast (not shown). The maxima reflect large magnitudes of the orthogonality of gradients between the Montgomery stream function and the relative vorticity primarily from northern Florida to eastern North Carolina, i.e., the intersection of the meridional temperature gradient and zonal ageostrophic relative vorticity (Table 15.2, Eq. 15.5). These maxima also reflect large magnitude values of the NCSU1 index (calculations not shown but index in Table 15.2, Eq. 15.4) because of the significant magnitude of inertial advection of ageostrophic v momentum in proximity to the gradient of ageostrophic along-stream relative vorticity and low Richardson number values in the frontal zone. The low Richardson number can be seen as elongated plumes below the .25 critical value extending from north central Florida to Delaware in Fig. 15.8c. The degree to which the frontal zone is being stretched downstream can be inferred from this Richardson number simulation consistent with strong along-stream advection of ageostrophic relative vorticity. The Charleston, South Carolina (CHS) sounding, reflects the vertical complexity of this baroclinic zone with the deep adiabatic layer between 475 and 350 hPa. The adiabatic layer is just above the convective outflow jet(let) resulting in the vertical wind shear zone evident in the vertical wind turning and increase in wind velocity at the base of this layer which also corresponds with the ageostrophy at 350 hPa evident in the NARR over South Carolina (Fig. 15.7a). Hence, horizontal and vertical indicators of microfronts and strong wind shear zones are in place close to the clusters of turbulence PIREPS along the south Atlantic coastal plain.

15.4 Summary and Synthesis of a Multi-scale Paradigm for the Organization of Severe Aviation Turbulence

Based upon the diverse literature on turbulence, there is no simple consensus as to why severe aviation turbulence exists. There are many possible dynamical and thermodynamical processes that have been linked to this phenomenon in this book, so therefore its causes are likely very complex. The research in this specific chapter indicates a certain scale-dependent sequence of events that one often observes and simulates in preturbulence environments. The end state for many environments in proximity to where the severe turbulence is reported by aircraft can be described as a mesoscale or microfront containing very substantial variation in ageostrophic vertical vorticity. This likely describes a scenario where mesoscale vortex tubes are rapidly changing and restructuring themselves. The following summary of key dynamical and thermodynamical processes represents our best estimate of how this atmospheric state occurs based on multi-scale observations and numerical simulations:

1. Curved flow develops at minimally the meso-α scale of atmospheric motion typically accompanying the confluence of two synoptic scale jet streaks and/or outflow from an upstream mesoscale convective system.
2. As the jets converge, there is a marked separation between the confluent geometry in the mass field and that in the wind field with the confluent streamlines typically leading the confluent isoheights downstream (separation between jet geostrophic wind entrance, and jet ageostrophic wind exit regions).
3. In this confluent streamline zone at the meso-β scale, the flow not only is highly ageostrophic but often highly supergradient with too much variation in centrifugal force for the cross- or along-stream pressure gradient and the Coriolis forces to balance.
4. In this same region, there is typically a significant temperature gradient initially oriented across the stream but which in time develops a finer typically meso-β scale along-stream bulge or microfront.
5. This microfront occurs where the downstream advection of ageostrophic relative vorticity is very large in the zone of supergradient flow and is also coincident with strong and primarily meridional shearing and stretching deformations.
6. The microfront is formed where the predominantly cross-stream temperature gradient is contracted in scale by the predominantly meridional shearing and stretching deformation. It is the colocation in space and time between meridional wind shear and meridional temperature gradients that is the result of the confluent horizontal advection of ageostrophic vertical vorticity downstream. This is facilitated by superagradient flow in proximity to a strong and primarily meridionally oriented cold front.

The result and implications for turbulence of this microfront is the concentration of 3-dimensional rotation accompanying vortex tubes in a location of reduced static stability due to the tilting of the isentropic surfaces. Therefore, rotation and low static stability are closely aligned which likely indicates a favored zone for atmospheric overturning in a strongly sheared environment resulting in turbulent vortex formation.

Acknowledgments This research has been funded under NASA Contract No. NAS1-99074 and Subcontract No. 82U-7473-008 from the Research Triangle Institute. The author acknowledges the support of Dr. Fred H. Proctor, the NASA-Langley Technical Monitor, and also thanks the North Carolina Supercomputing Center for providing supercomputing time. Benjamin J. Hatchett, of the University of Nevada Reno, prepared many of the figures in this manuscript.

References

Ellrod, G.P., Knapp, D.I.: An objective clear-air turbulence forecasting technique: verification and operational use. Weather Forecasting **7**, 150–165 (1992)

Hamilton, D.W., Lin, Y.-L., Weglarz, R.P., Kaplan, M.L.: Antecedent jetlet formation prior to the Palm Sunday 1994 Tornado Outbreak in Alabama and Georgia. Mon. Weather Rev. **136**, 2061–2089 (1998)

Kalnay, E., Kanamitsu, M., Kistler, R., Collins, W., Deaven, D., Gandin, L., Iredell, M., Saha, S., White, G., Woollen, J., Zhu, Y., Leetmaa, A., Reynolds, R., Chelliah, M., Ebisuzaki, W., Higgins, W., Janowiak, J., Mo, K.C., Ropelewski, C., Wang, J., Jenne, R., Joseph, D.: The NCEP/NCAR 40-Year Reanalysis Project. Bull. Am. Meteorol. Soc. **77**, 437–471 (1996)

Kaplan, M.L., Lin, Y.-L., Hamilton, D.W., Rozumalski, R.A.: A numerical simulation of an unbalanced jetlet and its role in the Palm Sunday 1994 Tornado outbreak in Alabama and Georgia. Mon. Weather Rev. **136**, 2133–2165 (1998)

Kaplan, M.L., Lin, Y.-L., Charney, J.J., Pfeiffer, K.D., Ensley, D.B., DeCroix, D.S., Weglarz, R. P.: A terminal area PBL Prediction System at Dallas-Fort Worth and its application in simulating diurnal PBL jets. Bull. Am. Meteorol. Soc. **81**, 2179–2204 (2000)

Kaplan, M.L., Charney, J.J., Waight, K.T. III, Lux, K.M., Cetola, J.D., Huffman, A.W., Slusser, S. D., Riordan, A.J., Lin, Y.-L.: Characterizing the severe turbulence environments associated with commercial aviation accidents. A Real-Time Turbulence Model (RTTM) designed for the operational prediction of moderate-severe aviation turbulence environments. NASA/CR-2004-213025, 50 pp. (2004)

Kaplan, M.L., Huffman, A.W., Lux, K.M., Charney, J.J., Cetola, J.D., Riordan, A.J., Lin, Y.-L.: Characterizing the severe turbulence environments associated with commercial aviation accidents. Part 1: A 44 case study synoptic observational analyses. Meteorol. Atmos. Phys. **88**, 129–152 (2005a)

Kaplan, M.L., Huffman, A.W., Lux, K.M., Cetola, J.D., Charney, J.J., Riordan, A.J., Lin, Y.-L., Waight III, K.T.: Characterizing the severe turbulence environments associated with commercial aviation accidents. Part 2: Numerical simulation of meso-beta scale supergradient wind flow and streamwise ageostrophic frontogenesis. Meteorol. Atmos. Phys. **88**, 153–175 (2005b)

Kaplan, M.L., Charney, J.J., Waight III, K.T., Lux, K.M., Cetola, J.D., Huffman, A.W., Slusser, S. D., Riordan, A.J., Lin, Y.-L.: Characterizing the severe turbulence environments associated with commercial aviation accidents. A Real-Time Turbulence Model (RTTM) designed for the operational prediction of moderate-severe aviation turbulence environments. Meteorol. Atmos. Phys. **94**, 235–270 (2006)

Knox, J.A.: Possible mechanisms of clear-air turbulence in strongly anticyclonic flows. Mon. Weather Rev. **125**, 1251–1259 (1997)

Mesinger, F., et al.: North American regional reanalysis. Bull. Am. Meteorol. Soc. **87**(758), 343–360 (2006)

Miller, J.E.: On the concept of frontogenesis. J. Meteorol. **5**, 169–171 (1948)

Proctor, F.H.: Personal communication (2000)
Stone, P.H.: On non-geostrophic baroclinic stability. J. Atmos. Sci. **23**, 390–400 (1966)
Wolf, B.J., Johnson, D.R.: The mesoscale forcing of a mid-latitude upper-tropospheric jet streak by a simulated convective system. Part I: Mass circulation and ageostrophic processes. Mon. Weather Rev. **123**, 1059–1087 (1995)

Chapter 16
Processes Underlying Near-Cloud Turbulence

Todd Lane

Abstract Thunderstorms generate turbulence both within cloudy air and in the clear air surrounding them. Both in-cloud and out-of-cloud turbulence are an aviation hazard, and the latter can pose a risk of unexpected encounters because it is invisible and is unable to be detected using standard on-board radar technology. In recent years there has been a significant improvement in the understanding of out-of-cloud turbulence generation mechanisms, including identifying the importance of breaking atmospheric gravity waves. In this chapter, numerical modelling studies are used to explain the processes leading to turbulence outside of cloud, including near the cloud boundary and above storms. Turbulence near the cloud boundary is explained by local instabilities associated with storm-induced wind shear. Turbulence further aloft is related to gravity wave breaking that is well explained by linear theory. Very high-resolution simulations are also used to quantify the intensity of turbulence and how it varies spatially. These simulations demonstrate that, although the main convective regions of storms possess the most intense turbulence, notable regions of turbulence can extend significant distances vertically and horizontally from the cloud boundary.

16.1 Introduction

Convective clouds form due to moist instabilities that arise as a consequence of the environmental thermodynamic profile, which may contain convective available potential energy (CAPE), and surface heating and/or a lifting mechanism sufficient to overcome convective inhibition (CIN). The result is convective updrafts that derive their buoyancy from latent heating associated with condensation. The convective updrafts are inherently turbulent, behaving similarly to plumes or thermals that have been studied in the laboratory (e.g. Morton et al. 1956), with the largest eddies corresponding to the scale of the updrafts and smaller eddies arising due to a

T. Lane (✉)
School of Earth Sciences, The University of Melbourne, Melbourne, VIC, Australia
e-mail: tplane@unimelb.edu.au

R. Sharman, T. Lane (eds.), *Aviation Turbulence*,
DOI 10.1007/978-3-319-23630-8_16

turbulent cascade and entrainment processes at the cloud boundary. Likewise, the formation of convective downdrafts is associated with diabatic cooling caused by the evaporation of precipitation. Depending on the organisational behaviour of the convective clouds, a convective system may be composed of numerous updrafts and downdrafts that are embedded within a contiguous cloud volume. For long-lived convective systems, like squall lines or multicell cloud clusters, numerous convective updrafts and downdrafts may exist concurrently in different stages of their life cycle.

The turbulence within convective clouds poses an established risk to aircraft. As demonstrated by Bryan et al. (2003) and Lane and Sharman (2014), the length scale of updrafts embedded within mesoscale convective systems can be about 2 km or less. These scales closely correspond to those that elicit the strongest turbulent response for large commercial aircraft (see Chap. 1; Lane et al. 2012). Thus, convective clouds pose a significant aviation turbulence hazard, and for this reason, along with other associated hazards like icing, thunderstorms are routinely avoided to minimise the risk of encountering turbulence within cloud. However, in addition to in-cloud turbulence, thunderstorms are known to induce turbulence in the clear air surrounding them. This turbulence is invisible, and the volume of hazardous air can extend substantial distances from the cloud boundary.

The US Federal Aviation Administration (FAA) states in their Thunderstorm Advisory Circular (AC 00-24C; FAA 2013) that 'potentially hazardous turbulence is present in all thunderstorms' and 'outside the cloud, shear turbulence is encountered several thousand feet above and up to 20 miles laterally from a severe storm. Additionally, clear-air turbulence may be encountered 20 or more miles from the anvil cloud edge'. Accordingly, the FAA has established a set of thunderstorm avoidance guidelines that are designed to facilitate en route tactical avoidance of thunderstorm hazards. Other international aviation administrations and airlines use similar guidelines. The relevant parts of the FAA guidelines (FAA 2015) state:

- 'Do avoid by at least 20 miles any thunderstorm identified as severe or giving an intense radar echo. This is especially true under the anvil of a large cumulonimbus'.
- 'Don't trust the visual appearance to be a reliable indicator of the turbulence inside a thunderstorm'.
- 'Don't attempt to fly under the anvil of a thunderstorm. There is a potential for severe and extreme clear air turbulence'.
- 'Do regard as extremely hazardous any thunderstorm with tops 35,000 feet or higher whether the top is visually sighted or determined by radar'.

Aspects of these guidelines can be traced back many decades, to the (now superceded) 1968 FAA Thunderstorm Advisory Circular (AC 00-24; FAA 1968).

The 1968 Circular (FAA 1968) also discusses the occurrence of turbulence above storm tops and its relation to cloud-top wind speed. Consistent with that discussion, for many years the FAA had a guideline that stated: 'Do clear the top of a known or suspected severe thunderstorm by at least 1,000 feet altitude for each 10 knots of wind speed at the cloud top' (FAA 2010). However, a number of recent studies (e.g. Lane and Sharman 2008; Trier et al. 2012; Lane et al. 2012; Kim and Chun 2012) showed that this particular guideline was inconsistent with the physics underlying turbulence generation above storms (especially gravity wave processes—see discussion later in this chapter) and in some observed cases provided inadequate vertical separations. This particular guideline has since been deleted, and the current Aeronautical Information Manual (FAA 2015) does not provide a guideline for vertical separation from thunderstorms to avoid turbulence at all.

On-board radar and visual identification can be used to minimise thunderstorm penetrations by aircraft, and the above guidelines can help reduce the risk of encountering turbulence in the clear air surrounding thunderstorms. However, many recent case studies (Lane et al. 2003, 2012; Trier et al. 2010, 2012; Fovell et al. 2007) have demonstrated that turbulence that was ultimately caused by thunderstorms can occur at distances well beyond the recommended separations. These studies have also provided significant fundamental advances in our understanding of turbulence generation by thunderstorms, especially in the clear air surrounding them. This chapter focuses on some aspects of the processes that generate this 'near-cloud turbulence'.

Near-cloud turbulence (NCT) is often defined as turbulence caused by thunderstorms that occurs outside of cloud, though this definition would not include turbulence occurring in thin cirrus associated with anvils, other storm outflows or large-scale ascent. Due to the size of the ice particles and their concentrations, such cirrus would be invisible to the on-board radar that is used to avoid convective clouds. Thus, here an alternate more practical definition is employed, with NCT being used to describe turbulence caused by thunderstorms, which occurs outside of the convective regions of storms.

As described in Lane et al. (2012; see their Fig. 3), in situ turbulence observations from commercial aircraft demonstrate that, on average, the risk of turbulence decreases with distance from the radar-detected cloud boundary. This decrease is relatively slow, with the risk of encountering moderate or greater turbulence being twice the background value as far as 70 km (laterally) from the storm and ten times the background value at 3.6 km above echo tops. These results underline the importance of near-cloud turbulence as an aviation hazard.

Figure 16.1 identifies regions of thunderstorms that would normally be expected to have an increased risk of NCT, based on a similar schematic in Lester (1993). These include the upper-level outflow (or anvil) region, the clear air under the anvil, the region directly adjacent to overshooting convective updrafts and the gravity wave region above the storm. The processes underlying the turbulence generation in those regions are considered in the following sections.

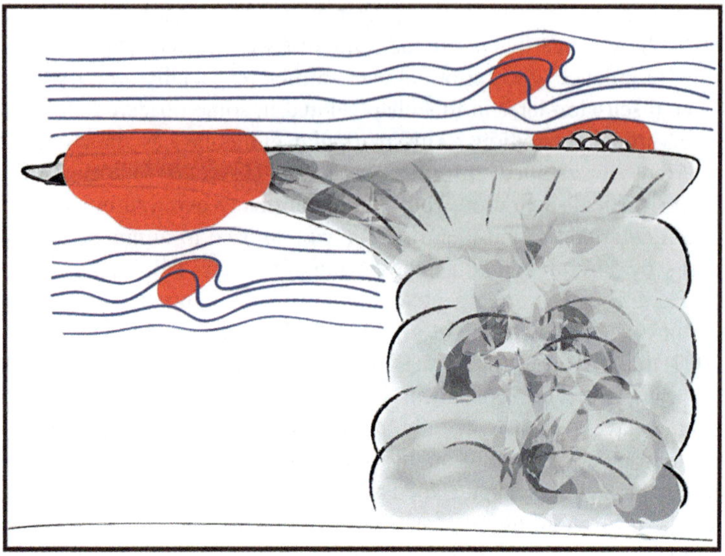

Fig. 16.1 Schematic depicting regions of potential near-cloud turbulence (*red shading*) near a thunderstorm. *Grey shading* depicts cloud, *black lines* represent the cloud boundary and *blue lines* represent isentropes

16.2 Turbulence Near the Cloud Boundary

On 10 July 1997, a commercial aircraft en route from Seattle, Washington, to New York City, New York, USA, encountered severe turbulence at 37,000 ft immediately above a rapidly developing deep convective cell near Dickinson, North Dakota. The aircraft suffered peak vertical accelerations of 1.01 and −1.75 perturbation *g*'s, resulting in 22 minor injuries. Witness statements placed the aircraft above the uppermost cloud boundary when it encountered turbulence. However, the actual distance above cloud was impossible to determine, making it difficult to unambiguously establish the cause. The case, hereafter referred to as the Dickinson case, is examined in detail in Lane et al. (2003), and one possible explanation for the turbulence event was turbulence generated at the cloud boundary. This section describes the mechanisms responsible for NCT directly adjacent to overshooting convection in this case.

To help explain the turbulence generation process in this case, we consider results from one of the numerical simulations reported in Lane et al. (2003). This cloud model simulation was configured in two spatial dimensions with 16.7-m grid spacing and initialized with a sounding representative of the observed conditions. Figure 16.2 contains three snapshots of a single overshooting updraft from the simulation, illustrating its 10-min evolution. Here the term 'overshoot' is used to describe when an updraft extends vertically beyond its equilibrium level, which in many cases of deep storms is near the tropopause.

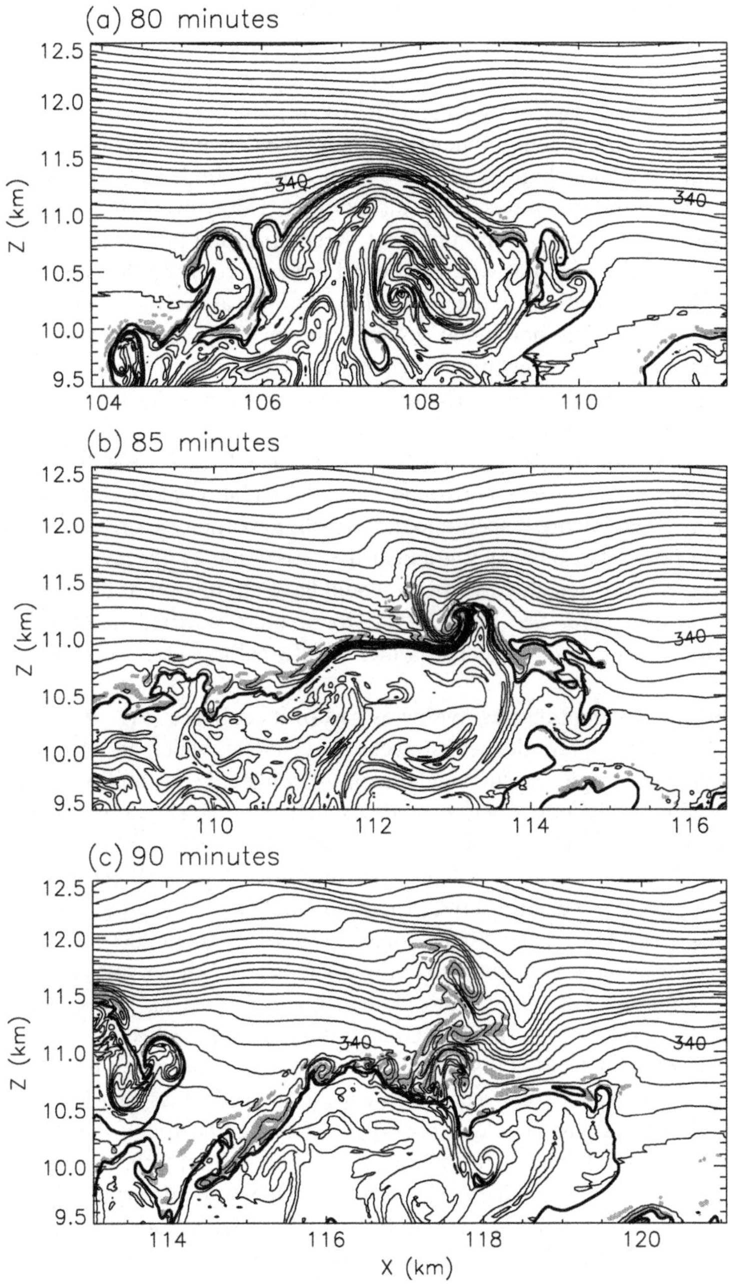

Fig. 16.2 Time evolution of the upper part of an overshooting convective updraft in idealised two-dimensional simulations. Potential temperature (*thin lines*) at 1-K intervals, cloud outline (total cloud mixing ratio of 0.05 g kg^{-1}, *thick line*) and shading in regions of convective instability outside of cloud. Adapted from Lane et al. (2003)

As indicated by the potential temperature, in the early stages of the overshoot (Fig. 16.2a), the uppermost cloud boundary is relatively smooth. The air aloft is also smooth and laminar, though with evidence of a gravity wave signature above the cloud with vertical displacements of a few 100 m. The complex structure within the cloud denotes mixing and turbulence owing to the convective instability.

Five minutes later (Fig. 16.2b) the overshooting updraft has begun to collapse, with the stable lower stratospheric air exerting a downward buoyancy force to return the updraft to its equilibrium level. Accompanying this collapse is a distinct horizontal vortex structure, with embedded convective instability and an implied anticlockwise circulation, on the right side of and just above the cloud at around $(x, z) = (113, 11)$ km. This horizontal vortex is noteworthy as it is about 500 m in horizontal scale and its accompanying velocity could thereby induce a significant commercial aircraft response. There are a few possible causes of this vortex. It could be a breaking small-scale gravity wave (the dynamics of which are described in Sect. 16.3), and indeed the vortex appears linked to a vertically propagating gravity wave with negative horizontal phase propagation aloft. Alternatively, the vortex may be an intense response to the baroclinic vorticity source associated with the strong horizontal buoyancy gradient that arises during the overshoot process. This baroclinically generated vorticity would have circulation of the same sense as the vortex seen here. See Lane (2008) for a more detailed description of the formation of this vortical response, which is referred to as the 'secondary circulation' in that paper.

As the updraft continues to evolve as part of its collapse (Fig. 16.2c), there are two main regions of turbulence that occur above cloud, which are identified by the complicated isentropic structure and regions of convective instability (shaded). The first is narrow in the horizontal, extends about 1.5 km above cloud and represents the continued evolution and breakdown of the horizontal vortex discussed above. The second region forms in a shallow (~200-m-deep) layer along the uppermost edge of the cloud (especially between approximately $x = 114$ and $x = 118$ km). This region features a few resolved scale horizontal vortices about 200 m in diameter, as well as other unstable structures that are less coherent. These vortices appear like Kelvin-Helmholtz billows and are akin to those formed as part of the 'cloud-interfacial instability' described by Grabowski and Clark (1991).

Grabowski and Clark (1991) used idealised simulations of thermals, as an analogue to deep moist convection, to expose the processes that lead to mixing and entrainment along the cloud boundary. They demonstrated that the thermal-induced flow would generate a shallow shear layer, with the shear vector oriented parallel to the thermal boundary, which was unstable to small perturbations. In their study the growth rate of the instabilities was consistent with Kelvin-Helmholtz instability (KHI). The flow deformation by the thermal generated both horizontal (along the sides of the thermal) and vertical shears (along the top of the thermals), which caused the Richardson number to be sufficiently small to induce instability. Here the full form of the Richardson (Ri) number is used to take into account horizontal shears such that

$$\mathrm{Ri} = \frac{g}{\theta}\frac{\partial \theta}{\partial z}/\mathrm{Def}^2, \tag{16.1}$$

where θ is the potential temperature, g is gravity and Def ^{2} is the square of the total deformation, which in two dimensions is

$$\mathrm{Def}^2 = \frac{1}{2}\left(D_{11}^2 + D_{22}^2\right) + D_{12}^2 \tag{16.2}$$

and D_{ij} is the deformation tensor defined as

$$D_{ij} = \frac{\partial u_i}{\partial x_j} + \frac{\partial u_j}{\partial x_i} - \delta_{ij}\frac{\partial u_k}{\partial x_k}, \tag{16.3}$$

where here the tensor notation refers to $u = u_1$, $w = u_2$, $x = x_1$ and $z = x_2$. Note that the form of Ri shown in Eq. 16.1 reduces to the regular form in the absence of horizontal shear. Grabowski and Clark (1991) showed that for their thermals Ri was between 0 and 1 along the upper edge of the thermal-environment interface and negative at the edge of the thermal itself. This thermal-induced shear led to billows forming along the edge of the thermal, which was argued to be an important process governing entrainment.

The cloud-interfacial instability explains the behaviour of the simulation shown in Fig. 16.2. In particular, at the earlier times (Fig. 16.2a, b), there is a thin layer of negative Ri along the cloud interface shown by shading. Though not shown here, there is also a layer immediately outside the cloud about 50–100 m wide/deep of $0 < \mathrm{Ri} < ¼$, which is supportive of KHI. Accordingly, the billows in Fig. 16.2c form along that low-Ri layer. Although it is apparent from the isentropes in Fig. 16.2a, b that there is an increase in static stability immediately above the cloud boundary, this stability increase is overwhelmed by increases in horizontal and vertical wind shear in the same locations, which leads to small Ri. Thus, KHI along the cloud boundary of overshooting updrafts via the cloud-interfacial instability is a potential cause of NCT, albeit close to the cloud boundary. This region is highlighted in the schematic in Fig. 16.1.

The above example demonstrates that a number of processes conspire to create a turbulent region within the geometric volume that was at some time occupied by the updraft during the overshoot process. The turbulence also extends beyond the maximum height of the updraft (see more discussion in Sect. 16.3). Other three-dimensional simulations of the Dickinson case also show a turbulent region *left behind* after the updraft overshoot (see Lane et al. 2003). Moreover, Kim and Chun (2012) identify the processes following updraft collapse as a potential cause of an observed severe turbulence encounter over Japan and highlighted the potential hazard posed by dissipating convection.

Shear instabilities have also been shown to be important along the cloud boundary, but away from the main convective regions of storms. For example, Zovko-Rajak and Lane (2014) showed that supercell thunderstorms were prone to

KHI above and below their main region of upper-level outflow. Their idealised storms consistently produced strong upper-level outflow jets centred below the tropopause, with sufficiently large vertical shears that Ri < ¼ and KHI occurred. Such instabilities occurred more than 100 km from the main convective region of the storms at the upper and lower margins of the anvil (similar to the anvil region highlighted in Fig. 16.1). Some of this turbulence was outside of cloud and some within cirrus. These results had some similarities to the results of Trier and Sharman (2009), who showed that MCS upper-tropospheric outflows could enhance the environmental shear and encourage KHI (see Chap. 17 for more details); though the supercells in Zovko-Rajak and Lane's (2014) study were able to induce KHI without any environmental shear at the level of the anvil cloud.

Thus KHI induced by convective cloud-induced circulations is an important source of turbulence near the cloud boundary. In the next section, processes that generate turbulence further aloft are explained as a consequence of gravity wave instabilities.

16.3 Turbulence Above Convection

The Dickinson case study reported in Lane et al. (2003) identified the role of gravity wave breaking above convection as a potentially important source of turbulence. This region is highlighted in Fig. 16.1. Deep convective clouds generate a rich spectrum of gravity waves that can propagate horizontally and vertically within the troposphere, stratosphere and further aloft. The generation of these waves can be linked to any process that induces vertical motion of the stable air surrounding the (unstable part of the) storms, which includes diabatic heating within the cloud (e.g. Holton et al. 2002; Chun and Baik 1998), shear effects (e.g. Clark et al. 1986; Beres et al. 2002) and the so-called mechanical oscillator effect (e.g. Fovell et al. 1992; Lane et al. 2001). This latter mechanism is most relevant here because it best explains the high-frequency waves with phase lines that are closest to vertical and influence the air directly above storms.

As described by Pierce and Coroniti (1966), when a convective updraft extends vertically above its level of neutral buoyancy (LNB), it will experience a downward buoyancy force, which may force it back below its LNB, and an oscillation will ensue. This process is analogous to a simple harmonic oscillator, with the restoring force being buoyancy, and is the fundamental process underlying gravity wave formation. As shown by Lane (2008), this oscillation can occur in any stable environment, though when an updraft impinges on a more stable layer, like the tropopause or an inversion, the amplitude of the wave response can be larger. Fovell et al. (1992) and Lane et al. (2001) described this 'mechanical oscillator effect' in detail using numerical cloud model simulations.

Linear theory of gravity waves can be used to show that a simplified condition for vertically propagating gravity waves in a two-dimensional non-rotating fluid is

$$0 < |\omega - Uk| < N, \tag{16.4}$$

where ω is the wave frequency, U is the background wind, k is the horizontal wave number (2π/wavelength) and $N = \left(\frac{g}{\theta}\frac{\partial \theta}{\partial z}\right)^{1/2}$ is the Brunt-Väisälä frequency. The term $(\omega - Uk)$ is known as the intrinsic frequency, which is equivalent to the frequency of the wave measured by an observer moving with the wind U; accordingly, ω is the frequency measured by a stationary observer.

The two limits on the intrinsic frequency defined by Eq. 16.4 correspond to two different wave behaviours. If $(\omega - Uk) = 0$, a wave critical level can occur, which is equivalent to stating $(U - c) = 0$, with c equal to the horizontal phase speed. Gravity waves can become unstable and break when they encounter a critical level, which can lead to turbulence. Alternatively, if $|\omega - Uk| = N$, then the wave would be evanescent, which means its amplitude would decay exponentially with the height. Evanescence does not necessarily cause turbulence, but gravity waves can be partially reflected off an evanescent layer leading to nonlinear interactions and instabilities and/or horizontal ducting of waves, which allows them to propagate significant horizontal distances. See Nappo (2002) for more details.

As described by Lane et al. (2001), overshooting updrafts are efficient generators of high-frequency gravity waves. Typically, such waves have intrinsic frequencies at their source altitude that are close to the upper-tropospheric Brunt-Väisälä frequency, resulting in intrinsic periods close to approximately 15 min, and horizontal wavelengths of about 5–20 km. These characteristics correspond to horizontal propagation speeds that range from about 5 to 20 m s^{-1}, relative to the wind at the source, i.e. the upper troposphere. The waves with the shortest horizontal wavelengths propagate slowest. Provided the vertical shear is not too large, there is horizontal symmetry to the wave generation, and waves will propagate in all horizontal directions relative to their source. This generation property leads to circular wavefronts in the horizontal plane above the convection. However, this circular pattern can be disrupted by filtering of the vertically propagating waves, as per Eq. 16.4, by shear and stability changes.

Of particular relevance to the problem of turbulence generation above convection is the fact that for a given wave frequency, the shortest horizontal wavelength waves have the slowest horizontal phase speed magnitude. It is those waves that are most susceptible to changes in horizontal wind (due to vertical wind shear) above the wave source (viz. the overshooting updrafts). For example, if the wave has a horizontal phase speed of 5 m s^{-1}, it would encounter a critical level if the wind were to change by 5 m s^{-1} in the direction of wave propagation. Also, a wave propagating in the opposite direction to the change in wind could become evanescent (see Lane et al. 2003; Lane and Sharman 2008 for more details). These wave properties have two important implications for (usual) conditions with wind shear above the cloud top: first, they imply that the shortest wavelength gravity waves would be confined to near the cloud top with longer and faster waves further aloft; and, second, there will be a predominance of wave breaking due to critical-level interactions for those waves propagating in the same direction as the above-cloud wind shear (i.e. downshear) relative to the wind at the cloud top. It is this downshear breaking that is of relevance here because it initiates a turbulent cascade from the relatively long-wave scale down to the scales of motion that affect aircraft. Moreover, even for modest wind shear, such as a change in the horizontal wind speed of ~20 m s^{-1} above the cloud top, a significant portion of the high-frequency wave spectrum would experience a critical level.

As an example, consider results from a high-resolution three-dimensional cloud-resolving model simulation (Figs. 16.3 and 16.4). This simulation was motivated by the Dickinson encounter and uses an approximation to the observed environment to initialise the model. The model grid has 150-m spacing (in all directions) and is able to resolve the wave breaking and part of the turbulent cascade. Of relevance is that the model wind is unidirectional and has negative vertical wind shear (directed to the left of both figures) above the cloud top. See Lane and Sharman (2006) for full details of the simulation.

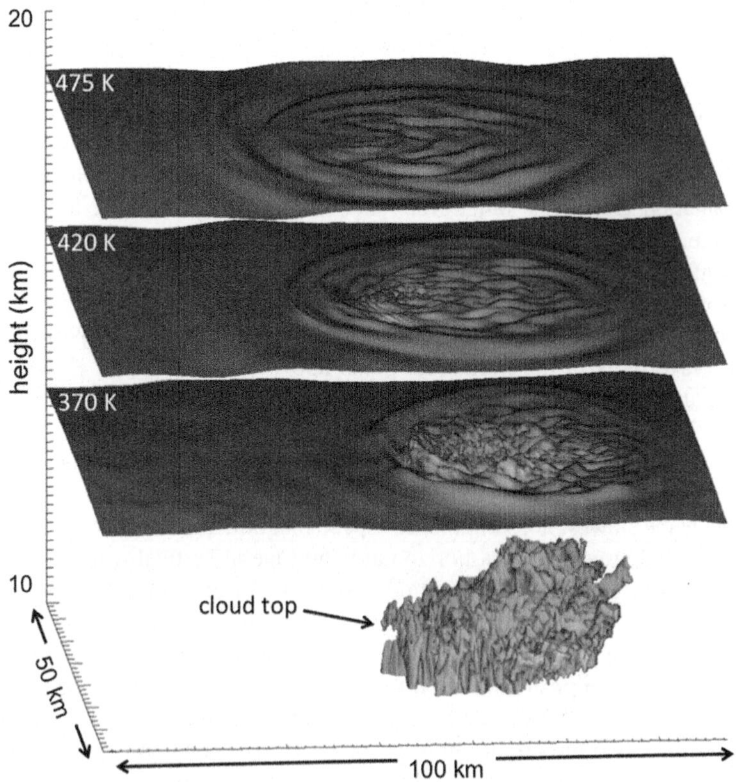

Fig. 16.3 Rendering of three-dimensional surfaces from an idealised cloud-resolving model simulation. The lower surface is cloud that extends above 10-km altitude. The three layers aloft represent the 370-, 420- and 475-K surfaces of potential temperature, which in this case have mean altitudes of approximately 13, 16 and 19 km, respectively

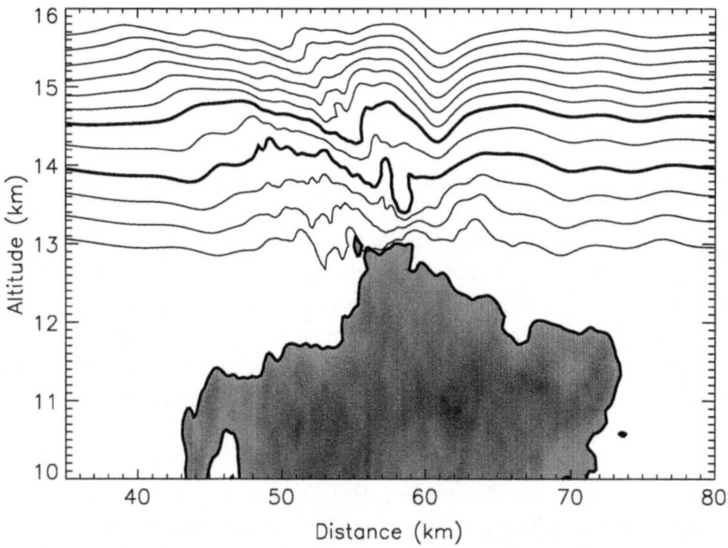

Fig. 16.4 Vertical cross-section from the idealised numerical simulation of a deep convective cloud shown in Fig. 16.3, which depicts gravity wave breaking above convection. Here cloud is shaded at mixing ratios greater than 1 g kg^{-1}; *line contours* are potential temperature at 4-K intervals with the lowest value equal to 362 K. Thicker potential temperature contours (at 374 and 382 K) are chosen to highlight regions of wave breaking

First consider the three-dimensional structure of the waves above the simulated storm (Fig. 16.3). Immediately above the storm ($\theta = 370$ K), the vertical displacements are dominated by large amplitude small-scale gravity wave signals—to the left of the wave region the surfaces are steepened, which is related to wave instability. At higher altitudes the wave signals are dominated by longer horizontal wavelengths as the vertical shear effectively filters the smaller-scale signals. At upper levels ($\theta = 475$ K) the wavefronts are approximately circular, which is consistent with the discussion above. There is a segment with reduced amplitude towards the left of the wave region; this reduced amplitude is caused by dissipation/breaking of the waves by critical-level interactions below this altitude.

A vertical cross-section of the simulation (Fig. 16.4) identifies this wave breaking immediately above the storm. Isentropes are steepened, with small regions of overturning between 13- and 15-km altitude. The steepening occurs primarily on the left (downshear) side of wavefront, and it is those downshear-propagating waves that break. This behaviour is consistent with the linear gravity wave theory discussed above and in Lane et al. (2003).

The above discussion, as well as that in Sect. 16.2, highlights the importance of the above-cloud wind shear and stability in providing a dynamical control over the occurrence of turbulence above convection. Lane and Sharman (2008) used a set of idealised model simulations to explore the sensitivity of above-cloud turbulence to these environmental conditions. Not surprisingly, they found that the volume of turbulence above cloud increased when the above-cloud static stability was reduced; this reduction leads to reduced Ri near the cloud boundary as well as more prevalent wave breaking aloft. Increasing the vertical shear increased the occurrence of turbulence within 500 m above the cloud top, which was also related to reduced Ri. However, further aloft the relationship between turbulence and shear was more complicated, with the maximum turbulence extent occurring at intermediate shears. Weak shears reduced the occurrence of wave breaking. On the other hand, strong shears lowered the height of the critical level considerably, which confined the turbulence to a shallower region above the cloud top. Among other things, these results highlighted inadequacies in the use of simple wind-speed-based guidelines for above-cloud turbulence avoidance.

It is important to note that the above discussion of above-cloud NCT considered altitudes higher than most commercial aircraft would normally fly. This is typical of warm-season storms, such as that encountered during the Dickinson case, where it would be unusual or even impossible for an aircraft to intentionally fly over the convective region of an intense system. However, in the cooler months, when the tropopause is lower and hence so are the storm tops, the above-cloud gravity wave-breaking mechanism is highly relevant. For example, Trier et al. (2012) examined cool-season above-cloud turbulence that was explained by this breaking mechanism occurring above a lowered tropopause, which confined the storm tops to well below flight level. This case is described in more detail in Chap. 17 of this book.

16.4 Estimates of Turbulence Intensity

The discussion in Sects. 16.2 and 16.3 has focused primarily on the *occurrence* of turbulence above convective clouds, that is, the processes by which instabilities are created by the storm-induced circulations (viz. shear and gravity waves). However, for a true understanding of the risks of turbulence related to convection, estimates need to be made about the *intensity* of turbulence as well as its spatial occurrence. In particular, determining the intensity of turbulence outside of cloud relative to that within cloud is of critical value for the development of improved turbulence guidelines and to quantify the spatial variation in turbulence risk. Observations from commercial or even research aircraft are insufficient to quantify this spatial variation because they are too sparse, especially within the convective regions of storms that are normally avoided for safety reasons.

Lane and Sharman (2014) conducted very high-resolution simulations of a linear mesoscale convective system with the aim of quantifying the turbulence intensity within and outside of cloud. The simulations were configured with 75-m grid spacing in all spatial directions and were therefore able to resolve part of the spectrum of turbulence that affects large aircraft (see Chap. 1). The model domain used $8000 \times 1200 \times 334$ grid points in the across-line, along-line and vertical directions, respectively, which made it highly computationally intensive. The size of the domain allowed the resolution of the mesoscale storm structure as well as the

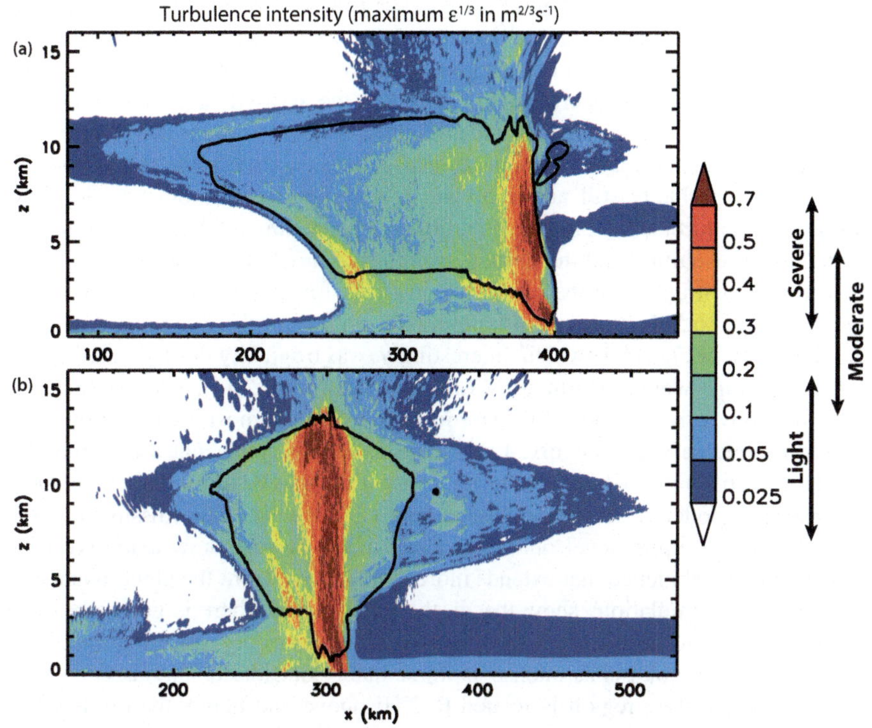

Fig. 16.5 Turbulence intensity (*coloured shading*), in terms of peak values of $\varepsilon^{1/3}$, derived from high-resolution three-dimensional simulations of idealised squall lines. The cloud boundary (*black line*) is defined as the 0.1 g kg^{-1} contour of line-averaged total cloud (liquid water and ice) mixing ratios. Two systems are simulated that used different low-level wind profiles (see text): (**a**) the tilted system and (**b**) the upright system. Also shown are subjective turbulence categories using values guided by Sharman et al. (2014)

small-scale turbulence. The simulation reported in Lane and Sharman (2014) used an idealised thermodynamic profile and a simplified wind profile, with 4 m s^{-1} km^{-1} of wind shear in the lowest 2.5-km and zero shear further aloft. This simulation produces a canonical squall line with notable upshear tilt—hereafter referred to as the ‘tilted system’. Another simulation is presented here, which uses the same model configuration but with a wind profile with a deeper shear layer: 4 m s^{-1} km^{-1} of wind shear in the lowest 5 km. This deeper shear simulation produces a storm that is more upright (referred to as the ‘upright system’), consistent with established theory of squall lines (e.g. Weisman and Rotunno 2004). The along-line-averaged structure of the cloud boundary associated with these two storms is shown in Fig. 16.5.

To quantify the turbulence in these two simulations, an approach similar to that used in the automated in situ turbulence reporting system (Chap. 5, Sharman et al. 2014) is employed here. At each height and across-line position, the model domain is separated into 19 overlapping along-line segments that are 9 km long (the model domain is 90 km long in the along-line direction). Each of these segments is analysed spectrally (see Lane and Sharman 2014 for details) to determine the cube root of the eddy dissipation rate, $\varepsilon^{1/3}$, and the maximum value of $\varepsilon^{1/3}$ is recorded for each location. This approach is analogous to obtaining the 1-min peak $\varepsilon^{1/3}$ along a flight leg, which is used operationally. Note that the across-line component of the velocity is used here because it provides slightly larger $\varepsilon^{1/3}$ values than the vertical velocity (which is normally used operationally) because even at the smallest scales the model is probably still not resolving three-dimensional turbulence properly. Moreover, the turbulence intensity is probably underestimated here because of the finite model resolution and the limited sampling of rare but intense events.

The spatial structure of the turbulence intensity for the two simulated storms is shown in Fig. 16.5. First, it is worth noting the similarities between Fig. 16.5a and the schematic in Fig. 16.1 (which, interestingly, was originally drafted years before the calculations were performed by Lane and Sharman 2014). Both simulations show the convective regions of the storms have by far the most intense turbulence, exceeding the 'severe' category. Large regions within cloud, but away from the main convective region, have 'moderate' intensities as well. Above the cloud, there is light and moderate turbulence associated with gravity wave dissipation (see Sect. 16.3). The upper-level outflow regions of the storms show extensive light and moderate turbulence that extends more than 100 km from the cloud boundary. Moreover, the simulations show that in the anvil region, there is generally not a distinct change in turbulence intensity at the cloud boundary, meaning the turbulence within the cloud is as intense as immediately outside the cloud. The turbulence in this outflow region is related to KHI above and below the outflow jets (e.g. see Zovko-Rajak and Lane 2014) and other instabilities within the outflow (e.g. see Chap. 17).

The lateral extent of the upper-tropospheric turbulence is sensitive to the storm structure and the tilt of the main updraft core, with the two simulated storms showing distinctly different patterns of turbulence. The tilted storm shows limited turbulence in the upper outflow ahead (to the right) of the main convective region, with turbulence dominant to the rear, whereas the upright system has a more symmetric pattern of turbulence in its upper outflow. Finally, the tilted storm (Fig. 16.5a) shows moderate to severe turbulence in a layer below the anvil, which was also highlighted in Fig. 16.1. This turbulence may be related to shear instabilities associated with the storm's rear-flank mesoscale downdraft or caused by thermodynamic instabilities associated with sublimation/evaporation (similar to those instabilities that may help produce mammatus clouds, e.g. Schultz et al. 2006).

16.5 Summary

This chapter has considered the processes underlying the generation of near-cloud turbulence (NCT), with specific focus on the turbulence generated by KHI near the cloud boundary and gravity wave breaking further aloft. The regions adjacent to storms are also prone to turbulence, as demonstrated by Fig. 16.5, which is linked to KHI and gravity waves. Turbulence adjacent to convection is considered in more detail in the Chap. 17 of this book. Identification of these processes represents a significant advance in recent understanding of NCT generation, which has been made possible by improved observing systems and numerical modelling capabilities.

Recent studies on NCT have, among other things, demonstrated inadequacies in current methods for near-cloud turbulence tactical avoidance. The ultimate goals of work in this area are to improve turbulence avoidance guidelines and also improve predictions of the spatial extent and intensity of turbulence related to thunderstorms. Better guidelines require enhanced knowledge of how the turbulence generation is related to conditions like storm intensity, environmental wind shear and environmental stability. Better forecasts require a combination of ongoing development of numerical weather prediction systems to improve the prediction of deep convection and new methods to relate aspects of the numerically predicted flow to (unresolved) aircraft turbulence.

The ongoing challenge to study NCT and convectively induced turbulence in general is related to the broad range of scales involved. Modelling studies, like those shown in Figs. 16.2–16.5, must resolve the mesoscale motions associated with the convective systems as well as those turbulent scales of motion that affect aircraft. This still poses a challenge, even for research studies, and numerical model experiments can push the limits of modern high-performance computing. Nonetheless, numerical modelling and state-of-the-art observations offer much promise for ongoing improvements to understanding and characterisation of convectively induced turbulence, which should eventually lead to improved avoidance and prediction methods.

Acknowledgments Parts of this work were supported by the Australian Research Council through their Discovery Projects (DP0770381), Future Fellowships (FT0990892) and Centres of Excellence (CE110001028) schemes. Supercomputing resources from the National Computational Infrastructure Facility in Canberra, Australia, are also gratefully acknowledged.

References

Beres, J.H., Alexander, M.J., Holton, J.R.: Effects of tropospheric wind shear on the spectrum of convectively generated gravity waves. J. Atmos. Sci. **59**, 1805–1824 (2002)

Bryan, G.H., Wyngaard, J.C., Fritsch, J.M.: Resolution requirements for the simulation of deep moist convection. Mon. Weather Rev. **131**, 2394–2416 (2003)

Chun, H.-Y., Baik, J.-J.: Momentum flux by thermally induced internal gravity waves and its approximation for large-scale models. J. Atmos. Sci. **55**, 3299–3310 (1998)

Clark, T.L., Hauf, T., Kuettner, J.P.: Convectively forced internal gravity waves: Results from two-dimensional numerical experiments. Quart. J. Roy. Meteorol. Soc. **112**, 899–925 (1986)

Federal Aviation Administration: Advisory Circular AC 00-24 – Thunderstorms. Available from: www.faa.gov or 800 Independence Avenue, SW Washington, DC 20591 (1968)

Federal Aviation Administration: Aeronautical Information Manual (version dated 11 February 2010). Available from: www.faa.gov 800 Independence Avenue, SW Washington, DC 20591 (2010)

Federal Aviation Administration: Advisory Circular AC 00-24C – Thunderstorms. Available from: www.faa.gov or 800 Independence Avenue, SW Washington, DC 20591 (2013)

Federal Aviation Administration: Aeronautical Information Manual (version dated 8 January 2015). Available from: www.faa.gov or 800 Independence Avenue, SW Washington, DC 20591 (2015)

Fovell, R., Durran, D., Holton, J.R.: Numerical simulations of convectively generated stratospheric gravity waves. J. Atmos. Sci. **49**, 1427–1442 (1992)

Fovell, R.G, Sharman, R.D., Trier, S.B.: A case study of convectively-induced clear-air turbulence. Proceedings of the 12th Conference on Mesoscale Processes, Amer. Meteor. Soc., Waterville Valley, N.H., 6–9 Aug 2007, Paper 13.4, 4pp (2007)

Grabowski, W.W., Clark, T.L.: Cloud environment interface instability: rising thermal calculations in two spatial dimensions. J. Atmos. Sci. **48**, 527–546 (1991)

Holton, J.R., Beres, J.H., Zhou, X.: On the vertical scale of gravity waves excited by localized thermal forcing. J. Atmos. Sci. **59**, 2019–2023 (2002)

Kim, J.-H., Chun, H.-Y.: A numerical simulation of convectively induced turbulence above deep convection. J. Appl. Meteor. Climatol. **51**, 1180–1200 (2012)

Lane, T.P.: The vortical response to penetrative convection and the associated gravity-wave generation. Atmos. Sci. Lett. **9**, 103–110 (2008)

Lane, T.P., Sharman, R.D.: Gravity wave breaking, secondary wave generation, and mixing above deep convection in a three-dimensional cloud model. Geophys. Res. Lett. **33**, L23813 (2006). doi:10.1029/2006GL027988

Lane, T.P., Sharman, R.D.: Some influences of background flow conditions on the generation of turbulence due to gravity wave breaking above deep convection. J. Appl. Meteor. Climatol. **47**, 2777–2796 (2008)

Lane, T.P., Sharman, R.D.: Intensity of thunderstorm-generated turbulence revealed by large-eddy simulation. Geophys. Res. Lett. **41**, 2221–2227 (2014). doi:10.1002/2014GL059299

Lane, T.P., Reeder, M.J., Clark, T.L.: Numerical modeling of gravity wave generation by deep tropical convection. J. Atmos. Sci. **58**, 1249–1274 (2001)

Lane, T.P., Sharman, R.D., Clark, T.L., Hsu, H.-M.: An investigation of turbulence generation mechanisms above deep convection. J. Atmos. Sci. **60**, 1297–1321 (2003)

Lane, T.P., Sharman, R.D., Trier, S.B., Fovell, R.G., Williams, J.K.: Recent advances in the understanding of near-cloud turbulence. Bull. Am. Meteorol. Soc. **93**, 499–515 (2012)

Lester, P.F.: Turbulence: A New Perspective for Pilots. Jeppeson Sanderson, Englewood, CO (1993)

Morton, B.R., Taylor, G., Turner, J.S.: Turbulent and gravitational convection from maintained and instantaneous sources. Proc. Roy. Soc. London **A234**, 1–23 (1956)

Nappo, C.J.: An Introduction to Atmospheric Gravity Waves. Academic Press, San Diego, CA (2002)

Pierce, A.D., Coroniti, S.C.: A mechanism for the generation of acoustic-gravity waves during thunderstorm formation. Nature **210**, 1209–1210 (1966)

Schultz, D.M., Kanak, K.M., Straka, J.M., Trapp, R.J., Gordon, B.A., Zrnić, D.S., Bryan, G.H., Durant, A.J., Garrett, T.J., Klein, P.M., Lilly, D.K.: The mysteries of mammatus clouds: observations and formation mechanisms. J. Atmos. Sci. **63**, 2409–2435 (2006)

Sharman, R.D., Cornman, L.B., Meymaris, G., Pearson, J., Farrar, T.: Description and derived climatologies of automated in situ eddy-dissipation-rate reports of atmospheric turbulence. J. Appl. Meteor. Climatol. **53**, 1416–1432 (2014)

Trier, S.B., Sharman, R.D.: Convection-permitting simulations of the environment supporting widespread turbulence within the upper-level outflow of a mesoscale convective system. Mon. Weather Rev. **137**, 1972–1990 (2009). doi:10.1175/2008MWR2770.1

Trier, S.B., Sharman, R.D., Fovell, R.G., Frehlich, R.G.: Numerical simulation of radial cloud bands within the upper-level outflow of an observed mesoscale convective system. J. Atmos. Sci. **67**, 2990–2999 (2010)

Trier, S.B., Sharman, R.D., Lane, T.P.: Influences of moist convection on a cold-season outbreak of clear-air turbulence (CAT). Mon. Weather Rev. **140**, 2477–2496 (2012)

Weisman, M.L., Rotunno, R.: A theory for strong long-lived squall lines. Revisited. J. Atmos. Sci. **61**, 361–382 (2004)

Zovko-Rajak, D., Lane, T.P.: The generation of near-cloud turbulence in idealized simulations. J. Atmos. Sci. **71**, 2430–2451 (2014)

Chapter 17
Modeling Studies of Turbulence Mechanisms Associated with Mesoscale Convective Systems

Stanley B. Trier

Abstract High-resolution numerical simulations have aided recent understanding of mechanisms responsible for convectively induced aviation turbulence (CIT) spatially removed from its parent moist convection. Anticyclonic jets occurring near the tropopause, which are often significantly enhanced by mesoscale convective systems or other forms of organized convection, influence a variety of different mechanisms of turbulence at commercial aviation cruising altitudes. Herein, we describe different CIT mechanisms that are possible near these anticyclonic outflow jets and illustrate them with examples from recent simulations.

17.1 Introduction

Climatological studies indicate that commercial aviation turbulence encounters frequently occur in environments near deep convection (e.g., Kaplan et al. 2005; Wolff and Sharman 2008). In situations where deep convection has experienced upscale growth into mesoscale convective systems, related turbulence is typically more widespread and therefore more difficult to avoid.

17.1.1 Mesoscale Convective Systems

A mesoscale convective system (MCS) is defined as "a cloud system that occurs in connection with an ensemble of thunderstorms and produces a contiguous precipitation area on the order of 100 km or more in horizontal scale in at least one

S.B. Trier (✉)
Mesoscale and Microscale Meteorology Laboratory, National Center for Atmospheric Research, Boulder, CO, USA
e-mail: trier@ucar.edu

R. Sharman, T. Lane (eds.), *Aviation Turbulence*,
DOI 10.1007/978-3-319-23630-8_17

direction" (American Meteorological Society 2014). Particularly large and intense MCSs, termed mesoscale convective complexes (MCCs), are recognizable by their nearly circular satellite signatures of cold cloud top temperature (Maddox 1980). Large MCSs (including MCCs) typically have three distinct mesoscale circulations that occupy different layers of the troposphere. These circulations include both near-surface and upper-tropospheric anticyclones and a region of midtropospheric cyclonic relative vorticity (Maddox 1983; Cotton et al. 1989).

When present, the MCS-induced midtropospheric vortex (MCV) is often the longest lasting of these three circulation components and can influence aviation turbulence through its role in either allowing deep convection to persist (Raymond and Jiang 1990) or reinitiating deep convection after the dissipation of the parent MCS (Trier and Davis 2007). However, the typically shorter-lived lower- and upper-tropospheric anticyclones have greater direct impacts on aviation turbulence.

The lower-tropospheric anticyclone is a hydrostatic consequence of the cool, dense air mass driven by evaporative cooling underneath thunderstorms. The most significant hazards with this mesoscale feature are (1) strong convection located horizontally within a few kilometers of the leading edge of the cold outflow and (2) Kelvin-Helmholtz instability (Sect. 17.3.1.1) along the upper-surface of the cold outflow, where the vertical shear can be particularly strong. The cold outflow is usually confined to the lowest few kilometers above the surface and is therefore typically most hazardous to commercial aviation during takeoffs and landings.

The upper-tropospheric anticyclone occurs at the MCS detrainment levels (roughly 250–175 hPa or 34–42 kft), which are near the height of the warm-season midlatitude tropopause. Of the three different MCS circulation components, the upper-level anticyclone has the largest horizontal scale, approaching up to 1000 km in diameter for the largest MCSs and MCCs (e.g., Fritsch and Maddox 1981). Turbulence can occur anywhere within the upper-level outflow, and, though it is often convectively induced, it may be located several hundreds of kilometers away from active deep convection. This latter aspect of the turbulence can make it difficult to avoid and can lead to it being classified (perhaps mistakenly) as clear-air turbulence (CAT). In the remainder of this chapter, we focus on different mechanisms leading to convectively induced turbulence (CIT) within upper-tropospheric/lower-stratospheric (UTLS) anticyclonic outflows of simulated warm- and cold-season convective systems, which often coincide with commercial aviation cruising altitudes.

17.1.2 Research Simulations

Recent advances in understanding mechanisms responsible for turbulence generation within convectively induced UTLS outflows have come from high-resolution simulations (Lane et al. 2012; Sharman et al. 2012). Such research simulations are computationally demanding since model domains need be large enough to adequately represent the synoptic environments conducive to MCS formation while simultaneously having horizontal and vertical grid spacings small enough to explicitly resolve deep convection. Still, present convection-permitting simulations are typically too coarse to resolve all but the largest horizontal scales of motion felt by commercial aircraft, which lie within the range of 100 m to 2 km (Lane et al. 2012, Chap. 16). Fortunately, the finest horizontal grid spacings in such convection-permitting simulations of $\Delta = 300$ m to $\Delta = 3$ km are often capable of resolving some of the different types of structures that can directly lead to the turbulence that affects aircraft.

As yet, though such models can produce realistic simulations that agree qualitatively with observed events, they do not consistently produce reliable deterministic forecasts of deep convection, which constitutes a major limitation in their use for operational forecasting of turbulence and suggests that ensembles are needed for future prediction efforts. However, by containing explicit deep convection at high resolution, these models are capable of simulating observed turbulence mechanisms and may therefore be used in a diagnostic capacity for physical process studies that further understanding of turbulence generation and when it may occur. This is the approach taken in the remainder of this chapter where we use several different simulations to illustrate a variety of CIT mechanisms associated with both warm- and cold-season episodes of organized midlatitude deep convection. Knowledge of these mechanisms and the more predictable larger-scale environmental conditions under which they are favored can be used in development of statistical or empirically based turbulence forecasting systems (Ellrod and Knapp 1992; Sharman et al. 2006).

17.2 MCS Anticyclonic Outflows and the Turbulence Environment

17.2.1 Climatological Aspects

Fritsch and Maddox (1981) showed how the MCS upper-level anticyclonic outflow leads to an enhanced jet north and northeast of its origination region and diminished winds to its south and southwest for a composite of ten cases in the central USA. This relationship between the MCS and its upper-level outflow within the midlatitude northern hemisphere is illustrated schematically in Fig. 17.1.

It is commonly observed that the strongest winds in MCS upper-level outflows (gray shading in Fig. 17.1a) occur up to several hundreds of kilometers from their origin in organized deep convection. Such increases in wind speed along outflow trajectories result from initially unbalanced horizontal pressure gradient forces (e.g., Fritsch and Brown 1982; Wolf and Johnson 1995; Trier and Sharman 2009) that are influenced by detrainment of cold air above the level of neutral buoyancy in deep convection, which helps produce the upper-level anticyclone. Significant directional changes in the outflow are influenced by Coriolis effects acting over

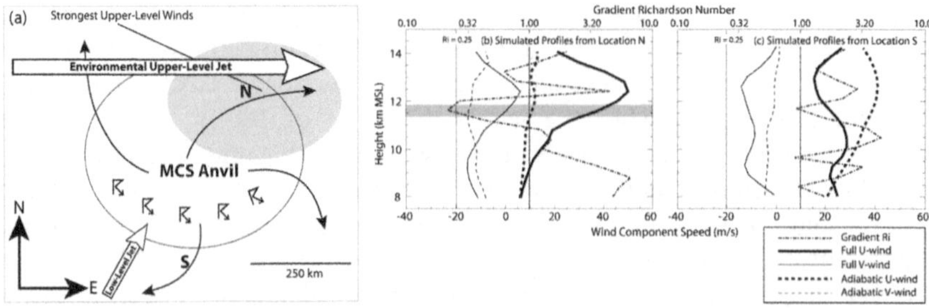

Fig. 17.1 (**a**) Schematic diagram illustrating spatial relationships among the organized regions of deep convection (*thunderstorm symbols*), the MCS anvil cloud (*thin ellipse*) and divergent upper-tropospheric outflow streamlines (*thick curves* with *arrowheads*), the environmental flow at the upper-tropospheric outflow level and low-level jet, and the location of the strongest net upper-level winds (*shading*). (**b**), (**c**) Ninety-minute time-averaged vertical profiles of zonal and meridional wind components from a full-physics simulation (*solid lines*), a simulation where temperature changes due to latent heating are eliminated (*dashed lines*), and the gradient Richardson number from the full-physics simulation (*dash-dotted line*) from locations on the north and south side of the MCS upper-tropospheric outflow marked on part (**a**) by N and S, respectively. The vertical profiles in parts (**b**) and (**c**) are from a simulation of the 16–17 June 2005 convectively induced turbulence case, and their locations relative to the observed MCS and turbulence reports are also indicated on Fig. 17.3b by the annotations N and S. Adapted from Trier and Sharman (2009)

several hours. For southerly outflow trajectories emanating from the line of thunderstorms depicted in Fig. 17.1a, the Coriolis effect acting alone will produce westerly flow over a timescale of $T/4 = \pi/4\Omega \sin\phi$, where T, ϕ, and Ω are the inertial period, latitude, and the earth's rotation rate, respectively.[1] The westerly component of the outflow may be further augmented by its superposition with the background westerly flow frequently observed near the tropopause (Fig. 17.1a).

The details of how the outflow and base-state horizontal winds superpose in individual cases depend on the synoptic pattern and MCS structural characteristics. However, the consistency of the scenario in Fig. 17.1a in an overall climatological sense is supported by the spatiotemporal relationship between rainfall and 200-hPa zonal wind for the heaviest 25 % of rain events for four warm seasons over the central USA (Fig. 17.2a), which are more generally dominated by large nocturnal MCSs (Fritsch et al. 1986). Here, the maximum westerlies are located ≈5° latitude north of the heaviest rainfall, for which peak values precede the maximum winds by several hours. The maximum vertical shear through the 200-hPa-deep layer immediately below the 200-hPa jet is similarly displaced north of the heaviest precipitation (Fig. 17.2b).

Environmental vertical shear of the horizontal wind, which may impact aviation turbulence in a variety of different ways (Sect. 17.3), has a temporal maximum in its zonal component that lags the heaviest rainfall by even a few hours more than the maximum in zonal wind (Fig. 17.2b) does. This phase lag between the maximum

[1] At 40° latitude $T/4 = 4.6$ h.

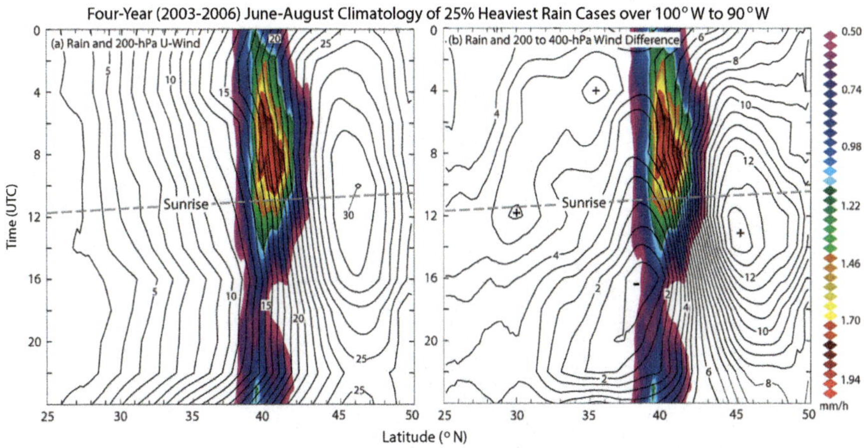

Fig. 17.2 Time-longitude diagram of the diurnal cycle of rainfall and (**a**) Rapid Update Cycle (RUC, Benjamin et al. 2004) analysis 200-hPa zonal winds (1-m s^{-1} contour interval) and (**b**) magnitude of RUC analysis 200–400-hPa vector wind difference (0.5-m s^{-1} contour interval) for the 25th percentile of heaviest rainfall cases during the warm seasons (1 June–31 August) of 2003 through 2006. The rainfall is derived using the WSI Corporation NOWrad national radar composite from the US National Weather Service (WSR-88D) Doppler radar network at 15-min frequency and applying the reflectivity-rain relationship ($Z = 300R^{1.5}$), where Z is the reflectivity factor ($mm^6\ m^{-3}$) and R is the rainfall rate ($mm\ h^{-1}$). Both the rainfall amounts and RUC-analyzed fields are longitudinally averaged between 90° and 100°W and are then shifted in latitude for individual rainfall cases so that longitudinally averaged rainfall maximum occurs at 40°N latitude. From Trier and Sharman (2009). (**c**) American Meteorological Society. Reprinted with permission

deep convection and vertical shear strength is potentially significant for commercial aviation over the central USA since the relatively strong vertical shear persists well after sunrise (Fig. 17.2b), when air traffic is increasing.

17.2.2 An Example Case (16–17 June 2005)

Figure 17.3a is time-latitude Hovmöller diagram that illustrates similar spatiotemporal relationships between the 100 and 90°W averaged rainfall and the magnitude of the vector vertical shear between 200 and 300 hPa for individual nocturnal MCSs on successive nights. However, the magnitude of the maximum vertical shear ($24\,\mathrm{ms}^{-1}(100\,\mathrm{hPa})^{-1}$ or $\sim 1 \times 10^{-2}\ \mathrm{s}^{-1}$) is significantly stronger than during the 4-year climatological period (Fig. 17.2b). This is partly related to artifacts from averaging in the climatology since near the edge of MCS anvils (where the vertical shear is maximized) the outflow jet and associated vertical shear typically occupy a depth of only a few km, which is comparable to the depth of variation in the altitude of this jet. The intensity and horizontal scale of the outflows in the MCSs of

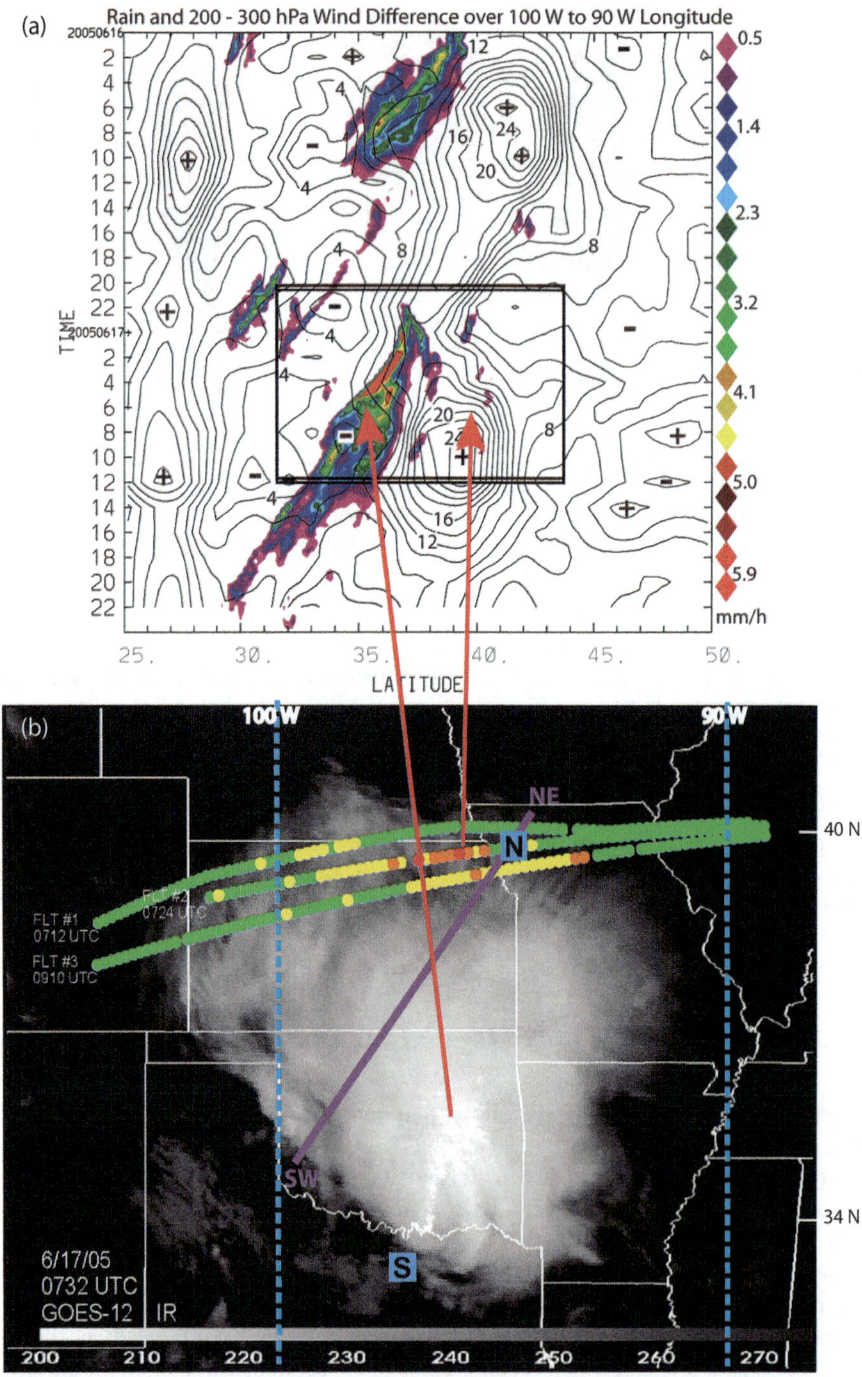

Fig. 17.3 (**a**) Time-latitude diagram of rainfall (color scale) and Rapid Update Cycle (RUC, Benjamin et al. 2004) analyzed 200–300-hPa vector wind difference magnitude (2-m s^{-1} contour interval), each longitudinally averaged from 100° to 90°W (*dashed blue lines* in part **b**). The

Fig. 17.3a and the additional vertical shear contributions from the meridional wind component also influence the comparison with climatology (Fig. 17.2b).

Commercial aviation turbulence occurred north of the MCS precipitation on both nights and is illustrated during the second night in Fig. 17.3b. Moderate-to-severe turbulence (indicated by the orange and red dots in Fig. 17.3b), diagnosed from in-flight estimates of eddy dissipation rate (EDR; Sharman et al. 2014, Chap. 5), is evident at the northern edge of the MCS cloud shield and is situated close to the region of the strongest longitudinally averaged vertical shear (Fig. 17.3a). Note that the turbulence is located several hundred kilometers north of where the heaviest hourly rain rates occur and lags the period of most intense rainfall (associated with the strongest deep convection) by several hours (Fig. 17.3). This is consistent with the turbulence being connected to earlier deep convection through physical processes occurring in its anticyclonic upper-level outflow.

Trier and Sharman (2009) isolated the role of the MCS on the vertical shear near the reported turbulence with a high-resolution numerical model having horizontal grid spacings of 3 km, using both "adiabatic" (latent heating effects were excluded) and full-physics simulations for the spatiotemporal region indicated by the rectangle in Fig. 17.3a. Comparison of Fig. 17.1b with 17.1c for the full-physics simulation reveals much stronger westerlies near the 12.5-km outflow jet level along the northeast edge of the MCS cloud shield (location N in Figs. 17.1a and 17.3b) than at its southern edge (location S in Figs. 17.1a and 17.3b). Furthermore, comparison of the bold-solid and bold-dashed curves in Fig. 17.1b indicates that the strong westerlies at this level result almost entirely from moist-convective processes (i.e., latent heating) excluded in the "adiabatic" run. Strong zonal and meridional vertical shears are maximized beneath the outflow jet in this full-physics simulation and contribute to small gradient Richardson numbers (Ri $\approx$ 0.25) at altitudes where the observed turbulence is indicated by the shading in Fig. 17.1b. In contrast, Ri (defined in Sect. 17.3.1.1) is larger on the south side of the MCS cloud shield where horizontal winds (solid curves) and their associated vertical shears are weaker (Fig. 17.1c).

Ri also depends on the moist static stability N_{m}^2 (Sect. 17.3), which is reduced in MCS anvils from that of the background environment. At the altitude of the outer

Fig. 17.3 (continued) annotations N and S indicate positions of the simulated time-averaged vertical profiles presented in Fig. 17.1b, c, respectively. The transect SW-NE indicates the position of the vertical cross section in Fig. 17.4, which is averaged for 75 km on each side of this line. Hourly rainfall is calculated as in Fig. 17.2. (**b**) Eddy dissipation rate (EDR, $\varepsilon^{1/3}$) automated turbulence measurements along the tracks of three commercial aircrafts (see Trier and Sharman 2009 for details) superposed on Geostationary Operational Environmental Satellite-12 (*GOES-12*) infrared satellite imagery from 0732 UTC 17 June 2005. Each *circle* is a color-coded peak EDR measurement during a 1-min interval of flight time, where *green* is EDR < 0.1, *yellow* 0.1–0.2, *orange* 0.2–0.3, and *red* 0.3–0.4 $\mathrm{m}^{2/3}\ \mathrm{s}^{-1}$, which are categorized as smooth, light, moderate, and moderate-to-severe turbulence, respectively. The *red lines* with *arrows* indicate the latitudinal positions of the longitudinally averaged quantities in (**a**) corresponding to heavy convective rainfall (southern tail) and observed turbulence (northern tail). Adapted from Trier and Sharman (2009)

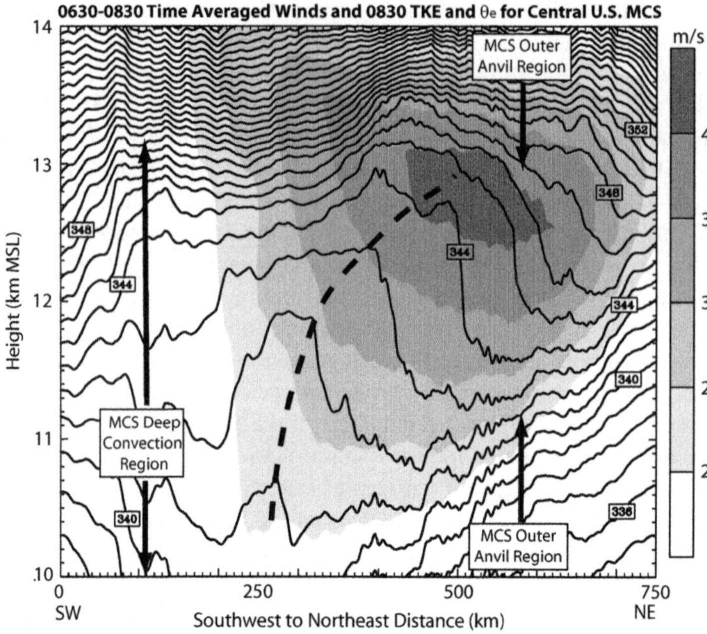

Fig. 17.4 Vertical cross section from the full-physics numerical simulation reported in Trier and Sharman (2009) averaged for 75 km on each side of line SW-NE in Fig. 17.3b. The simulated winds parallel to the cross section are time averaged between 0630 and 0800 UTC 17 June 2005 and are shaded. The simulated 0830 UTC 17 June 2005 equivalent potential temperature is contoured in 1-K intervals. The *bold-dashed line* highlights the vertically and horizontally displaced equivalent potential temperature surfaces discussed in the text. Adapted from Lane et al. (2012)

MCS anvil, N_m^2 is approximately proportional to the lapse rate of equivalent potential temperature $\partial\theta_e/\partial z$.[2] Trier and Sharman (2009) describe how θ_e surfaces are displaced upward by vertical motions in the anvil that lag pulsations in the strength of upstream convection. Figure 17.4 shows evidence of these vertical displacements beginning slightly above $z = 10$ km near $x = 250$ km. Such θ_e perturbations were noted by Trier and Sharman (2009) to have periodic temporal variations (see their Fig. 16) and may be associated with thermally (e.g., Pandya and Durran 1996) or mechanically (e.g., Fovell et al. 1992) forced mesoscale gravity waves related to the upstream convection. The upward displaced isentropes indicated by the bold-dashed line in Fig. 17.4 are then steepened by differential advection on their downshear (northeast) side, which is facilitated by the strong vertical shear in the outer portion of the MCS anvil (Fig. 17.4). There, the combination of strong shear and small static stability contributes to low Ri (cf. Fig. 17.1b),

[2] Durran and Klemp (1982) emphasize that this approximation is valid only in the upper troposphere/lower stratosphere, where dry- and moist-adiabatic lapse rates are nearly equal.

representing one of the ways that strong vertical shear and reduced static stability act cooperatively in producing mesoscale environments favorable for turbulence.

17.3 Turbulence Mechanisms in Convectively Induced UTLS Anticyclonic Outflows

In the previous section, we demonstrated how convectively induced UTLS outflows can modify the mesoscale environment in a manner that supports turbulence in regions that are sometimes significantly removed from active deep convection. We now summarize mechanisms that can be directly responsible for turbulence in such environments (Sect. 17.3.1) and illustrate them using examples from high-resolution simulations of observed CIT during the midlatitude warm (Sect. 17.3.2) and cold (Sect. 17.3.3) seasons. These mechanisms consist of instabilities and gravity waves and combinations of the two in some cases. More complete treatments of these phenomena are found in other chapters of this volume that are dedicated to waves and instabilities.

17.3.1 Overview of CIT Mechanisms

17.3.1.1 Instability Mechanisms

Kelvin-Helmholtz instability (KHI) may be the most common explanation for classical CAT (e.g., Dutton and Panofsky 1970). KHI, which has additionally been linked to CIT in numerical simulations (e.g., Trier et al. 2012; Zovko-Rajak and Lane 2014), is possible in three-dimensional flow when values of the gradient Richardson number,

$$\mathrm{Ri} = N^2/|\partial\mathbf{V}/\partial z|^2, \tag{17.1}$$

are $0 \leq \mathrm{Ri} \leq 1$ (Miles 1986). It is clear from Eq. (17.1) that strong vertical shear $|\partial\mathbf{V}/\partial z| \gg 0$ forces this instability, while strong static stability $N^2 = g/\theta(\partial\theta/\partial z) \gg 0$ opposes it. Knox (1997) illustrates that for a given geostrophic flow strength, the vertical shear is enhanced in gradient-balanced anticyclones (see his Fig. 1). Though Knox (1997) appropriately cautions that particularly strong and curved anticyclonic jets are likely to violate gradient balance, there have been numerous observations of strong vertical shear near anticyclonic MCS outflow jets (e.g., Fig. 17.3). Thus, it seems a reasonable expectation that the outer regions of large MCS anvils that are in proximity to anticyclonic UTLS outflow jets could be favorably predisposed to KHI. Under atmospheric conditions of ice or water saturation, overturning billow clouds (Scorer 1969) that are oriented normal to the vertical shear may provide visual evidence of KHI. In this situation, the instability criterion is obtained by replacing the dry static stability in Eq. (17.1) with the moist static stability N_m^2 (Durran and Klemp 1982).

Figure 17.4 from the previous section (Sect. 17.2.2) illustrated conditions of near moist static instability $N_m^2 \approx g/\theta_e(\partial\theta_e/\partial z) < 0$ in the layer underneath the jet maximum within the outer region of the MCS anvil cloud from 11.5- to 12.5-km MSL. In such layers, which are typically only a few km deep, shallow convective overturning that leads to turbulence can occur.

Such instabilities, where N^2 or $N_m^2 < 0$, can be either entirely convective or mixed dynamic convective. When statically unstable conditions coincide with significant vertical shear, the mixed dynamic-convective instability known as *thermal-shear instability* may occur (e.g., Asai 1970, 1972), which manifests as horizontal convective roll (HCR) vortices whose longitudinal axis is aligned along the vertical shear vector.

Thermal-shear instability is common in the strongly heated daytime PBL, with shallow cumulus "cloud streets" along the updraft zone between two such counterrotating vortices providing visual evidence of the instability when sufficient moisture is present (e.g., Houze 2014, their Fig. 1.1.0). Unlike for billow clouds arising from KHI, HCR cloud bands are aligned along, rather than normal to, the vertical shear. The horizontal spacing between HCR bands is typically on the order of, or slightly greater than, the depth of the neutral to statically unstable layer in which they form (LeMone 1973). Recent high-resolution simulations (Trier et al. 2010; Kim et al. 2014) discussed in Sect. 17.3.2 point toward thermal-shear instability as a possible explanation for turbulence occurring near or within banded cirrus often seen at outer edges of MCS anvils (Lenz et al. 2009).

In inertially stable regions, horizontally displaced air parcels return to their original position due to the conservation of absolute momentum, $M = fy - u$. In the northern hemisphere (where $f > 0$), the criterion for *inertial instability* is

$$\left(-\frac{\partial V}{\partial n}+\frac{V}{R_S}\right)+f<0, \tag{17.2}$$

where the term in the parentheses is the vertical component of relative vorticity and f is the planetary vorticity. In Eq. (17.2), which is expressed in natural coordinates, V is the tangential wind speed, the n direction is normal to and left of the flow, and the radius of curvature of the streamlines R_S is negative (positive) for flows with anticyclonic (cyclonic) curvature and is infinite for perfectly straight flow. Thus, in the northern (southern) hemisphere, inertial instability is most likely on the right (left) side of UTLS MCS outflow jets when the horizontal shear and/or flow curvature is particularly strong. Knox (1997) noted the linkage in earlier observational investigations between CAT and such strongly anticyclonic flows.

However, the physical processes through which inertial instability may lead to aviation turbulence have not been well documented and are, at present, poorly understood. Inertia-gravity waves emitted as part of an adjustment process to restore balance in inertially unstable flows represent one possible link in such a physical process. Ciesielski et al. (1989) found disturbances having horizontal wavelengths of ≈400 km that propagated along the outer edge of a broader cirrus cloud canopy in inertially unstable anticyclonic flow conditions. These horizontal wavelengths are one to two orders of magnitude larger than those typically associated with the shearing ($0 < \mathrm{Ri} < 1$) and convective or mixed dynamic-convective ($\mathrm{Ri} < 0$) instabilities discussed earlier.

17.3.1.2 Internal Gravity Waves

Convectively generated internal gravity waves can contribute, through a variety of different mechanisms, to the development of turbulence at locations remote from their source region within MCS circulations. Internal gravity waves that help produce turbulence have been simulated in high-resolution models and may exhibit either vertical- or horizontal-energy propagation depending on environmental conditions.

Horizontally propagating internal gravity waves require an environmental wave duct in order to maintain their coherence as they propagate away from their convective source region. This can occur when

$$\frac{N^2}{(U-c)^2} - \frac{U_{zz}}{U-c} - k^2 \tag{17.3}$$

becomes negative within a nearby layer, which limits vertical propagation of the wave. In (17.3) N is the Brunt-Väisälä frequency, U is the horizontal flow in the plane of the wave propagation, k is the horizontal wave number, and c is the horizontal phase speed of the wave. Such waves, which are often described as "vertically trapped," have been hypothesized to aid turbulence generation outside of deep convection by locally reducing Ri in simulated environments that are initially close to permitting KHI (e.g., Lane et al. 2012, Chap. 15).

Vertically propagating internal gravity waves can similarly help excite KHI in adjacent low Ri environments (e.g., Trier et al. 2012). These vertically propagating waves may also play a more direct role in the onset of turbulence by breaking upon reaching a critical level, where $U = c$. Such waves are often triggered by deep convective updrafts impinging on the stable stratosphere above (e.g., Fovell et al. 1992; Lane et al. 2003). Lane and Sharman (2008) illustrate how turbulence intensity arising from the breaking of these vertically propagating waves depends on the strength and depth of vertical shear layers and static stability and possibly other factors, making development of simple turbulence guidelines elusive.

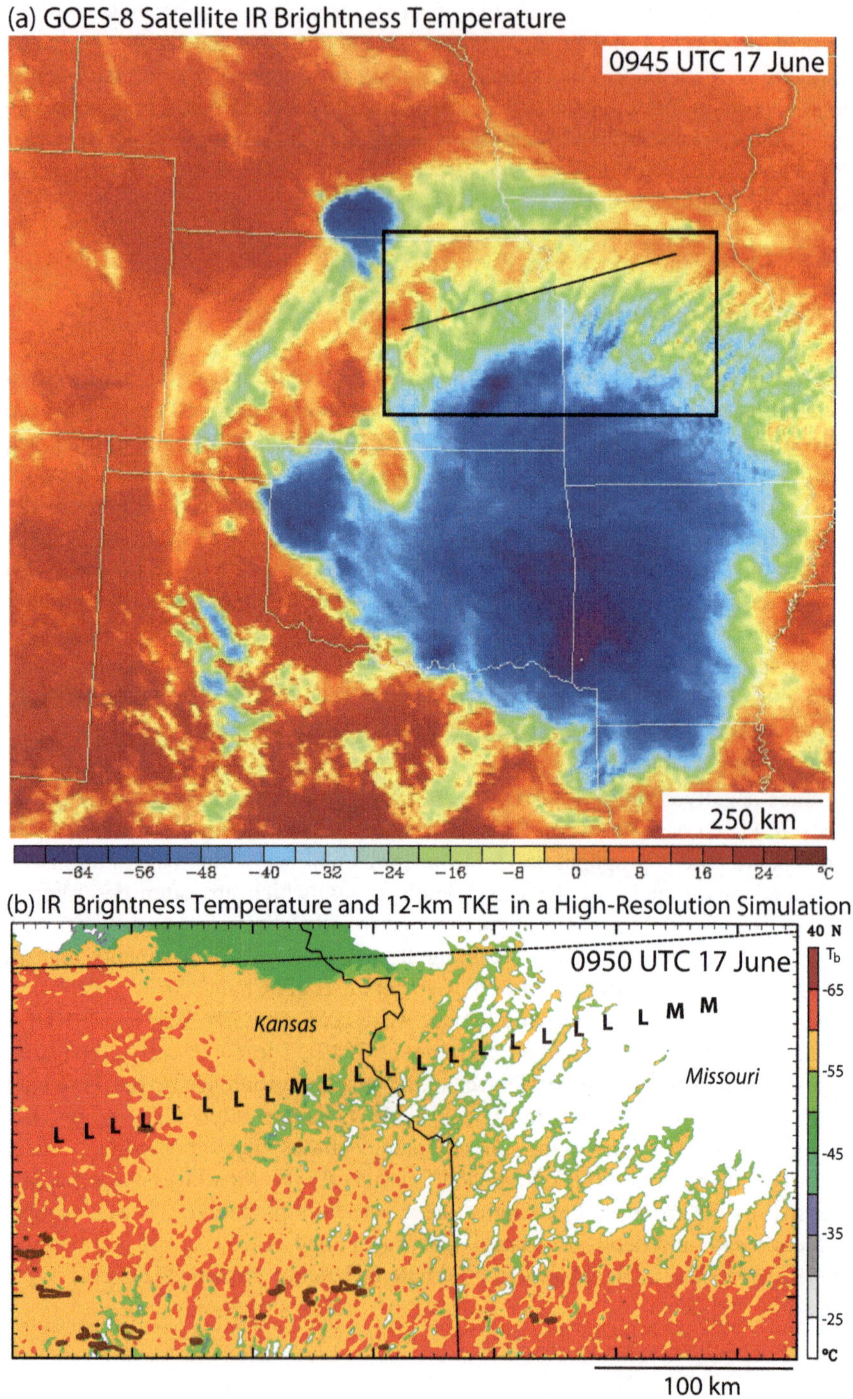

Fig. 17.5 (**a**) Infrared Geostationary Operational Environmental Satellite-8 (*GOES-8*) brightness temperature and (**b**) simulated brightness temperature and turbulence kinetic energy (*brown*

17.3.2 Warm-Season Examples of CIT (16–17 June 2005 and 8–9 September 2010)

The 17 June 2005 turbulence event shown in Fig. 17.3b occurred within a region of banded cirrus clouds at the northern edge of the outer portion of the MCS anvil, which is most evident in color-enhanced IR satellite imagery about 2 h later (Fig. 17.5a). Such bands extending radially outward from the anvil edge are common in the mature-to-dissipating stages of MCSs and are frequently colocated with reports of aviation turbulence.

Lenz et al. (2009) found such banding in 41 % (54) of 131 organized convective systems over the continental USA during May–August 2006. Furthermore, EDR observations that were available for the majority of cases with banding indicated at least one observation of light (moderate) turbulence in 93 % (44 %) of their cases. Like in the current case, Lenz et al. (2009) found that the banding was typically restricted to a particular half or quadrant of the anvil cloud edge (their Fig. 5). However, these locations were not always along the northern edge.

Trier et al. (2010) simulated radial cirrus bands in the 17 June 2005 case by adding an interior domain with 600-m horizontal grid spacing to the model configuration employed by Trier and Sharman (2009). The simulated bands of colder brightness temperature T_b are situated close to their counterparts in the satellite observations (Fig. 17.5a), have similar ~10-km horizontal separation, and are located in the vicinity of in situ EDR turbulence reports (Fig. 17.5b). Though it can be intermittently severe in some cases, the turbulence within such banded regions is more often light to moderate. However, it can be quite extensive, often spanning mesoscale regions of several hundreds of kilometers. In the current example shown in Fig. 17.5b, the recorded turbulence along the flight track was primarily light. However, simulated vertical velocities at the height of the bands (not shown) suggest it was likely more intense closer to the band origination region south of the flight track (Fig. 17.5a).

The simulated bands emanate from a region of near neutral to moist static instability $N_m^2 < 0$ (Fig. 17.6c) occurring within the upper-level anticyclonic MCS outflow (Fig. 17.6a, dashed inset), several hundreds of kilometers north of the heavy MCS precipitation (Fig. 17.6a). The bands are oriented approximately parallel to the vertical shear through the depth of anvil near its outer edge (Fig. 17.6c) and are elongated beyond the region of $N_m^2 < 0$ due to horizontal advection.

Fig. 17.5 (continued) contours denoting values greater than 0.75 $m^2 s^{-2}$). The *rectangle* in part (**a**) indicates the region shown in part (**b**), *and the line* shows the approximate length of the latest (southernmost) of the three flight tracks on which turbulence was recorded in Fig. 17.3b. The locations and strength of the recorded observed turbulence (L = light and M = moderate) are annotated in part (**b**). From Lane et al. (2012). © American Meteorological Society. Reprinted with permission

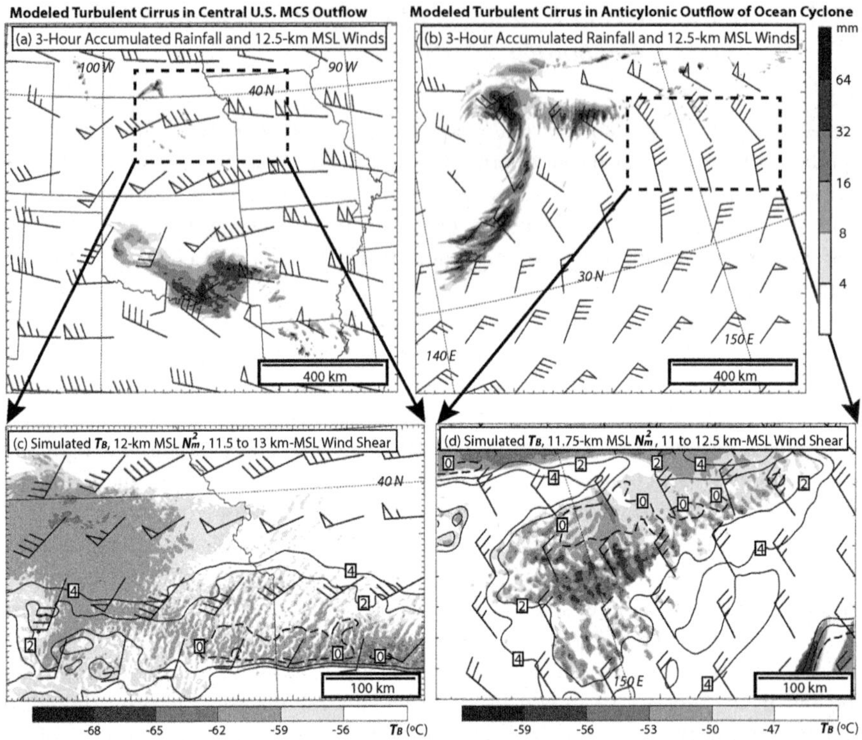

Fig. 17.6 *Top*: Simulated 12.5-km MSL horizontal winds (standard meteorological wind plotting convention with half barb = 5 knots, full barb = 10 knots, flag = 50 knots; 1 knot = 0.5144 m s^{-1}) and 3-h accumulated precipitation (*shading*) ending the time of the wind plots for the (**a**) 16–17 June 2005 MCS and (**b**) 8–9 September 2010 oceanic cyclone cases. *Bottom*: Simulated brightness temperature (*shading*), smoothed moist static stability (contoured in intervals of 2×10^{-5} s^{-1} with negative values *dashed*) and vector wind differences between indicated heights (meteorological wind plotting convention as in parts **a** and **b**) for the simulated (**c**) 16–17 June 2005 MCS and (**d**) 8–9 September 2010 oceanic cyclone cases over the regions indicated by the dashed insets in the above panels

The origination of the bands in a neutral to statically unstable region, together with the shear-parallel orientation of the bands, indicates they are likely a manifestation of thermal-shear instability (Sect. 17.3.1.1). Trier et al. (2010) demonstrated that the banding was enhanced by cloud-radiative feedbacks. However, such effects are unable to account for the regionalization of the banding on the north side of the MCS anvil. Recall that the mesoscale region of lower-static stability toward the northern edge of the anvil in Fig. 17.4 was linked to the differential advection of θ_e surfaces by the vertical shear underneath the outflow jet. In this way the vertical shear not only organizes the pattern of the banding (via thermal-shear instability) but also plays an important role in generating the regional thermodynamic conditions that supports the shallow convection that leads to the bands within the outer MCS anvil.

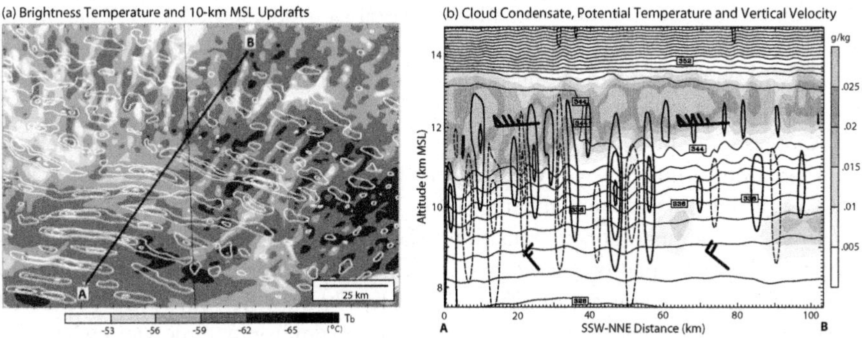

Fig. 17.7 (**a**) Simulated brightness temperature (*shading*) and 10-km MSL updrafts (*white contours* starting at 0.75 m s^{-1} with 1.5-m s^{-1} contour intervals). (**b**) Vertical cross section averaged for 3 km on both sides of line AB in part (**a**) of total cloud condensate (*shading* with scale at the *right*), horizontal winds (meteorological wind plotting convention as in Fig. 17.6a, b), potential temperature (*thin solid lines* 2-K contour intervals), and vertical velocity with positive values contoured with *bold-solid lines* (1-m s^{-1} contour interval starting at 0.5 m s^{-1}) and negative values contoured with *medium-dashed lines* (−1-m s^{-1} contour interval starting at −0.5 m s^{-1}). Adapted from Trier et al. (2010)

Satellite-based observational studies have provided visual evidence of the coexistence of the radial bands at the outer anvil edge with wavelike cloud features that propagate outward from regions of active deep convection within the interior of the storm (e.g., Lenz et al. 2009, upper left in their Fig. 3). In the current high-resolution simulation of 17 June 2005, seemingly analogous features correspond to short-wavelength horizontally propagating internal gravity waves (Fig. 17.7a), which have maximum vertical motions of 2–3 m s^{-1} near 10 km, and are oriented approximately perpendicular to the radial bands of T_b located above.

The simulated gravity waves (Fig. 17.7a) are partially trapped within the layer of strong vertical wind shear and increased static stability situated beneath the layer of near-neutral stability within the anvil where the radial bands reside (Fig. 17.7b). Trier et al. (2010) point out that the coexistence of radial bands within the anvil with trapped gravity waves below is not surprising since both are favored by the near-neutral conditions in the anvil. They further note that most of the updraft maxima in the anvil cloud are situated above upwardly displaced potential temperature surfaces associated with the underlying trapped gravity waves (Fig. 17.7b). This led to the hypothesis that the gravity waves beneath the anvil may play a role in radial band formation in some cases by helping to excite the thermal-shear instability, which itself is generated by other processes including differential radiative forcing and mesoscale differential advection in the strongly vertically sheared MCS upper-level outflow.

Lenz et al. (2009) also referred to these radial bands in the upper-level MCS outflow as transverse bands. Whitney et al. (1966) was among the first to use this terminology when referring to bands oriented transverse to the major axis of the cloud shield in jet stream cirrus. The association of such transverse bands with aviation turbulence was subsequently discussed by Anderson et al. (1973). Subsequent works (e.g., Knox et al. 2010) have illustrated the diverse environmental settings in which cirrus banding occurs, though their formation mechanisms and association with turbulence may vary.

Another example of upper-tropospheric cirrus banding associated with significant aviation turbulence was described and simulated by Kim et al. (2014). This case occurred on 8–9 September 2010 in conjunction with a midlatitude ocean cyclone that had undergone a transition from a tropical storm (Malou) several days earlier. These simulations used five interacting model domains with different horizontal grid spacings ranging from 30 km to 370 m, which collectively resolved horizontal scales ranging from that of the synoptic environment down to what may be the largest scales of turbulence that affected the aircraft.

The Kim et al. (2014) simulations (Fig. 17.6b, d) exhibit several aspects in common with the previously discussed continental MCS case (Fig. 17.6a, c). First, the moderate and severe turbulence, observed off the coast of Japan (not shown), occurred within a strong upper-tropospheric region of anticyclonic flow that was removed by several hundreds of kilometers (Fig. 17.6b, dashed inset) from large areas of significant deep convective precipitation. Similar to the continental MCS case (Fig. 17.6c), banded features in T_b emanate from a zone where $N_m^2 < 0$ and the resulting bands are aligned along the vertical shear vector (Fig. 17.6d), indicating the likely dominant role of thermal-shear instability in their formation.

However, along with these similarities, there are several important differences between this oceanic cyclone case and the previous MCS case. Cloud-radiative feedbacks were found to be crucial to cirrus banding in the current case (Kim et al. 2014), whereas they only enhanced the banding in the continental MCS case (Trier et al. 2010). The spacing between the bands is about twice as large in the current case, which allows them to be largely resolved (Kim et al. 2014) within a domain having 3.3-km horizontal grid spacing (Fig. 17.6d), which was too coarse to resolve the bands in the continental MCS case. Unlike for the continental MCS case, the strong upper-level anticyclone (within which the turbulence occurred) was also present in a dry simulation where cloud microphysical processes were disabled (Kim et al. 2014). Here, the upper troposphere was strongly perturbed prior to the event by the multiday presence of the weakening tropical cyclone Malou over this portion of the western Pacific Ocean, and outflow from the MCS (Fig. 17.6b) located closer to the surface oceanic cyclone (not shown) had only a minor contribution to the upper-level flow near the location of observed turbulence (dashed inset in Fig. 17.6b). Kim et al. (2014) noted that mesoscale pockets of simulated inertial instability (Sect. 17.3.1.1) were also present near the banding, but emphasized that the banded structure was more consistent with thermal-shear instability. Further research is needed to investigate how these different instability mechanisms might act synergistically to influence aviation turbulence in strong UTLS anticyclones.

17.3.3 A Cold-Season Example of CIT (9–10 March 2006)

Though deep convection is less common over continental midlatitude regions during the cold season than during late spring and summer, it is sometimes found within regions of strong warm advection along and ahead of strong surface cold fronts and tropopause-based disturbances. We now discuss turbulence-producing mechanisms in simulations of an observed widespread CIT case that occurred during the cold season (9–10 March 2006) over the central USA reported in Trier et al. (2012). Here, the convective precipitation (Fig. 17.8a) occurs within portions of a large-scale midlatitude cyclone instead of within an individual MCS. However, similar to the previously discussed warm-season cases that contained individual MCSs, the majority of EDR observations and pilot reports of moderate-or-greater turbulence are located outside of deep moist convection.

In this late winter case, the turbulence occurred mostly above a strong upper-tropospheric jet (Fig. 17.8b). Trier et al. (2012) found the intensity of the southerly jet was enhanced by about 30 % by the anticyclonic UTLS circulation resulting from the aggregate moist convection. The anticyclonic sense of this convectively induced circulation is illustrated schematically by the arrows on Fig. 17.8a.

While the percentage contribution from the convectively induced anticyclone to the upper-level jet is much less than for the MCS case of 17 June 2005, the total vertical shear near the wintertime jet is greater (Trier et al. 2012, their Fig. 9). Moreover, the convectively induced contribution to the total flow and its associated shear increases are necessary for the simulation to produce significant values of parameterized TKE (not shown) above the jet in the $\Delta = 3.3$-km domain along the transect SN (Fig. 17.8a, b). The most intense observed turbulence along this north-south transect was clustered in two regions including one ($x = 640$–760 km) located directly above the southerly jet maximum and another ($x = 300$–420 km) located above the jet entrance region and relatively shallow but strong moist convection (Fig. 17.8b).

Trier et al. (2012) was able to resolve different turbulence-producing mechanisms in these two locations on a higher-resolution model grid with $\Delta = 667$-m horizontal spacing. Overturning billows evident in the potential temperature field (Fig. 17.8e), which are characteristic of KHI (Sect. 17.3.1.1), are simulated in the region of the strongest vertical shear directly above the southerly jet maximum (Fig. 17.8b). In contrast, the convection below the jet entrance region excites vertically propagating gravity waves when it impinges on the region of larger-static stability beginning near 8 km MSL (Fig. 17.8c). These internal gravity waves break upon reaching a critical level (Sect. 17.3.1.2) where $U = c$ (Fig. 17.8d), similar to the mechanism discussed in Chap. 16.

As noted earlier, intense, moist convection, which can trigger gravity waves, is considerably more common during the midlatitude warm season than in winter. However, when this mechanism occurs in the winter, it may pose an even greater threat of aviation turbulence since flight levels are more likely to be located above

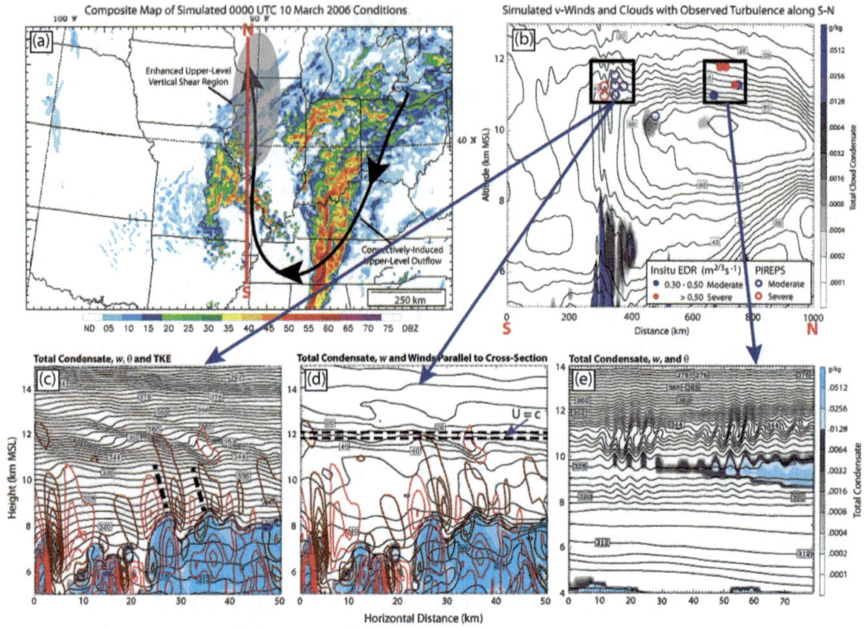

Fig. 17.8 (**a**) Composite diagram of the relationship between rainfall and convectively induced UTLS anticyclonic outflow (indicated schematically by *bold curve* with *arrows*) from simulations described in Trier et al. (2012) using a model domain with 3.3-km horizontal grid spacing. (**b**) Vertical cross section of simulated meridional wind (5-m s^{-1} contour intervals) and simulated total cloud condensate (*color shading*; scale at the *right*) averaged for 16.67 km on each side of line SN in part (**a**) with time-space corrected reports of observed moderate-or-greater turbulence within 33.3 km of SN projected onto the cross section and described in the legend. (**c**), (**d**) Vertical cross sections from higher-resolution simulations with 667-m horizontal grid spacing representing the left inset region in part (**b**) of total cloud condensate (*color shading*; scale at the *right*), vertical velocity (1-m s^{-1} contour interval; *red lines* ≥ 1 m s^{-1}; *brown lines* ≤ 1 m s^{-1}), where (**c**) has potential temperature in increments of 2 K contoured in *black solid lines* and (**d**) has winds parallel to cross section in increments of 5 m s^{-1} contoured in *black solid lines*. The *dashed vertically tilted lines* in part (**c**) indicate example phase lines of vertically propagating gravity waves, and the *dashed horizontal lines* in part (**d**) indicate a range of estimates for the location of critical levels for vertically propagating waves of different frequencies. (**e**) Vertical cross section from higher resolution with 667-m horizontal spacing representing the right inset region in part (**b**) of total cloud condensate (*color shading*; scale at *right*) and potential temperature (2-K contour intervals)

the lower wintertime tropopause and, thus, at levels where wave breaking can occur.

In some circumstances convectively induced gravity waves and enhanced vertical shear, the two primary factors influencing CIT in this particular case, are closely linked. In particular, Trier et al. (2012) also found locations where convectively induced internal gravity waves are able to excite KHI in the strong vertical shear layer above (their Fig. 17).

17.4 Summary

The generation of turbulence experienced at commercial aviation cruising altitudes often results from organized deep moist convection. However, in many cases the turbulence is removed from the region of active convection, either existing several kilometers or more above it or being laterally displaced by up to several hundreds of kilometers for the most organized mesoscale convective systems. These aspects of the turbulence make its avoidance more difficult and may contribute to it sometimes being mistaken for more classical clear-air turbulence typically associated with large-scale jet streams.

Many of the recent advances in understanding turbulence related to areas of organized moist convection have come from the diagnosis of high-resolution research numerical simulations. Horizontal scales of motion that contain the turbulence that affects aviation are aircraft dependent but are typically quite small (100–2000 m), which does not allow them to be properly resolved by most simulation models. However, when multiple horizontal grid refinements are used, such models are able to adequately represent scales ranging from those that control important aspects of the synoptic environment that may favor turbulence down to those which contain atmospheric structures that may be directly responsible for its initiation.

Convectively induced turbulence (CIT) is often associated with strongly anticyclonic upper-tropospheric/lower-stratospheric (UTLS) outflow jets. There are numerous mechanisms that may directly influence turbulence in UTLS anticyclones. For instance, recent numerical simulations point toward thermal-shear instability as an organizing factor governing the turbulent cloud bands occupying the outer edges of horizontally extensive cirrus anvils of mesoscale convective systems (MCSs). This is the same instability mechanism that organizes shallow cumulus into cloud streets in the strongly heated daytime planetary boundary layer (PBL). However, the moist static instability in anvil cirrus bands arises from different forcings, including differential cloud-radiative forcing and differential temperature advection owing to the strong vertical shear at altitudes adjacent to the outflow jet. This latter aspect may be a primary factor that often restricts cirrus banding to a preferred side of midlatitude MCS anvils.

MCSs are a warm-season phenomenon in midlatitudes, but organized moist convection can also play a role in the generation of turbulence at commercial aviation flight levels in the cold season. Because of a more baroclinic background atmosphere, the maximum vertical shear is typically stronger in winter, which supports the Kelvin-Helmholtz instability (KHI) found in high-resolution simulations. However, KHI onset in these simulations is clearly dependent on upstream moist convection unlike in classical clear-air turbulence. Wintertime scenarios are also susceptible to vertically propagating gravity waves excited by convection impinging on the stable tropopause. Since the tropopause is often significantly lower in portions of wintertime synoptic systems and the vertical shear is stronger at this time of the year, related wave breaking may be more likely to result in turbulence at commercial aviation flight levels. Such vertically propagating gravity waves may also be more likely to induce KHI in the strong shear.

High-resolution simulations are clearly an important resource in developing hypotheses for mechanisms influencing CIT and aviation turbulence in general. However, it is important that both observations from multiscale meteorological field experiments and routine in situ estimates of turbulence on equipped aircraft become more readily available for analysis to help confirm (or refute) these model-inspired hypotheses. Combined approaches of model diagnostics, data analysis, and empirical methods are likely to be most effective in advancing understanding and ultimately real-time prediction and avoidance of aviation turbulence.

References

American Meteorological Society: "mesoscale convective system". In: Glossary of Meteorology (2014). Available online at http://glossary.ametsoc.org/wiki/"mesoscale convective system"

Anderson, R.K., Veltishshev, N.F. (eds.): The use of satellite pictures in weather analysis and forecasting. Tech Note 124, WMO 333, World Meteorological Organization, Geneva, Switzerland (1973)

Asai, T.: Stability of a plane parallel flow with variable vertical shear and unstable stratification. J. Meteor. Soc. Japan **48**, 129–138 (1970)

Asai, T.: Thermal instability of a shear flow turning the direction with height. J. Meteor. Soc. Japan **50**, 525–532 (1972)

Benjamin, S.G., et al.: An hourly assimilation-forecast cycle: the RUC. Mon. Weather Rev. **132**, 495–518 (2004)

Ciesielski, P.E., Stevens, D.E., Johnson, R.H., Dean, K.R.: Observational evidence for asymmetric inertial instability. J. Atmos. Sci. **46**, 817–831 (1989)

Cotton, W.R., Lin, M.-S., McAnelly, R.L., Tremback, C.J.: A composite model of mesoscale convective complexes. Mon. Weather Rev. **117**, 765–783 (1989)

Durran, D.R., Klemp, J.B.: On the effects of moisture on the Brunt-Väisälä frequency. J. Atmos. Sci. **39**(10), 2152–2158 (1982)

Dutton, J., Panofsky, H.A.: Clear air turbulence: a mystery may be unfolding. Science **167**, 937–944 (1970)

Ellrod, G.P., Knapp, D.L.: An objective clear-air turbulence forecasting technique: verification and operational use. Weather Forecasting **7**, 150–175 (1992)

Fovell, R.G., Durran, D.R., Holton, J.R.: Numerical simulation of convectively generated stratospheric gravity waves. J. Atmos. Sci. **49**, 1427–1442 (1992)

Fritsch, J.M., Brown, J.M.: On the generation on convectively driven mesohighs aloft. Mon. Weather Rev. **110**, 1554–1563 (1982)

Fritsch, J.M., Maddox, R.A.: Convectively driven mesoscale weather systems aloft. Part I: Observations. J. Appl. Meteor. **20**, 9–19 (1981)

Fritsch, J.M., Kane, R.J., Chelius, C.R.: The contribution of mesoscale convective weather systems to the warm-season precipitation in the United States. J. Clim. Appl. Meteorol. **25**, 1333–1345 (1986)

Houze Jr., R.A.: Cloud Dynamics, 2nd edn. Academic Press, Oxford (2014). 432 pp

Kaplan, M.L., Huffman, A.W., Lux, K.M., Charney, J.J., Riordan, A.J., Lin, Y.-L.: Characterizing the severe turbulence environments associated with commercial aviation accidents. Part 1: A 44-case study synoptic observational analysis. Meteor. Atmos. Phys. **88**, 129–152 (2005)

Kim, J.H., Chun, H.-Y., Sharman, R.D., Trier, S.B.: The role of vertical shear on aviation turbulence within cirrus bands of a simulated western Pacific cyclone. Mon. Weather Rev. **142**, 2794–2813 (2014)

Knox, J.A.: Possible mechanisms for clear-air turbulence in strongly anticyclonic flows. Mon. Weather Rev. **125**, 1251–1259 (1997)

Knox, J.A., Bachmeier, A.S., Carter, W.M., Tarantino, J.E., Paulik, L.C., Wilson, E.N., Bechdol, G.S., Mays, M.J.: Transverse cirrus bands in weather systems: a grand tour of an enduring enigma. Weather **5**, 35–41 (2010)

Lane, T.P., Sharman, R.D.: Some influences of background flow conditions on the generation of turbulence due to gravity-wave breaking above deep convection. J. Appl. Meteor. Climatol. **47**, 2777–2796 (2008)

Lane, T.P., Sharman, R.D., Clark, T.L., Hsu, H.-M.: An investigation of turbulence generation mechanisms above deep convection. J. Atmos. Sci. **60**, 1297–1321 (2003)

Lane, T.P., Sharman, R.D., Trier, S.B., Fovell, R.G., Williams, J.K.: Recent advances in the understanding of near-cloud turbulence. Bull. Amer. Meteor. Soc. **93**, 499–515 (2012)

LeMone, M.A.: The structure and dynamics of horizontal roll vortices in the planetary boundary layer. J. Atmos. Sci. **30**, 1077–1091 (1973)

Lenz, A., Bedka, K., Feltz, W., Ackerman, S.: Convectively-induced transverse band signatures in satellite imagery. Weather Forecasting **24**, 1362–1373 (2009)

Maddox, R.A.: Mesoscale convective complexes. Bull. Am. Meteorol. Soc. **61**, 1374–1387 (1980)

Maddox, R.A.: Large-scale meteorological conditions associated with mesoscale convective complexes. Mon. Weather Rev. **111**, 1475–1493 (1983)

Miles, J.: Richardson's criterion for the stability of stratified shear flow. Phys. Fluids **29**, 3470–3471 (1986)

Pandya, R.E., Durran, D.R.: The influence of convectively generated thermal forcing on the mesoscale circulation around squall lines. J. Atmos. Sci. **53**, 2924–2951 (1996)

Raymond, D.J., Jiang, H.: A theory for long-lived mesoscale convective systems. J. Atmos. Sci. **47**, 3067–3077 (1990)

Scorer, R.S.: Billow mechanics. Radio Sci. **4**, 1299–1308 (1969)

Sharman, R., Tebaldi, C., Wiener, G., Wolff, J.: An integrated approach to mid- and upper-level turbulence forecasting. Weather Forecasting **21**(3), 268–287 (2006)

Sharman, R.D., Trier, S.B., Lane, T.P., Doyle, J.D.: Sources and dynamics of turbulence in the upper troposphere and lower stratosphere: a review. Geophys. Res. Lett. **39**, L12803 (2012)

Sharman, R.D., Cornman, L.B., Meymaris, G., Pearson, J., Farrar, T.: Description and derived climatologies of automated *in situ* eddy dissipation rate reports of atmospheric turbulence. J. Appl. Meteor. Climatol. **53**, 1416–1432 (2014)

Trier, S.B., Davis, C.A.: Mesoscale convective vortices observed during BAMEX. Part II: Influences on secondary deep convection. Mon. Weather Rev. **135**, 2051–2075 (2007)

Trier, S.B., Sharman, R.D.: Convection-permitting simulations of the environment supporting widespread turbulence within the upper-level outflow of a mesoscale convective system. Mon. Weather Rev. **137**, 1972–1990 (2009)

Trier, S.B., Sharman, R.D., Fovell, R.G., Frehlich, R.G.: Numerical simulation of radial cloud bands within the upper-level outflow of an observed mesoscale convective system. J. Atmos. Sci. **67**, 2990–2999 (2010)

Trier, S.B., Sharman, R.D., Lane, T.P.: Influences of moist convection on a cold-season outbreak of clear-air turbulence (CAT). Mon. Weather Rev. **140**, 2477–2496 (2012)

Whitney Jr., L.F., Timchalk, A., Gray Jr., T.I.: On locating jet streams from TIROS photographs. Mon. Weather Rev. **94**, 127–138 (1966)

Wolf, B.J., Johnson, D.R.: The mesoscale forcing of a midlatitude upper-level jet streak by a simulated convective system. Part I: Mass circulation and ageostrophic processes. Mon. Weather Rev. **123**, 1059–1087 (1995)

Wolff, J.K., Sharman, R.D.: Climatology of upper-level turbulence over the continental Unites States. J. Appl. Meteor. Climatol. **47**, 2198–2214 (2008)

Zovko-Rajak, D., Lane, T.P.: The generation of near-cloud turbulence in idealized simulations. J. Atmos. Sci. **71**, 2430–2451 (2014)

Chapter 18
Numerical Modeling and Predictability of Mountain Wave-Induced Turbulence and Rotors

James D. Doyle, Qingfang Jiang, and P. Alexander Reinecke

Abstract A survey is provided of a series of studies that demonstrate the capabilities of modern nonhydrostatic numerical models to simulate and predict the occurrence of mountain waves, wave-induced turbulence in the upper troposphere and stratosphere, and rotors in the lower troposphere generated by flow over topography. Field campaign measurement from research aircraft, turbulence reports from commercial aviation, and numerical simulations demonstrate that flow over larger-scale topography (e.g., Greenland) and three-dimensional complex terrain (e.g., Alps, Sierra Nevada Range) frequently generates upper-level wave breaking and turbulence in a variety of conditions including in the presence of environmental critical levels. In some situations, the boundary layer can strongly influence wave launching, and may limit wave amplitudes and impact the altitude and likelihood of wave breaking and turbulence. Near the surface, rotors occur when strong downslope flow in the boundary layer along the lee slopes separate from the surface as a turbulent vortex sheet, which is lifted aloft within the lee wave. Because of their strong turbulent flow, rotors are often significant aeronautical hazards.

The predictive skill of numerical simulations of mountain wave-induced turbulence observed in nature is limited by errors in initial conditions, boundary conditions (e.g., for limited area models), and the models themselves (e.g., parameterizations, dynamical methods). Ultimately, high-resolution ensemble methods that are capable of explicitly resolving mountain waves should be used to provide probabilistic forecasts of turbulence needed for aviation hazard mitigation.

J.D. Doyle (✉) • Q. Jiang • P.A. Reinecke
Marine Meteorology Division, Naval Research Laboratory, 7 Grace Hopper Avenue, Monterey, CA, USA
e-mail: james.doyle@nrlmry.navy.mil

R. Sharman, T. Lane (eds.), *Aviation Turbulence*,
DOI 10.1007/978-3-319-23630-8_18

18.1 Introduction

Stratified airflow that passes over a topographic barrier generates atmospheric gravity waves that can propagate horizontally and/or vertically. These waves, typically referred to as mountain waves, have a profound influence on the atmosphere on a variety of spatial and temporal scales. These range from turbulence scales, associated with downslope windstorms and mountain wave overturning or breaking, to the aggregate effects of mountain wave drag and vertical flux of horizontal momentum, which contributes to the momentum balance of the atmospheric general circulation and climate (e.g., see reviews from Smith 1989; Fritts and Alexander 2003).

Vertically propagating mountain waves often increase in amplitude with height because of decreasing air density (Hines 1960; Lindzen 1967) or environmental conditions such as reverse (or negative) vertical wind shear layers (Smith 1989; Shen and Lin 1999). Both of these conditions, either independently or operating in concert, may lead to wave steepening, overturning, and subsequent turbulent breakdown, particularly in the lower stratosphere (Bacmeister and Schoeberl 1989; Fritts and Alexander 2003) at commercial aircraft cruise altitudes. Mountain waves may overturn and break as they approach a critical level (Clark and Peltier 1984; Dörnbrack 1998), a level at which the wave phase speed (which is often zero for mountain waves) is equivalent to the wind component projected along the horizontal wave vector. An increase in the atmospheric stability, such as the abrupt transition across the tropopause, can reduce the vertical wavelength and increase the potential for wave breaking (VanZandt and Fritts 1989). Wave breaking characterized by overturning of isentropic surfaces often involve nonlinear interactions that occur in the transition to turbulence (e.g., Fritts et al. 1996).

Mountain wave breaking and turbulence generation are important for several reasons: impact of orographic drag on the large-scale circulation (Bretherton 1969; Palmer et al. 1986); development of severe downslope windstorms (Peltier and Clark 1979; Durran 1986); potential vorticity generation (Schär and Smith 1993) and associated upscale forcing (Thorpe et al. 1993; Aebischer and Schär 1998); vertical mixing of water vapor, aerosols, and chemical species (Dörnbrack and Dürbeck 1998); and, of most relevance for this discussion, clear-air turbulence that poses a hazard to aviation (Lilly and Kennedy 1973; Lilly 1978; Ralph et al. 1997; Clark et al. 2000). Mountains are one of the most prolific generators of clear-air turbulence at cruise altitudes of commercial aircraft (e.g., Wolff and Sharman 2008), but also pose a hazard at lower altitudes as well.

Direct research observations of mountain wave breaking and turbulence are relatively rare, particularly at upper levels of the troposphere and lower stratosphere, although there have been some noteworthy studies that document mountain waves (e.g., Shutts 1992; Vosper and Mobbs 1996; Doyle and Smith 2003). The remarkable in situ aircraft observations of the downslope windstorm and wave breaking in the lee of the Front Range of the Rockies on 11 January 1972 (Lilly and Zipser 1972; Lilly 1978) are one of the most remarkable set of observations of large-amplitude mountain wave breaking to date and provide a clear link between resonant mountain wave amplification positioned beneath upper-level wave breaking regions that contain "wave-induced" critical levels and downslope windstorm and gravity wave drag enhancement (Peltier and Clark 1979). The downslope windstorm has been shown to be consistent with internal hydraulic theory based on theoretical considerations (Smith 1985) and numerical simulations (Durran 1986). Other more recent observational evidence of mountain wave breaking has been documented, for example, above the Front Range of the Rockies (Clark et al. 2000), the Alps (Jiang and Doyle 2004), and the Welsh Mountains (Worthington 1998).

The lee side of prominent terrain barriers, such as the Sierra Nevada and Front Range of the Rocky Mountains, are well known for topographically forced atmospheric phenomena such as downslope windstorms, trapped lee waves, and rotors. In these situations, severe downslope winds near the surface, occasionally in excess of 50 m s^{-1}, decelerate rapidly in the lee and give way to an unsteady return flow back toward the mountain crest that is the lower branch of a turbulent and intense horizontal circulation, referred to as a rotor. Figure 18.1 schematically illustrates the conceptual relationship between rotors and mountain lee waves. Rotor or roll clouds typically cap the rotor circulation and are often positioned beneath the lee wave crests, with layers of lenticular clouds present aloft in the middle or upper troposphere above the more turbulent rotor circulation. The circulation in the rotor is characterized by unsteady and turbulent flow (e.g., Doyle et al. 2009; Kühnlein et al. 2013). A number of early studies made the remarkable link between stationary orographic cloud formations, mountain waves, and turbulent rotor-like circulations (e.g., Koschmieder 1920; Kuettner 1938; Grubišić and Orlić 2007). Rotors have been observed in a number of mountainous regions including the Rockies (Lester and Fingerhut 1974; Ralph et al. 1997), Sierra Nevada and Owens Valley (e.g., Holmboe and Klieforth 1957; Kuettner 1959; Grubišić and Lewis 2004; Grubišić and Billings 2007), Adriatic coast of Croatia (Grubišić and Orlić 2007), Falkland

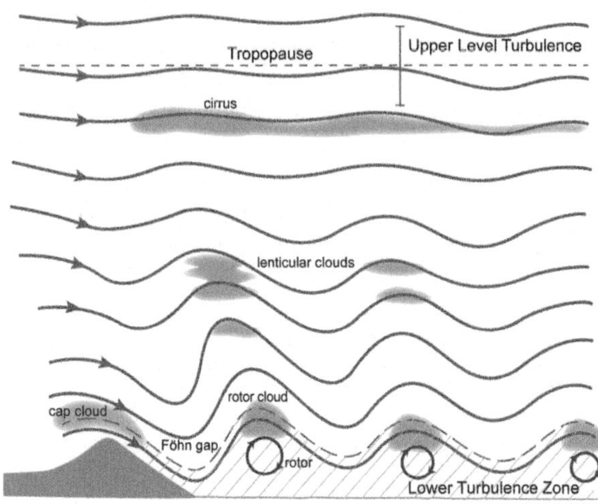

Fig. 18.1 Schematic of mountain wave phenomena including lee waves and rotors. Significant visible weather phenomenon are shown, including lenticular clouds, rotor clouds, and cap cloud on the windward side of the mountain range (Adapted from Kuettner and Jenkins 1953, Fig. 5)

Islands (Mobbs et al. 2005; Sheridan and Vosper 2006), and United Kingdom (Scorer 1955). Rotors can be severe aeronautical hazards and have been associated with severe turbulence incidents that have contributed to aviation accidents (Carney et al. 1996; Darby and Poulos 2006). For example, Ágústsson and Ólafsson (2014) discuss a commercial aircraft encounter of severe turbulence over Iceland. Their numerical simulations highlight the salient features including severe downslope winds, lee waves, and a rotor aloft that featured strong shear turbulence at the interface of the lee wave and the rotor.

In this chapter, we provide an overview of numerical modeling of mountain wave turbulence, rotors, and predictability issues. We focus on model simulations to illustrate the characteristics, predictability, and dynamics of breaking mountain waves and turbulence, as well as topographically induced rotor circulations. Highlights of mountain wave-induced turbulence generated by large topographic obstacles such as Greenland are described in Sect. 18.2. The role of complex terrain in generating mountain waves and wave breaking is discussed in Sect. 18.3. The predictability of mountain waves and associated turbulent breakdown is addressed in Sect. 18.4. The dynamics and characteristics of rotors are discussed in Sect. 18.5. Section 18.6 contains the summary and future directions.

18.2 Mountain Wave Breaking and Turbulence over Large Topographic Obstacles

Stratified flow over large topographic obstacles, such as Greenland, generates mountain waves and is a prominent source of mountain wave-induced turbulence in the lower stratosphere. The topography of Greenland is characterized by steep terrain near the coastlines and an ice-covered plateau exceeding 3000 m with an average ice thickness of 1790 m and a thickness maximum of 3400 m (Dolgushin and Osipova 1989). The terrain of Greenland is complex and characterized by a spectrum of mountain shapes ranging from half-widths of less than 150 km in the south to greater than 500 km in the central portion. Mountain wave-induced turbulence in the lower stratosphere often occurs above Greenland, which is of particular significance since southern and central Greenland is a frequently traveled commercial aviation corridor (e.g., see Lester et al. 1989). Here, we provide several examples of the characteristics of mountain wave turbulence (MWT) induced by the large-scale topographic barrier, Greenland.

A large-amplitude mountain wave generated by strong southwesterly flow over southern Greenland was observed during the Fronts and Atlantic Storm Track Experiment (FASTEX) (Joly et al. 1997) on 29 January 1997 by the NOAA G-IV research aircraft (Doyle et al. 2005). Dropwindsondes deployed every 50 km combined with flight-level data depict a vertically propagating large-amplitude wave with deep convectively unstable layers, potential temperature perturbations of 25 K that deformed the tropopause and lower stratosphere (Fig. 18.2a, b), and a

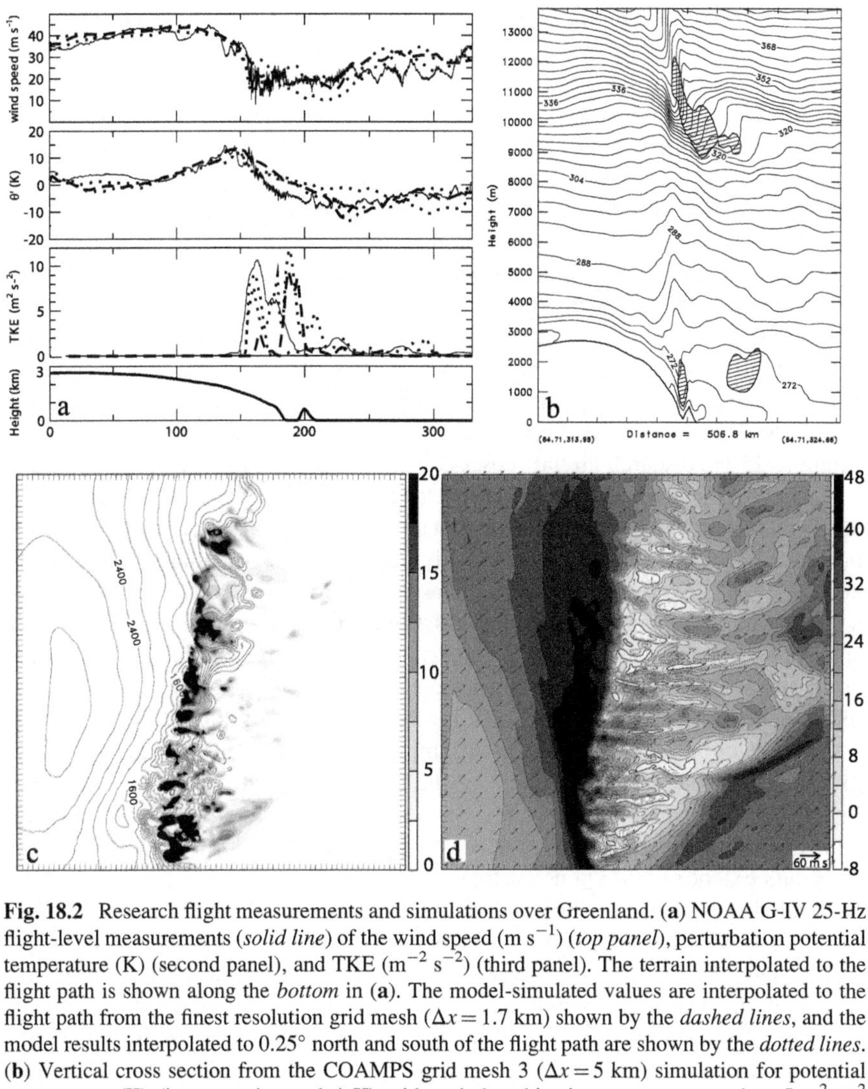

Fig. 18.2 Research flight measurements and simulations over Greenland. (**a**) NOAA G-IV 25-Hz flight-level measurements (*solid line*) of the wind speed (m s^{-1}) (*top panel*), perturbation potential temperature (K) (second panel), and TKE (m^{-2} s^{-2}) (third panel). The terrain interpolated to the flight path is shown along the *bottom* in (**a**). The model-simulated values are interpolated to the flight path from the finest resolution grid mesh ($\Delta x = 1.7$ km) shown by the *dashed lines*, and the model results interpolated to 0.25° north and south of the flight path are shown by the *dotted lines*. (**b**) Vertical cross section from the COAMPS grid mesh 3 ($\Delta x = 5$ km) simulation for potential temperature (K) (isentrope interval 4 K) with turbulent kinetic energy greater than 5 m^2 s^{-2} hatched. Horizontal plots of model simulated: (**c**) TKE (gray scale every 2.5 m^2 s^{-2}) and (**d**) wind vectors and cross-mountain wind component (gray scale every 4 m s^{-1}) valid at 1130 UTC 29 January 1997 (11.5-h simulation time) at 12 km ASL for the finest resolution grid mesh ($\Delta x = 1.7$ km). The model topography is shown in (**c**) with a 200 m contour interval (Modified from Doyle et al. 2005) © American Meteorological Society. Reprinted with permission

vertical velocity maximum of nearly 10 m s^{-1} in the stratosphere (not shown). The Naval Research Laboratory's (NRL's) Coupled Ocean-Atmosphere Mesoscale Prediction System (COAMPS®) with four nested grid meshes with a minimum horizontal resolution of 1.7 km was applied (see Doyle et al. 2005; Hodur 1997) (using the atmospheric model only), and the simulated amplitude, location, and timing of the mountain wave and turbulent breakdown were performed with a reasonable degree of accuracy. The simulated wind speed, potential temperature, and turbulence kinetic energy (TKE) derived from the 1.7-km resolution grid mesh were interpolated to a portion of the research aircraft flight transect at the nearest available model output time (1130 UTC 29 January 1997) and are displayed in Fig. 18.2a along with the in situ aircraft observations. In order to illustrate the spread and variability of the simulated flow in the vicinity of the flight track, the model results interpolated to 0.25° north and south of the flight path are shown by the dotted lines in Fig. 18.2a. The highest-resolution grid mesh accurately captures the gross large-scale features of the wave. For example, the model and observations both exhibit a wind speed perturbation of nearly 30 m s^{-1} associated with the gravity wave (Fig. 18.2a top panel) and a potential temperature perturbation greater than 25 K (Fig. 18.2a second panel). The TKE predicted by the model and derived from the 25 Hz research aircraft observations is shown in Fig. 18.2a (third panel). The model-simulated TKE shows two distinct peaks. One maximum occurs in the observed region of minimum wind speed and highest frequency activity; however, the model-simulated TKE maximum of 2 m^2 s^{-2} is substantially less than the derived TKE maximum of 10.5 m^2 s^{-2} based on the aircraft data. In the simulation, a second maximum exists 25-km downstream of the initial breaking zone, associated with strong buoyancy and shear contributions. The surrounding grid points just north and south of the flight track indicate considerable variability and appear to match better with the flight-level data for the western TKE maximum.

Vertical cross section of potential temperature, based on the 5-km resolution mesh and oriented along a portion of the research aircraft transect, is shown in Fig. 18.2b. The simulated upper-level wave breaking, as diagnosed by the region with TKE in excess of 5 m^2 s^{-2}, vertically spans the 8–11 km layer above the lee slope and extends ~100 km downstream (Fig. 18.2b). The isentropic cross section has a number of characteristics in common with the observation-based analysis (see Fig. 8 in Doyle et al. 2005), which includes the weak stability layer in the upper troposphere and the deformed tropopause associated with the wave amplification and breaking. The analysis contains a layer of weaker static stability in the middle and upper troposphere upstream of the Greenland crest relative to the simulation. The implication of this static stability difference is that the simulated hydrostatic vertical wavelength, $2\pi U/N$, may be underestimated. The static stability differences in the 5–9 km layer imply a larger vertical wavelength based on the observations relative to the simulation.

The TKE at 12 km ASL (above sea level), obtained from the fourth grid mesh (1.7-km horizontal resolution) for 1130 UTC 29 January is shown in Fig. 18.2c. The simulated TKE indicates a widespread region of wave breaking occurs above the steepest portion of the lee-side topography and decays with distance downstream. The cross-mountain wind component at 12 km ASL, shown in Fig. 18.2d, suggests that a strong horizontal gradient in the wind speed is present associated with the wave breaking. The region of wind speed in excess of 40 m s^{-1} upstream of the breaking transitions over a relatively short distance of 5–10 km to weak or reversed flow downstream. Downstream of the wave breaking, the cross-mountain wind field contains rich structure comprised of flow-aligned streams of weak and strong flow. The sources of the weak or reversed flow plumes are linked with the more intense regions of dissipation-associated wave breaking, as opposed to the stronger flow maxima that are correlated with regions of relatively weak breaking and smaller values of dissipation (Fig. 18.2c).

Lane et al. (2009) examined records of aviation turbulence encounters above Greenland for the period from 2000 to 2006. They analyzed commercial aircraft turbulence reports, meteorological analyses and numerical simulations in the vicinity of Greenland, and identified processes that contribute to the occurrence of turbulence encounters. The turbulence was found to be more common over Greenland during the winter months and more likely to occur when the low-level wind near Greenland's southern tip is from the southeast quadrant. The passage of extratropical cyclones to the south of the turbulent regions is one source of low-level southeasterly flow and also influences Greenland's southern tip more often in winter than summer. The incident southeasterly low-level flow often results in the generation of vertically propagating mountain waves. Often the wind rotates toward the east with height. In these situations, mountain waves interact with the directional wind shear and break due to the existence of a critical level. As mountain waves approach a layer with zero wind velocity component parallel to the wave vector, often referred to as a critical level, the vertical wavelength approaches zero and wave action accumulates beneath the critical layer, leading to wave breaking (Dörnbrack et al. 1995; Grubišić and Smolarkiewicz 1997). Gravity wave reflection may occur at a critical level when the Richardson number, $\mathrm{Ri} = N^2/(\partial U/\partial z)^2$, is less than 2 (e.g., Breeding 1971; Wang and Lin 1999).

The long-term turbulence statistics and two specific cases are examined in more detail by Lane et al. (2009). The results highlight the importance of the critical levels in contributing to the occurrence of enhanced levels of turbulence encounters near Greenland. The COAMPS model provided realistic simulations of the mountain wave breaking and parameterized turbulence very close to the locations of the commercial aircraft turbulence reports for the two events studied. Turbulence was closely associated with large-amplitude mountain wave breaking below an environmental critical level. Waves were generated by the western lee slope of the Greenland ice sheet. Transient wind structures characterized by northward moving middle-tropospheric jets contributed to the temporal variability of turbulence occurrence and intensity. Forecasting the presence of these jets and their interaction with the terrain below may be crucial for turbulence avoidance strategies.

The Lane et al. (2009) study noted that westerly and northwesterly wind directions were less likely to produce turbulence (reports) than expected from a climatological perspective. The westerly flow regime might be expected to produce the largest wave response (i.e., Fig. 18.2) and as shown by Wolff and Sharman (2008) is the most common scenario over the Rocky Mountains. Doyle et al. (2005) did identify wave breaking in the lower stratosphere over Greenland under westerly flow, as highlighted previously, but the absence of a critical level in such cases means that the altitude of the upper-level wave breaking is dependent on the wave amplitude, which was shown to be sensitive to a variety of upstream conditions. Thus, in many cases of west or northwesterly flow, the upper-level wave breaking may occur above the usual flight level of most commercial air carriers (~10 km), leaving lower altitudes relatively quiescent and free of turbulence (with the exception of the low levels above the terrain). The results elucidated by the turbulence statistics of the Lane et al. (2009) study can only be applied to turbulence near 10 km, i.e., the part of the atmosphere sampled by commercial air travel.

Sharman et al. (2012) utilized digital flight data recorder (DFDR) analyses and high-resolution numerical simulations to investigate a severe clear-air turbulence (CAT) encounter over western Greenland by a Boeing 777 aircraft at 10-km altitude on 25 May 2010. The environmental flow was characterized by an extratropical cyclone to the southeast of the Greenland tip, resulting in easterly flow at all levels, similar to the regime highlighted in the Lane et al. (2009) study. The synoptic conditions with deep easterly flow observed at all levels are similar to another study of turbulence over western Greenland by Ólafsson and Ágústsson (2009). The results of the Sharman et al. (2012) analyses indicate that the CAT encounter was related to mountain wave breaking on the western lee (downslope) of the Greenland plateau. The simulations indicate large-amplitude lee waves and overturning in reasonably good agreement with the encounter location as indicated by the DFDR. The COAMPS simulation captured the timing and location of the turbulence event, even for the relatively long lead time analyzed (13-h), although the shorter wavelength disturbances found in the DFDR analysis were not resolved. The model-derived sub-grid TKE and eddy dissipation ($\varepsilon^{1/3}$) values that were higher than those inferred from the DFDR possibly due to the relatively coarse resolution of the model and possible underestimates of the eddy dissipation rate for short period discrete events.

The severe turbulence and associated abrupt change in wind speed encountered by a commercial jet flying over southern Greenland, while cruising at approximately the 200 hPa level, was reasonably reproduced by Ólafsson and Ágústsson (2009) using the MM5 model in a two-level nested mode with grid resolutions of 9 and 3 km. They found that the 3-km grid simulated stronger wave breaking and MWT and revealed more complex wave patterns than the 9-km grid, suggestive of the value of high-resolution real-time modeling for aviation MWT forecasts over Greenland.

The success of the Sharman et al. (2012) simulation and the other numerical modeling studies in capturing and characterizing MWT events suggest that operational implementation of high-resolution nonhydrostatic simulation models, possibly an ensemble of models (see Sect. 18.4), over MWT-prone areas could produce more reliable forecasts of MWT than are routinely available currently, such as those using gravity wave drag or post-processing MWT algorithms derived from relatively coarse-resolution global weather prediction models.

18.3 Mountain Wave Breaking over Complex Terrain

The Greenland cases provide examples of highly complex 3D transient events, which are difficult to understand physically, so many numerical model-based studies reported on in the literature have used idealized terrain shapes with idealized initial states to establish the characteristics of gravity wave breaking (e.g., Clark and Peltier 1984; Bacmeister and Schoeberl 1989; Ólafsson and Bougeault 1996, 1997; Epifanio and Durran 2001). Relatively few studies use complex terrain and real-atmospheric initial states to simulate wave breaking. Exceptions include the Clark et al. (2000) study that used a high-resolution model to simulate a downslope windstorm and clear-air turbulence event over the Rockies and documented the generation of horizontal vortex tubes within wave breaking regions along a jet stream in the upper troposphere. Schmid and Dörnbrack (1999) simulated lower stratospheric wave breaking over the Alps associated with a critical level above the jet stream. Leutbecher and Volkert (2000) found that the simulation of mountain waves generated by Greenland was sensitive to the horizontal model resolution and surface friction representation. The multi-scale terrain of the Alps was shown by Jiang and Doyle (2004) to promote wave breaking and increase wave drag. Although several of these previous real-data simulations have replicated the limited observations of the wave breaking evolution with reasonable accuracy, the predictability of wave breaking is not well established. For example, simulations of the 1972 Boulder windstorm using 11 different numerical models, all with identical initial states and two-dimensional terrain, exhibit diverse wave breaking evolution in some cases (Doyle et al. 2000).

During the Terrain-Induced Rotor Experiment (T-REX) (Grubišić et al. 2008), the National Science Foundation (NSF)/National Center for Atmospheric Research (NCAR) Gulf Stream V (G-V) aircraft measured mountain wave properties along repeated transects across the central and southern Sierra Nevada Range, separated by a distance of approximately 50 km. Racetrack flight patterns executed across the Sierra ridge at several different stratospheric altitudes, and repeated over a number of different cases, provide insight into the three-dimensionality of stratospheric gravity waves generated by the complex terrain of the Sierra Nevada Range. Doyle et al. (2011) present observations from several of the G-V research flights that indicate vertical velocities in the primary mountain wave exhibit variations up to a factor of two between the southern and northern portion of the racetrack flight segments in the lower stratosphere, with the largest amplitude waves most often occurring over the southern flight leg, which has terrain that is 800 m lower than the northern leg. Multiple racetracks at the 11.7 km and 13.1 km altitudes indicate that these differences were repeatable, which is suggestive that the deviations were likely due to vertically propagating mountain waves that varied systematically in amplitude rather than associated with transients. The cross-mountain horizontal velocity perturbations are also a maximum above the southern portion of the Sierra ridge. Nonhydrostatic numerical model simulations from COAMPS attribute the variability in the wave amplitude and characteristics in the along-barrier direction to be a combination of blocking by the three-dimensional Sierra Range and the Coriolis effect.

Several studies highlight the consequence of the wave launching and vertical propagation up through the lower stratosphere. Smith et al. (2008) used the T-REX G-V measurements to compute vertical energy fluxes and confirmed the dynamical relationship between the momentum and energy flux that Eliassen and Palm (1961) first hypothesized. In a follow-on study, Woods and Smith (2010) discovered collocated long up-going and short down-going waves in the lower stratosphere in at least four cases, suggesting secondary generation of gravity waves by wave breaking aloft. In two other cases, they noted trapped waves along the tropopause inversion layer. Distinguishing and predicting the quasi-laminar mountain wave events from ones that produce significant turbulence through upper-level wave breaking still remains a major challenge. Here, we discuss one of the few T-REX cases in which mountain wave-induced turbulence was observed in the lower stratosphere.

A high-resolution simulation of the 25 March 2006 mountain wave event during T-REX, referred to as Intensive Observing Period (IOP) 6, was conducted using four nested grid meshes of COAMPS with a horizontal resolution of 1 km on the finest grid mesh and 80 vertical levels, as described by Doyle et al. (2011). A vertical cross section of the cross-mountain wind component and potential temperature oriented along the more active southern leg is shown in Fig. 18.3a, b, which correspond to 2100 UTC (21-h forecast) and 2300 UTC (23-h forecast) on 25 March. The simulation contains a region of wave steepening, positioned above the jet in a reversed shear region, indicated by the hatched region in Fig. 18.3 corresponding to areas with TKE greater than 20 $m^2\ s^{-2}$. The region of wave steepening increases in amplitude and area, and progressively occurs at lower altitudes in the 2100–2300 UTC time period. The simulation is consistent with the timing and spatial location above Owens Valley and downstream from the Sierra crest in which the G-V encountered turbulence and a large-amplitude mountain wave (see Fig. 9 in Doyle et al. 2011). It is noteworthy that the other flight legs at the 13.1-km altitude indicated the presence of large-amplitude waves, but the signatures of wave breaking were only observed at the 13.7-km level.

The T-REX IOP 6 was also simulated by Mahalov et al. (2011) using a high-resolution microscale model driven by 3-km Weather Research and Forecasting (WRF) model fields (see also Chap. 20). With an increased vertical resolution, their microscale model was able to resolve some fine-scale dynamical processes such as wave breaking and shear instability in the upper troposphere and lower stratosphere. The simulated small-scale perturbations were attributed to Kelvin-Helmholtz instability associated with strong wind shear induced by the mountain waves. They believed that the vertical resolution refinement of the microscale model was critical to resolve both waves and wave-induced shear instability properly.

In complex terrain, the boundary layer can play an important role in wave launching and ultimately acts as a control on the wave amplitudes and likelihood of breaking. Smith et al. (2002) describe a case of a stationary mountain wave generated by southwesterly flow over Mont Blanc in the Alps that was observed

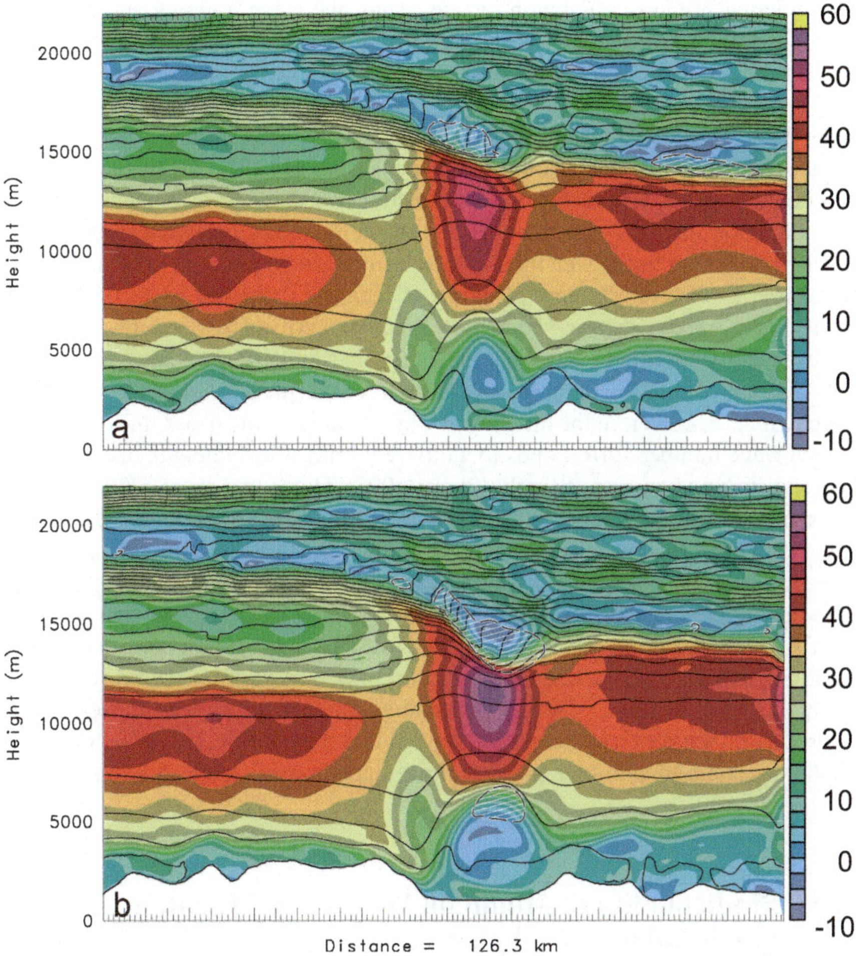

Fig. 18.3 Vertical cross section of the model simulated along-section wind speed component shaded in color (every 2.5 m s^{-1}) and potential temperature (every 8 K) valid at (**a**) 2100 UTC 25 March and (**b**) 2300 UTC 25 March 2006. The vertical cross section is oriented along a flight segment over the southern Sierra. The TKE in excess of 20 m^2 s^{-2} is shown by the hatch regions (From Doyle et al. 2011) © American Meteorological Society. Reprinted with permission

simultaneously by research aircraft and remote sensing instruments during the Mesoscale Alpine Programme (MAP). Only the highest peaks such as Mount Blanc extended above a layer of blocked stagnant flow near the ground that reduced the amplitude of the wave generation. The strong wind and weak stability in the upper troposphere resulted in evanescent waves; however, no lee waves were observed due to a stagnant layer below 3 km that absorbed downward-reflected waves, preventing resonance and wave trapping. The authors demonstrate the significance of this boundary layer absorption effect using theoretical models and full numerical weather prediction models. Thus, the wave trapping and boundary layer absorption, which was observed and simulated for several MAP cases, reduced the amount of vertically propagating wave energy and ultimately upper-level breaking.

Another issue that may impact the mountain wave amplitudes aloft and the likelihood of breaking and clear-air turbulence is related to filtering by the wind in a directionally sheared environment. A directional critical level occurs when the wind direction changes with altitude and the flow at any level is parallel to the gravity wave phase lines, even if the wind speed does not reach a zero velocity. This is the case for the easterly flow cases over Greenland described in Sect. 18.2. The gravity wave packets stagnate below the directional critical level, and wave action is subsequently advected away from the mountain (Shutts 1998; Shutts and Gadian 1999). In the unidirectional case, all waves have a critical level at the zero wind velocity line. However, in the directional wind shear case, the phase line orientation of each wave number corresponds to a different critical level height, and the wave drag is distributed correspondingly (Shutts 1995). Doyle and Jiang (2006) discuss an example of a directional critical level that influences small-amplitude gravity waves generated by flow over the southwestern French Alps observed during MAP. Three research aircraft and backscatter from two lidars showed a rapid decrease in the wave amplitude with height due to a directional critical layer that partially absorbs wave energy in the lower to middle troposphere. Numerical simulations using COAMPS with a horizontal resolution of 556 m confirm the presence of a directional critical level. These results highlight the important impact of the absorption of wave energy associated with a directional critical layer on the wave amplitudes and ultimately the potential for mountain wave-induced turbulence.

18.4 Predictability of Mountain Waves and Mountain Wave-Induced Turbulence

The science of predictability seeks to quantify the fundamental limitations of numerical weather prediction given uncertainties in the numerical model, the initial and boundary conditions, and the observations. In seminal work, Lorenz (1969) argued that the rapid upscale propagation of uncertainties in the specification of initial conditions would limit the predictability of mesoscale motions with spatial scales on the order of 10 km to time scales on the order of 1 h. While early practical experience with numerical weather prediction (NWP) led many to believe that known mesoscale forcing, such as terrain or lateral boundary conditions, may act to constrain error growth and extend the predictability of certain motions (e.g., Anthes et al. 1985; Paegle et al. 1990; Mass et al. 2002), more recent work has suggested that mesoscale predictability is not only limited by rapid upscale growth of initially small-scale errors as originally suggested by Lorenz, but even faster downscale growth of synoptic-scale errors (Reinecke and Durran 2009; Rotunno and Snyder

2008; Durran and Gingrich 2014). We begin this section by highlighting several examples which demonstrate how initial condition uncertainty can limit the predictability of terrain-induced gravity wave breaking and turbulence and then address the role of model error in mesoscale predictability.

The focus of this section is on the role of initial condition error on the predictability of mesoscale terrain-induced flows. Initial condition errors arise from observational uncertainty, and from uncertainties in the numerical model and the data assimilation system. One way to examine the sensitivity of model forecasts to errors in initial conditions is through an ensemble forecast. Given that the initial condition of each ensemble member represents an equally likely set of states, the ensemble forecast gives a probabilistic view of the likely outcomes, as well as a quantitative estimate of the forecast sensitivity to the initial conditions.

Doyle and Reynolds (2008) used a suite of high-resolution two-dimensional ensemble simulations, in which spatially uncorrelated perturbations with amplitude typical of observational uncertainty was added to the initial fields, to investigate the predictability of mountain waves and wave breaking. They progressively increased the mountain height to induce a more nonlinear mountain wave response. As would be expected, for a small mountain in the linear regime, the perturbation growth was relatively small. However, as the mountain height was increased to a point near wave breaking, the perturbations grew rapidly and evoked a bimodal response. Several ensemble members exhibit a trapped wave response, and others reveal a hydraulic jump and large-amplitude breaking in the stratosphere. These results indicate that the theoretical transition across the regime boundary for gravity wave breaking can be interpreted as a blurred transition zone from a practical predictability standpoint.

As an illustration of some of these threshold predictability issues, the ensemble mean and variance for potential temperature and u-wind component are shown in Fig. 18.4 (from Doyle and Reynolds 2008) for a simulation performed with a mean state critical level at 10 km using a mountain height of 2000 m. The potential temperature ensemble mean indicates a train of large-amplitude lee waves that extend downstream from the mountain (Fig. 18.4a). The potential temperature and wind velocity (Fig. 18.4b) variance is larger in the lower troposphere than in the ensemble without a mean state critical level (see Fig. 4k, l in Doyle and Reynolds 2008). The spread in the maximum wind speed at the surface is 13 m s^{-1} and similar to the ensemble without a mean state critical level. However, the u-variance maximum near the surface is similar to that of the variance maximum in the ensemble that exhibited bifurcation characteristics (some members show large-amplitude breaking waves; other members contain laminar trapped waves). The variance is reduced considerably in the lower stratosphere as a result of the critical level absorption present near the tropopause. Thus, the critical level is enhancing predictability of mountain waves at higher altitudes above the critical level, but decreasing it near the surface.

While the results from Doyle and Reynolds (2008) focus on the role of initially small-amplitude and small-scale error on the predictability of mountain wave breaking, a number of studies suggest that in NWP relatively small-amplitude

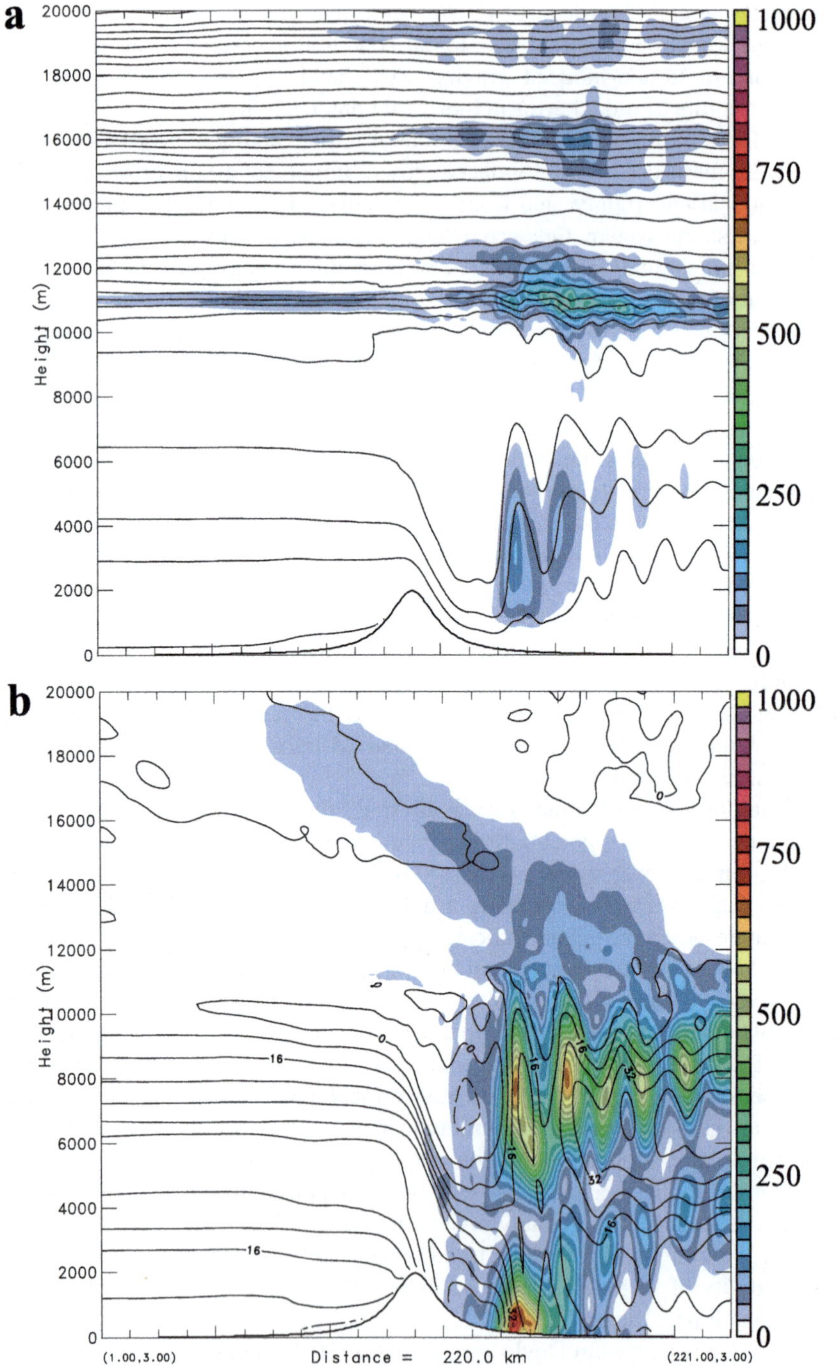

Fig. 18.4 The ensemble mean (**a**) potential temperature (contours every 8 K) and variance (color shading interval 25 K^2) and (**b**) ensemble mean u (contours every 8 m s^{-1}) and variance (color shading interval 25 $m^2\ s^{-2}$) are shown for a mountain height of 2000 m with a mean state critical

large-scale errors can quickly propagate downscale and contaminate mesoscale forecasts. For example, mesoscale predictions of landfalling fronts were found to be very sensitive to small changes in incident flow, as deduced through simulations made with small modifications to the topography orientation by Nuss and Miller (2001).

In another example, Reinecke and Durran (2009) considered a 70-member ensemble simulation of a large-amplitude mountain wave and downslope wind event that occurred as part of IOP 6 during the T-REX (Grubišić et al. 2008). From the 70-member ensemble, they analyze the differences between the strongest and weakest ten members, as characterized by the intensity of the downslope wind in the immediate lee of the Sierra Nevada mountains, for a forecast valid on 00 UTC, 26 March 2006. Figure 18.5a, b shows a composite of zonal wind and forecasted TKE along a cross section across the Sierra Nevada for the weak- and strong-member subset, respectively. It is clear that the strong-member subset (Fig. 18.5b) contains a large-amplitude breaking mountain wave with an extensive region of turbulent mixing directly above and to the lee of the Sierra Nevada. In contrast, wave breaking and turbulence are limited to a small region in the upper troposphere/lower stratosphere in the weak-member subset (Fig. 18.5a). Figure 18.5c, d shows the 500 hPa geopotential height and wind speed for the weak- and strong-member subsets, respectively. Surprisingly, the differences in the synoptic-scale forcing flow are relatively small, suggesting that small-amplitude large-scale errors can lead to large mesoscale error growth.

In addition to initial condition uncertainty, model error can play an important role in the predictability of terrain forced flows. One way to explore the contribution of the model error component on predictability is through a model intercomparison where an ensemble of numerical models, which differ only in numerical formulation and parameterized processes, are integrated using identical initial and lateral boundary conditions. Doyle et al. (2000) performed a model intercomparison of the 11 January 1972 Boulder downslope windstorm using a two-dimensional model framework over an idealized terrain profile. While every model predicted upper-level wave breaking in a qualitatively similar location, there were significant differences in the details and timing of the wave breaking and turbulent region. Furthermore, the tropospheric wave structure differed between the models with the majority of simulations producing a hydraulic jump-type feature and the remaining simulations predicting a large-amplitude mountain wave response.

To further explore the sensitivity of mountain wave breaking and turbulence to the model formulation, Doyle et al. (2011) performed an additional set of model intercomparison using 11 different two-dimensional nonhydrostatic numerical models with a free-slip lower boundary condition. The initial state is derived from a sounding sampled upstream of the Sierra Nevada mountain range on

Fig. 18.4 (continued) level specified at 10 km (From Doyle and Reynolds 2008) © American Meteorological Society. Reprinted with permission

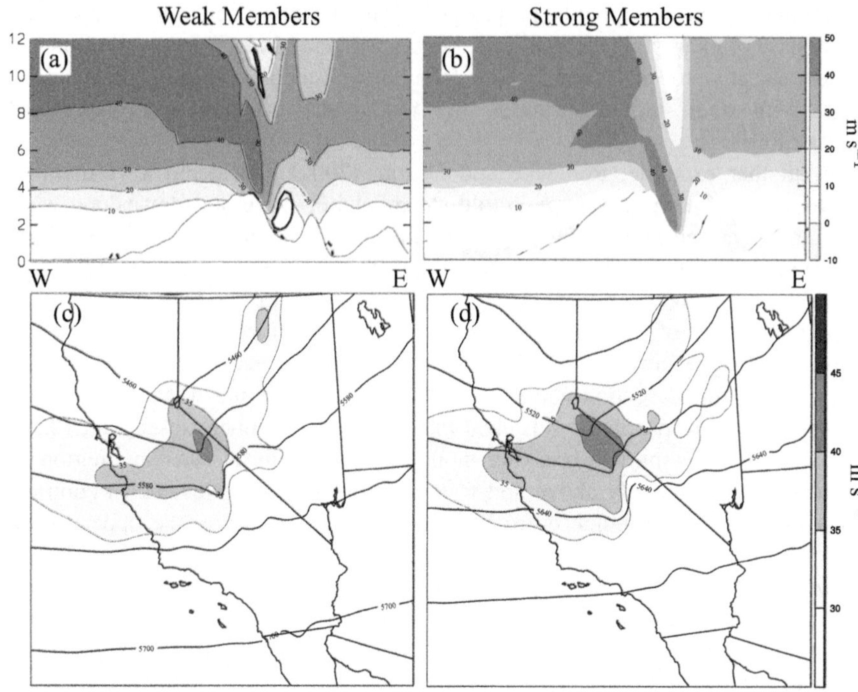

Fig. 18.5 Composite plots of a forecast valid 00 UTC 26 March 2006 of the (**a**, **c**) weak and (**b**, **d**) strong 10-member subsets as characterized by the downslope wind intensity in the Owens Valley for the forecast valid 00 UTC 26 March 2006. (**a**, **b**) The zonal wind (*shaded*) and predicted TKE (*heavy line*) along a W-E cross-section across the Sierra-Nevada mountains. (**c**, **d**) The 500 hPa geopotential height (contour) and wind speed (*shading*) (Adapted from Reinecke and Durran 2009) © American Meteorological Society. Reprinted with permission

25 March 2006 in which a large-amplitude mountain wave was observed during IOP 6 of the T-REX. The results from Doyle et al. (2011) indicate a surprisingly diverse spectrum of simulated mountain wave characteristics including lee waves, hydraulic-like jump features, and gravity wave breaking. As an example, consider the experiment that makes use of a terrain transect across the Sierra and a free-slip lower boundary condition shown in Fig. 18.6. The simulated potential temperature and horizontal wind speed perturbation for the 4-h time are shown. Broad similarities are apparent between all of the simulations; for example, they contain significant wave activity in the lee of the Inyo Range (to the east of the Sierra), as well as large-amplitude waves in the stratosphere. However, substantial variations are also apparent among the 11 model simulations (see Doyle et al. 2011 for details of the model configurations). Some models simulations develop a large-amplitude mountain wave and a strong wind storm along the lee slope of the Sierra. Several models simulate large-amplitude wave breaking and turbulence in the stratosphere, in contrast to other models that feature smaller amplitude waves in the lower troposphere. The results imply relatively low predictability of key characteristics of

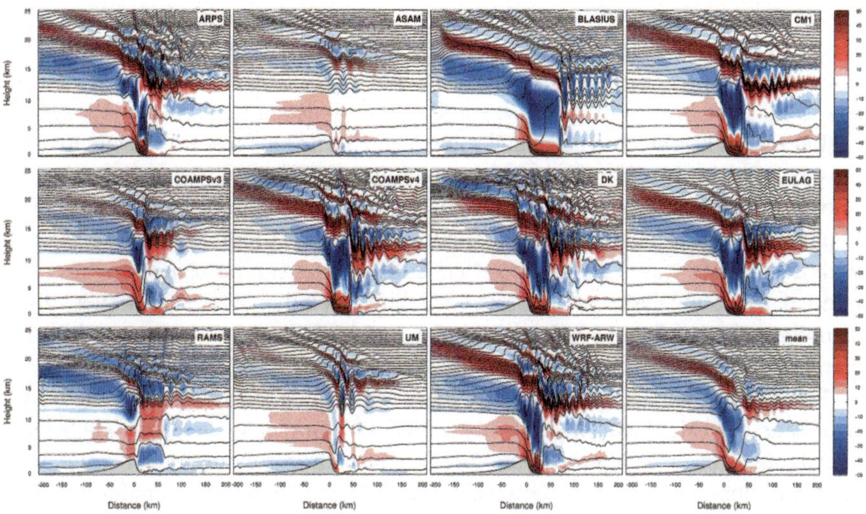

Fig. 18.6 Horizontal perturbation wind component (color, interval 5 m s^{-1}) and potential temperature (*black contours*, interval 10 K) for the Sierra terrain, free-slip case at the final time (4 h) for all 11 intercomparison models and the mean (*lower right panel*) (From Doyle et al. 2011) © American Meteorological Society. Reprinted with permission

topographically forced flows such as the stratospheric wave breaking and low-level severe downslope winds. The diversity among the various model simulations, all initialized with identical initial states, suggests that model dynamical cores may be an important component of diversity for the design of mesoscale ensemble systems for topographically forced flows. Doyle et al. (2011) found that the inter-model differences are significantly larger than sensitivity experiments within a single modeling system.

The model intercomparisons of the 11 January 1972 Boulder downslope windstorm and the 25 March 2006 T-REX IOP 6 case suggest that mountain wave breaking exhibits a strong sensitivity to the numerical formulation of the model dynamical core as well as sensitivity to the parameterized physics, such as sub-grid-scale mixing and the surface layer processes. The diversity of solutions between the dynamical cores suggests that model error is an important component of quantifying the predictability of mountain wave turbulence and should be captured in the design of mesoscale ensemble systems for topographically forced flows.

18.5 Rotors

The nomenclature referring to layers and regions of turbulence associated with rotors and mountain waves varies considerably. Lester and Fingerhut (1974) refer to a lower turbulent zone (LTZ) to describe a turbulent layer downwind of elevated terrain beneath an undulating inversion and coincident with lee waves (e.g., Fig. 18.1). The aviation community often uses the term "rotor" to more closely refer to the LTZ (e.g., Hertenstein 2009), while turbulence within lee waves and rotors also has been referred to as rotor-zone turbulence (e.g., WMO 1973). The spatial scale of turbulent eddies and rotors can vary from a few kilometers to a few tens of meters. Intense small-scale rotors within the LTZ have been called subrotors (Doyle and Durran 2007). Based on observations, two types of rotors have been hypothesized (Kuettner 1959; Lester and Fingerhut 1974): the first type associated with trapped lee waves and a second category having similar characteristics as an internal hydraulic jump or low-level breaking mountain wave. However, direct sampling of rotors is relatively rare to date and insufficient to distinguish between rotor types in a statistically meaningful sense.

Rotors are often associated with strong turbulence in the lower troposphere, and their characteristics have only recently been more completely understood in the past decade as a result of high-resolution turbulent eddy-resolving model simulations and new observations (e.g., Grubišić et al. 2008). Numerical simulations have suggested that a key aspect of rotor development involves the synergistic interaction between the lee wave and boundary layer, as discussed, for example, in Doyle and Durran (2002). Their simulations highlight lee waves that develop with attendant strong downslope winds (~25 m s^{-1}), along with rotors present near the surface beneath the lee wave crests, which feature regions of reversed cross-mountain winds. A sheet of horizontal vorticity develops in the region of high shear within the strong downslope winds in the boundary layer and forms in part due to surface friction processes along the lee slope (e.g., Doyle and Durran 2002, 2007). In the idealized simulations of Doyle and Durran (2002), the vortex sheet separates from the surface, ascends into the crest of the first lee wave, and remains aloft as it is advected downstream by the undulating flow in the lee waves. They found that boundary layer separation is facilitated by the adverse pressure gradients associated with trapped mountain lee waves. The leading edge of the rotor is a region of strong wind shear and turbulence production along an elevated sheet of horizontal vorticity, in general agreement with aircraft observations over the Sierra Nevada during the Sierra Wave Project (Holmboe and Klieforth 1957) and T-REX (Doyle et al. 2009), as well as in the lee of the Rocky Mountain Front Range (Lester and Fingerhut 1974).

Rotor characteristics, such as the turbulence intensity, are thought to be closely dependent on the boundary layer processes. As an illustration, simulations conducted by Doyle and Durran (2002) indicate that an increase in the surface roughness weakens the reversed flow, decreases the depth of the mean circulation, and impacts the downstream location due to an influence on the boundary layer separation. Föhn clouds upstream and over the mountains are often present in these events, which leads to an asymmetry in the solar radiation reaching the ground, with a cloud-free lee side due to descending air motion. Idealized simulations with an increased lee-side surface heat flux acts to further deepen the rotor circulation and increase the turbulence intensity (Doyle and Durran 2002), in general agreement with measurements (e.g., Kuettner 1959). In the absence of upstream Föhn clouds, surface heating deepens the upstream boundary layer and decreases the static stability, resulting in weaker downslope winds and an absence of rotors (Smith and Skyllingstad 2009).

High-resolution eddy-resolving simulations of mountain waves and rotors indicate the presence of small-scale vortices or subrotors (Doyle and Durran 2002, 2007) and highly turbulent flow (e.g., Hertenstein and Kuettner 2005) embedded within the parent or larger-scale mean rotor circulation. These subrotors may actually pose the greatest hazard to aviation. The evolution of the y-component of the horizontal vorticity for flow over an idealized three-dimensional ridge using a 60-m horizontal resolution mesh carried out by Doyle and Durran (2007) is shown in Fig. 18.7. A vertical section normal to the ridge along the centerline of the grid at 1 min intervals beginning at 3 h 30 min simulation time is displayed. Surface friction generates a sheet of horizontal vorticity along the lee slope that is lifted aloft by the mountain lee wave at the boundary layer separation point. Parallel shear instability breaks this vortex sheet into small intense vortices or subrotors. In two-dimensional simulations, the subrotors are less intense and are entrained into the larger-scale rotor circulation, where they dissipate and contribute their vorticity toward the maintenance of a vortex inside the larger-scale rotor. In three-dimensional simulations, the subrotors are more intense and are primarily advected downstream, with weaker vorticity in the interior of the main rotor and more chaotic flow. Doyle and Durran (2007) found that when an isolated peak is added to a ridge, systematic along-ridge velocity perturbations create regions of preferential vortex stretching at the leading edge of the rotor. Subrotors passing through such regions are intensified by stretching and achieve extreme intensities in some cases.

Doyle et al. (2009) present high-resolution simulations and observations from scanning Doppler and aerosol lidars, wind profiler radars, and surface and aircraft measurements during IOP 13 of T-REX. They document subrotor vortices with characteristic length scales of ~500–1000 m, which overturn and intensify, in some cases, to a maximum spanwise vorticity greater than 0.2 s^{-1}. Radar wind profiler observations document a series of vortices, with strong updrafts/downdraft couplets and regions of enhanced reversed flow, which are generated in a layer of strong vertical wind shear and subcritical Richardson number. A three-dimensional simulation (using a 60-m horizontal resolution) of the downslope flow, vortex sheet, and subrotor vortices is shown in Fig. 18.8 for the 3-h simulation time (from Doyle et al. 2009). The model was initialized with an upstream radiosonde at 2100 UTC 16 April 2006 (IOP 13) and uses terrain interpolated from a section across and normal to the Sierra Nevada crest near Independence, CA. A series of vortices form at the interface between the stronger flow aloft and weak flow positioned beneath the lee wave. An elevated zone of strong turbulence near the crest level of the Sierra occurs as vortices detach from the vortex sheet and subsequently advect downstream. The vortices also contain considerable three-dimensional structure, including tubes of horizontal vorticity with maxima greater than 0.15 s^{-1} oriented in the y-direction, approximately normal to the mean flow. As the subrotors break from the vortex sheet, intensification occurs through vortex stretching, and in some cases,

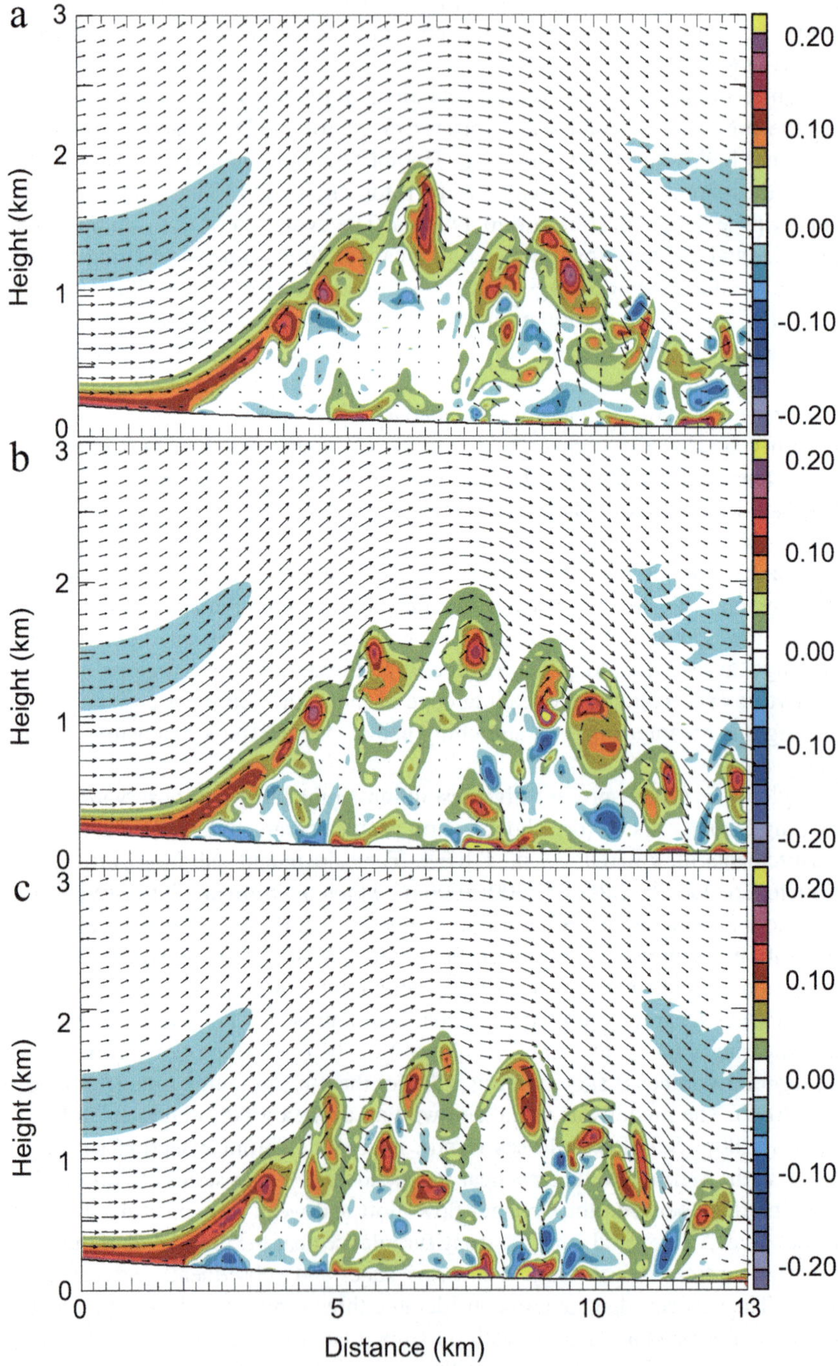

Fig. 18.7 Vertical cross section of the *y*-component of horizontal vorticity and wind vectors in the plane of the cross section for the 60-m resolution mesh of a 3D simulation with an infinitely long uniform ridge at the (**a**) 3 h 30 min, (**b**) 3 h 31 min, and (**c**) 3 h 32 min times. The horizontal

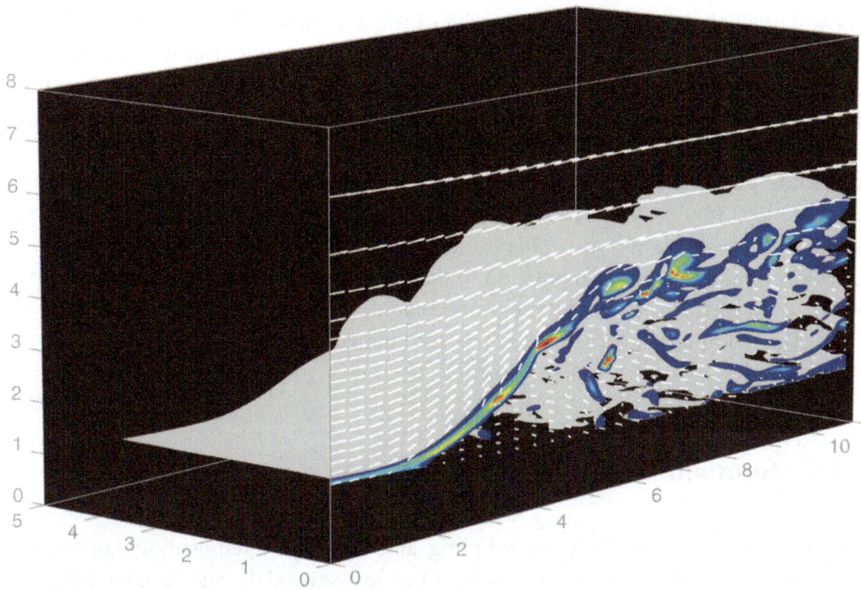

Fig. 18.8 Three-dimensional view of the downslope flow and subrotor vortices. The isosurface of the 0.02 s^{-1} y-component of the horizontal vorticity is shown by the *gray shading*, with the *red* isosurface corresponding to the 0.15 s^{-1} vorticity. The *gray* isosurface that intersects the $y = 0$ km plane (at approximately 3.2 km or 53 grid cells from the northern edge) is shaded corresponding to positive values of the y component of the horizontal vorticity (*red* is 0.15 s^{-1}). Wind vectors in the $y = 0$ plane are shown. Distances are shown along the axes in km (From Doyle et al. 2009) © American Meteorological Society. Reprinted with permission

tilting processes related to three-dimensional turbulent mixing. Beneath the lee wave within the broader rotor-scale circulation, the flow is more chaotic with episodic periods of localized reversed flow directed back toward the mountains.

Hertenstein (2009) found the internal hydraulic jump type of rotor forms more often in the presence of a strong near-mountaintop inversion. The importance of mountaintop inversions for lee wave dynamics has been known for decades (e.g., Kuettner 1938, 1939; Corby and Wallington 1956). The sign and magnitude of vertical shear across the upstream inversion have a strong influence on rotors, with positive shear leading to the formation of the trapped-wave type of rotors (e.g., Hertenstein and Kuettner 2005). Hertenstein (2009) found that the presence of stronger inversions leads to higher-altitude and more turbulent mountain waves and rotors. Vosper (2004) used a series of idealized two-dimensional simulations to

Fig. 18.7 (continued) vorticity is represented by the color shading with interval of 0.02 s^{-1} (scale on the *right*). The cross section displays a portion of the 60-m resolution mesh. *Tick marks* along the abscissa are shown every 250 m with major *tick marks* every 1 km (From Doyle and Durran 2007) © American Meteorological Society. Reprinted with permission show that when the upwind shallow-water Froude number ($\overline{U}/\sqrt{g' z_i}$, where $\bar{U}$ is the cross barrier wind speed, g' is a reduced gravitational acceleration and z_i is the depth of the lowest layer or inversion height) is less than a threshold value, rotors form beneath the crests of resonant lee wave. With a further reduction of the Froude number, a hydraulic jump forms. Vosper (2004) notes a strong dependence of the rotor characteristics on the inversion strength and height. Similarly, Doyle et al. (2009) noted with a modest change in the upstream inversion strength, the subrotors and ambient vortex sheet intensify significantly. The strong dependence of rotors on the strength and position of the inversion and shear across the inversion layer described in the aforementioned studies highlights the challenges for predicting rotor characteristics, such as turbulence severity.

18.6 Summary and Future Directions

The rapid progress in our understanding and ability to predict mountain waves, wave-induced turbulence, and rotors over the past several decades can be attributed to a large degree to the advancement of high-resolution nonhydrostatic numerical weather prediction models, along with focused field campaigns. These campaigns have provided in situ and remotely sensed measurements of mountain waves, which has motivated a thorough evaluation of the models and elucidated new gravity wave dynamics and processes. In this chapter, we have highlighted aspects of upper-level mountain wave-induced turbulence associated with large topographic mountains such as Greenland and complex three-dimensional terrain and associated processes, rotors, as well as the predictability of mountain waves, wave breaking, and turbulence.

Large topographic obstacles such as Greenland have been shown to be an important source of mountain wave-induced turbulence. Field campaign measurement, turbulence reports from commercial aviation, and numerical simulations have highlighted the proclivity of flow over Greenland to generate upper-level wave breaking and turbulence in a variety of environmental conditions. The meteorological situation that most frequently generates wave breaking and turbulence of significance to aircraft in the upper troposphere and lower stratosphere involves low-level airflow over Greenland from the southeast. The southeasterly low-level flow results in the generation of mountain waves that propagate vertically up through the upper troposphere or lower stratosphere where the wind direction may change with height or even become westerly. In these situations, mountain waves may break, and turbulence will occur associated with a (mean state, directional) critical level. Other situations highlight the potential of strong westerly flow that impinges on Greenland and generates large-amplitude gravity waves, which overturn above the jet stream in regions of reverse wind shear (and reduced density of the air) and result in strong turbulence.

Airflow over more complex and three-dimensional terrain (relative to a large barrier such as Greenland) is capable of generating mountain wave-induced turbulence as well. After all, mountains are one of the most frequent generators of clear-air turbulence and are therefore of great importance for aviation. Measurements and numerical modeling based on T-REX, MAP, and other field campaign cases highlight the importance of the boundary layer and directional critical levels that have an important influence on the wave amplitudes that eventually reach the upper troposphere and stratosphere. Stagnant boundary layer winds that arise due to the blocking effects of three-dimensional terrain can result in the absorption of downward propagating gravity waves that reduce the amplitudes and downstream extent of trapped mountain lee waves and ultimately reduce the likelihood of wave energy leaking into the upper troposphere and lower stratosphere and breaking into turbulence. Mountain waves in the presence of directional wind shear can be absorbed or filtered by directional critical levels that effectively reduce the mountain wave energy and the potential for wave breaking and turbulence aloft. The role of upstream blocking by complex terrain can lead to situations where lower terrain near the end of a three-dimensional ridge launches larger-amplitude mountain waves relative to higher terrain that is located closer to the center of the ridge, as was the case over the Sierra Nevada Range during T-REX.

A great deal has been learned about the internal structure of rotors in the past two decades, in part because of the great advances in the realism of numerical models and more sophisticated observations. Numerical models and observations have documented the fine-scale structure of rotors and subrotors and their turbulent nature, as well as their dynamical links to upstream atmospheric conditions, mountain lee waves, and downslope winds. However, forecast guidance for rotors has yet to be established despite their importance for aviation safety. Recent numerical simulations highlight the meteorological conditions favorable for the formation of strong rotors, which are broadly similar to those correlated with the development of trapped lee waves and downslope winds including: (1) a significant cross-mountain wind speed component of at least 10 m s^{-1}, (2) an elevated inversion near crest level, and (3) a significant increase in the cross-mountain wind speed above crest level. More extensive theoretical, numerical modeling, and observational research is needed to map out the governing parameters and refine these general guidelines. Forecasting of mountain waves and turbulence for operational aviation applications currently primarily uses empirical and statistical techniques, which could be extended to include rotors using more sophisticated methods. Additionally, eddy-resolving, three-dimensional, high-resolution, ensemble models may be able to ultimately provide probabilistic predictions of the nature of the turbulence associated with rotors in real time, as greater computational resources become available.

Numerical prediction of mountain wave-induced turbulence observed in nature is fraught with difficulty due to errors in the initial conditions, boundary conditions (e.g., for limited area models), and the model parameterizations and dynamics. Initial condition errors can grow rapidly in some circumstances and lead to large errors in the flow impinging on the orography, which may result in poor prediction of the timing or location of mountain wave launching. Similarly, errors in the predicted static stability, winds, or moisture fields can result in dramatically different wave responses (which have implications for rotor formation). For example, subtle changes in vertical stability can lead to downstream wave responses that evolve from trapped states to vertically propagating waves that may break in the upper levels and may be an aviation hazard. The mountain top stability likely needs to be predicted with great fidelity for accurate prediction of mountain waves and related turbulence, as well as rotors. Errors in the wind profile above a mountain may filter waves or allow more wave energy to propagate to higher altitudes, subsequently impacting the likelihood of wave breaking and turbulence and the formation of rotors. These predictability issues underscore the need to apply high-resolution ensemble models to predict mountain wave-induced turbulence in a probabilistic manner. Application of high-resolution ensembles that are capable of explicitly resolving mountain waves will allow for probabilistic forecasts of turbulence, which are ultimately needed for aviation hazard avoidance. As these mesoscale ensemble approaches are still in their infancy, probabilistically based tools and metrics are needed for aviation applications. The communication of uncertainty in the aviation turbulence forecast process remains a challenge.

Quantifying the practical predictability limits of mountain waves and wave-induced turbulence including the sensitivity to the initial state, boundary conditions, and model components remains unresolved. Predictability tools such as high-resolution adjoint and ensemble models need to be applied to topographically forced flows to address basic questions related to the predictability of mountain waves and turbulence forced by the lower boundary. These models need to be capable of simulating and predicting orographically forced phenomena with a high degree of skill and realism in order to test predictability theories and simulate conditions found in nature. The numerical weather prediction community may face important questions regarding the point of diminishing returns as model horizontal resolutions approach 1 km and less because of predictability issues. Higher horizontal and vertical resolutions should ultimately produce more physically realistic predictions of turbulence relevant to aviation. However, the balance between high-resolution deterministic models and number of ensemble members needs to carefully assessed for topographically forced flows including mountain wave-induced turbulence.

Acknowledgments We acknowledge a number of collaborations and helpful discussions related to this work with Adrian Broad, Andreas Dörnbrack, Dale Durran, Steve Eckermann, David Fritts, Sasa Gabersek, Vanda Grubišić, Todd Lane, Carolyn Reynolds, Melvyn Shapiro, Robert Sharman, and Ronald Smith. We appreciate the thorough review and constructive comments from Teddie Keller, Todd Lane, and Robert Sharman. Thanks to Lara Ziady for constructing Fig. 18.1. The authors acknowledge support from the Chief of Naval Research through the NRL Base Program, PE 0601153N. We acknowledge computational support from a grant of High Performance Computing (HPC) time from the Navy Defense Resource Center (DSRC) at Stennis, MS. COAMPS is a trademark of the Naval Research Laboratory.

References

Aebischer, U., Schär, C.: Low-level potential vorticity and cyclogenesis to the lee of the Alps. J. Atmos. Sci. **55**, 186–207 (1998)

Ágústsson, H., Ólafsson, H.: Simulations of observed lee waves and rotor turbulence. Mon. Weather Rev. **142**, 832–849 (2014)

Anthes, R.A., Kuo, Y., Baumhefner, D.P., Errico, R.M., Bettge, T.W.: Prediction of mesoscale atmospheric motions. Adv. Geophys. **28B**, 159–202 (1985)

Bacmeister, J.T., Schoeberl, M.R.: Breakdown of vertically propagating two-dimensional gravity waves forced by orography. J. Atmos. Sci. **46**, 2109–2134 (1989)

Breeding, R.J.: A nonlinear investigation of critical levels for internal atmospheric gravity waves. J. Fluid Mech. **50**, 545–563 (1971)

Bretherton, F.P.: Momentum transport by gravity waves. Quart. J. Roy. Meteorol. Soc. **95**, 213–243 (1969)

Carney, T.Q., Bedard, A.J., Jr., Brown, J.M., McGinley, J., Lindholm, T., Kraus, M.J.: Hazardous Mountain Winds and Their Visual Indicators. Handbook. Department of Commerce, NOAA, Boulder, Colorado, 80 pp. [Federal Aviation Administration Advisory Circular AC00-57.] (1996)

Clark, T.L., Peltier, W.R.: Critical level reflection and the resonant growth of nonlinear mountain waves. J. Atmos. Sci. **41**, 3122–3134 (1984)

Clark, T.L., Hall, W.D., Kerr, R.M., Middleton, D., Radke, L., Ralph, F.M., Nieman, P.J., Levinson, D.: Origins of aircraft-damaging clear air turbulence during the 9 December 1992 Colorado downslope windstorm: numerical simulations and comparison to observations. J. Atmos. Sci. **57**(8), 1105–1131 (2000)

Corby, G.A., Wallington, C.E.: Airflow over mountains: the lee-wave amplitude. Quart. J. Roy. Meteor. Soc. **82**, 266–274 (1956)

Darby, L.S., Poulos, G.S.: The evolution of mountain wave/rotor activity in the lee of Pike's Peak under the influence of a cold frontal passage: implications for aircraft safety. Mon. Weather Rev. **134**, 2857–2876 (2006)

Dolgushin, L.D., Osipova, G.B.: Glaciers. Mysl Publishing, Moscow (1989). 444 pp

Dörnbrack, A.: Turbulent mixing by breaking gravity waves. J. Fluid Mech. **375**, 113–141 (1998)

Dörnbrack, A., Dürbeck, T.: Turbulent dispersion of aircraft exhausts in regions of breaking gravity waves. Atmos. Environ. **32**, 3105–3112 (1998)

Dörnbrack, A., Gerz, T., Schumann, U.: Turbulent breaking of overturning gravity waves below a critical level. Appl. Sci. Res. **54**, 163–176 (1995)

Doyle, J.D., Durran, D.R.: The dynamics of mountain-wave-induced rotors. J. Atmos. Sci. **59**, 186–201 (2002)

Doyle, J.D., Durran, D.R.: Rotor and subrotor dynamics in the lee of three-dimensional terrain. J. Atmos. Sci. **64**, 4202–4221 (2007)

Doyle, J.D., Jiang, Q.: Observations and numerical simulations of mountain waves in the presence of directional wind shear. Quart. J. Roy. Meteor. Soc. **132**, 1877–1905 (2006)

Doyle, J.D., Reynolds, C.A.: Implications of regime transitions for mountain wave breaking predictability. Mon. Weather Rev. **136**, 5211–5223 (2008)

Doyle, J.D., Smith, R.B.: Mountain waves over the Hohe Tauern. Quart. J. Roy. Meteor. Soc. **129**, 799–823 (2003)

Doyle, J.D., Durran, D.R., Colle, B.A., Chen, C., Georgelin, M., Grubišić, V., Hsu, W.R., Huang, C.Y., Landau, D., Lin, Y.-L., Poulos, G.S., Sun, W.Y., Weber, D.B., Wurtele, M.G., Xue, M.: An intercomparison of model predicted wave breaking for the 11 January 1972 Boulder windstorm. Mon. Weather Rev. **128**, 901–914 (2000)

Doyle, J.D., Shapiro, M.A., Jiang, Q., Bartells, D.: Large-amplitude mountain wave breaking over Greenland. J. Atmos. Sci. **62**, 3106–3126 (2005)

Doyle, J.D., Grubišić, V., Brown, W.O.J., De Wekker, S.F.J., Dörnbrack, A., Jiang, Q., Mayor, S. D., Weissmann, M.: Observations and numerical simulations of subrotor vortices during T-REX. J. Atmos. Sci. **66**, 1229–1249 (2009)

Doyle, J.D., Gaberšek, S., Jiang, Q., Bernardet, L., Brown, J.M., Dörnbrack, A., Filaus, E., Grubišić, V., Kirshbaum, D., Knoth, O., Koch, S., Stiperski, I.M., Vosper, S., Zhong, S.: An intercomparison of T-REX mountain wave simulations and implications for mesoscale predictability. Mon. Weather Rev. **139**, 2811–2831 (2011)

Durran, D.R.: Another look at downslope windstorms. Part I: The development of analogs to supercritical flow in an infinitely deep, continuously stratified fluid. J. Atmos. Sci. **43**, 2527–2543 (1986)

Durran, D.R., Gingrich, M.A.: Atmospheric predictability: why butterflies are not of practical importance. J. Atmos. Sci. **71**, 2476–2488 (2014)

Eliassen, A., Palm, E.: On the transfer of energy in stationary mountain waves. Geofys. Publ. **22** (3), 1–23 (1961)

Epifanio, C.C., Durran, D.R.: Three-dimensional effects in high-drag-state flows over long ridges. J. Atmos. Sci. **58**, 1051–1065 (2001)

Fritts, D.C., Alexander, M.J.: Gravity wave dynamics and effects in the middle atmosphere. Rev. Geophys. **41**(1), 1003 (2003). doi:10.1029/2001RG000106

Fritts, D.C., Garten, J.F., Andreassen, Ø.: Wave breaking and transition to turbulence in stratified shear flows. J. Atmos. Sci. **53**, 1057–1085 (1996)

Grubišić, V., Billings, B.J.: The intense lee-wave rotor event of Sierra Rotors IOP 8. J. Atmos. Sci. **64**, 4178–4201 (2007)

Grubišić, V., Lewis, J.M.: Sierra Wave Project revisited: 50 years later. Bull. Am. Meteorol. Soc. **85**, 1127–1142 (2004)

Grubišić, V., Orlić, M.: Early observations of rotor clouds by Andrija Mohorovičić. Bull. Am. Meteorol. Soc. **88**, 693–700 (2007)

Grubišić, V., Smolarkiewicz, P.K.: The effect of critical levels on 3D orographic flows: linear regime. J. Atmos. Sci. **54**, 1943–1960 (1997)

Grubišić, V., Doyle, J.D., Kuettner, J., Mobbs, S., Smith, R.B., Whiteman, C.D., Dirks, R., Czyzyk, S., Cohn, S.A., Vosper, S., Weissman, M., Haimov, S., De Wekker, S., Pan, L., Chow, F.K.: The terrain-induced rotor experiment: an overview of the field campaign and some highlights of special observations. Bull. Am. Meteorol. Soc. **89**, 1513–1533 (2008)

Hertenstein, R.F.: The influence of inversions on rotors. Mon. Weather Rev. **137**, 433–446 (2009)

Hertenstein, R.F., Kuettner, J.P.: Rotor types associated with steep lee topography: Influence of the wind profile. Tellus **57A**, 117–135 (2005)

Hines, C.O.: Internal atmospheric gravity waves at ionospheric heights. Can. J. Phys. **38**, 1441–1481 (1960)

Hodur, R.M.: The Naval Research Laboratory's Coupled Ocean/Atmosphere Mesoscale Prediction System (COAMPS). Mon. Weather Rev. **125**, 1414–1430 (1997)

Holmboe, J., Klieforth, H.: Investigations of mountain lee waves and airflow over the Sierra Nevada. Final Rep., Contract AF19(604)-728, University of California ADNo. 133606, Dept. of Meteorology, University of California, Los Angeles, CA, 290 pp. (1957)

Jiang, Q., Doyle, J.D.: Gravity wave breaking over the central Alps: role of complex terrain. J. Atmos. Sci. **61**, 2249–2266 (2004)

Joly, A., et al.: The Fronts and Atlantic Storm-Track Experiment (FASTEX): scientific objectives and experimental design. Bull. Am. Meteor. Soc. **78**, 1917–1940 (1997)

Koschmieder, H.: Zwei bemerkenswerte Beispiele horizontaler Wolkenschlaeuche. Beitr. Phys. Frei Atmos. **9**, 176–180 (1920)

Kuettner, J.P.: Moazagotl und Föhnwelle. Beitr. Phys. Frei Atmos. **25**, 79–114 (1938)

Kuettner, J.P.: Zur Entstehung der Föhnwelle. Beitr. Phys. Frei Atmos. **25**, 251–299 (1939)

Kuettner, J., Jenkins, C.F.: Flight aspects of the mountain wave. Air Force Surveys in Geophysics No. 35. AFCRC Technical Report 53-36, Air Force Cambridge Research Center, 20 pp. 19 pp. (1953)

Kuettner, J.P.: The rotor flow in the lee of the mountains. Geophysics Research Directorate (GRD) Research Notes 6, AFCRC-TN-58-626, Air Force Cambridge Research Center, 20 pp. (1959)
Kühnlein, C., Dörnbrack, A., Weissmann, M.: High-resolution doppler lidar observations of transient downslope flows and rotors. Mon. Weather Rev. **141**, 3257–3272 (2013)
Lane, T.P., Doyle, J.D., Sharman, R.D., Shapiro, M.A., Watson, C.D.: Statistics and dynamics of aircraft encounters of turbulence over Greenland. Mon. Weather Rev. **117**, 2687–2702 (2009)
Lester, P.F., Fingerhut, W.A.: Lower turbulent zones associated with lee waves. J. Appl. Meteorol. **13**, 54–61 (1974)
Lester, P.F., Sen, O., Bach, R.E.: The use of DFDR information in the analysis of a turbulence incident over Greenland. Mon. Weather Rev. **137**, 1103–1107 (1989)
Leutbecher, M., Volkert, H.: The propagation of mountain waves into the stratosphere: quantitative evaluation of three-dimensional simulations. J. Atmos. Sci. **57**, 3090–3108 (2000)
Lilly, D.K.: A severe downslope windstorm and aircraft turbulence event induced by a mountain wave. J. Atmos. Sci. **35**, 59–77 (1978)
Lilly, D.K., Kennedy, P.J.: Observations of a stationary mountain wave and its associated momentum flux and energy dissipation. J. Atmos. Sci. **30**, 1135–1152 (1973)
Lilly, D.K., Zipser, E.J.: The Front Range windstorm of January 11, 1972. Weatherwise **25**, 56–63 (1972)
Lindzen, R.S.: Thermally driven diurnal tide in the atmosphere. Quart. J. Roy. Meteor. Soc. **93**, 18–42 (1967)
Lorenz, E.N.: The predictability of a flow which possesses many scale of motion. Tellus **21**, 289–307 (1969)
Mahalov, A., Moustaoui, M., Grubišić, V.: A numerical study of mountain waves in the upper troposphere and lower stratosphere. Atmos. Chem. Phys. **11**, 4487–4532 (2011)
Mass, C.F., Ovens, D., Westrick, K., Colle, B.A.: Does increasing horizontal resolution produce more skillful forecasts? Bull. Am. Meteorol. Soc. **83**, 407–430 (2002)
Mobbs, S.D., Vosper, S., Sheridan, P., Cardoso, R.M., Burton, R.R., Arnold, S.J., Hill, M.K., Horlacher, V., Gadian, A.M.: Observations of downslope winds and rotors in the Falkland Islands. Quart. J. Roy. Meteor. Soc. **131**, 329–351 (2005)
Nuss, W.A., Miller, D.K.: Mesoscale predictability under various synoptic regimes. Nonlinear Process Geophys. **8**, 429–438 (2001)
Ólafsson, H., Ágústsson, H.: Gravity wave breaking in easterly flow over Greenland and associated low-level barrier- and reverse tip-jets. Meteorol. Atmos. Phys. **104**, 191–197 (2009)
Ólafsson, H., Bougeault, P.: Nonlinear flow past an elliptical ridge. J. Atmos. Sci. **53**, 2465–2489 (1996)
Ólafsson, H., Bougeault, P.: The effect of rotation and surface friction on orographic drag. J. Atmos. Sci. **54**, 193–210 (1997)
Paegle, J.: Predictability of flows over complex terrain. Atmospheric Processes over Complex Terrain, Meteor. Monogr., No. 45. Am. Meteor. Soc., 285–299 (1990)
Palmer, T.N., Shutts, G.J., Swinbank, R.: Alleviation of systematic westerly bias in general circulation and numerical weather prediction models through an orographic gravity wave drag parameterization. Quart. J. Roy. Meteor. Soc. **112**, 1001–1039 (1986)
Peltier, W.R., Clark, T.L.: The evolution and stability of finite-amplitude mountain waves. Part II: Surface wave drag and severe downslope windstorms. J. Atmos. Sci. **36**, 1498–1529 (1979)
Ralph, F.M., Neiman, P.J., Levinson, D.: Lidar observations of a breaking mountain wave associated with extreme turbulence. Geophys. Res. Lett. **24**, 663–666 (1997)
Reinecke, P.A., Durran, D.R.: Initial condition sensitivities and the predictability of downslope winds. J. Atmos. Sci. **66**, 3401–3418 (2009)
Rotunno, R., Snyder, C.: A generalization of Lorenz's model for the predictability of flows with many scales of motion. J. Atmos. Sci. **65**, 1063–1076 (2008)
Schär, C., Smith, R.B.: Shallow-water flow past isolated topography. Part I: Vorticity production and wake formation. J. Atmos. Sci. **50**, 1373–1400 (1993)

Schmid, H., Dörnbrack, A.: Simulation of breaking gravity waves during the south foehn of 7–13 January 1996. Beitr. Phys. Atmos. **72**, 287–303 (1999)
Scorer, R.S.: The theory of airflow over mountains–IV. Separation of flow from the surface. Q. J. R. Meteor. Soc. **81**, 340–350 (1955)
Sharman, R.D., Doyle, J.D., Shapiro, M.A.: An investigation of a commercial aircraft encounter with severe clear-air turbulence over western Greenland. J. Appl. Meteor. Climatol. **51**(1), 42–53 (2012)
Shen, B.-W., Lin, Y.-L.: Effects of critical levels on two-dimensional back-sheared flow over an isolated mountain ridge on an *f* plane. J. Atmos. Sci. **56**, 3286–3302 (1999)
Sheridan, P.F., Vosper, S.B.: A flow regime diagram for forecasting lee waves, rotors and downslope winds. Meteor. Appl. **13**, 179–195 (2006)
Shutts, G.: Observations and numerical model simulations of a partially trapped lee wave over the Welsh mountains. Mon. Weather Rev. **120**, 2056–2066 (1992)
Shutts, G.J.: Gravity-wave drag parameterization over complex terrain: the effect of critical-level absorption in directional wind-shear. Quart. J. Roy. Meteor. Soc. **121**, 1005–1021 (1995)
Shutts, G.J.: Stationary gravity-wave structure in flows with directional wind shear. Quart. J. Roy. Meteor. Soc. **124**, 1421–1442 (1998)
Shutts, G.J., Gadian, A.: Numerical simulation of orographic gravity waves in flows which back with height. Quart. J. Roy. Meteor. Soc. **125**, 2743–2765 (1999)
Smith, R.B.: On severe downslope winds. J. Atmos. Sci. **43**, 2597–2603 (1985)
Smith, R.B.: Hydrostatic flow over mountains. Adv. Geophys. **31**, 1–41 (1989)
Smith, C.M., Skyllingstad, E.D.: Investigation of upstream boundary layer influence on mountain wave breaking and lee wave rotors using a large-eddy simulation. J. Atmos. Sci. **66**, 3147–3164 (2009)
Smith, R.B., Skubis, S.T., Doyle, J.D., Broad, A., Kiemle, C., Volkert, H.: Mountain waves over Mt. Blanc: influence of a stagnant boundary layer. J. Atmos. Sci. **59**, 2073–2092 (2002)
Smith, R.B., Woods, B.K., Jensen, J., Cooper, W.A., Doyle, J.D., Jiang, Q., Grubišić, V.: Mountain waves entering the stratosphere. J. Atmos. Sci. **65**, 2543–2562 (2008)
Thorpe, A.J., Volkert, H., Heimann, D.: Potential vorticity of flow along the Alps. J. Atmos. Sci. **50**, 1573–1590 (1993)
VanZandt, T.E., Fritts, D.C.: A theory of enhanced saturation of the gravity wave spectrum due to increases in atmospheric stability. Pure Appl. Geophys. **130**, 399–420 (1989)
Vosper, S.B.: Inversion effects on mountain lee waves. Quart. J. Roy. Meteor. Soc. **130**, 1723–1748 (2004)
Vosper, S.B., Mobbs, S.D.: Lee waves over the English Lake District. Quart. J. Roy. Meteorol. Soc. **122**, 1283–1305 (1996)
Wang, T.-A., Lin, Y.-L.: Wave ducting in a stratified shear flow over a two-dimensional mountain. Part I: General linear criteria. J. Atmos. Sci. **56**, 412–436 (1999)
WMO: The airflow over mountains: Research 1958–1972. World Meteorological Organization Tech. Note 127, Geneva, Switzerland, 72 pp. (1973)
Wolff, J., Sharman, R.: Climatology of upper-level turbulence over the continental Unites States. J. Appl. Meteor. Climatol. **47**, 2198–2214 (2008)
Woods, B.K., Smith, R.B.: Energy flux and wavelet diagnostics of secondary mountain waves. J. Atmos. Sci. **67**, 3721–3738 (2010)
Worthington, R.M.: Tropopausal turbulence caused by the breaking of mountain waves. J. Atmos. Solar-Terr. Phys. **60**, 1543–1547 (1998)

Chapter 19
Gravity Waves Generated by Jets and Fronts and Their Relevance for Clear-Air Turbulence

Riwal Plougonven and Fuqing Zhang

Abstract Jets and fronts are known from observations and modeling studies to be an important source of gravity waves, primarily waves with low intrinsic frequencies. Waves in the jet-exit regions have recurrently been documented and are now rather well understood. Nonetheless, other waves (e.g., involving convection) are also present in the jet/front regions, and there is no simple model available yet to quantitatively predict excited waves from the knowledge of the large-scale flow. Only a handful of case studies have analyzed how jet-generated gravity waves contribute to the occurrence of clear-air turbulence (CAT) in the vicinity of the upper tropospheric jet. They have confirmed that the modulation of vertical shear and stability by strong, low-frequency gravity waves produces localized bands of turbulence. Further studies would be required to discriminate between this and other mechanisms (convectively generated gravity waves, inertial instability) that lead to CAT in the vicinity of jet streaks.

19.1 Introduction

Jets and fronts in the mid-latitudes produce intense motions on scales smaller than the synoptic scale, and part of these motions are internal gravity waves. This has been known from observations of the surface pressure (Uccelini and Koch 1987) and has become a topic of research motivated by the contribution of gravity waves to the circulation of the middle atmosphere (Fritts and Alexander 2003). The strong horizontal and vertical shear and the discontinuity in static stability at the

R. Plougonven (✉)
Laboratoire de Météorologie Dynamique, Ecole polytechnique, Palaiseau, France
e-mail: plougon@lmd.ens.fr

F. Zhang
Department of Meteorology, Pennsylvania State University, University Park, PA, USA
e-mail: fzhang@psu.edu

R. Sharman, T. Lane (eds.), *Aviation Turbulence*,
DOI 10.1007/978-3-319-23630-8_19

tropopause provide a favorable environment to reflect, capture, and break the jet-/front-generated gravity waves which may contribute to clear-air turbulence (Koch et al. 2005). Now, upper-level jets are also regions of strong shears (Kennedy and Shapiro 1980), which are favorable to events of clear-air turbulence (CAT). It has been highlighted for a long time that gravity waves generated from fronts will contribute to fluctuations of the shear and stability on small scales and can hence contribute to the appearance of CAT in these regions (Mancuso and Endlich 1966; Knox 1997). Indeed, for waves with large enough amplitudes, wave breaking produces mixing and turbulence (e.g., Fritts et al. 2003). It is of particular importance to predict occurrences of CAT, and the tropopause region near the jet stream is a major source of CAT events (e.g., Kim and Chun 2011; Sharman et al. 2012).

Despite considerable progress being made on the emission of gravity waves from jets and fronts over the past few decades (Plougonven and Zhang 2014 and references therein), the mechanisms responsible for the generation of gravity waves by jets and fronts remain somewhat elusive, making the prediction of CAT events tied to jet-generated gravity waves difficult. A fundamental reason is the theoretical difficulty in describing gravity waves excited by motions (jets and fronts) which have been understood through balanced approximations which precisely filter out gravity waves. In other words, the emission of gravity waves by balanced models is generally weak enough that the appropriate simplification made to model jets and fronts is to exclude gravity wave motions. Modeling the gravity waves emitted by motions that were initially balanced (hence referred to as *spontaneously* emitted waves) is a challenging problem which has motivated numerous theoretical studies (Vanneste 2013 and references therein). Moreover, gravity waves present in the vicinity of jet/front systems can be forced, modified, or enhanced by other sources. For instance, the contribution of moist processes has been emphasized in recent studies (Wei and Zhang 2014; Mirzaei et al. 2014; Plougonven et al. 2015).

The typical characteristics of waves emitted near jets and fronts (low-frequency waves, amplitudes up to 10 m s^{-1} for the horizontal wind perturbations) are such that one can expect a significant contribution from them to shear and stability. The induced fluctuations are expected to lead to conditions locally conducive to turbulence. This has been confirmed by the existing case studies of CAT due to gravity waves generated by jet/front systems (Lane et al. 2004; Koch et al. 2005).

Below we first recall certain relevant properties of internal gravity waves (Sect. 19.2). We then provide elements on the observational evidence of gravity waves excited by jets and fronts (Sect. 19.3) and give an overview of the current understanding of gravity wave generation mechanisms relevant in the vicinity of jets and fronts (Sect. 19.4). Case studies of CAT due to gravity waves generated from upper-level jets are described in Sect. 19.5. The main results and current challenges are summarized in Sect. 19.6.

19.2 Some Considerations on Gravity Wave Characteristics

It is useful to begin by recalling certain fundamental characteristics of gravity waves and certain key relations between physical variables within a wave. Internal gravity waves in a stratified atmosphere have intrinsic frequencies bounded above by the buoyancy frequency (or Brunt–Väisälä frequency) $N = \sqrt{\frac{g}{\theta_0}\frac{d\theta}{dz}}$, where θ is the potential temperature with θ_0 as a reference value and g is gravity, and bounded below by the Coriolis frequency $f = 2\Omega \sin\varphi$, where φ is the latitude and Ω is the Earth's rotation rate. Typical values of the buoyancy frequency are ~10^{-2} s^{-1} in the troposphere and about double in the lower stratosphere (Gettelman et al. 2011). In mid-latitudes, typical values of the Coriolis parameter f are about ~10^{-4} s^{-1}, implying that intrinsic frequencies span about two orders of magnitudes. The dispersion relation (Gill 1982; Fritts and Alexander 2003), disregarding compressibility, is

$$\omega^2 = \frac{(k^2 + l^2)N^2 + m^2 f^2}{k^2 + l^2 + m^2}, \tag{19.1}$$

where k, l, and m are the three components of the wave number and ω is the intrinsic frequency. The dispersion relation reveals that intrinsic frequency depends only on the slope of the tilt of the wave number relative to the vertical. In the atmosphere, horizontal wavelengths are found to vary on a wider range of scales (roughly from thousands of kilometers to a kilometer) than vertical wavelengths (roughly from 1 to 10 km). In consequence, although the dispersion relation does not depend on the actual scale of the wave, atmospheric gravity waves display a contrast between slow waves (low intrinsic frequencies, close to f) which have large horizontal scales (and rather small vertical scales) and fast waves (frequencies intermediate between f and N, or close to N) which are deeper in the vertical and have shorter horizontal scales (such as those described in Chap. 16).

The contribution of gravity waves to turbulence in the upper troposphere and lower stratosphere region comes from the possible initiation of dynamic instabilities (shear or convective) due to the wave-induced fluctuations. Hence it is of interest to estimate the possible contribution of gravity waves to the vertical shear and to the vertical gradient of potential temperature. As explained in Lane et al. (2004), linear theory allows us to calculate the contribution of a gravity wave to the vertical shear and to the vertical gradient of potential temperature as

$$\frac{A_{u_h} 2\pi}{\lambda_z} \quad \text{and} \quad \frac{A_\theta 2\pi}{\lambda_z}, \tag{19.2}$$

respectively, where A_{u_h} is the amplitude of the wave perturbation on the horizontal wind and A_θ is the amplitude on potential temperature. Again disregarding compressibility, the relation between the two can be shown to be (Fritts and Alexander 2003)

$$|A_\theta| = \frac{\theta_0 N^2}{g} \left|\frac{\lambda_z}{\lambda_h \omega}\right| |A_{u_h}|. \tag{19.3}$$

Hence, if we consider typical values of background parameters for the lower stratosphere and mid-latitudes ($f = 10^{-4}$ s^{-1} and $N = 0.02$ s^{-1}) and inject values typical of large-amplitude inertia–gravity waves described in case studies ($\lambda_h \sim 200$ km, $\lambda_z \sim 2$ km, $A_{u_h} = 5$ m s^{-1}, yielding $\omega = 2.2f$), we find that the wave's contribution to the vertical shear is of the order of 0.016 s^{-1}, and its contribution to the vertical gradient of potential temperature is about 9 K km^{-1}, to be compared with 12 K km^{-1} typically for the background stratification. In contrast, a high-frequency wave with a deeper vertical wavelength (e.g., associated to convection) would, for the same significant signature in the horizontal wind, have weaker impacts on the modulation of stratification and shear. Taking $\lambda_h \sim 40$ km and $\lambda_z \sim 10$ km (which yields $\omega = 48.5f = 0.25$ N), for example, yields contributions of about 0.003 s^{-1} for the shear and 2 K km^{-1} for potential temperature. While still significant, this is, for a given amplitude of perturbation, considerably weaker than in the case of the low-frequency inertia–gravity wave.

The main points to retain from the above relations (19.1–19.3) are as follows:

1. Low-frequency waves (often called "inertia–gravity waves") correspond to waves with nearly horizontal phase lines and a signature which is relatively strong in horizontal velocities; in practice they are found to have vertical wavelengths typically of 1–4 km, while horizontal scales range from a hundred to several hundreds of km (Sato 1994; Guest et al. 2000). Amplitudes of 5–10 m s^{-1} have commonly been described in case studies (see Sect. 19.3), implying that such waves have a potential to contribute significant modulations of the shear and stratification.
2. High-frequency waves correspond to waves with phase lines which have horizontal scales that are comparable or shorter than the vertical scales. These are for instance convectively generated waves, for which the vertical wavelengths typically scale with the troposphere depth (Piani et al. 2000; Beres et al. 2004). These waves contribute relatively strongly to vertical displacements and hence to potential temperature fluctuations. Convectively generated gravity waves have been shown to contribute to CAT events, as discussed in Chap. 16 and Trier et al. (2012).

19.3 Observational Evidence

Gravity waves in the vicinity of jets and fronts have been identified in observations for several decades, both in case studies and in more statistical investigations (Uccelini and Koch 1987). The focus has expanded, over the years, from

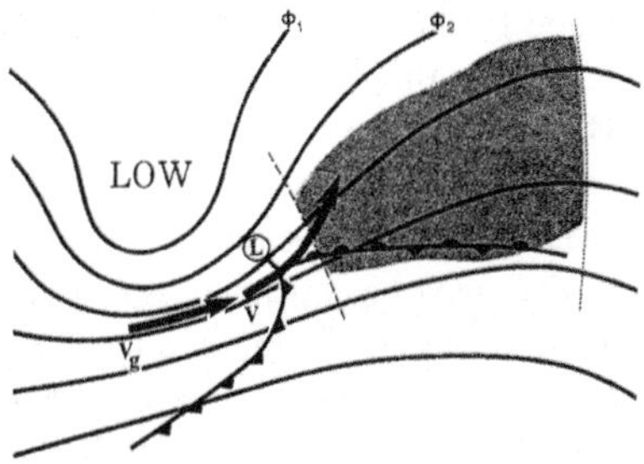

Fig. 19.1 Schematic of the configuration of the flow identified by Uccelini and Koch (1987) as conducive to intense gravity waves in the exit region of the upper-level jet: lines of geopotential height in the mid-troposphere and surface fronts are indicated. Intense inertia–gravity waves have recurrently been found in the shaded region, just downstream of the inflection axis (*dashed line*). Adapted from Koch and O'Handley (1997)

tropospheric waves, typically detected in networks of surface barographs, to upper tropospheric and stratospheric waves, detectable from a variety of different measurements (radar, radiosondes, superpressure balloons, satellites) (Fritts and Alexander 2003; Plougonven and Zhang 2014). Early case studies have identified a configuration of the flow regularly present in events of intense gravity waves in the troposphere. In a mid-latitude baroclinic wave, it consists of the region downstream of an upper-level jet, cold-air side of the surface front, and close to the ridge of geopotential (Uccelini and Koch 1987, see Fig. 19.1). The relevance of this paradigm has been confirmed numerous times in subsequent case studies (e.g., Bosart et al. 1998; Guest et al. 2000). Such large-amplitude waves in the lower layers can have impacts on the organization and enhancement of moist convection but are challenging to be accurately simulated in convection-permitting models (Zhang et al. 2001, 2003). Several other studies also identified the exit region of the upper-level jet as the source for stratospheric gravity waves using sounding or satellite measurements in combination with numerical models and/or ray-tracing techniques (Guest et al. 2000; Hertzog et al. 2001; Wu and Zhang 2004; Zhang et al. 2013). Other regions of the flow have also been highlighted as significant from observational case studies. First, jet-exit regions upstream of a trough also appear favorable for enhanced gravity wave activity (Plougonven et al. 2003). In several studies of waves from upper-level jets, gravity waves have been found above and below the jet, propagating upward and downward, respectively (Hirota and Niki 1986; Thomas et al. 1999). This is a clear evidence that the jet is the source. These waves typically have intrinsic frequencies close to the Coriolis parameter. In contrast to this, the vicinity of a cold front has also been highlighted as a source, in particular when significant wind impinges on the front (e.g., Ralph et al. 1999). Complementary to case studies, observational evidence showing gravity waves from jet/front systems also comes from statistical studies, i.e., studies that examine small-scale motions in a large number of observations and relate those to likely sources (e.g., Fritts and Nastrom 1992; Koppel et al. 2000; Guest et al. 2000). Systematic in situ measurements of gravity waves associated with jet/front are very scarce, although Nastrom and Fritts (1992) and Fritts and Nastrom (1992) did infer that jet/front systems were significant producers of gravity waves using commercial aircraft measurements. Many other types of observations can be used to identify gravity waves, from in situ measurements from high-resolution radiosondes to global remote-sensing observations from satellites. Each type of observation will see only a part of the gravity wave spectrum (Alexander et al. 2010). Presently, there is an ongoing effort to combine the knowledge acquired from the different datasets available, complemented by the outcomes of numerical models, in order to obtain a global estimation of the gravity wave field (Geller et al. 2013).

Global descriptions of the gravity wave field obtained from satellite observations typically highlight several large-scale regions with enhanced gravity wave activities: regions of orography, the tropics, and the winter mid-latitudes associated, respectively, with mountain waves, convective waves, and waves from jets and fronts. Often, maps will tend to highlight the hotspot regions of orography on the path of jet streams (Hoffmann et al. 2013). It matters to distinguish between orographic and non-orographic gravity waves (Plougonven et al. 2013), the first being the most conspicuous but the latter also being important. Indeed, relatively intense gravity waves can also be found in connection to jets and fronts (Guest et al. 2000; Plougonven et al. 2003). Regarding the circulation of the middle atmosphere, the contribution from non-orographic gravity waves is likely at least comparable to that from orographic sources (Hertzog et al. 2008; Plougonven et al. 2013). Positive gravity wave momentum forcing in the summer extratropical middle atmosphere, which is required for driving observed wind and temperature in the middle atmosphere, comes completely from non-orographic gravity waves (Kim et al. 2003). Regarding their contribution to turbulence, Pavelin et al. (2001) describes a persistent inertia–gravity wave observed by radar and radiosonde in the lowermost stratosphere above the upper tropospheric jet. A vertical wavelength of about 2 km and amplitudes of nearly 10 m s^{-1} for the wind perturbations were such that the Richardson number associated to these waves was locally weak and that turbulence due to wave breaking was present. In another case study using airborne measurements, the development of Kelvin–Helmholtz billows from a region of strong shear near the jet stream (but not necessarily involving gravity waves) was captured as a sawtooth pattern in potential temperature (Whiteway et al. 2004).

Gravity waves are not solely described from a time-averaged amplitude. An important characteristic of gravity waves is their intermittency (Hertzog et al. 2008; Alexander et al. 2010). Different approaches have been used to quantify this (Hertzog et al. 2008, 2012; Plougonven et al. 2013; Wright et al. 2013). Consistently across datasets, orographic waves are found to be significantly more intermittent than non-orographic gravity waves, meaning that rare, extremely intense gravity waves above mountains are more likely than such events near jets and fronts.

Finally, observational case studies have highlighted the complexity of the flow and processes involved in a life cycle of a gravity wave. Starting in the 1990s, case studies have frequently included numerical simulations with a mesoscale meteorological model as complementary to observations (e.g., Powers and Reed 1993; Zhang et al. 2001). The more complete description of the three-dimensional, time-dependent flow has revealed that for a given wave event there are generally several processes occurring, each of which may influence the emission and maintenance of waves (emission from the jet and/or front, interaction with topography and/or moist processes leading to amplification of the waves, ducting, etc.). In consequence, there can be several possible interpretations for a given gravity wave event.

In summary, observations have provided compelling evidence regarding the importance of jets and fronts as sources of gravity waves and have identified several configurations particularly favorable to gravity waves, in particular jet-exit regions. Mechanisms of generation, however, have not been clearly identified from observations alone. A more extensive review of this subject can be found in Plougonven and Zhang (2014).

19.4 Generation Mechanisms of Gravity Waves Generated by Jets and Fronts

Understanding which mechanisms excite gravity waves from jets and fronts is a crucial first step to determine characteristics of gravity waves likely to be excited and whether they may produce CAT. Ideally, one wishes to have a conceptual model that would determine, from a large-scale description of the flow, the gravity waves that are excited and how they propagate away (as for orographic waves, Gill 1982; Queney 1948). However, understanding the generation of gravity waves near jets and fronts has proven a challenging problem, and no such model is available. Nonetheless, theoretical studies, modeling, and observational studies have proposed several generation mechanisms and have identified regions of the flow and wave types that are likely to be found in the vicinity of jets. Below we first explain why this is a difficult problem, outline some confusions which should be avoided, and finally describe the current understanding of generation mechanisms for gravity waves near jets and fronts.

The essential difficulty in describing the emission of gravity waves excited by jets and fronts comes from the fact that the waves are generated by the internal dynamics of the atmosphere and more precisely by features (jet/front systems) which have themselves been a challenge to model and are tied to essentially horizontal motions. This contrasts with orographic gravity waves and even with convectively generated waves, for both of which the forcing (flow over an obstacle, diabatic heating) directly forces vertical motion. Now, mid-latitude low-pressure systems have been understood in approximations like the quasi-geostrophic approximation, which provide a framework for analyzing baroclinic instability (Charney 1948; Eady 1949; Holton 1992). The mesoscale circulations leading to frontogenesis have required more sophisticated models, such as the semi-geostrophic approximation (Hoskins and Bretherton 1972; Hoskins 1982). Both of these approximations are balanced approximations: they simplify the flow by describing only the part of the flow associated with potential vorticity (Hoskins et al. 1985), and they hence filter out gravity waves by construction. This assumption has been remarkably efficient and is justified by the weakness of the coupling between balanced motions and gravity waves, which has even led to the conjecture that balanced motions did not radiate any energy into gravity wave motions (Leith 1980; Lorenz 1980) and remained on a "*slow manifold*" (see Vanneste 2013 and references therein).

Many studies of gravity waves generated by jets and fronts have invoked geostrophic adjustment as the mechanism responsible for the emission of the waves (e.g., Uccelini and Koch 1987; O'Sullivan and Dunkerton 1995). Classical geostrophic adjustment theories typically describe the evolution of either small perturbations to a fluid at rest or of perturbations to a flow that has a symmetry (zonally symmetric flow or axisymmetric flow) (Rossby 1938; Blumen 1972). Although this picture of adjustment may guide our intuition, it is not quantitatively helpful: first, it circumvents the main difficulty of understanding how the evolution of the balanced flow itself produces regions of imbalance. Second, theoretical studies have described the propagation of emitted waves only in simple flows, whereas in practice the waves are emitted in a complex, three-dimensional flow which strongly influences their characteristics. Third, the underlying balance in the classical adjustment hypothesis is the large-scale geostrophic balance that may not be accurate to describe balance motions associated with jets and fronts. As a generalization of geostrophic adjustment, Zhang (2004) proposed the concept of the spontaneous balance adjustment to explain the gravity waves generated from flow imbalance in the jet-exit region in which (1) the nonlinear balance is used as the underlying balance and (2) the mostly balanced larger-scale background (baroclinic) flow may continuously generate imbalance while the flow is continuously adjusted toward balance (removal of imbalance) through forcing gravity waves. A heuristic semi-analytical derivation of the spontaneous balance adjustment concept is presented in Plougonven and Zhang (2007), with a proof-of-concept demonstration through numerical simulations in Wang and Zhang (2010).

An important non-dimensional number which indicates the relevance of balanced dynamics is the Rossby number. Consider a flow with L a typical length scale and U a typical velocity. The Rossby number is the ratio of the inertial timescale, $1/f$, where f is the Coriolis parameter, and the advective timescale, L/U: $\mathrm{Ro} = U/fL$. When Ro is <1, this indicates that advection occurs on timescales much longer than the fastest of the gravity waves (inertia–gravity waves with periods close to the inertial period). The dynamics is then well captured by approximations which describe the part of the flow that is tied to the advection of potential vorticity, i.e., the balanced or vortical part of the flow.

Therefore, determining what gravity waves are generated from jets and fronts requires determining how the evolution of flows that are initially balanced leads to emission of gravity waves. This spontaneous emission (or spontaneous adjustment emission, SAE) is quite distinct from classical geostrophic adjustment problems, in which the imbalance is prescribed in the initial condition. Quantifying spontaneous emission is an arduous problem which has motivated numerous studies over the past three decades.

Several pathways for SAE have been studied and quantified theoretically. The first has been Lighthill radiation, which is analogous to the emission of acoustic waves from turbulent motions, as was described by James Lighthill (1952). It describes the emission of waves which have the same vertical length scale and the same timescale as the balanced flow that is emitting them. It has been successfully applied to quantify waves emitted from vortices and jets in the shallow water model (e.g., Ford 1994a, b; Sugimoto et al. 2008) and in a stratified fluid (e.g., Plougonven and Zeitlin 2002; Schecter 2008). The assumptions (same vertical wavelength) and the dispersion relation of gravity waves impose that the emitted waves occur in the far field, with horizontal length scales that are larger than those of the balanced flow. This scale separation is what makes an analytical description possible (Ford et al. 2000), but it also makes Lighthill radiation clearly distinct from the emission from jets and fronts, where waves are found within the balanced flow and with shorter horizontal and vertical wavelengths (Plougonven et al. 2009).

Two other mechanisms illustrating the emission of gravity wave motions from balanced flow have been described analytically:

- Unbalanced instabilities are instabilities which combine motions that are balanced (in a region of the flow) to motions that are gravity waves (in another region), the two being coupled through shear (Plougonven et al. 2005; Vanneste and Yavneh 2007) or differential rotation (Ford 1994c).
- Transient generation in shear is the emission of waves from sheared anomalies of potential vorticity and has been described in horizontal (Vanneste and Yavneh 2004) and vertical shear (Lott et al. 2010).

Both mechanisms emphasize the importance of a background flow (shear) which acts to couple motions that have different intrinsic timescales but which may share the same absolute frequency, thanks to Doppler shift. In such simple but nontrivial background flows, the standard separation between balanced motions and gravity waves (which underlies the classical scenario of geostrophic adjustment) no longer strictly holds. Direct application to observed or simulated atmospheric flow is however not straightforward as the background flows considered remain highly simplified relative to the three-dimensional, time-evolving flow found in developing baroclinic waves. Moreover, the gravity waves explainable from both types of the SAE hypotheses are likely too weak to be relevant to CAT.

Much understanding of baroclinic waves and frontogenesis has come from idealized simulations, and such simulations have also constituted key contributions to the understanding of gravity waves generated from fronts (Snyder et al. 1993) and jets (O'Sullivan and Dunkerton 1995; Zhang 2004). In many cases, both from observations and from idealized case studies, jet-exit regions have been emphasized as a favored locus for the presence of intense inertia–gravity waves. Consideration of the propagation of inertia–gravity waves in a flow characterized by strong deformation and vertical shear, akin to what is found in a jet-exit region, was shown to lead to waves with a specific intrinsic frequency and with phase lines oriented along the extensional axis of the deformation field. This is consistent with the orientation and intrinsic frequencies of waves simulated and observed in such regions. This phenomenon has been called "wave capture" and has been emphasized as providing a route for the background flow to lead a gravity wave to dissipation (as the wavelength contracts) (Bühler and McIntyre 2005). The relevance of these "propagation effects" has been verified and illustrated in a number of idealized studies (Plougonven and Snyder 2005; Lin and Zhang 2008; Wang et al. 2010). The similarity between jet-exit region waves simulated in idealized baroclinic life cycles and those observed in case studies (Uccelini and Koch 1987; Guest et al. 2000; Plougonven et al. 2003) have motivated further idealized studies of the phenomenon, using dipoles as a background flow. Dipoles retain essential features (deformation, vertical shear, along-jet deceleration) but are quasi-steady in the appropriate frame of reference, providing a simpler configuration to quantify the waves. A robust phenomenology has emerged from the multiple simulations carried out by different groups using very diverse models (Snyder et al. 2007; Viudez 2007, 2008; Wang et al. 2009): the dipole continuously emits waves that have low intrinsic frequencies and are partly captured in the jet-exit region, with characteristics consistent with wave capture. A successful framework for predicting the emission of waves has been applied to this problem: it consists of approximating the dipole with a balanced model, linearize the equations around this balanced dipole, and force these linear equations with the residual tendencies resulting from the difference between the balanced and the full equations (Plougonven and Zhang 2007; Snyder et al. 2009; Wang and Zhang 2010). This is analogous to approaches described in earlier works on frontogenesis (Snyder et al. 1993; Reeder and Griffiths 1996).

The waves predicted to be present in jet-exit regions are low-frequency waves, with intrinsic frequencies typically between 1 and $2f$ and with vertical scales typically between 1 and 4 km (Plougonven and Snyder 2005). As such, they are likely to significantly contribute to an enhancement of the shear already present. Moreover, a number of observational case studies have reported amplitudes of such waves on the order of 5–10 m s^{-1} (e.g., Pavelin et al. 2001; Guest et al. 2000; Plougonven et al. 2003), stronger than what is found in idealized simulations. These factors are favorable to a contribution of these waves to CAT. The pathways for such contribution will be discussed based on case studies in Sect. 19.5.

The emphasis on jet-exit region-emitted (JEREmi) waves should not hide that other waves and generation mechanisms are also expected to play a role near jet/front systems: surface fronts are also emitting gravity waves (Zhang 2004), possibly at higher intrinsic frequencies. The attribution to the jet or to surface fronts can be delicate given the complex nature of the flow (Lin and Zhang 2008). Yet there are cases for which the emission from the fronts is unambiguous (Plougonven and Snyder 2007) and strongly reminiscent of orographic waves, with the front playing the role of an obstacle for an impinging surface wind (Ralph et al. 1999). Such effects are present in dry-idealized simulations and lead to waves with intrinsic frequencies that are somewhat larger (a few times the Coriolis parameter,

Plougonven and Snyder 2007). In moist-idealized baroclinic life cycles (Wei and Zhang 2014; Mirzaei et al. 2014) as well as in real atmospheric flows (Wu and Zhang 2004; Zhang et al. 2013; Plougonven et al. 2015), recent studies have emphasized the contribution of moist convection to the emission of gravity waves. Although weaker than in the tropics, the convection is efficient at forcing gravity waves because the effect of the diabatic heating directly forces vertical motions. Here again, these mechanisms will tend to produce waves with shorter horizontal scales than JEREmi waves and larger vertical wavelengths.

Finally, it is worth mentioning that the emission of gravity waves from small-scale shear instability has long been considered as a potential mechanism relevant near jets (e.g., Mastrantonio et al. 1976; Lalas and Einaudi 1976; Fritts 1982). However, the emission is a priori weak, hindered by fundamental reasons (McIntyre and Weissman 1978), requiring nonlinear processes (Chimonas and Grant 1984; Fritts 1984), and has been difficult to model because of the wide range of scales involved (Scinocca and Ford 2000). This pathway for gravity wave generation is however the opposite (upscale transfer of energy from turbulence to gravity waves) to the one that interests us here and will not be discussed further.

19.5 Case Studies of Clear-Air Turbulence and Gravity Waves

The severe clear-air turbulence colliding with air traffic (SCATCAT) experiment, conducted in 2001 using the National Oceanographic and Atmospheric Administration's Gulfstream IV (G-IV) reconnaissance aircraft, was a key observational campaign dedicated to the observation of CAT at altitudes relevant to air traffic. On the flight of 17–18 February 2001, an intense jet/front system was sampled over the Pacific at approximately 40N, 200E, including regions of clear-air turbulence in the strongly sheared region above the jet (see Fig. 19.2). This case was studied extensively, using mesoscale simulations with several models (Lane et al. 2004; Koch et al. 2005). The highest resolution run, using a nested cloud-resolving model (Clark 1977), had a resolution of $dx = 1$ km in the horizontal and $\Delta z = 50$ m in the vertical (Fig. 19.3). The simulations showed that strong gravity waves were emitted from the jet core from a region of strong imbalance, with phase lines aligned with the strong northwesterly flow. This location and orientation is reminiscent of waves found in idealized experiments downstream of the jet-exit region, along the northwesterly flow going into the trough (Zhang 2004; Plougonven and Snyder 2007). These waves had a vertical wavelength of the order of 2.5 km, horizontal wavelength on the order of 120–200 km, and signatures of about 3 K on temperature and 5 m s^{-1} on the wind (Lane et al. 2004; Koch et al. 2005). These waves have amplitudes sufficient to contribute significantly to vertical shear (~1.3×10^{-2} s^{-1}) and to the vertical gradient of θ (~7 K km^{-1}), modulating the Richardson number Ri and producing alternating bands of enhanced and decreased Richardson number.

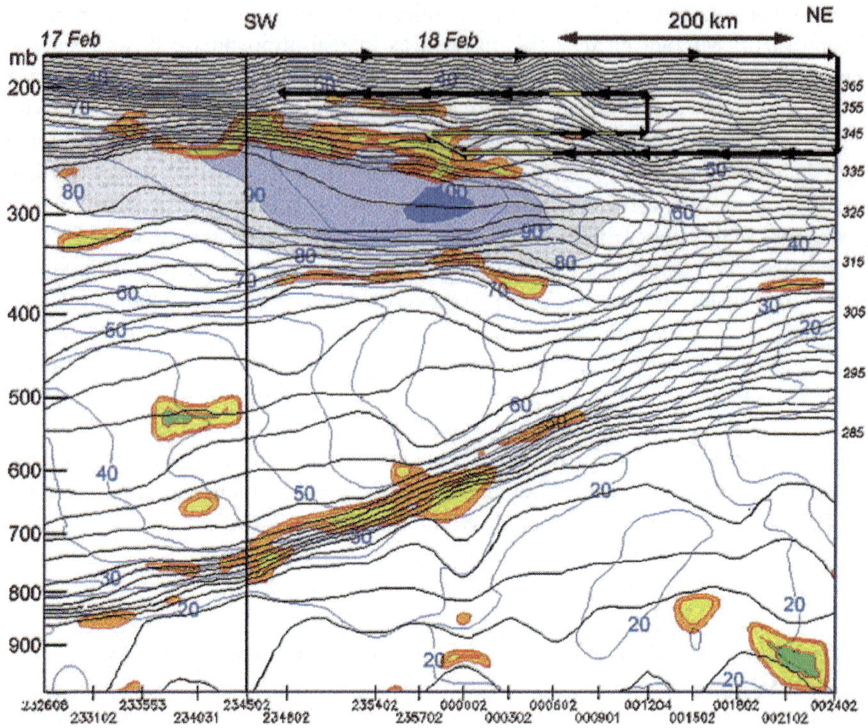

Fig. 19.2 Vertical cross section depicting the jet/front system, regions of expected turbulence, and track of the research aircraft, as obtained from dropsondes: wind speed (*blue lines*, 5 m s^{-1} isotachs), potential temperature (*black lines*, 2K isolines), and DTF3 turbulence diagnostic (shaded; see Koch et al. 2005 for details) computed from the dropsondes (release times at the bottom of the figure). Also shown are the four stacked legs of the G-IV aircraft (*thick black lines*), with *yellow* highlight indicating segments for which moderate or greater turbulence was encountered. Note the wavy patterns in the isentropes above the tropopause, indicative of gravity waves emanating from the jet core. Adapted from Koch et al. (2005)

The shear associated to the jet was such that the background Richardson number had values around 2. Bands of reduced Ri coincide with regions of enhanced turbulent kinetic energy in the high-resolution nested runs, supporting the interpretation that the modulation of the shear and stability by the gravity waves led to at least some of the CAT encountered by the G-IV. Other indicators of turbulence were also investigated, with horizontal maps again yielding banded distributions emphasizing the contribution from the jet-generated gravity waves (Koch et al. 2005). Dropsondes from the G-IV were analyzed for gravity waves, albeit only on a limited vertical range, and revealed waves consistent with the simulations (Lane et al. 2004).

Aircraft measurements were analyzed and showed that a wide spectrum of high-frequency gravity waves were also present at smaller scales, with horizontal wavelengths between 1 and 20 km (Koch et al. 2005). Regions of turbulence were found to be highly correlated with the appearance of packets of gravity

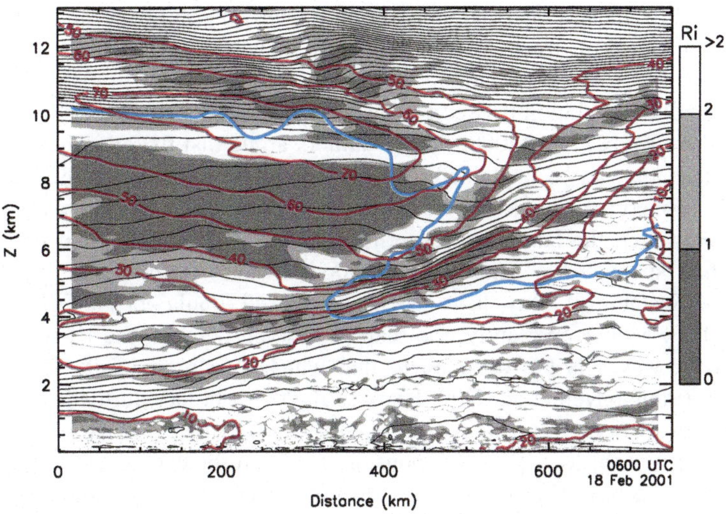

Fig. 19.3 Vertical cross section from the model simulation described by Lane et al. (2004) for which observations are shown in Fig. 19.2: Richardson number (Ri, shaded), potential temperature (*black lines*, 2K isolines), wind speed (*red lines*, 5 m s^{-1} isotachs), and the tropopause, defined as the 2 PVU contour of potential vorticity (*blue line*). The cross section is oriented southwest to northeast, with its center located at approximately 43N, 205E; the horizontal grid spacing is 3 km and the vertical grid spacing is 100 m. Adapted from Lane et al. (2004)

waves with similar characteristics. Analysis of third-order structure functions were used to analyze energy transfers between gravity waves and turbulence. This provided evidence for a downscale cascade of energy with gravity waves creating conditions conducive to turbulence on scales of several hundred meters (Koch et al. 2005).

Further analysis using spectral methods and third-order structure functions was carried out with particular attention to distinguishing flight segments with and without turbulence (Lu and Koch 2008). A clear k^{-2} spectrum was found for flight segments that did not include turbulence, consistent with expectations. In flight segments that sampled turbulence, a narrow region of upscale energy transfer was found for scales between 100 m and ~1 km, suggesting that gravity wave generation from turbulent motions (Kelvin–Helmholtz billows from shear instability typically) may be occurring, as studied theoretically (Bühler et al. 1999; Scinocca and Ford 2000).

In the case studies described above, the connection between the flow and the waves is not quantitatively related to one of the mechanisms described in Sect. 19.4. The flow in the vicinity of jets and fronts is complex, three-dimensional and time varying, making a quantitative analysis difficult. Parameterizations of gravity waves in climate models have been elaborated with tentative relations between the flow and the generated waves, based on qualitative arguments (Charron and Manzini 2002; Richter et al. 2010). Quantitative comparison has been attempted in only a few cases (Zülicke and Peters 2006, 2008). In practice, tying the emitted waves to an indicator of frontogenesis introduces intermittency and locates the sources in the storm tracks, which has been shown to have a positive effect in climate models.

A quantitative relation between jet-generated gravity waves and CAT has been explored by Knox et al. (2008). These authors argue that the emission of gravity waves near jets can be explained as Lighthill emission, their argument referring strongly to laboratory experiments (Lovegrove et al. 2000; Williams et al. 2005, 2008). However, the laboratory experiments were controversial, and phenomena other than spontaneous generation of gravity waves (unbalanced instabilities, Holmboe instability) may have been overlooked (Gula et al. 2009; Scolan et al. 2011). More importantly, the scale separation between the jet motions and the gravity waves is the opposite to that which is a prerequisite for Lighthill radiation (Plougonven et al. 2009; Knox et al. 2009). In fact, further investigation of this case (Trier et al. 2012) has recently showed that gravity waves due to convection were at least partly responsible for the turbulence events analyzed by Knox et al. (2008). The turbulence indicator set up by these authors may have relevance, but the interpretation of its success as evidence of Lighthill radiation is erroneous.

The relation of clear-air turbulence to jet/front systems has also been investigated over South Korea in case studies (Kim and Chun 2010) and in statistics over 5 years of pilot reports (Kim and Chun 2011). The strong shear associated to the upper-level jet can be associated to clear-air turbulence (Kim and Chun 2010), and turbulence events associated to the tropospheric jet account for the majority of the reported cases of clear-air turbulence (twice as frequent as mountain wave cases in the location and sample studied (Kim and Chun 2011)). The mechanisms responsible for the turbulence can include shear instability, inertial instability, and gravity waves generated from the jet/front system but cannot be identified in this statistical analysis.

One of the objectives of the recent Stratosphere–Troposphere Analyses of Regional Transport 2008 (START08) experiment (Pan et al. 2010) was to observe the properties of gravity waves generated by jets and fronts. In particular, one of the research flights was dedicated to measuring gravity waves associated with a strong upper tropospheric jet/front system (Zhang et al. 2015). Using a combination of spectral and wavelet analyses of the in situ aircraft observations, along with a diagnosis of the polarization relationships, Zhang et al. (2015) concluded that there are clear signals of significant mesoscale variations with wavelengths ranging from 50 to 500 km and amplitude ranging from 0.01 m s^{-1} to 1.0 m s^{-1} in vertical motion over a wide range of background conditions including near the jet core, a jet over the high mountains, and the exit region of the jet. They also found prevalence of transient smaller-scale wavelike oscillations below 50 km wavelengths which may be signals of wave-induced turbulence or aircraft measurement error: indeed, aircraft measurements of several flight segments are dominated by signals with sampled periods of 20–60 s and wavelengths of 5–15 km (assuming that the typical flight speed is approximately 250 m s^{-1}). The spectral analysis of horizontal winds and temperature produce power spectra which, averaged over many START08 flight segments, generally follow the −5/3 power law which was consistent with past observational (Nastrom and Gage 1985; Lindborg 1999) and modeling studies (e.g., Skamarock 2004; Waite and Snyder 2012; Bei and Zhang 2014).

19.6 Summary and Perspectives

Jets and fronts are significant sources of gravity waves in the mid-latitudes, and many studies in the past two decades have contributed to advance our understanding of the occurrence and generation of gravity waves near jets and fronts. The main motivation has been the role of gravity waves in driving the circulation of the middle atmosphere (Fritts and Alexander 2003) and the need for a physical foundation to the specification of non-orographic sources in parameterizations (Kim et al. 2003). Many observational case studies have identified and emphasized jet-exit regions as a favorable location for large-amplitude gravity waves (e.g., Uccelini and Koch 1987; Guest et al. 2000; Pavelin et al. 2001; Plougonven et al. 2003).

Idealized simulations, first of baroclinic life cycles (O'Sullivan and Dunkerton 1995; Zhang 2004; Plougonven and Snyder 2007) and then, simply, of dipoles (Snyder et al. 2007; Viudez 2007; Wang et al. 2009), have provided a theoretical framework for explaining these jet-exit region waves. The dipole is rather well described by a balanced approximation, but this nevertheless remains an approximation so that there is a residual between the balanced tendency and the full tendency. The spontaneously emitted waves constitute the response, in a diffluent, vertically sheared flow, to the discrepancy between the full equations of motion and the balanced approximation of the motion (Snyder et al. 2009; Wang et al. 2010; Wang and Zhang 2010). The fact that this response occurs in a background flow is more important than the specific choice of the balanced approximation yielding the residuals. In other words, the horizontal deformation and vertical shear present in the jet-exit region are crucial to influence and determine the characteristics of the gravity waves that are present, and this is consistent with the wave-capture mechanism (Bühler and McIntyre 2005; Plougonven and Snyder 2005). This constitutes an explanation for many waves observed in jet-exit regions, with phase lines normal to the flow and an intrinsic frequency close to the inertial frequency (Pavelin et al. 2001; Plougonven and Teitelbaum 2003; Plougonven and Snyder 2005).

This successful explanation of one conspicuous type of gravity wave encountered near jets and fronts should not overshadow the fact that other gravity waves are also present in the vicinity of jet/front systems. Surface fronts themselves are a source of gravity waves (Snyder et al. 1993; Ralph et al. 1999; Lin and Zhang 2008), with intrinsic frequencies somewhat larger (Plougonven and Snyder 2007). The flow is generally complex, and real case studies have often emphasized that a complex combination of processes (excitation, amplification, ducting) can play a role in shaping the gravity waves (Zhang et al. 2001).

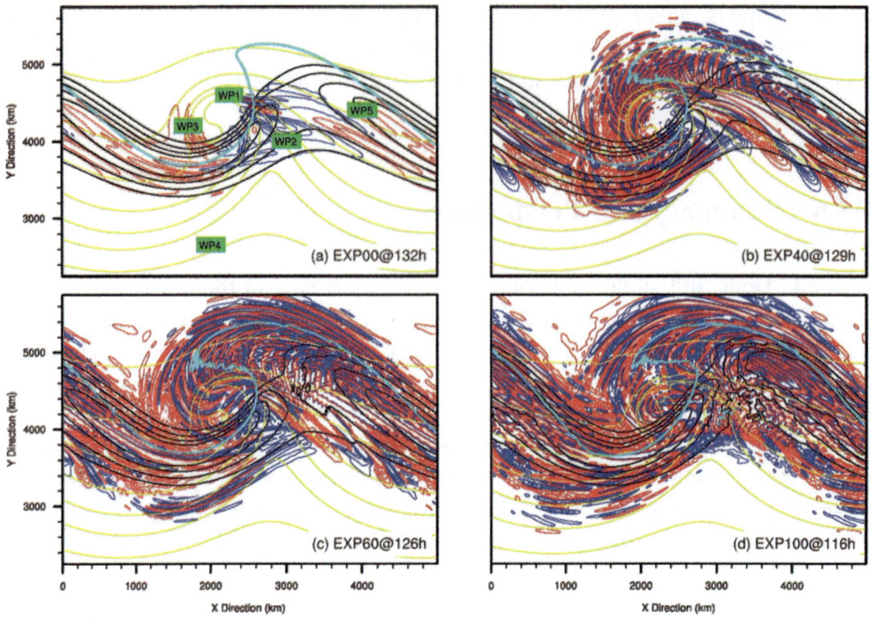

Fig. 19.4 Gravity waves in similar baroclinic jet/front systems but with different degree of convective instability from dry to fully moist. Shown are simulated 1-km temperature (*yellow lines*), 8-km horizontal wind (*black lines*), and 12-km horizontal divergence (*blue lines*, positive; *red lines*, negative) for (**a**) completely dry experiment EXP00 at 132 h, weakly to moderately convective (**b**) EXP40 at 129 h and (**c**) EXP60 at 126 h, and (**d**) strongly convective EXP100 at 116 h of the simulations. The *turquoise lines* denote the 7-km dynamic tropopause where potential vorticity equals 1.5 PVU. WP1–WP5 marked in panel (**a**) denote five different types of wave packets that can be generated in the dry baroclinic jet/front systems as discussed in Wei and Zhang (2014). Adapted from Wei and Zhang (2014)

Two additional remarks from idealized simulations need to be made: first, it has often been found that, although at early times in a cleanly prepared simulation it is possible to tie emerging gravity waves to a certain region of the flow, different sets of simulations have repeatedly shown how low-frequency waves, which have a slow, nearly horizontal group velocity, quickly occupy most of the jet region at later times (Plougonven and Snyder 2005; Waite and Snyder 2009, 2012; Wei and Zhang 2014). Second, recent simulations have included moisture in order to increase the realism and have emphasized how moist convection enhances and complicates the emitted gravity waves (Wei and Zhang 2014; Mirzaei et al. 2014). Figure 19.4 exemplifies the drastic difference in amplitude and wavelength of gravity waves among different baroclinic jet/front systems that have similar amplitude in baroclinicity but with a wide range of convective instability. The effect of moisture on gravity waves in the jet/front systems are also confirmed by real-data case studies (Wu and Zhang 2004; Zhang et al. 2013; Plougonven et al. 2015).

The waves that are emitted from jets and fronts can a priori have impacts on the turbulence occurring in the vicinity of the upper tropospheric jet. This is demonstrated by case studies (Lane et al. 2004; Koch et al. 2005) which have analyzed observations of a campaign dedicated to the issue of clear-air turbulence (SCATCAT; see Sect. 19.5). It was shown from observations and several numerical simulations that inertia–gravity waves were present on the flank of an intense jet stream. The waves modulated the shear and stratification, and their contribution was sufficient to produce turbulence in banded regions associated to a certain phase of the wave. This was shown through the analysis of different turbulence indicators (notably the Richardson number), the analysis of resolved and subgrid turbulent kinetic energy, and was consistent with the airborne observations of patches of turbulence. Other mechanisms leading to CAT in the vicinity of jets are also active (Kim and Chun 2011), including gravity waves that are tied to convection (Lane et al. 2012; Trier et al. 2012; Wei and Zhang 2014; Chaps. 16 and 17). A more systematic quantification of the link between jet-generated gravity waves and turbulence will require further case studies. While this research effort will likely benefit from research on non-orographic gravity waves that is motivated by parameterizations in climate models (Alexander et al. 2010), the specific issue of turbulence calls for specific observations. While pilot reports are a precious tool in this regard, dedicated research campaigns such as SCATCAT or the more recent START08 campaign provide unique opportunities for such investigations.

References

Alexander, M.J., Geller, M., McLandress, C., Polavarapu, S., Preusse, P., Sassi, F., Sato, K., Eckermann, S., Ern, M., Hertzog, A., Kawatani, Y., Pulido, M., Shaw, T.A., Sigmond, M., Vincent, R., Watanabe, S.: Recent developments in gravity-wave effects in climate models and the global distribution of gravity-wave momentum flux from observations and models. Q. J. Roy. Meteorol. Soc. **136**, 1103–1124 (2010)

Bei, N., Zhang, F.: Mesoscale predictability of moist baroclinic waves: variable and scale dependent error growth. Adv. Atmos. Sci., p. 9951008 (2014) doi:10.1007/s00376-014-3191-7

Beres, J., Alexander, M., Holton, J.: A method of specifying the gravity wave spectrum above convection based on latent heating properties and background wind. J. Atmos. Sci. **61**, 324–337 (2004)

Blumen, W.: Geostrophic adjustment. Rev. Geophys. Space Phys. **10**(2), 485–528 (1972)

Bosart, L., Bracken, W., Seimon, A.: A study of cyclone mesoscale structure with emphasis on a large-amplitude inertia-gravity wave. Mon. Weather Rev. **126**, 1497–1527 (1998)

Bühler, O., McIntyre, M.: Wave capture and wave-vortex duality. J. Fluid Mech. **534**, 67–95 (2005)

Bühler, O., McIntyre, M., Scinocca, J.: On shear-generated gravity waves that reach the mesosphere. Part I: Wave generation. J. Atmos. Sci. **56**, 3749–3763 (1999)

Charney, J.: On the scale of atmospheric motions. Geophys. Publ. Oslo **17**(2), 1–17 (1948)

Charron, M., Manzini, E.: Gravity waves from fronts: parameterization and middle atmosphere response in a general circulation model. J. Atmos. Sci. **59**, 923–941 (2002)

Chimonas, G., Grant, J.: Shear excitation of gravity waves. Part II: Upscale scattering from Kelvin-Helmholtz waves. J. Atmos. Sci. **41**, 2278–2288 (1984)

Clark, T.: A small-scale dynamical model using a terrain following coordinate transformation. J. Comput. Phys. **24**, 186–215 (1977)

Eady, E.: Long waves and cyclone waves. Tellus **1**, 33–52 (1949)

Ford, R.: Gravity wave radiation from vortex trains in rotating shallow water. J. Fluid Mech. **281**, 81–118 (1994a)
Ford, R.: The response of a rotating ellipse of uniform potential vorticity to gravity wave radiation. Phys. Fluids **6**(11), 3694–3704 (1994b)
Ford, R.: The instability of an axisymmetric vortex with monotonic potential vorticity in rotating shallow water. J. Fluid Mech. **280**, 303–334 (1994c)
Ford, R., McIntyre, M.E., Norton, W.A.: Balance and the slow quasi-manifold: some explicit results. J. Atmos. Sci. **57**, 1236–1254 (2000)
Fritts, D.: Shear excitation of atmospheric gravity waves. J. Atmos. Sci. **39**, 1936–1952 (1982)
Fritts, D.: Shear excitation of atmospheric gravity waves. 2: Nonlinear radiation from a free shear-layer. J. Atmos. Sci. **41**, 524–537 (1984)
Fritts, D., Alexander, M.: Gravity wave dynamics and effects in the middle atmosphere. Rev. Geophys. **41**(1), 1003 (2003)
Fritts, D., Nastrom, G.: Sources of mesoscale variability of gravity waves. Part II: Frontal, convective, and jet stream excitation. J. Atmos. Sci. **49**(2), 111–127 (1992)
Fritts, D., Bizon, C., Werne, J., Meyer, C.: Layering accompanying turbulence generation due to shear instability and gravity-wave breaking. J. Geophys. Res. **108**(D8), 8452 (2003)
Geller, M., Alexander, M., Love, P., Bacmeister, J., Ern, M., Hertzog, A., Manzini, E., Preusse, P., Sato, K., Scaife, A., Zhou, T.: A comparison between gravity wave momentum fluxes in observations and climate models. J. Climate **26**, 6383–6405 (2013). doi:10.1175/JCLI-D-1200545.1
Gettelman, A., Hoor, P., Pan, L., Randel, W., Hegglin, M., Birner, T.: The extratropical upper troposphere and lower stratosphere. Rev. Geophys. **49**(RG3003), 2011RG000,355 (2011)
Gill, A.E.: Atmosphere-Ocean Dynamics. Academic Press/Harcourt Brace, San Diego, CA (1982)
Guest, F., Reeder, M., Marks, C., Karoly, D.: Inertia-gravity waves observed in the lower stratosphere over Macquarie Island. J. Atmos. Sci. **57**, 737–752 (2000)
Gula, J., Zeitlin, V., Plougonven, R.: Instabilities of two-layer shallow-water flows with vertical shear in the rotating annulus. J. Fluid Mech. **638**, 27–47 (2009)
Hertzog, A., Souprayen, C., Hauchecorne, A.: Observation and backward trajectory of an inertia-gravity wave in the lower stratosphere. Ann. Geophys. **19**, 1141–1155 (2001)
Hertzog, A., Boccara, G., Vincent, R., Vial, F., Coquerez, P.: Estimation of gravity-wave momentum fluxes and phase speeds from long-duration stratospheric balloon flights. 2. Results from the Vorcore campaign in Antarctica. J. Atmos. Sci. **65**, 3056–3070 (2008)
Hertzog, A., Alexander, M., Plougonven, R.: On the probability density functions of gravity waves momentum flux in the stratosphere. J. Atmos. Sci. **69**, 3433–3448 (2012)
Hirota, I., Niki, T.: Inertia-gravity waves in the troposphere and stratosphere observed by the MU radar. J. Meteor. Soc. Japan **64**, 995–999 (1986)
Hoffmann, L., Xue, X., Alexander, M.: A global view of stratospheric gravity wave hotspots located with atmospheric infrared sounder observations. J. Geophys. Res. **118**, 416–434 (2013)
Holton, J.R.: An Introduction to Dynamic Meteorology, 3rd edn. Academic Press, London (1992)
Hoskins, B.J.: The mathematical theory of frontogenesis. Annu. Rev. Fluid Mech. **14**, 131–151 (1982)
Hoskins, B.J., Bretherton, F.P.: Atmospheric frontogenesis models: mathematical formulation and solution. J. Atmos. Sci. **29**, 11–37 (1972)
Hoskins, B., McIntyre, M., Robertson, A.: On the use and significance of isentropic potential vorticity maps. Q. J. Roy. Meteorol. Soc. **111**(470), 877–946 (1985)
Kennedy, P., Shapiro, M.: Further encounters with clear air turbulence in research aircraft. J. Atmos. Sci. **37**, 986–993 (1980)
Kim, J.-H., Chun, H.-Y.: A numerical study of Clear-Air Turbulence (CAT) encounters over South Korea on 2 April 2007. J. Appl. Meteor. Climatol. **49**, 2381–2404 (2010)
Kim, J.-H., Chun, H.-Y.: Statistics and possible sources of aviation turbulence over South Korea. J. App. Meteor. Climatol. **50**, 311–324 (2011)

Kim, Y.-J., Eckermann, S., Chun, H.-Y.: An overview of the past, present and future of gravity-wave drag parameterization for numerical climate and weather prediction models. Atmos Ocean **41**, 65–98 (2003)

Knox, J.: Possible mechanisms of clear-air turbulence in strongly anticyclonic flows. Mon. Weather Rev. **125**, 1251–1259 (1997)

Knox, J., McCann, D., Williams, P.: Application of the Lighthill-Ford theory of spontaneous imbalance to Clear-Air Turbulence forecasting. J. Atmos. Sci. **65**, 3292–3304 (2008)

Knox, J., McCann, D., Williams, P.: Reply. J. Atmos. Sci. **66**, 2511–2516 (2009)

Koch, S., O'Handley, C.: Operational forecasting and detection of mesoscale gravity waves. Weather Forecasting **12**, 253–281 (1997)

Koch, S., Jamison, B., Lu, C., Smith, T., Tollerud, E., Girz, C., Wang, N., Lane, T., Shapiro, M., Parrish, D., Cooper, O.: Turbulence and gravity waves within an upper-level front. J. Atmos. Sci. **62**, 3885–3908 (2005)

Koppel, L., Bosart, L., Keyser, D.: A 25-yr climatology of large-amplitude hourly surface pressure changes over the conterminous United States. Mon. Weather Rev. **128**(1), 51–68 (2000)

Lalas, D., Einaudi, F.: On the characteristics of waves generated by shear layers. J. Atmos. Sci. **33**, 1248–1259 (1976)

Lane, T.P., Doyle, J.D., Plougonven, R., Shapiro, M.A., Sharman, R.D.: Observations and numerical simulations of inertia-gravity waves and shearing instabilities in the vicinity of a jet stream. J. Atmos. Sci. **61**(22), 2692–2706 (2004)

Lane, T.P., Sharman, R.D., Trier, S.B., Fovell, R.G., Williams, J.K.: Recent advances in the understanding of near-cloud turbulence. Bull. Am. Meteorol. Soc. **93**(4), 499–515 (2012). doi:10.1175/BAMS-D-11-00062.1

Leith, C.: Nonlinear normal mode initialization and quasi-geostrophic theory. J. Atmos. Sci. **37**, 958–968 (1980)

Lighthill, J.M.: On sound generated aerodynamically. I. General theory. Proc. Roy. Soc. London **211**(A), 564–587 (1952)

Lin, Y., Zhang, F.: Tracking gravity waves in baroclinic jet-front systems. J. Atmos. Sci. **65**, 2402–2415 (2008)

Lindborg, E.: Can the atmospheric kinetic energy spectrum be explained by two-dimensional turbulence? J. Fluid Mech. **388**, 259–288 (1999)

Lorenz, E.: Attractor sets and quasi-geostrophic equilibrium. J. Atmos. Sci. **37**, 1685–1699 (1980)

Lott, F., Plougonven, R., Vanneste, J.: Gravity waves generated by sheared potential vorticity anomalies. J. Atmos. Sci. **67**, 157–170 (2010). doi:10.1175/2009JAS3134.1

Lovegrove, A., Read, P., Richards, C.: Generation of inertia-gravity waves in a baroclinically unstable fluid. Q. J. Roy. Meteorol. Soc. **126**, 3233–3254 (2000)

Lu, C., Koch, S.: Interactions of upper-tropospheric turbulence and gravity waves as obtained from spectral and structure function analyses. J. Atmos. Sci. **65**, 2676–2690 (2008). doi:10.1175/2007JAS2660.1

Mancuso, R., Endlich, R.: Clear air turbulence frequency as a function of wind shear and deformation. Mon. Weather Rev. **94**, 581–585 (1966)

Mastrantonio, G., Einaudi, F., Fua, D., Lalas, D.: Generation of gravity waves by jet streams in the atmosphere. J. Atmos. Sci. **33**, 1730–1738 (1976)

McIntyre, M., Weissman, M.: On radiating instabilities and resonant overreflection. J. Atmos. Sci. **35**, 1190–1196 (1978)

Mirzaei, M., Zuelicke, C., Moheballojeh, A., Ahmadi-Givi, F., Plougonven, R.: Structure, energy and parameterization of inertia-gravity waves in dry and moist simulations of a baroclinic wave life cycle. J. Atmos. Sci. **71**, 2390–2414 (2014)

Nastrom, G., Fritts, D.: Sources of mesoscale variability of gravity waves. Part I: Topographic excitation. J. Atmos. Sci. **49**(2), 101–109 (1992)

Nastrom, G., Gage, K.: A climatology of atmospheric wavenumber spectra of wind and temperature observed by commercial aircraft. J. Atmos. Sci. **42**(9), 950–960 (1985)

O'Sullivan, D., Dunkerton, T.: Generation of inertia-gravity waves in a simulated life cycle of baroclinic instability. J. Atmos. Sci. **52**(21), 3695–3716 (1995)

Pan, L.L., Bowman, K.P., Atlas, E.L., Wofsy, S.C., Zhang, F., Bresch, J.F., Ridley, B.A., Pittman, J.V., Homeyer, C.R., Romashkin, P., Cooper, W.A.: The stratosphere–troposphere analyses of regional transport 2008 experiment. Bull. Am. Meteor. Soc. **91**, 327–342 (2010)

Pavelin, E., Whiteway, J., Vaughan, G.: Observation of gravity wave generation and breaking in the lowermost stratosphere. J. Geophys. Res. **106**(D6), 5173–5179 (2001)

Piani, C., Durran, D., Alexander, M., Holton, J.: A numerical study of three-dimensional gravity waves triggered by deep tropical convection and their role in the dynamics of the QBO. J. Atmos. Sci. **57**(22), 3689–3702 (2000)

Plougonven, R., Snyder, C.: Inertia-gravity waves spontaneously generated by jets and fronts. Part I: Different baroclinic life cycles. J. Atmos. Sci. **64**, 2502–2520 (2007)

Plougonven, R., Snyder, C.: Gravity waves excited by jets: propagation versus generation. Geophys. Res. Lett. **32**(L18892), (2005). doi:10.1029/2005GL023,730

Plougonven, R., Teitelbaum, H.: Comparison of a large-scale inertia-gravity wave as seen in the ECMWF and from radiosondes. Geophys. Res. Lett. **30**(18), 1954 (2003)

Plougonven, R., Zeitlin, V.: Internal gravity wave emission from a pancake vortex: an example of wave-vortex interaction in strongly stratified flows. Phys Fluids **14**(3), 1259–1268 (2002)

Plougonven, R., Zhang, F.: On the forcing of inertia-gravity waves by synoptic-scale flows. J. Atmos. Sci. **64**, 1737–1742 (2007)

Plougonven, R., Zhang, F.: Internal gravity waves from atmospheric jets and fronts. Rev. Geophys. **52**, 33–76 (2014)

Plougonven, R., Teitelbaum, H., Zeitlin, V.: Inertia-gravity wave generation by the tropospheric mid-latitude jet as given by the FASTEX radio soundings. J. Geophys. Res. **108**(D21), 4686 (2003)

Plougonven, R., Muraki, D., Snyder, C.: A baroclinic instability that couples balanced motions and gravity waves. J. Atmos. Sci. **62**, 1545–1559 (2005)

Plougonven, R., Snyder, C., Zhang, F.: Comments on 'Application of the Lighthill-Ford theory of spontaneous imbalance to clear-air turbulence forecasting'. J. Atmos. Sci. **66**, 2506–2510 (2009)

Plougonven, R., Hertzog, A., Guez, L.: Gravity waves over Antarctica and the Southern Ocean: consistent momentum fluxes in mesoscale simulations and stratospheric balloon observations. Q. J. Roy. Meteorol. Soc. **139**, 101–118 (2013)

Plougonven, R., Hertzog, A., Alexander, M.: Case studies of non-orographic gravity waves over the Southern Ocean emphasize the role of moisture. J. Geophys. Res. **120**, 1278–1299 (2015)

Powers, J., Reed, R.: Numerical simulation of the large-amplitude mesoscale gravity wave event of 15 December 1987 in the Central United States. Mon. Weather Rev. **121**, 2285–2308 (1993)

Queney, P.: The problem of air flow over mountains: a summary of theoretical studies. Bull. Am. Meteorol. Soc. **29**, 16–26 (1948)

Ralph, F., Neiman, P., Keller, T.: Deep-tropospheric gravity waves created by leeside cold fronts. J. Atmos. Sci. **56**, 2986–3009 (1999)

Reeder, M.J., Griffiths, M.: Stratospheric inertia-gravity waves generated in a numerical model of frontogenesis. Part II: Wave sources, generation mechanisms and momentum fluxes. Q. J. Roy. Meteorol. Soc. **122**, 1175–1195 (1996)

Richter, J., Sassi, F., Garcia, R.: Toward a physically based gravity wave source parameterization in a general circulation model. J. Atmos. Sci. **67**, 136–156 (2010). doi:10.1175/2009JAS3112.1

Rossby, C.: On the mutual adjustment of pressure and velocity distributions in certain simple current systems II. J. Mar. Res. **1**, 239–263 (1938)

Sato, K.: A statistical study of the structure, saturation and sources of inertia-gravity waves in the lower stratosphere observed with the MU radar. J. Atmos. Terr. Phys. **56**(6), 755–774 (1994)

Schecter, D.: The spontaneous imbalance of an atmospheric vortex at high Rossby number. J. Atmos. Sci. **65**, 2498–2521 (2008)

Scinocca, J., Ford, R.: The nonlinear forcing of large-scale internal gravity waves by stratified shear instability. J. Atmos. Sci. **57**, 653–672 (2000)
Scolan, H., Flor, J.-B., Gula, J.: Frontal instabilities and waves in a differentially rotating fluid. J. Fluid Mech. **685**, 532–542 (2011)
Sharman, R.D., Trier, S.B., Lane, T.P., Doyle, J.D.: Sources and dynamics of turbulence in the upper troposphere and lower stratosphere: a review. Geophys. Res. Lett. **39**, L12803 (2012). doi:10.1029/2012GL051996
Skamarock, W.C.: Evaluating mesoscale NWP models using kinetic energy spectra. Mon. Weather Rev. **132**, 3019–3032 (2004)
Snyder, C., Skamarock, W., Rotunno, R.: Frontal dynamics near and following frontal collapse. J. Atmos. Sci. **50**(18), 3194–3211 (1993)
Snyder, C., Muraki, D., Plougonven, R., Zhang, F.: Inertia-gravity waves generated within a dipole vortex. J. Atmos. Sci. **64**, 4417–4431 (2007)
Snyder, C., Plougonven, R., Muraki, D.: Forced linear inertia-gravity waves on a basic-state dipole vortex. J. Atmos. Sci. **66**(11), 3464–3478 (2009)
Sugimoto, N., Ishioka, K., Ishii, K.: Parameter sweep experiments on spontaneous gravity wave radiation from unsteady rotational flow in an f-plane shallow water system. J. Atmos. Sci. **65**, 235–249 (2008)
Thomas, L., Worthington, R., McDonald, A.: Inertia-gravity waves in the troposphere and lower stratosphere associated with a jet stream exit region. Ann. Geophys. **17**, 115–121 (1999)
Trier, S.B., Sharman, R.D., Lane, T.P.: Influences of moist convection on a cold-season outbreak of clear-air turbulence (CAT). Mon. Weather Rev. **140**(8), 2477–2496 (2012). doi:10.1175/MWR-D-11-00353.1
Uccelini, L., Koch, S.: The synoptic setting and possible energy sources for mesoscale wave disturbances. Mon. Weather Rev. **115**, 721–729 (1987)
Vanneste, J.: Balance and spontaneous wave generation in geophysical flows. Annu. Rev. Fluid Mech. **45**, 147–172 (2013)
Vanneste, J., Yavneh, I.: Exponentially small inertia-gravity waves and the breakdown of quasi-geostrophic balance. J. Atmos. Sci. **61**, 211–223 (2004)
Vanneste, J., Yavneh, I.: Unbalanced instabilities of rapidly rotating stratified shear flows. J. Fluid Mech. **584**, 373–396 (2007)
Viudez, A.: The origin of the stationary frontal wave packet spontaneously generated in rotating stratified vortex dipoles. J. Fluid Mech. **593**, 359–383 (2007)
Viudez, A.: The stationary frontal wave packet spontaneously generated in mesoscale dipoles. J. Phys. Oceanogr. **38**, 243–256 (2008)
Waite, M.L., Snyder, C.: The mesoscale kinetic energy spectrum of a baroclinic life cycle. J. Atmos. Sci. **66**(4), 883–901 (2009)
Waite, M.L., Snyder, C.: Mesoscale energy spectra of moist baroclinic waves. J. Atmos. Sci. **70**(4), 1242–1256 (2012)
Wang, S., Zhang, F.: Source of gravity waves within a vortex-dipole jet revealed by a linear model. J. Atmos. Sci. **67**, 1438–1455 (2010)
Wang, S., Zhang, F., Snyder, C.: Generation and propagation of inertia-gravity waves from vortex dipoles and jets. J. Atmos. Sci. **66**, 1294–1314 (2009)
Wang, S., Zhang, F., Epifanio, C.: Forced gravity wave response near the jet exit region in a linear model. Q. J. Roy. Meteorol. Soc. **136**, 1773–1787 (2010)
Wei, J., Zhang, F.: Mesoscale gravity waves in moist baroclinic jet-front systems. J. Atmos. Sci. **71**, 929–952 (2014). doi:10.1175/JAS-D-13-0171.1
Whiteway, J., Klaassen, G., Bradshaw, N., Hacker, J.: Transition to turbulence in shear above the tropopause. Geophys. Res. Lett. **31**, L02,118 (2004)
Williams, P., Haine, T., Read, P.: On the generation mechanisms of short-scale unbalanced modes in rotating two-layer flows with vertical shear. J. Fluid Mech. **528**, 1–22 (2005)
Williams, P., Haine, T., Read, P.: Inertia gravity waves emitted from balanced flow: observations, properties, and consequences. J. Atmos. Sci. **65**(11), 3543–3556 (2008)

Wright, C., Osprey, S., Gille, J.: Global observations of gravity wave intermittency and its impact on the observed momentum flux morphology. J. Geophys. Res. **118**, 10,980–10,993 (2013). doi:10.1002/jgrd.50869

Wu, D. L., Zhang, F.: A study of mesoscale gravity waves over the North Atlantic with satellite observations and a mesoscale model. J. Geophys. Res. **109**(D22104) (2004). doi:10.1029/2004JD005090

Zhang, F.: Generation of mesoscale gravity waves in upper-tropospheric jet-front systems. J. Atmos. Sci. **61**(4), 440–457 (2004)

Zhang, F., Koch, S., Davis, C., Kaplan, M.: Wavelet analysis and the governing dynamics of a large amplitude mesoscale gravity wave event along the east coast of the United States. Q. J. Roy. Meteorol. Soc. **127**, 2209–2245 (2001)

Zhang, F., Koch, S.E., Kaplan, M.L.: Numerical simulations of a large-amplitude gravity wave event. Meteorol. Atmos. Phys. **84**, 199–216 (2003)

Zhang, F., Zhang, M., Wei, J., Wang, S.: Month-long simulations of gravity waves over North America and North Atlantic in comparison with satellite observations. Acta Meteorol. Sin. **27**, 446–454 (2013)

Zhang, F., Wei, J., Zhang, M., Bowman, P., Pan, L., Atlas, E., Wolfsy, S.: Aircraft measurements of gravity waves in the upper troposphere and lower stratosphere during the start08 field experiment. Atmos. Chem. Phys. Discuss. **15**, 4725–4766 (2015). doi:10.5194/acpd-15-4725-2015

Zülicke, C., Peters, D.: Simulation of inertia-gravity waves in a poleward breaking Rossby wave. J. Atmos. Sci. **63**, 3253–3276 (2006)

Zülicke, C., Peters, D.: Parameterization of strong stratospheric inertia gravity waves forced by poleward breaking Rossby waves. Mon. Weather Rev. **136**, 98–119 (2008)

Chapter 20
Turbulence and Waves in the Upper Troposphere and Lower Stratosphere

Alex Mahalov

Abstract The generation mechanisms and physical characteristics of jet stream turbulence, mountain, and inertia-gravity waves in the upper troposphere and lower stratosphere (UTLS) are investigated for real atmospheric conditions. To resolve multi-scale physical processes of wave breaking and laminated structures in the UTLS region, vertical nesting and adaptive vertical gridding have been developed and applied in nested, high-resolution, coupled mesoscale-microscale simulations. The fully three-dimensional, moist, compressible Navier-Stokes equations are solved with a stretched, adaptive grid in the vertical and improved resolution in the UTLS region. For verification purposes, real-case simulations are conducted for the Terrain-Induced Rotor Experiment (T-REX) campaign of measurements and selected cases from pilot reports (PIREPs). Comparisons with observational datasets highlight significant benefits of nested computational techniques that take into account the shear-stratified turbulence physics of the UTLS. Localized sharp shear layers characterized by stiff gradients of potential temperature and strong alternating vertical velocity patches are resolved in the tropopause region within the embedded microscale nest. We describe fully three-dimensional multi-scale dynamics of laminated structures and nonlinear processes in turbulent layers observed in the UTLS region. Depending on atmospheric conditions, the gravity waves might be trapped at the altitude of the jet stream and break or propagate into higher altitudes acquiring characteristics of inertia-gravity waves. Three-dimensional instabilities in nonparallel shear-stratified flows such as those induced by mountain and polarized inertia-gravity waves in UTLS are characterized by a polarized Richardson number.

A. Mahalov (✉)
School of Mathematical and Statistical Sciences, Arizona State University, Tempe, AZ, USA

School for Engineering of Matter, Transport and Energy, Arizona State University, Tempe, AZ, USA
e-mail: mahalov@asu.edu

R. Sharman, T. Lane (eds.), *Aviation Turbulence*,
DOI 10.1007/978-3-319-23630-8_20

20.1 Introduction

The extended region consisting of the bulk of the troposphere and the lower stratosphere represents a significant challenge for turbulence modeling and numerical prediction. The collusion between the stratification and shear in upper-level jet streams leads to many complex multi-scale physics phenomena, including the formation of vertically thin, laminated coherent structures in the UTLS and clouds in the upper troposphere. Associated irregularities and inhomogeneities of anisotropic, non-Kolmogorov, and patchy shear-stratified turbulence have a spatial range from tens of kilometers through meter scales. A wide variety of physical processes occur on these disparate scales, and this has posed a considerable challenge to the goal of a truly self-consistent, comprehensive physics-based understanding of turbulent dynamics and morphology in the UTLS (Sharman et al. 2012a, b; Mahalov et al. 2000, 2009, 2011).

Mountain waves and jet stream turbulence present a major challenge to the safety, controllability, and flight path optimization for commercial aircrafts and high-altitude unmanned aerial vehicles (UAVs). Flying through turbulence causes the autopilot to begin pitch oscillations that seriously degrade performance as well as put the platform itself at risk. The UTLS turbulence phenomena also play a major role in atmospheric sciences due to its important influence on the propagation of electromagnetic (EM) waves (radio waves, microwaves, lasers). High-impact UTLS environments can significantly impact communication, navigation, and imaging systems primarily through the development of laminated structures (layers) associated with shear-stratified turbulence, often induced by mountain/inertia-gravity waves and jet streams.

The observed fine vertical structure of the UTLS may present a severe limitation in resolving small vertical-scale processes such as optical and clear-air turbulence, thin adiabatic layers, and sharp gradient formations that develop near the tropopause. They are observed in the UTLS region during extreme events such as wave breaking, overshooting moist convection, and shear instability induced by jet streams or in the presence of gravity wave-critical level interactions. These small-scale upper-level processes are particularly sensitive to the vertical resolution, implying that the relatively coarse vertical grid spacing typically used in operational models is insufficient to resolve vertical scales. Computations of these processes require that a fine mesh in both the vertical and the horizontal is used to encompass all pertinent multi-scales of atmospheric phenomena in the UTLS environment. Effective resolution and prediction of strongly nonlinear multi-scale physical phenomena and laminated structures in the UTLS region is a significant challenge for real-time operational forecasting.

There are two main techniques that are used in atmospheric and oceanic models to improve resolution over limited areas. In dynamically adaptive methods, the spatial resolution is constantly changing with time by coarsening or refining the grid spacing depending on local conditions (e.g., Dietachmayer and Droegemeier 1992). The adaptive methods are not well established in the atmospheric modeling systems for several reasons: (1) adaptive techniques can incur massive overhead due to indirect data addressing and additional efforts for grid handling which increase the cost of real-time forecasting or long-term predictions; (2) physical parameterizations of subgrid processes are usually optimized for a specific grid resolution, making it difficult to use dynamically and temporally refined or coarsened grids. The other method uses nesting to improve spatial resolution over a limited area. Nesting techniques are widely used in atmospheric (e.g., Dudhia 1993; Clark and Hall 1996; Skamarock and Klemp 2008; Mahalov et al. 2009, 2011; Mahalov and Moustaoui 2009) and oceanic models (e.g., Shchepetkin and McWilliams 2005). Large domain models with coarse resolution are used to predict large-scale dynamics, while limited area models with boundary conditions interpolated from the coarse grids are used over small domains with finer resolution. The improvement allowed by nesting techniques is that small-scale processes which are not resolved in a coarse-grid model, and therefore need to be represented by using subgrid-scale parameterizations, may be explicitly resolved in the nested model.

In studying limited area atmospheric environments, a key challenge is the development of nesting methods that use robust computational techniques to control numerical errors such as the ones that are generated at the boundaries of the nested models. These errors are inevitable in any nested or downscaled simulation because the coarse-grid fields specified at the boundaries of the nested grid are not consistent with the finer fields simulated by the nest. If not treated effectively, these errors will propagate into the interior of the nested domain, resulting in poor quality of numerical simulations. These errors are particularly sensitive when instabilities dominate dynamics near the boundaries of the nested domain. Thus, numerical methods that control and reduce propagation of these errors are essential for reliable high-resolution regional atmospheric models.

Real-time forecasting methods for the UTLS region that can predict the strength and distribution of high-impact turbulence and cloud layers at fine scales require new capabilities that incorporate the shear-stratified turbulence physics of the UTLS. In this paper we review physics-based predictive modeling techniques that include improved subgrid-scale parameterizations for inhomogeneous shear-stratified turbulence (using a polarized Richardson number and spatially varying turbulent Prandtl number), microscale vertical nesting, and adaptive vertical gridding in multi-scale atmospheric physics codes with initial and boundary conditions from high-resolution global datasets such as T799L91 ECMWF analysis datasets (25 km horizontal resolution and 91 vertical levels over the entire globe). The computational method for the UTLS is based on nested mesoscale-microscale simulations coupled with global dynamics (Mahalov et al. 2009, 2010, 2011). The Weather Research and Forecasting (WRF) mesoscale code is coupled with a microscale code. The microscale code executes vertical nesting and adaptive vertical gridding to provide a finer resolution than the WRF code can provide. The WRF mesoscale code is not designed for microscale simulations in the UTLS region.

For the purpose of illustration of computational results, we define a 3D domain centered on a geographical region 100 km $\times$ 100 km horizontally and extending to

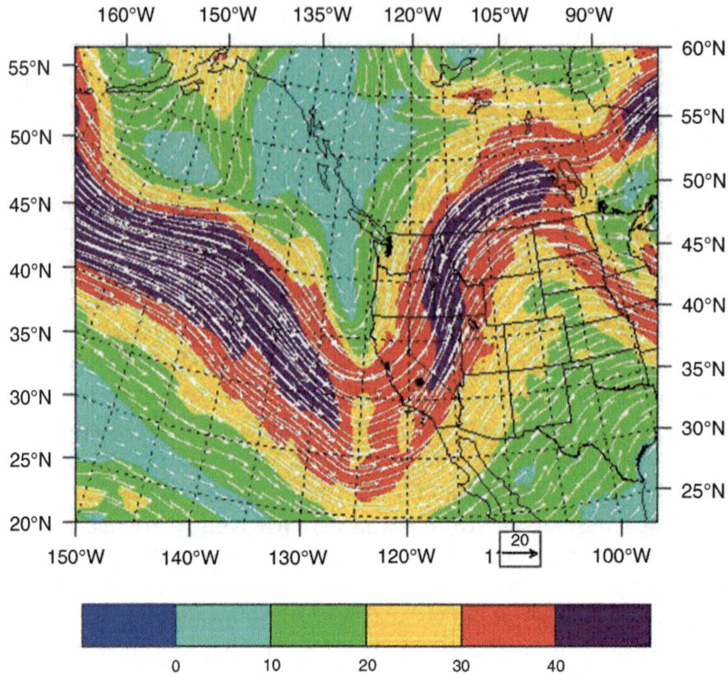

Fig. 20.1 Mesoscale variations of jet stream. Data from global ECMWF T799L91 model. Wind fields are shown for 320 K isentrope on 1 April 2006. The *dot* indicates the location of experimental campaign launching site

30 km altitude. Global atmospheric NWP data (NCEP GFS or ECMWF T799L91) is uploaded every 6 h and then zoomed into a specific geographical region with a sequence of nested mesoscale and microscale domains. To produce a sequence of finer-resolution nested domains, microscale simulations use both vertical and horizontal nesting until the required resolution is obtained. Figures 20.1, 20.2, and 20.3 show the results of coupled mesoscale/microscale real-case simulations of the 3D flux-form fully compressible equations for atmospheric dynamics conducted for the period of 48 h. The domains are centered on an area of Owens Valley, CA, where an intensive campaign of measurements was conducted to study topographically induced turbulent events (Grubišić et al. 2008). Figure 20.1 shows mesoscale variations of the upper-level jet stream. The size of the microscale domain in Fig. 20.2 is approximately 100 km × 100 km × 30 km but only a portion of it extending to 20 km altitude is presented. The microscale nest is implemented with 180 staggered vertical levels. A fifth innermost nesting implemented with 450 staggered vertical levels is not shown, but verifies that the results were unchanged.

Figure 20.2 are microscale cross sections of Fig. 20.1 showing the mountainous terrain at the bottom of the figure panels. The results are from the innermost

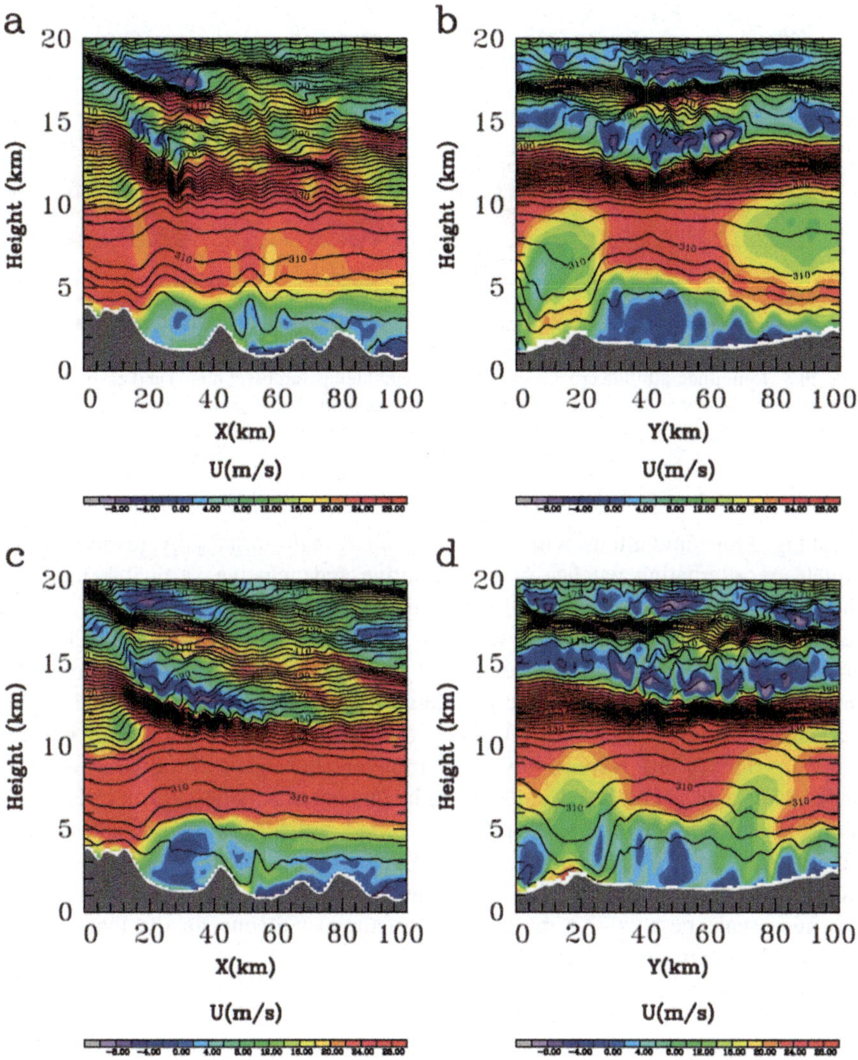

Fig. 20.2 Vertical cross sections of horizontal wind component transverse to the valley and potential temperature (K) on 1 April 2006 at 8 UTC: (**a**) across and (**b**) along the valley. (**c**) and (**d**) are the same as (**a**) and (**b**), respectively, but at 6 UTC. The horizontal axes X and Y indicate the distance with respect to the location (36.70 N, 118.50 W) and (36.29 N, 118.01 W), respectively

microscale domain with 333 m horizontal resolution and 180 vertical levels. The time is 8:00 UTC, 1 April 2006. Strongly laminated structures can be seen in the upper troposphere/lower stratosphere region in Fig. 20.2 (altitudes 8–20 km). These turbulent layers are characterized by steep vertical gradients and are located at the edges of relatively well-mixed regions produced by shear instabilities and wave

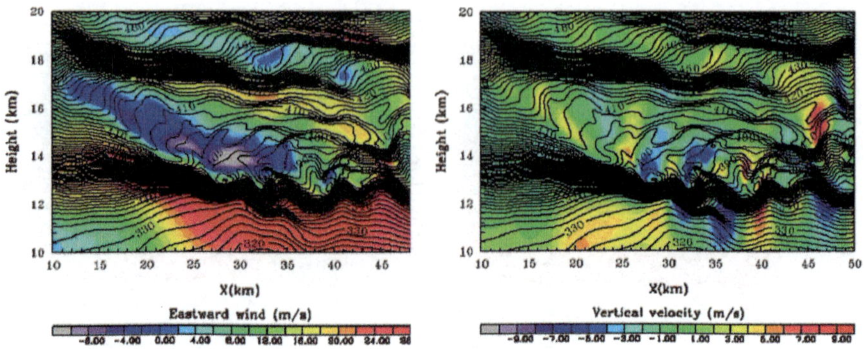

Fig. 20.3 Longitude-altitude cross sections of potential temperature (contour) and eastward wind (*left panel*) and potential temperature and vertical velocity (*right panel*) from the innermost microscale domain (333 m grid), 300 grid points in horizontal directions, 180 vertical levels. The time is 8:00 UTC, 1 April 2006

breaking. Our simulations with embedded microscale nests fully resolve rotors which are alternating patches of high positive and negative vertical velocity in the boundary layer. Dynamics of fluctuating dipoles of vertical velocity within turbulent layers in the UTLS is strongly nonlinear phenomena. Figure 20.3 are zoomed in views of the UTLS region for altitudes from 10 km to 20 km in the microscale domain. They show longitude-altitude cross sections of potential temperature with eastward wind (left panel) and potential temperature with vertical velocity (right panel), respectively. A sharp reversal of the eastward wind velocity component seen in Figs. 20.2 and 20.3 are in agreement with observations. Homogenized regions of potential temperature are found in the UTLS region. These well-mixed layers are associated with patches of high vertical velocity reaching 5–9 m/s in the UTLS. The potential temperature and vertical velocity fields calculated by the microscale nest exhibit multi-scale structures with fine details that are not resolved by the mesoscale nests.

20.2 Model Numerics, Multi-nesting, and Implicit Relaxation

The fully three-dimensional, moist, compressible Navier-Stokes equations are solved with a stretched, adaptive grid in the vertical and improved resolution in the UTLS region. For nesting, both lateral and vertical boundary conditions are treated via relaxation zones where the velocity and temperature fields are relaxed to those obtained from the mesoscale inner nest (Mahalov et al. 2009, 2011). The solver uses a time-split integration scheme. Slow or low-frequency modes are integrated using a third-order Runge-Kutta (RK3) time integration scheme, while the high-frequency acoustic modes are integrated over smaller time steps to maintain numerical stability. The acoustic-mode integration is cast in the form of a correction to the RK3 integration. Additionally, to increase the accuracy of the splitting, we integrate a perturbation form of the governing equations using smaller acoustic time steps within the RK3 large-time-step sequence. To form the perturbation equations for the RK3 time-split acoustic integration, we define small-time-step variables that are deviations from the most recent RK3 predictor ($\psi'' = \psi - \psi^t$). The spatial discretization in the solver uses a C grid staggering for the variables. That is, normal velocities are staggered one-half grid length from the thermodynamic variables. An adaptive time-stepping capability is introduced so that the RK3 time step is chosen based on the temporally evolving wind fields. The adaptively chosen time step is usually larger than the typical fixed time step; hence, the dynamics integrates faster and physics modules are called less often, and the time to completion of the simulation is substantially reduced.

The horizontal and vertical nesting allows resolution to be focused over a region of interest in the UTLS by introducing an additional grid (or grids) into the simulation. Vertical and horizontal refinement options have been developed. The nested grids are rectangular and are aligned with the parent (coarser) grid within which they are nested. Additionally, the nested grids allow any integer spatial and temporal refinement ratios of the parent grid (the spatial and temporal refinement ratios are usually 3, but not necessarily the same for each nest). This implementation allows for nesting with embedded nested domains in the horizontal as well as in the vertical directions and a more robust novel method to control errors near the boundaries (implicit relaxation). Nested-grid simulations can be produced using either one-way nesting or two-way nesting. The one-way and two-way nesting options refer to how a coarse grid and the fine grid interact. In both the one-way and two-way simulation modes, the fine-grid boundary conditions (i.e., the lateral and vertical boundaries) are interpolated from the coarse-grid forecast. In a one-way nest, this is the only information exchange between the grids (from the coarse grid to the fine grid). In the two-way nest integration, the fine-grid solution replaces the coarse-grid solution for coarse-grid points that lie inside the fine grid. This information exchange between the grids is now in both directions (coarse to fine for the fine-grid lateral, bottom, and top boundary computation and fine to coarse during the feedback at each coarse-grid time step). The one-way nest setup may be run in one of two different methods. One option is to produce the nested simulation as two separate simulations. In this mode, the coarse grid is integrated first and the coarse-grid forecast is completed. Output from the coarse-grid integration is then processed to provide boundary conditions for the nested run (usually at a much lower temporal frequency than the coarse-grid time step), and this is followed by the complete time integration of the fine (nested) grid. Hence, this one-way option is equivalent to running two separate simulations with a processing step in between. The second one-way option (lockstep with no feedback) is run as a traditional simulation with two (or more) grids integrating concurrently, except with the feedback runtime option shut off. This option provides lateral boundary conditions to the fine grid at each coarse-grid time step, which is an advantage of the concurrent one-way method (no feedback).

The model supports a strategy to refine a coarse-grid simulation with the introduction of a nested grid. When using concurrent one-way and two-way nesting, an option for initializing the fine grid is provided. The entire fine-grid variables are interpolated from the coarse grid. This option allows the fine grid to start at a later time in the coarse grid's simulation. A simulation involves one outer grid and may contain multiple inner nested grids. Each nested region is entirely contained within a single coarser grid, referred to as the *parent* grid. The finer, nested grids are referred to as *child* grids. Using this terminology, children are also parents when multiple levels of nesting are used. The fine grids may be telescoped to any depth (i.e., a parent grid may contain one or more child grids, each of which in turn may successively contain one or more child grids), and several fine grids may share the same parent at the same level of nesting. For both one-way and two-way nested-grid simulations, the ratio of the parent horizontal grid distance to the child horizontal grid distance (the spatial refinement ratio) must be an integer. For two-way and concurrent one-way nesting, this is also true for the time steps (the temporal refinement ratio). The model does allow the time step refinement ratio to differ from the spatial refinement ratio.

One of the main challenges faced in atmospheric as well as oceanic limited area modeling is the specification of the lateral boundary conditions. Usually, the prognostic fields at the lateral boundaries of the nested grid are specified from the large domain. These fields have coarse resolution and are interpolated in space and time to the nested grid. The inconsistencies between the limited and the large domain solutions create spurious reflections that may propagate and affect the solution in the interior of the nested domain. Several approaches are used to handle the lateral boundary conditions. The flow relaxation scheme (e.g., Davies 1983) is frequently used for atmospheric mesoscale forecasting models over a limited domain. Lateral open boundary conditions are often used in limited area ocean modeling. These conditions include radiation condition, combined radiation and prescribed condition depending on the inflow and outflow regime at the boundary, and a scale-selective approach. A review of these methods is given in Oddo and Pinardi (2008).

The relaxation method that we implemented for use in the nested UTLS simulations consists of progressively relaxing the fine-grid fields toward the coarse-grid fields. It is implemented using the operator splitting method applied in the acoustic steps where a prognostic variable is updated first without any relaxation. The corrected variable is then used to update the other prognostic variables. The normal velocity component located at the boundaries is treated differently from the tangential velocities and the thermodynamic variable which are located a half grid point inside the domain and adjacent to the boundary. The relaxation boundary scheme consists of smoothly constraining the main prognostic variables of the nested model to match the corresponding values from the coarse-grid model in a buffer zone next to the boundary called the "relaxation" zone. The flow relaxation scheme is a combination of Newtonian and diffusive relaxation that has the form:

$$\partial_t \psi = -N(x)(\psi - \psi^c) + D(x)\partial_{xx}(\psi - \psi^c)$$

where ψ is a prognostic variable of the limited area model that needs to be relaxed to the corresponding variable from the coarse-grid model ψ^c, x denotes the direction normal to the boundary, and $N(x)$ and $D(x)$ are the Newtonian and the diffusive relaxation factors. The choice of the profiles and the values of the coefficients $N(x)$ and $D(x)$ in the relaxation zones control reflections at the boundary.

The relaxation is implemented after each acoustic step as an implicit correction. Let $\widetilde{\psi}^{\prime\prime,n+1}$ denote the perturbation of the updated variable after each acoustic time step. The relaxation is then applied as a correction in a subsequent step using the following implicit flow relaxation equation:

$$\frac{\psi_i^{\prime\prime,n+1}-\widetilde{\psi}_i^{\prime\prime,n+1}}{\delta\tau} = -N_i\left(\psi_i^{\prime\prime,n+1}+\psi_i^t-\psi_i^{c,n+1}\right)+\frac{D_i}{\delta x^2}$$
$$\times\left\{\left(\psi_{i+1}^{\prime\prime,n+1}+\psi_{i+1}^t-\psi_{i+1}^{c,n+1}\right)-2\left(\psi_i^{\prime\prime,n+1}+\psi_i^t-\psi_i^{c,n+1}\right)\right.$$
$$\left.+\left(\psi_{i-1}^{\prime\prime,n+1}+\psi_{i-1}^t-\psi_i^{c,n+1}\right)\right\}$$

where $\delta\tau$ is the acoustic time step, δx is the grid spacing, and $\psi_i^{c,n+1}$ is the coarse-grid value interpolated in space and to the time step $(n+1)$. $\psi_i^{\prime\prime,n+1}=\psi_i^{n+1}-\psi_i^t$, where ψ_i^{n+1} is the total fine-grid value, ψ_i^t is the most updated value in the RK3 step, and $\psi_i^{\prime\prime,n+1}$ is the perturbation with respect to ψ_i^t. For the prognostic variables located at half grid points adjacent to the lateral boundary, this implicit equation is solved for $\psi_i^{\prime\prime,n+1}$ along the relaxation zone, subject to the boundary conditions:

$$\psi_{s+1}^{\prime\prime,n+1}=\widetilde{\psi}_{s+1}^{\prime\prime,n+1} \quad \text{and} \quad \psi_1^{\prime\prime,n+1}=\psi_1^{c,n+1}-\psi_i^t$$

where s is the index of the last relaxed point in the interior of the domain. For the normal velocities at the boundary, the same system is solved except that consistency between the coarser and the finer mass fluxes in the continuity equation is imposed, that is:

$$\nabla\cdot\vec{V}=\nabla\cdot\vec{V}^c$$

where $\vec{V}$ and $\vec{V}^c$ are the velocity vectors from the nested and the coarse grids. Since the tangential velocities adjacent to the boundary are imposed by the coarse-grid model, the above relation reduces to:

$$\frac{\partial U}{\partial x}=\frac{\partial U^c}{\partial x}$$

where U and U^c are the normal components of $\vec{V}$ and $\vec{V}^c$, respectively. This relation is imposed implicitly at the lateral boundaries, and the implicit equation is solved for $U^{\prime\prime,n+1}$ along the relaxation zone as above, but with the following implicit boundary conditions:

$$U_{s+1}^{\prime\prime,n+1} = \widetilde{U}_{s+1}^{\prime\prime,n+1} \quad \text{and} \quad U_2^{\prime\prime,n+1} - U_1^{\prime\prime,n+1} = \left(U_1^t - U_1^{c,n+1}\right) - \left(U_2^t - U_2^{c,n+1}\right)$$

The Newtonian and diffusive relaxation times are fixed by the choice of the coefficients N_i and D_i. Optimal profiles for the Newtonian relaxation coefficient N are proposed in Davies (1983). They are constructed in such a way that, under idealized conditions, the unwanted partial reflection of outgoing waves leaving the domain is minimized. Marbaix et al. (2003) conducted a detailed theoretical and numerical study on the choice of the Newtonian and diffusive relaxation coefficients, where different profiles and relaxation times were compared. They concluded that a diffusive relaxation combined with exponential or optimized profiles gives better results. Also, they presented a formula for what they called the leading coefficients N_2^* and D_2^* which they derived using the criterion of minimum reflection. These are non-dimensional coefficients at the first relaxed grid point. These coefficients together with the relaxation profiles determine the entire relaxation time in the relaxation zone through the relations:

$$N_i = -\frac{c}{2\Delta x} N_2^* \widetilde{N}_i \quad \text{and} \quad D_i = -\frac{c}{2\Delta x} D_2^* \widetilde{D}_i$$

In these relations c is the phase speed of the fastest wave, and $\tilde{N}_i$ and $\widetilde{D}_i$ are the normalized profiles of the coefficient ($\widetilde{N}_1 = 1$ and $\widetilde{D}_1 = 1$). We choose a five- or nine-point deep relaxation zone where Newtonian and diffusive relaxation are applied. Following Marbaix et al. (2003), we choose $N_2^* = 0.9$ and $D_2^* = 0.9$; c is the speed of sound, and an optimized profile is computed for $\tilde{N}_i$ and $\widetilde{D}_i$. These specifications determine the Newtonian and diffusive relaxation times. In our UTLS studies, it was found that implicit relaxation schemes with a five-point deep relaxation zone have optimal performance for computational speed and adequate accuracy.

20.3 Multi-scale Resolution of Jet Stream Turbulence and Mountain Waves in UTLS at Fine Scales: Verification Studies Using T-REX Datasets

Nonhomogeneous, anisotropic, shear-stratified flow computations in the UTLS require that a fine mesh be used to encompass all pertinent multi-scales. Coupled with stiff velocity and temperature gradient profiles, this presents significant challenges for nesting and adaptive gridding. Our approach is based on vertical nesting and adaptive vertical gridding using nested mesoscale WRF/microscale simulations. The inner nest of WRF (1 km grid in the horizontal) is coupled with a sequence of embedded microscale nests, both horizontally and vertically. The fully three-dimensional, moist, compressible, non-hydrostatic Navier-Stokes equations are solved with a stretched, adaptive grid in the vertical. An adaptive, staggered grid mesh is used in the vertical, with a grid spacing down to a few meters in the UTLS region. For nesting, both lateral and vertical boundary conditions are treated via relaxation zones where the velocity and temperature fields are relaxed to those obtained from the mesoscale WRF inner nest. Temporal discretization uses an adaptive time-split integration scheme and the Thompson microphysics parameterization scheme. The numerical code is written in Fortan90, which facilitates its portability to different platforms. The code is fully parallelized using MPI, and the memory used by the code is optimized.

The Terrain-Induced Rotor Experiment (T-REX) campaign (e.g., Grubišić et al. 2008) represents an important benchmark for real-case simulations of strongly nonlinear multi-scale dynamics associated with high activity of topographic gravity waves and upper-level jet stream dynamics. During this campaign, several radiosondes were launched from the upwind side of Owens Valley, CA. Owens Valley is located to the east of the Sierra Nevada range. This obstacle is nearly a two-dimensional mountain range. It extends 640 km long and 60–130 km wide, mostly in Eastern California. The ridge line of Sierra Nevada Mountain rises to approximately 3,500 m, and the tallest peak reaches 4,418 m in Mt. Whitney, the highest peak in the USA outside Alaska. Vertical profiles of temperature and wind components were measured with high vertical resolution from the ground up to 30 km. Many profiles exhibit large wave-like fluctuations in temperature and wind wave-like fluctuations in the UTLS, with different vertical wavelengths and localized adiabatic layers.

Figure 20.4 shows topography and wind vector fields at 12 km altitude from the coupled mesoscale-microscale simulations. The oval curves in Fig. 20.4a–d show the trajectory of the HIAPER research aircraft operated by the National Center for Atmospheric Research (NCAR) during the T-REX campaign. Panels (a)–(c) show mesoscale domains and (d) is the microscale domain. Our coupled mesoscale with embedded microscale vertical nest simulations is conducted for the period from 24 March 2006, 00 UTC, to 4 April 2006, 00 UTC, of the T-REX campaign of measurements. The mesoscale domains are centered at the location (36.49 N, 118.8 W) to cover the region of the extensive measurements. Three mesoscale domains (Fig. 20.4a–c) are used with horizontal resolutions of 9 km, 3 km, and 1 km and sigma pressure levels from the ground up to 10 mb (~30 km altitude). These levels are adjusted for better resolution of the jet stream and tropopause dynamics. Mesoscale WRF simulations are initialized with high-resolution ECMWF T799L91 analysis data, 25 km horizontal resolution, and 91 vertical levels.

Our microscale domain in Fig. 20.4d has 333 m resolution in the horizontal and much higher vertical resolution (180 staggered vertical levels). It is nested both horizontally and vertically, with initial and boundary conditions from the finest WRF mesoscale domain. Figure 20.4d shows topography for the microscale nest. The wind vector field at 12 km altitude from the microscale run on 25 March 2006 at 20:00 UTC is also superimposed in this figure. The wind directions are

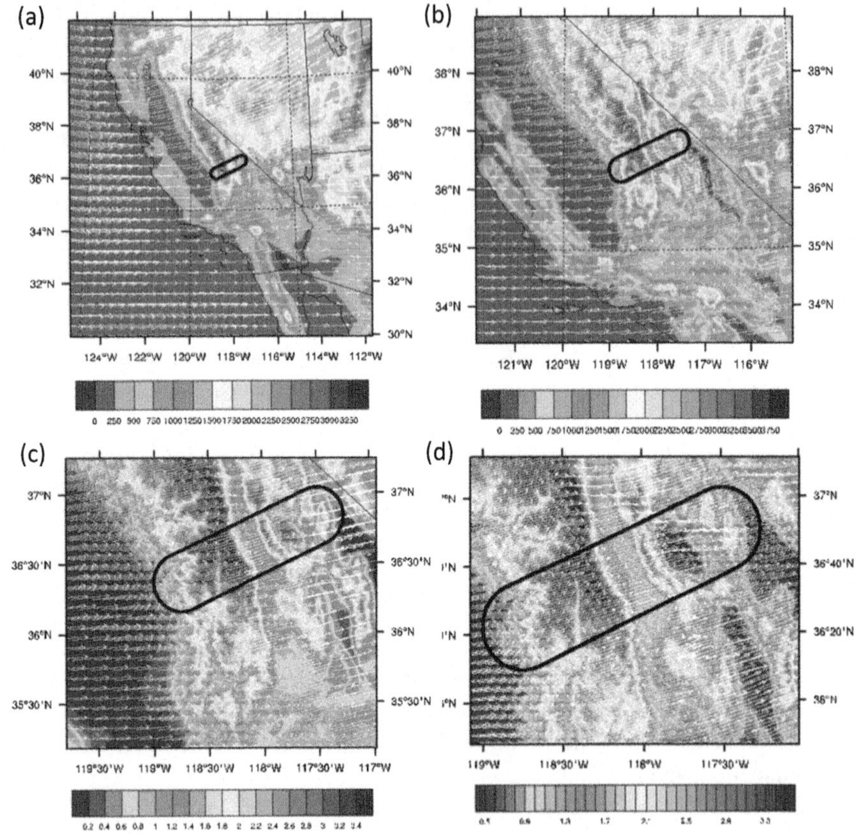

Fig. 20.4 Topography and the wind field at 12 km altitude on 25 March 2006 at 20 UTC from (**a**) mesoscale domain 1 (9 km grid spacing), (**b**) mesoscale domain 2 (3 km grid spacing), (**c**) mesoscale domain 3 (1 km grid spacing), and (**d**) microscale domain 4 (333 m horizontal grid spacing). The *oval curve* shows the path of the plane

dominated by south westerlies in agreement with observations. We note that the relaxation of the wind field is very smooth at the boundaries in the microscale nest. Also, regions with strong turbulent flow are found to be well resolved in the microscale nest. In these regions, the horizontal wind field shows strong drag, and the direction of the wind is complex.

Figure 20.5 shows a time series of (a) potential temperature, (b) vertical velocity, and (c) eastward (solid) and northward (dashed) winds from aircraft measurements; panels (d)–(f) are the same as (a)–(c) but from the high-resolution coupled mesoscale/microscale simulations. The time is relative to 15 UTC on 25 March 25 2006. The NCAR HIAPER research aircraft circled the Owens Valley at altitudes ~12 km. We note the excellent agreement in amplitudes of vertical velocity and timing of extreme vertical velocity peaks computed along the flight path.

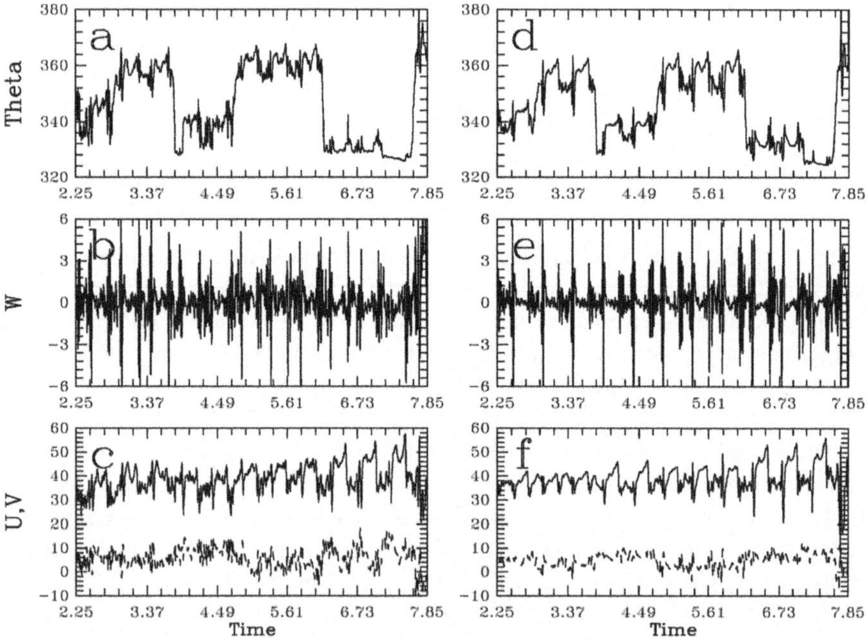

Fig. 20.5 Time series of (**a**) potential temperature, (**b**) vertical velocity, and (**c**) eastward (*solid*) and northward (*dashed*) winds from aircraft measurements. (**d**)–(**f**) are the same as (**a**)–(**c**) but from simulations. 25 March 2006. The time is relative to 15 UTC

Detailed results on the physics-based predictive modeling and ensemble forecasting of multi-scale UTLS dynamics can be found in Mahalov et al. (2008, 2009, 2011). Turbulence, waves, and rotational wind shears in the UTLS have been investigated for several intensive observational periods (IOPs) of the T-REX campaign. Verification of the UTLS modeling and computational results were obtained. These simulations use a UTLS microscale model that is driven by the finest mesoscale nest. During the IOPs, the simulation results reveal the presence of perturbations with short wavelengths in zones of strong vertical wind shear in the UTLS that cause a reversal of momentum fluxes. The spectral properties of these perturbations and the vertical profiles of heat and momentum fluxes show strong divergence near the tropopause indicating that they are generated by shear instability along shear lines locally induced by the primary mountain wave originating from the lower troposphere. The polarization relation between the horizontal wind components is elucidated by the hodograph of the horizontal wind vector. The polarized Richardson number and locally variable turbulent Prandtl number were introduced in Mahalov et al. (2008) to characterize the instabilities and turbulence of nonparallel shear-stratified flows.

The physical and dynamical processes associated with polarized instabilities are inherently three-dimensional. The horizontal velocity vector ($U(z)$, $V(z)$) rotates with the vertical coordinate z. Let $\alpha(z)$ denote the angle between the vector $d\mathbf{U}_h/dz = (dU/dz,\ dV/dz)$ and the horizontal wave vector. Then we obtain the following expression for the *polarized Richardson number*:

$$\mathrm{Ri}_p(z) = \frac{N^2(z)}{\left(\left(\frac{dU}{dz}\right)^2 + \left(\frac{dV}{dz}\right)^2\right)\cos^2(\alpha(z))}$$

Here $N(z)$ is the Brunt-Väisälä frequency profile. The polarized Richardson number takes into account horizontal anisotropy and the angle between the horizontal wave vectors and the velocity vectors at each vertical level. In the case of polarized wind fields in stably stratified environments, the polarized Richardson number was rigorously derived in Mahalov et al. (2008). For parallel shear flows $(V = 0)$, we recover definition of the classical Richardson number $\mathrm{Ri} = N^2/(dU/dz)^2$.

The instabilities and shear-stratified turbulence are induced by rotating winds and breaking polarized inertia-gravity waves in the UTLS. The shear-stratified UTLS turbulence is strongly inhomogeneous, patchy, and non-Kolmogorov. For IOP6 of the T-REX described in Grubišić et al. (2008), Mahalov et al. (2011) analyzed the distributions of trace gases in UTLS observed in aircraft measurements. They show small-scale fluctuations with amplitudes and phases that vary along the path of the flight. Detailed comparisons between these fluctuations from observations and numerical modeling provide further verification of the computational results. The observed vertical velocity shows that the behavior of these short fluctuations is due not only to the vertical motion but also to the local mean vertical gradients where the waves evolve, which are modulated by larger variations. The microscale model simulation results show favorable agreement with in situ radiosonde and aircraft observations of vertical velocity, horizontal winds, and temperature. Vertical nesting and improved physics-based predictive modeling of UTLS offered by the microscale model are found to be critical for the resolution of smaller-scale processes such as the formation of inversion layers associated with trapped lee waves in the troposphere and propagating mountain waves in the lower stratosphere. Localized sharp shear layers, wave breaking, and stiff gradients of potential temperature and vertical velocity are resolved in the UTLS using new physics-based modeling (variable turbulent Prandtl number, polarized Richardson number) within the embedded microscale nest. The three-dimensional character of the generated waves is investigated with analysis of co-spectra of the vertical and horizontal velocities in Mahalov et al. (2007). The polarization relation between the horizontal wind components is exposed by the hodograph of the horizontal wind vector, further confirming the upward energy propagation.

20.4 PIREP Cases: Turbulence, Waves, Rotational Wind Shear, and Polarized Richardson Number

The selection of cases from pilot reports (PIREPs) was based primarily on the height of (severe) turbulence being at least 30,000 ft AGL. Here we present an example from a catalogue of studied 2012–2014 PIREP cases. The application of microscale vertical nesting was performed in studies where PIREPs indicated extreme events taking place in the UTLS region. These extreme events are characterized by strongly nonlinear phenomena associated with high activity of mountain waves interacting with upper-level jet streams and formation of sharp gradients and adiabatic layers. In our vertically nested approach, 180 points with improved grid spacing in the UTLS region are used in the microscale nest. Embedded microscale nests resolve thin turbulent layers near the tropopause. They have complex streamwise-spanwise dependence influenced by topographic effects. The corresponding multi-scale nonhomogeneous laminated turbulent flow fields are characterized by alternating patches with high vertical velocity values.

Figure 20.6a shows location of a severe turbulence case reported over Montana during the time period from 12 UTC, 5 December 2012, to 00 UTC, 6 December 2012. Figure 20.6b and 20.6c shows plots of typical cross sections from microscale simulations zoomed to UTLS altitudes. Figure 20.6b shows the horizontal velocity U component, vertical velocity, and potential temperature (K) contours in cross section A (see Fig. 20.6a). Analyses of computational results indicated that the physical mechanism of turbulence is associated with high vertical shear of horizontal wind components and rapid changes in the direction of the horizontal wind vector with altitude in the UTLS zone. Examination of PIREP cases revealed that three-dimensional shear instabilities attributed to the breakdown of the polarized wind fields are common in the UTLS. These flows are characterized by wind profiles with intense variations in both eastward and westward components over a short vertical distance. These intense nonparallel shear-stratified flows lead to the development of 3D helical Kelvin-Helmholtz instabilities with characteristics that are different from the one found in parallel flows. The stability criterion for a general nonparallel polarized wind field was established and analyzed in Mahalov et al. (2008). Contrary to the parallel flows, this criterion is anisotropic and depends on the direction of the wave vectors of the unstable modes.

Figure 20.6c is an example cross section of the horizontal velocity, vertical velocity, and potential temperature at location E for two times. Turbulent patches in UTLS with positive/negative values of vertical velocities exceeding 10 m/s in magnitude were computed at these locations during a time interval spanning several hours. Analysis of PIREP cases and the example presented in Fig. 20.6 revealed that many physical mechanisms of instabilities and turbulence induced in UTLS by upper-level jet streams and mountain waves are similar to those previously studied in Joseph et al. (2003, 2004), Tse et al. (2003), and Mahalov et al. (2004, 2007, 2009, 2010, 2011). Complex turbulent dynamics is successfully resolved using the physics-based predictive modeling and microscale nested simulations with implicit

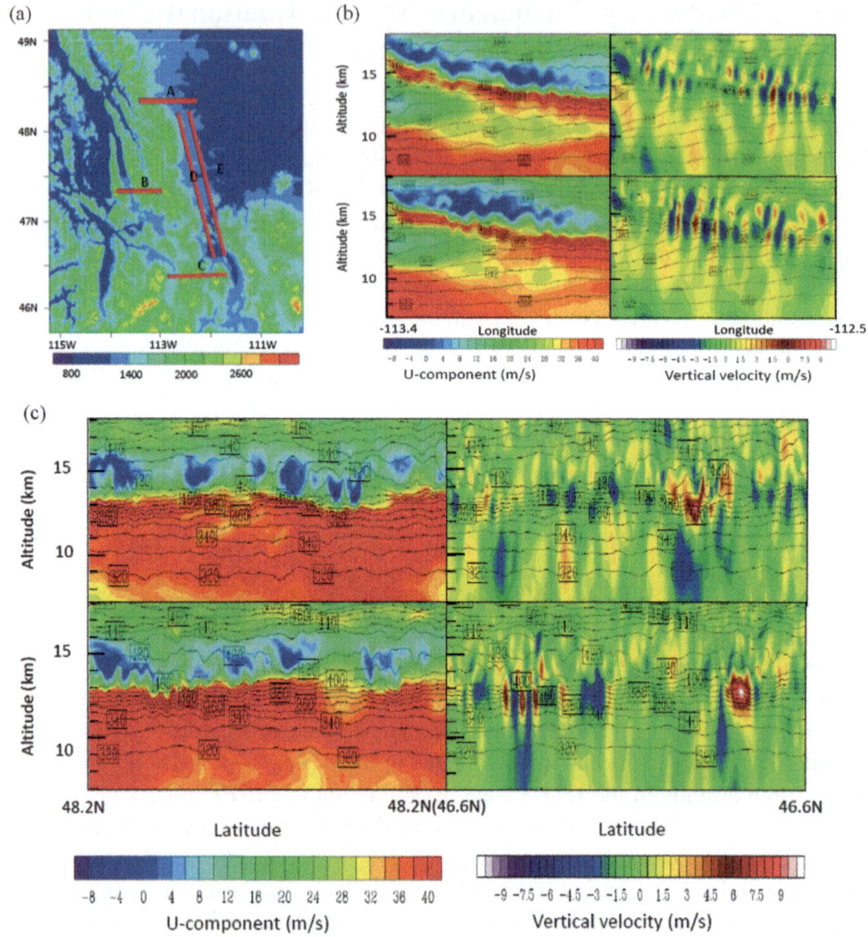

Fig. 20.6 Northwestern/western Montana case: (**a**) location of cross sections A, B, C, D, and E; (**b**) horizontal wind, vertical velocity, and potential temperature (K) in the cross section A at 17:00 (*top*) and 18:00 (*bottom*); and (**c**) horizontal wind, vertical velocity, and potential temperature in the cross section E at 19:00 (*top*) and 23:20 (*bottom*)

relaxation techniques and parameterizations based on the polarized Richardson number and variable turbulent Prandtl number.

20.5 Closing Comments

Research into enhanced vertical resolution to assess benefits and recommendations for operational use on new high-performance computing platforms is currently attracting considerable attention. For example, the UK Met Office is conducting investigations of the impact of increased vertical resolution in global configurations of the Unified Model (UM), in preparation for the anticipated step change in their supercomputer resources. Initial tests presented in Bushell et al. (2015) compare the current 70-level and 85-level sets used in global NWP forecast with a 120-level set with increased boundary layer and stratosphere resolution and a 180-level set with its resolution increased further in the stratosphere. Essentially all current operational models de-emphasize UTLS and they have low vertical resolution near the tropopause. The key aim is to identify where to concentrate additional resolution to improve model performance across all time scales.

Here we presented results of the high-resolution mesoscale/microscale modeling and forecasting system with advanced upper troposphere and lower stratosphere physics and fast computational algorithms based on vertical nesting and implicit relaxation techniques. The physics-based modeling approach incorporates rotational shear-stratified turbulence generation mechanisms including spatially varying turbulent Prandtl numbers and a polarized Richardson number that is new to UTLS turbulence forecasting products.

Coupled mesoscale/microscale simulations presented in this study are conducted typically for 48 h of physical time. The microscale code is run for 3 h covering observational periods and aircraft measurements. A large speedup is achieved if coupled mesoscale-microscale simulations are executed in a two-way concurrent mode. This eliminates the large slow down caused by the frequent input/output required when these two codes run separately. These simulations can be conducted in real-time forecasting, using a refined vertical grid in the microscale code with 120–180 vertical levels. Comparisons with observational datasets reveal significant benefits of nested computational techniques that take into account the shear-stratified turbulence physics of the UTLS.

Acknowledgments This work is sponsored in part by AFOSR contract FA9550-12-C-0067.

References

Bushell, A.C., Cullen, M.J.P., Walters, D.N., Milton, S.: Research into enhanced vertical resolution UM trials across timescales to assess benefits and recommendations for operational use on new HPC at NWP-seasonal timescales and for next-generation climate configurations. UK Met Office Technical Report, March 10 (2015)

Clark, T.L., Hall, W.D.: The design of smooth, conservative grids for interactive grid nesting with stretching. J. Appl. Meteorol. **35**, 1040–1046 (1996)

Davies, H.C.: Limitations of some common lateral boundary schemes used in regional NWP models. Mon. Weather Rev. **111**, 1002–1012 (1983)

Dietachmayer, G.S., Droegemeier, K.K.: Application of continuous dynamic grid adaptation techniques to meteorological modeling, Part I. Mon. Weather Rev. **120**, 1675–1706 (1992)

Dudhia, J.: A nonhydrostatic version of the Penn State NCAR mesoscale model. Mon. Weather Rev. **121**, 1493–1513 (1993)

Grubišić, V., Doyle, J.D., Kuettner, J., Mobbs, S., Smith, R.B., Whitman, D., Dirks, R., Czyzyk, S., Cohn, S.A., Vosper, S., Weissmann, M., Haimov, S., De Wekker, S., Pan, L.L., Chow, F.K.:

The terrain-induced rotor experiment: an overview of the field campaign and some highlights of special observations. Bull. Am. Meteorol. Soc. **89**, 1513–1533 (2008)

Joseph, B., Mahalov, A., Nicolaenko, B., Tse, K.L.: High resolution DNS of jet stream generated tropopausal turbulence. Geophys. Res. Lett. **30**(10), (2003). doi:10.1029/2003GL017252

Joseph, B., Mahalov, A., Nicolaenko, B., Tse, K.L.: Variability of turbulence and its outer scales in a nonuniformly stratified tropopause jet. J. Atmos. Sci. **41**, 524–537 (2004)

Mahalov, A., Moustaoui, M.: Vertically nested non-hydrostatic model for multi-scale resolution of flows in the upper troposphere and lower stratosphere. J. Comp. Phys. **228**, 1294–1311 (2009)

Mahalov, A., Pacheco, J.R., Fernando, H.J.S., Hunt, J.C.R.: Effects of rotation on fronts of density currents. Phys. Lett. **270**, 149–156 (2000)

Mahalov, A., Nicolaenko, B., Tse, K.-L., Joseph, B.: Eddy-mixing in jet-stream turbulence under stronger stratification. Geophys. Res. Lett. **23**, L23111–L23115 (2004)

Mahalov, A., Moustaoui, M., Nicolaenko, B.: Computational studies of inertia-gravity waves radiated from upper tropospheric jets. Theor. Comp. Fluid Dyn. **21**(6), 399–422 (2007)

Mahalov, A., Moustaoui, M., Nicolaenko, B.: Three-dimensional instabilities in non-parallel shear stratified flows. Kinetic Rel. Models **2**(1), 215–229 (2008)

Mahalov, A.: Atmospheric characterization and ensemble forecasting of multi-scale flows in the Upper Troposphere and Lower Stratosphere (UTLS). American Institute of Aeronautics and Astronautics, Paper AIAA 2009-110 (2008), invited paper

Mahalov, A., Moustaoui, M.: Characterization of atmospheric optical turbulence for laser propagation. Laser Photonics Rev. **4**, 144–159 (2010). Special Issue: 50 Years of Laser

Mahalov, A., Moustaoui, M., Grubišić, V.: A numerical study of mountain waves in the upper troposphere and lower stratosphere. Atmos. Chem. Phys. **11**, 5123–5139 (2011)

Marbaix, P., Gallée, H., Brasseur, O., Van Ypersele, J.-P.: Lateral boundary conditions in regional climate models: a detailed study of the relaxation procedure. Mon. Weather Rev. **111**, 461–479 (2003)

Oddo, P., Pinardi, N.: Lateral open boundary conditions for nested limited area models: a scale selective approach. Ocean Model. **10**, 134–156 (2008)

Sharman, R.D., Trier, S.B., Lane, T.P., Doyle, J.D.: Sources and dynamics of turbulence in the upper troposphere and lower stratosphere: a review. Geophys. Res. Lett. **39**, L12803 (2012a). doi:10.1029/2012GL051996

Sharman, R.D., Doyle, J.D., Shapiro, M.A.: An investigation of a commercial aircraft encounter with severe clear-air turbulence over western Greenland. J. Appl. Meteorol. Climatol. **51**(1), 42–53 (2012b)

Shchepetkin, A., McWilliams, J.C.: Regional ocean model system: a split-explicit ocean model with a free surface and topography-following vertical coordinate. Ocean Model. **9**, 347–404 (2005)

Skamarock, W.C., Klemp, J.B.: A time-split nonhydrostatic atmospheric model for weather research and forecasting applications. J. Comp. Phys. **227**, 1465–1485 (2008)

Tse, K.L., Mahalov, A., Nicolaenko, B., Fernando, H.J.S.: Quasi-equilibrium dynamics of shear-stratified turbulence in a tropospheric jet. J. Fluid Mech. **496**, 73–103 (2003)

Chapter 21
Similarity of Stably Stratified Geophysical Flows

Zbigniew Sorbjan

Abstract The article reviews the gradient-based similarity theory of shear-dominated, stably-stratified turbulent flows. The gradient-based similarity scales are classified as explicit or implicit. The explicit scaling employs the length scale as a specified function of height. Within the implicit type, the mixing length is locally related to various moments of turbulence. The analytical form of the explicit similarity functions of the Richardson number Ri is obtained based on experimental data collected in the atmospheric surface layer. The implicit similarity functions can be derived by renormalization of the explicit-type expressions. Since the implicit scales and similarity functions are not directly dependent on height, they are expected to be universally valid in shear-driven, stably-strafified turbulent flows, in the atmospheric boundary layer and in the upper atmosphere.

21.1 Introduction

Prediction of mesoscale flows in the atmosphere requires appropriate parameterization of small-scale turbulence. A number of approaches have been developed in the past, including the K-theory, higher-order closure modeling, as well as similarity theory approach (e.g., Sorbjan 2016a). The latter technique, introduced a century ago by Buckingham (1914), has been frequently applied in situations, when the complexity of physical processes prevents obtaining direct solutions. The classical milestones of the approach include the "2/3 law" for the second-order structure functions of Kolmogorov (1941), the "−5/3 law" for the energy spectrum of Obukhov (1941), as well as the surface-layer similarity theory of Monin and Obukhov (1954).

The Monin-Obukhov similarity provides a foundation for describing the near-surface atmospheric turbulence (e.g., Foken 2006). Weak turbulence in stable

Z. Sorbjan (✉)
Institute of Geophysics, Polish Academy of Sciences, Warsaw, Poland

Department of Physics, Marquette University, Milwaukee, WI 53201-1881, USA
e-mail: zbigniew.sorbjan@marquette.edu

R. Sharman, T. Lane (eds.), *Aviation Turbulence*,
DOI 10.1007/978-3-319-23630-8_21

conditions limits applications of the theory, due to uncertainty introduced by small values of fluxes, used to define similarity scales, and self-correlation errors (e.g., Klipp and Mahrt 2004; Baas et al. 2006). The fact that the Monin-Obukhov similarity scales are based on fluxes prevents applications of the theory in higher regions of the atmosphere, even within the local formulation of Nieuwstadt (1984). To overcome these difficulties, an alternate conceptual approach of the gradient-based similarity was introduced by Sorbjan (2008a, b), Sorbjan (2010), and subsequently elaborated by Sorbjan and Grachev (2010), Sorbjan (2012), Sorbjan and Czerwinska (2013), Grachev et al. (2015), Sorbjan (2014), and Sorbjan (2016). The approach was primarily developed for the shear-dominated, stable atmospheric boundary layer, but could also be applied in other stably-stratified turbulent flows. Such flows appear in a wide range of conditions, extending from small-scale engineering flows to large-scale geophysical motions in the upper atmosphere and oceans, controlled by the coupled relationship between kinetic and potential energy (e.g., Mater and Venayagamoorthy 2014).

21.2 Surface-layer similarity theories

The Monin-Obukhov similarity theory (1954) is limited to the surface layer, where the turbulent fluxes of momentum and temperature, τ_o, H_o, are nearly constant with height. The fluxes are chosen to construct three similarity scales: $u* = {\tau_o}^{1/2}$ as the velocity scale, $T* = H_o/u*$ as the temperature scale, and the Obukhov length, $L* = -\tau_o^{3/2}/(\kappa\beta H_o)$ as the length scale, where $\beta = g/T_o$ is the buoyancy parameter, g is the gravity acceleration, T_o is the reference absolute temperature, and $\kappa = 0.4$ is the von Karman constant. Subsequently, it can be concluded that nondimensional products of statistical moments of turbulence and the flux-based similarity scales are universal functions of the dimensionless height $z/L*$, where z is the height above the underlying surface. As pointed out by Nieuwstadt (1984), the assumption of flux constancy with height is not necessary, so that the flux-based scales can be local and defined analogously as $U{*}(z) = \tau^{1/2}$, $Y{*}(z) = H/U{*}$ and $\Lambda{*}(z) = -\tau^{3/2}/(\kappa\beta H)$, where τ, H are the local values of the momentum and temperature fluxes.

Within the alternative local gradient-based scaling (Sorbjan 2008a, b; 2010), the positive potential temperature gradient, $\Gamma = d\Theta/dz$, and the mixing length, l, are used to formulate the gradient-based scales:

$$\begin{aligned} &\text{for velocity :} && U_N = l\,N \\ &\text{for temperature :} && T_N = l\Gamma \\ &\text{for length :} && L_N(z) = l \end{aligned} \qquad (21.1)$$

where $N = (\beta\Gamma)^{1/2}$ is the Brunt–Väisälä frequency, which provides a useful description of static stability. In oceanic applications, the Brunt–Väisälä frequency

N is defined through the gradient $\Gamma = -d\rho/dz$ of the potential density ρ, which depends on both temperature and salinity, and the buoyancy parameter $\beta = g/\rho_o$, where ρ_o is the reference potential density. Since turbulent mixing is fundamentally related to overturning motions, the mixing length scale l is used above as the basis for construction of the scaling framework.

The definition of the length scale l is required in order to complete the system of scales (21.1). The mixing length l was originally introduced by Prandtl (1926), based on the analogy to the mean-free path of kinetic gas theory. Prandtl argued that l may be understood as a diameter of a moving parcel of fluid, before it blends into the neighboring environment. In neutral conditions, the concept allows expressing the momentum flux τ as the squared product of the mixing length l, characterizing dominant eddies, and shear $S : \tau = (lS)^2$, where the squared wind shear is defined as $S^2 = (dU/dz)^2 + (dV/dz)^2$. Since observations in the close proximity of the underlying surface show that $S = u*/(\kappa z)$, the mixing length in this region is a linear function of height:

$$l = \kappa z. \tag{21.2}$$

For the gradient-based scales (21.1), the similarity theory implies that dimensionless statistical moments of turbulence are functions of the local Richardson number $\text{Ri} = N^2/S^2$. Note that the local scales and the local Richardson number represent turbulence locally, within relatively thin layers of the atmosphere.

The gradient-based approach, based on Eqs. (21.1) and (21.2), was first examined in the atmospheric surface layer by using data collected during the SHEBA experiment (Sorbjan 2010; Sorbjan and Grachev 2010). The following gradient-based similarity functions were derived for the momentum flux, τ; the temperature flux, H; and standard deviations of the vertical velocity and temperature, σ_w and σ_θ, in the sub-critical range:

$$\begin{aligned}
\frac{\tau}{U_N^2} &= \frac{1}{\text{Ri}\left(1+300\text{Ri}^2\right)^{3/2}} \equiv G_t,\\
-\frac{H}{U_N T_N} &= \frac{1}{0.9\,\text{Ri}^{1/2}\left(1+250\text{Ri}^2\right)^{3/2}} \equiv G_h,\\
\frac{\sigma_w}{U_N} &= \frac{1}{0.85\,\text{Ri}^{1/2}\left(1+450\text{Ri}^2\right)^{1/2}} \equiv G_w,\\
\frac{\sigma_\theta}{T_N} &= \frac{5}{\left(1+2500\text{Ri}^2\right)^{1/2}} \equiv G_\theta.
\end{aligned} \tag{21.3}$$

The similarity curves for fluxes in (21.3) are reproduced in Fig. 21.1a, b, together with data collected during the SHEBA experiment. The figures show that both functions monotonically decrease with the increasing Richardson number Ri. In the near-neutral conditions, $G_t \sim \text{Ri}^{-1}$, and for larger values of the Richardson number, $G_t \sim \text{Ri}^{-4}$. On the other hand, $G_h \sim \text{Ri}^{-1/2}$ in the near-neutral range, and

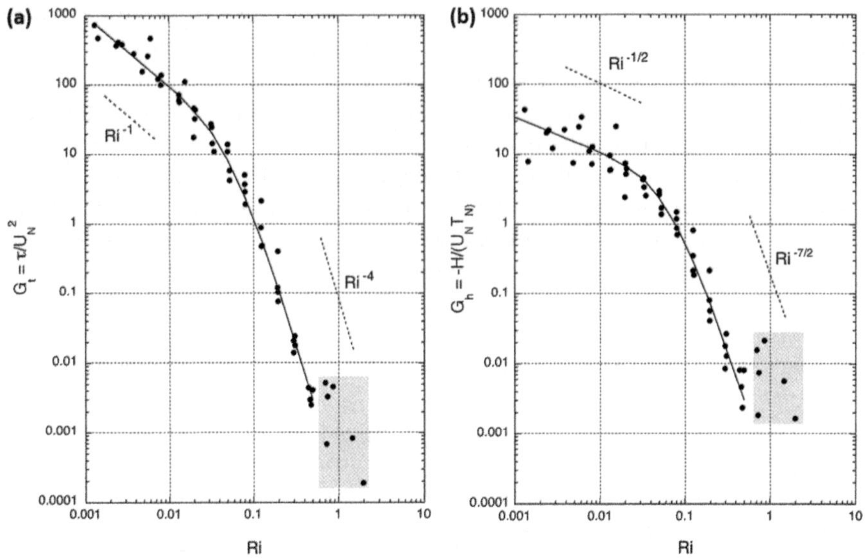

Fig. 21.1 A dependence of bin-averaged, 1-h medians of the dimensionless: (**a**) momentum flux $G_t = \tau/U_N^2$ and (**b**) temperature flux $G_h = -H/(U_N T_N)$, on the Richardson number in the atmospheric surface layer. The data points in the figure were evaluated based on observations during the SHEBA experiment at five levels: $z_1 = 2.2$ m, $z_2 = 3.2$ m, $z_3 = 5.1$ m, $z_4 = 8.9$ m, and $z_5 = 18.2$ m. The *shaded boxes* encompass data collected in the overcritical regime (based on Sorbjan and Grachev 2010)

subsequently, $G_h \sim \mathrm{Ri}^{-7/2}$. The data points in the figures are bin-averaged, 1-h medians (note that medians are not affected by excessively high or low values of data points), collected on the 20 m main tower at five levels, located at $z_1 \approx 2.2$ m, $z_2 \approx 3.2$ m, $z_3 \approx 5.1$ m, $z_4 \approx 8.9$ m, and $z_5 \approx 18.2$ m, but 14 m during most of the winter (e.g., Andreas et al. 1999; Persson et al. 2002). One might notice a scatter of observational points within each bin, related to multi-level observations. Details of the turbulent and profile measurements, accuracy, calibration, data processing, and data-quality criteria for the SHEBA dataset can be found in the papers of Grachev et al. (2005, 2013, 2015) and references therein.

The similarity expressions for the eddy viscosity $K_m = \tau/S$, and eddy diffusivity $K_h = -H/\Gamma$, take the form (Sorbjan 2014):

$$\frac{K_m}{L_N U_N} = \mathrm{Ri}^{1/2} G_t \qquad \frac{K_h}{L_N U_N} = G_h, \tag{21.4}$$

To obtain approximate expressions for the dissipation rate of the turbulent kinetic energy ε, one can consider simplified, quasi-steady state budget for the turbulent kinetic energy: $\varepsilon = K_m S^2(1 - \mathrm{Rf})$, received under assumption that the divergence of the vertical transport terms can be neglected in stable conditions (e.g., Wyngaard and Coté 1972; Champagne et al. 1977; Caughey et al. 1979), where $\mathrm{Rf} = \mathrm{Ri}^{1/2} G_h/G_t$ is the flux Richardson number (Sorbjan 2010). Consequently, using Eqs. (21.3) and (21.4), the following expressions can be obtained (Sorbjan 2010):

$$\frac{\varepsilon}{{U_N}^3/L_N} = \frac{1}{\mathrm{Ri}^{3/2}(1+300\mathrm{Ri}^2)^{3/2}}\left[1 - \frac{\mathrm{Ri}}{0.9}\frac{(1+250\mathrm{Ri}^2)^{3/2}}{(1+300\mathrm{Ri}^2)^{3/2}}\right] \equiv G_\varepsilon, \tag{21.5}$$

21.3 Explicit Scaling Systems Above the Surface Layer

Farther from the surface, the growth of the mixing length with height is expected to be nonlinear. This fact can be taken into consideration by adopting the inverse linear approximation of Delage (1974), $1/l = 1/(\kappa z) + 1/\lambda_0$, between values of the mixing length near the underlying surface (Eq. 21.2) and above the surface ($= \lambda_0$), which yields

$$l = \frac{\kappa z}{1 + \frac{\kappa z}{\lambda_0}}, \tag{21.6}$$

where the quantity λ_0 is an external parameter. Blackadar (1962) suggested that $\lambda_0 = 0.009\, u*/f$, where u* is the friction velocity, and f is the Coriolis parameter. For the CASES-99 data (observations collected within the first 50 m above the underlying surface), Sorbjan (2012) found that $\lambda_0 = 12$ in stable conditions. Sorbjan and Czerwinska (2013) showed the modification of the similarity scales, by using the mixing length defined by Eq. (21.6), does not alter the dependence of similarity functions on the Richardson number Ri. Huang et al. (2013) employed an ad hoc modification of the mixing length in the form $1/l = 1/(\kappa z) + 1/\lambda_m$, where $1/\lambda_m = 1/\lambda_0 + 1/\lambda_B$, λ_B is a parameter, $\lambda_B = \lambda_S/\mathrm{Ri}$, and the parameter λ_S was assumed equal to 1 m by Sorbjan (2014). This yields

$$l = \frac{\kappa z}{1 + \frac{\kappa z}{\lambda_0} + \frac{\kappa z}{\lambda_B}}. \tag{21.7}$$

21.4 Implicit scaling systems in stably-stratified flows

The form of the mixing length (21.7) introduces discrepancy, implied by the fact that both similarity scales (21.1) and similarity functions (21.3) depend on the Richardson number. This indicates that the general validity of the explicit scaling is limited. An analysis made by Sorbjan and Czerwinska (2013), based on experimental data described by Hunt et al. (1985), showed that the explicit scaling fails in more complex meteorological conditions. Both findings signal the need for replacing the explicit length scale by a length scale defined in terms of various statistical moments of turbulence, such as the vertical velocity variance σ_w^2; the temperature variance σ_θ^2; or the dissipation rate of turbulent kinetic energy ε. The choice of the vertical velocity variance σ_w^2 as the scaling parameter, instead of the kinetic turbulent energy E, is due to our focusing on mixing, caused by vertical overturning motions within shear-dominated turbulence in subcritical flows, with $\mathrm{Ri} < \mathrm{Ri}_{\mathrm{cr}}$, where $\mathrm{Ri}_{\mathrm{cr}}$ is the critical Richardson number. Turbulence, with overcritical Richardson numbers, $\mathrm{Ri} > \mathrm{Ri}_{\mathrm{cr}}$, supports internal waves and anisotropic turbulence, which makes the parameterization of stratified flows much more difficult.

For the mentioned three quantities, σ_w^2, σ_θ^2, ε, and the Brunt–Väisälä frequency N, three length scales can be proposed: (1) the turbulent length scale $L_w = \sigma_w/N$, which involves the turbulent kinetic energy of the vertical motions, (2) the Ellison length scale $L_\theta = \beta\sigma_\theta/N^2$, as a measure of the vertical distance traveled by fluid parcels before returning toward an equilibrium position (Ellison 1957), and (3) the Dougherty–Ozmidov length scale $L_\varepsilon = \left(\varepsilon/N^3\right)^{1/2}$, which marks the range of small eddies $\lambda < L_\varepsilon$, which are not influenced by effects of buoyancy in stably stratified turbulence (Dougherty 1961; Ozmidov 1965).

It can be noted that close to the underlying surface, the considered mixing lengths, $L_w, L_\theta, L_\varepsilon$, are proportional to Eq. (21.2). Indeed, taking into consideration that in near-neutral conditions: $\sigma_w \sim u^*$, $\sigma_\theta \sim T^*, \varepsilon \sim u^{*3}/(kz), N^2 \sim \beta T^*/(\kappa z)$, and $S = u^*/(\kappa z)$, we obtain $L_w \sim \kappa z\, \mathrm{Ri}^{-1/2}, L_\theta \sim \kappa z, L_\varepsilon \sim \kappa z \mathrm{Ri}^{-3/4}$. Because N is used instead of S in the definitions of scales L_w and L_ε, the dependence on the Richardson number appears in the resulting expressions.

In the next sections, we will consider three implicit scaling systems, based on the listed above length scales, L_w, L_θ, and L_ε, and derive gradient-based similarity functions for these scales through renormalization (i.e., rescaling). The renormalization eliminates explicit scales and replaces them with implicit ones. The resulting similarity functions depend only on the Richardson number Ri, and the explicit scales are not directly dependent on height. This fact implies universality of the approach, which can be applied not only in the atmospheric surface layer but also in a wide range of geophysical flows.

21.4.1 The σ_w-Scaling

Let us first consider the σ_w-based scaling of the following form (Sorbjan 2010):

$$\begin{aligned} &\text{for velocity :} && U_w = \sigma_w \\ &\text{for temperature :} && T_w = \frac{\sigma_w N}{\beta} \\ &\text{for length :} && L_w = \frac{\sigma_w}{N} \end{aligned} \tag{21.8}$$

Taking into consideration that

$$\frac{U_w}{U_N} = \frac{T_w}{T_N} = \frac{L_w}{L_N} = \frac{\sigma_w}{U_N} = G_w(R_i) \tag{21.9}$$

where G_w is described by expression (21.3c), the gradient-based similarity functions for momentum and temperature fluxes, and for the temperature standard deviation, can be found by renormalization of expressions (21.3a, b, d) (Sorbjan 2010):

$$\begin{aligned} \frac{\tau}{U_w^2} &= G_t/G_w^2 = 0.72\frac{\left(1+450\mathrm{Ri}^2\right)}{\left(1+300\mathrm{Ri}^2\right)^{3/2}}, \\ -\frac{H}{U_w T_w} &= G_h/G_w^2 = 0.8\mathrm{Ri}^{1/2}\frac{\left(1+450\mathrm{Ri}^2\right)}{\left(1+250\mathrm{Ri}^2\right)^{3/2}}, \\ \frac{\sigma_\theta}{T_w} &= G_\theta/G_w = 4.25\mathrm{Ri}^{1/2}\frac{\left(1+450\mathrm{Ri}^2\right)^{1/2}}{\left(1+2500\mathrm{Ri}^2\right)^{1/2}} \end{aligned} \tag{21.10}$$

The first two similarity functions, for the dimensionless momentum and temperature fluxes, are plotted in Fig. 21.2, together with data collected during the SHEBA experiment. In the near-neutral conditions, the dimensionless momentum flux is constant, $\tau/U_w{}^2 \sim \mathrm{Ri}^0$, and decreases as Ri^{-1} for larger values of the Richardson number. The dimensionless temperature flux increases exponentially as $\mathrm{Ri}^{1/2}$ in the near-neutral range, reaches a maximum at $\mathrm{Ri} \sim 0.1$, and subsequently decreases as $\mathrm{Ri}^{-1/2}$.

The similarity expressions for the eddy viscosity and eddy diffusivity are of the form

$$\frac{K_m}{L_w U_w} = \mathrm{Ri}^{1/2}\frac{G_t}{G_w^2}, \qquad \frac{K_h}{L_w U_w} = \frac{G_h}{G_w^2}. \tag{21.11}$$

It could be mentioned that the renormalization changes the dependence of the similarity functions (21.3) on the Richardson number, but does not introduce any additional corrections. Note that the similarity scales can be understood as representing external forcing conditions, while the similarity functions express dependence on thermodynamic stability, which ideally should be universal. The gradient-based similarity functions (21.11), obtained for implicit scales, can be

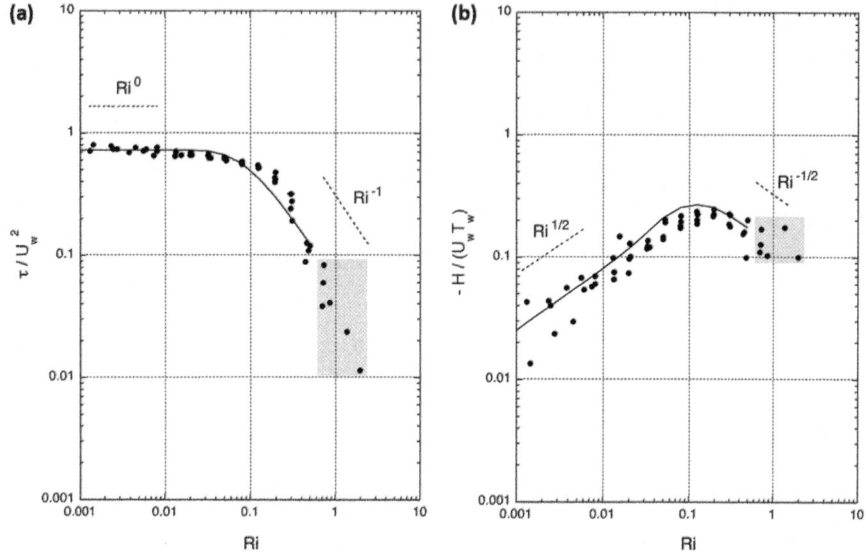

Fig. 21.2 A dependence of bin-averaged, 1-h medians of the dimensionless: (**a**) momentum flux $\tau/U_w^{\ 2}$ and (**b**) temperature flux $-H/(U_wT_w)$ on the Richardson number. The data points in the figure were evaluated based on observations at five levels: $z_1 = 2.2$ m, $z_2 = 3.2$ m, $z_3 = 5.1$ m, $z_4 = 8.9$ m, and $z_5 = 18.2$ m (but at 14 m during most of the winter) during the SHEBA experiment. The *curves* are described by Eq. (21.10a, b) in the range. The *shaded boxes* encompass data collected in the overcritical regime (based on Sorbjan and Grachev 2010)

expected to be generally valid within stable-stratified subcritical geophysical flows. The same comments apply to other scaling systems described below.

21.4.2 The σ_θ-Scaling

For the σ_θ-based scaling, one can obtain (Sorbjan 2010)

$$\begin{aligned} &\text{for velocity}: && U_\theta = \frac{\beta\sigma_\theta}{N} \\ &\text{for temperature}: && T_\theta = \sigma_\theta \\ &\text{for length}: && L_\theta = \frac{\beta\sigma_\theta}{N^2} \end{aligned} \tag{21.12}$$

Taking into consideration that

$$\frac{U_\theta}{U_N} = \frac{T_\theta}{T_N} = \frac{L_\theta}{L_N} = \frac{\sigma_\theta}{T_N} = G_\theta \tag{21.13}$$

where G_θ is described by expression (21.3d), the gradient-based similarity functions in this case can be found by renormalization of expressions (21.3a–c) (Sorbjan 2010)

$$\begin{aligned}\frac{\tau}{U_\theta^2} &= G_t/G_\theta^2 = 0.04\frac{(1+2500\mathrm{Ri}^2)}{\mathrm{Ri}(1+300\mathrm{Ri}^2)^{3/2}}\\ -\frac{H}{U_\theta T_\theta} &= G_h/G_\theta^2 = 0.044\frac{(1+2500\mathrm{Ri}^2)}{\mathrm{Ri}^{1/2}(1+250\mathrm{Ri}^2)^{3/2}}\\ \frac{\sigma_w}{U_\theta} &= G_w/G_\theta^2 = 0.24\frac{(1+2500\mathrm{Ri}^2)^{1/2}}{\mathrm{Ri}^{1/2}(1+450\mathrm{Ri}^2)^{1/2}}\end{aligned} \tag{21.14}$$

The first two similarity functions, for the dimensionless momentum and temperature fluxes, are plotted in Fig. 21.3, together with bin-averaged, 1-h medians, collected during the SHEBA experiment. The dimensionless momentum flux $\tau/U_\theta{}^2$ monotonically decreases in the near-neutral conditions as Ri^{-1}, and as Ri^{-2} for larger values of the Richardson number. The negative dimensionless temperature flux decreases as $\mathrm{Ri}^{-1/2}$ in the near-neutral conditions, reaches a local maximum at $\mathrm{Ri} \sim 0.05$, and subsequently decreases as $\mathrm{Ri}^{-3/2}$.

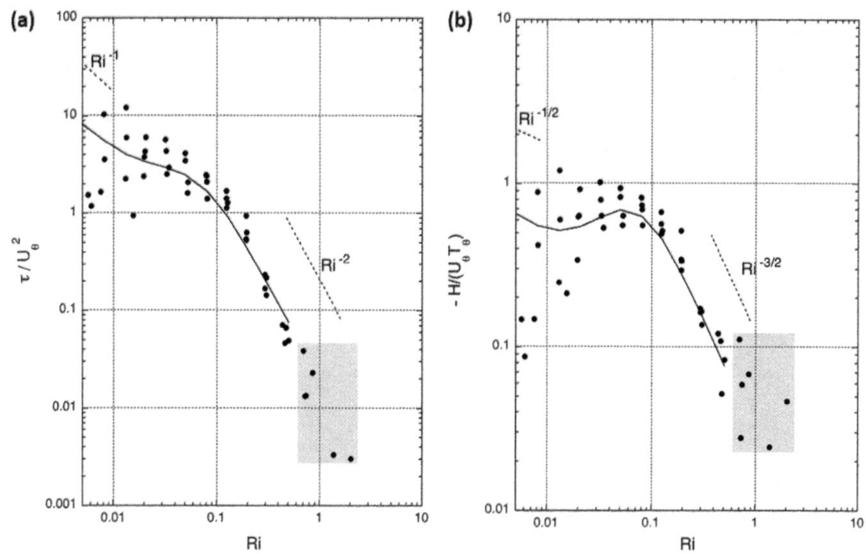

Fig. 21.3 A dependence of bin-averaged, 1-h medians of the dimensionless: (**a**) momentum flux $\tau/U_\theta{}^2$, (**b**) temperature flux $-H/(U_\theta T_\theta)$ on the Richardson number. The data points in the figure were evaluated based on observations collected at five levels: $z_1 = 2.2$ m, $z_2 = 3.2$ m, $z_3 = 5.1$ m, $z_4 = 8.9$ m, and $z_5 = 18.2$ m (but at 14 m during most of the winter) during the SHEBA experiment. The *curves* are described by Eq. (21.14a, b). The *shaded boxes* encompass data collected in the overcritical regime (based on Sorbjan and Grachev 2010)

The similarity expressions for the eddy viscosity and eddy diffusivity are of the form

$$\frac{K_m}{L_\theta U_\theta} = \mathrm{Ri}^{1/2}\frac{G_t}{G_\theta^2}, \qquad \frac{K_h}{L_\theta U_\theta} = \frac{G_h}{G_\theta^2}. \tag{21.15}$$

21.4.3 The ε-Scaling

The following three scales can be proposed within the ε-based scaling (Sorbjan and Balsley 2009; Grachev et al. 2015; Sorbjan 2016):

$$\begin{aligned} &\text{for velocity :} && U_\varepsilon = \left(\tfrac{\varepsilon}{N}\right)^{1/2} \\ &\text{for temperature :} && T_\varepsilon = \frac{(\varepsilon N)^{1/2}}{\beta} \\ &\text{for length :} && L_\varepsilon = \left(\tfrac{\varepsilon}{N^3}\right)^{1/2} \end{aligned} \tag{21.16}$$

where L_ε is the Dougherty–Ozmidov length scale. Taking into consideration that

$$\frac{U_\varepsilon}{U_N} = \frac{T_\varepsilon}{T_N} = \frac{L_\varepsilon}{L_N} = \left(\frac{\varepsilon}{U_N^2 N}\right)^{1/2} = G_\varepsilon^{1/2} \tag{21.17}$$

where G_ε is described by Eq. (21.5), one can obtain the following dimensionless similarity function (Sorbjan 2016):

$$\frac{\tau}{U_\varepsilon^2} = G_t/G_\varepsilon = \frac{\mathrm{Ri}^{1/2}}{\left[1 - \dfrac{\mathrm{Ri}(1+300\mathrm{Ri}^2)^{3/2}}{0.9(1+250\mathrm{Ri}^2)^{3/2}}\right]}$$

$$-\frac{H}{U_\varepsilon T_\varepsilon} = G_h/G_\varepsilon = \frac{\mathrm{Ri}}{0.9(1+250\mathrm{Ri}^2)^{3/2}} \frac{(1+300\mathrm{Ri}^2)^{3/2}}{\left[1 - \dfrac{\mathrm{Ri}(1+300\mathrm{Ri}^2)^{3/2}}{0.9(1+250\mathrm{Ri}^2)^{3/2}}\right]}$$

$$\frac{\sigma_w}{U_\varepsilon} = G_w/G_\varepsilon^{1/2} = \frac{\mathrm{Ri}^{1/4}}{0.85(1+450\mathrm{Ri}^2)^{1/2}} \frac{(1+300\mathrm{Ri}^2)^{3/4}}{\left[1 - \dfrac{\mathrm{Ri}}{0.9}\dfrac{(1+300\mathrm{Ri}^2)^{3/2}}{(1+250\mathrm{Ri}^2)^{3/2}}\right]^{1/2}}$$

$$\frac{\sigma_\theta}{T_\varepsilon} = G_\theta/G_\varepsilon^{1/2} = \frac{5\mathrm{Ri}^{3/4}}{(1+2500\mathrm{Ri}^2)^{1/2}} \frac{(1+300\mathrm{Ri}^2)^{3/4}}{\left[1 - \dfrac{\mathrm{Ri}(1+300\mathrm{Ri}^2)^{3/2}}{0.9(1+250\mathrm{Ri}^2)^{3/2}}\right]^{1/2}} \tag{21.18}$$

For the dimensionless eddy viscosity and eddy diffusivity, the renormalization yields

$$\frac{K_m}{U_\varepsilon L_\varepsilon} = \mathrm{Ri}^{1/2}\frac{G_t}{G_\varepsilon}, \qquad \frac{K_h}{U_\varepsilon L_\varepsilon} = \frac{G_h}{G_\varepsilon} \tag{21.19}$$

The similarity functions (21.18a, b) are compared with observational data in Fig. 21.4a, b. The data points in the figures were obtained by Grachev et al. (2015), based on the observations made during the SHEBA experiment. The authors imposed several methodological restrictions of observations, based on the values of the Richardson numbers Ri and Rf. Specifically, observational points were excluded, when both the Richardson numbers, Ri and Rf, exceeded the critical value, assumed to be 0.2 for "Kolmogorov turbulence". Such prerequisites differed from those employed earlier in Sorbjan and Grachev (2010) to derive Eqs. (21.3)–(21.6) and to draw lines in Figs. 21.1–21.3 in the range Ri < 0.7.

Figure 21.4a shows the dimensionless flux $\tau/U_\varepsilon{}^2$, as a function of the Richardson number Ri. The data points are displayed for five levels, denoted as 1–5, and marked by various symbols. The lowest level, z_1, is marked by black dots, while data points for the highest level, z_5, are gray. One might notice that there is a scatter of observational points within each bin, due to multi-level observations. The curve in the figure is plotted based on Eq. (21.4a), which was originally evaluated for Ri $\leq$ 0.7 (Sorbjan and Grachev 2010). To indicate the discrepancy of the domain, in which the empirical points were evaluated ($Ri < 0.2$), and the similarity functions drawn (Ri < 0.7), the section of the curve for Ri < 0.2 is marked as a continuous line, while the remaining portion is indicated by a broken line.

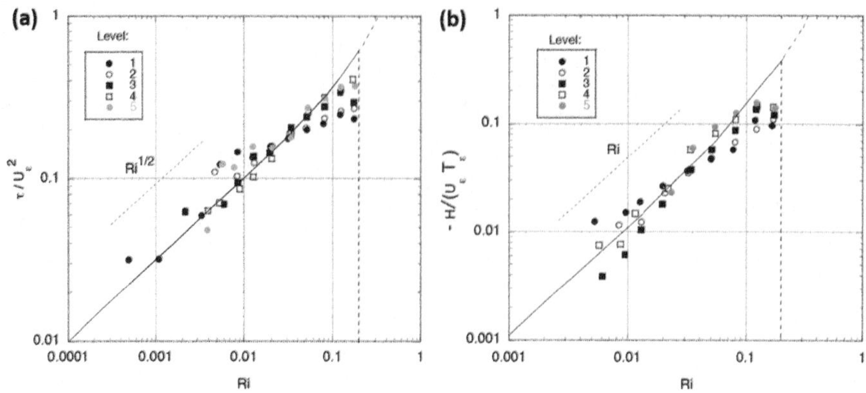

Fig. 21.4 A dependence of bin-averaged, 1-h medians of the dimensionless: (**a**) momentum flux $\tau/U_\varepsilon^{\,2}$, (**b**) temperature flux $-H/(U_\varepsilon T_\varepsilon)$ on the Richardson number. The data points in the figure were obtained by Grachev et al. (2015) for $\text{Ri} \leq 0.2$, based on observations collected during the SHEBA experiment, at five levels of the surface layer. The levels $z_1 = 2.2$ m, $z_2 = 3.2$ m, $z_3 = 5.1$ m, $z_4 = 8.9$ m, and $z_5 = 18.2$ m (but at 14 m during most of the winter) are marked by various symbols. The curves based on Eq. (21.18a, b) are plotted as continuous lines for $\text{Ri} \leq 0.2$ and then continued as a *broken line* for $\text{Ri} \leq 0.7$. The *broken vertical line* marks the abscissa of $\text{Ri} = 0.2$ (based on Sorbjan 2015)

In the near-neutral range in Fig. 21.4a, the data points follow the line $\tau/U_\varepsilon^{\,2} \sim \text{Ri}^{1/2}$ up to the value of $\text{Ri} \sim 0.1$. Above this value, the observational points seem to decrease with the increasing values of Ri. Accepting the notion that the critical Richardson number $\text{Ri}_{\text{cr}} = 0.2$ implies that the dimensionless flux should be represented by a continuous line for $\text{Ri} < 0.2$, and subsequently, it should drop to zero at $\text{Ri} = 0.2$, as marked by the vertical broken line. The remaining portion of the curve, for $\text{Ri} > 0.2$, could be understood as representing "non-Kolmogorov turbulence."

Figure 21.4b depicts the dimensionless temperature flux $-H/(U_\varepsilon T_\varepsilon)$ as functions of the Richardson number Ri. The SHEBA data are marked analogously as in Fig. 21.4a and plotted for $\text{Ri} \leq 0.2$. The curve is plotted based on Eq. (21.4b). As in Fig. 21.4a, the section of the curve for $\text{Ri} \leq 0.2$ is marked by a continuous line, while the remaining portion is indicated by a broken line. The data points follow the line Ri up to the value of $\text{Ri} \sim 0.5$ and, subsequently, turn slightly toward larger values of the ordinate at larger values of Ri. The scatter of the observational points is larger for small values of Ri. The analytical curve and the observational points coincide in the range $\text{Ri} < 0.08$.

21.5 Final Remarks

Applications of the presented approach in the upper atmosphere could be possible through high-resolution soundings, using tethered balloons or GPS-controlled unmanned aerial vehicles (UAVs). Another, simpler and less precise approach could involve the evaluation of the length scale L_T, proposed by Thorpe (1977) for ocean applications. The scale is a measure of vertical mixing in a stably stratified environment, which can also be obtained based on high-resolution soundings in the atmosphere (e.g., Thorpe 1977; Sorbjan and Balsley 2009). The method involves reordering of a non-monotonic instantaneous profile of the potential temperature $\Theta(z)$ into a monotonic profile. The reordering can be accomplished by an evaluation of "overturns" ΔZ, which occur at those levels, where the potential temperature profile decreases with increasing height. The resulting, reordered profile is a proxy of the profile that would take place if the turbulence were allowed to relax adiabatically to a motionless state. The Thorpe scale is defined as the root mean square of the vertical displacements ΔZ.

Various measurements in the atmosphere (e.g., Gavrilov et al. 2005; Clayson and Kantha 2008; Basha 2009; Alappattu and Kunhikrishnan 2010; Haack et al. 2014; Luce et al. 2014), oceans, and lakes (e.g., Itsweire 1984; Oakey 1982; Mater et al. 2013) and also direct numerical simulations (e.g., Mater et al. 2013) have shown that the Thorpe scale L_T is approximately proportional to (or is at least of the same order as) the Dougherty–Ozmidov scale L_ε. Based on maritime observations, e.g., Dillon (1982) found that $L_T \sim 1.25L_\varepsilon$. Crawford (1986) obtained a larger coefficient $L_T \sim 1.52L_\varepsilon$ for oceanic thermocline, while Ferron et al. (1998) received $L_T \sim 1.05L_\varepsilon$ for an abyssal region. A strong correlation is also found between the Thorpe and Ellison scales, $L_T \sim L_\theta$ (e.g., Cimatoribus et al. 2014).

Acknowledgment This work was partially supported within statutory activities No 3841/E-41/S/2015 of the Ministry of Science and Higher Education of Poland and the US National Science Foundation grant AGS-1500900.

References

Alappattu, D.P., Kunhikrishnan, P.K.: First observations of turbulence parameters in the troposphere over the Bay of Bengal and the Arabian Sea using radiosonde. J. Geophys. Res. **115**, D06105 (2010). doi:10.1029/2009JD012916

Andreas, E.L., Fairall, C.W., Guest, P.S., Persson, P.O.G.: An overview of the SHEBA atmospheric surface flux program. 13th Symposium on Boundary Layers and Turbulence, Dallas, TX. Am. Meteorol. Soc. Proc. 550–555 (1999)

Baas, P., Steeneveld, G.J., van de Wiel, B.J.H., Holtslag, A.A.M.: Exploring self-correlation in flux-gradient relationships for stably stratified conditions. J. Atmos. Sci. **63**, 3045–3054 (2006)

Basha, H.A.: High resolution observations of turbulence in the troposphere and lower stratosphere over Gadanki. Ann. Geophys. **27**, 2407–2415 (2009)

Blackadar, A.K.: The vertical distribution of wind and turbulent exchange in neutral atmosphere. J. Geophys. Res. **67**, 3095–3103 (1962)

Buckingham, E.: On physically similar systems; illustrations of the use of dimensional equations. Phys. Rev. **4**, 345–376 (1914)

Caughey, S.J., Wyngaard, J.C., Kaimal, J.C.: Turbulence in the evolving stable boundary layer. J. Atmos. Sci. **36**(6), 1041–1052 (1979)

Champagne, F.H., Friehe, C.A., Larue, J.C.: Flux measurements, flux estimation techniques, and fine-scale turbulence measurements in the unstable surface layer over land. J. Atmos. Sci. **34**, 515–530 (1977)

Cimatoribus, A., van Haren, A.H., Gostiaux, L.: Comparison of Ellison and Thorpe scales from Eulerian ocean temperature observations. J. Geophys. Res. Oceans **119**, 7047–7065 (2014). doi:10.1002/2014JC010132

Clayson, C.A., Kantha, L.: On turbulence and mixing in the free atmosphere inferred from high-resolution soundings. J. Atmos. Oceanic Technol. **25**, 833–852 (2008)

Crawford, W.R.: A comparison of length scales and decay times of turbulence in stably stratified flows. J. Phys. Oceanogr. **16**(11), 2147–2153 (1986)

Delage, Y.: A numerical study of the nocturnal atmospheric boundary layer. Q. J. Roy. Meteorol. Soc. **100**, 351–364 (1974)

Dillon, T.M.: Vertical overturns: a comparison of Thorpe and Ozmidov length scales. J. Geophys. Res. **87**(C12), 9601–9613 (1982). doi:10.1029/JC087iC12p09601

Dougherty, J.P.: The anisotropy of turbulence at the meteor level. J. Atmos. Terr. Phys. **21**, 210–213 (1961)

Ellison, T.H.: Turbulent transport of heat and momentum from an infinite rough plane. J. Fluid Mech. **2**, 456–466 (1957)

Ferrron, B., Mercier, H., Speer, K., Gargett, A., Polzin, K.: Mixing in the romance fracture zone. J. Phys. Oceanogr. **28**, 1929–1946 (1998)

Foken, T.: 50 Years of the Monin-Obukhov similarity theory. Boundary-Layer Meteorol. **2**, 7–29 (2006)

Gavrilov, N.M., Luce, H., Crochet, M., Dalaudier, F., Fukao, S.: Turbulence parameter estimations from high-resolution balloon temperature measurements of the MUTSI-2000 campaign. Ann. Geophys. **23**, 2401–2413 (2005)

Grachev, A.A., Fairall, C.W., Persson, P.O.G., Andreas, E.L., Guest, P.S.: Stable boundary-layer scaling regimes: the SHEBA data. Boundary-Layer Meteorol. **116**(2), 201–235 (2005). doi:10.1007/s10546-004-2729-0

Grachev, A.A., Andreas, E.L., Fairall, C.W., Guest, P.S., Persson, P.O.G.: The critical Richardson number and limits of applicability of local similarity theory in the stable boundary layer. Boundary-Layer Meteorol. **147**(1), 51–82 (2013). doi:10.1007/s10546-012-9771-0

Grachev AA, Andreas EL, Fairall CW, Guest, PS, Persson OG.: Similarity theory based on the Dougherty–Ozmidov length scale. Q. J. R. Meteorol. Soc. (2015) doi: 10.1002/qj.2488

Haack, A.M., Gerding, M., Lübken, F.-J.: Characteristics of stratospheric turbulent layers measured by LITOS and their relation to the Richardson number. J. Geophys. Res. Atmos. **119**(10), 10,605–10,618 (2014). doi:10.1002/2013JD021008#_blank#Link

Huang, J., Bou-Zeid, E., Golaz, J.C.: Turbulence and vertical fluxes in the stable atmospheric boundary layer. Part II: A novel mixing-length model. J. Atmos. Sci. **70**, 1528–1542 (2013)

Hunt, J.C.R., Kaimal, J.C., Gaynor, J.E.: Some observations of turbulence structure in stable layers. Q. J. Roy. Meteorol. Soc. **111**, 793–815 (1985)

Itsweire, E.C.: Measurements of vertical overturns in a stably stratified turbulent flow. Phys. Fluids **27**(4), 764–766 (1984). doi:10.1063/1.864704

Klipp, C.L., Mahrt, L.: Flux-gradient relationship, self-correlation and intermittency in the stable boundary layer. Q. J. Roy. Meteorol. Soc. **130**, 2087–2103 (2004)

Kolmogorov, A.N.: The local structure of turbulence in an incompressible fluid at very high Reynolds numbers. Doklady AS USSR **30**, 299–303 (1941) (in Russian)

Luce, H., Wilson, R., Dalaudier, F., Hashiguchi, H., Nishi, N., Shibagaki, Y., Nakajo, T.: Simultaneous observations of tropospheric turbulence from radiosondes using Thorpe analysis and the VHF MU radar. Radio Sci. **49**, 1106–1123 (2014). doi:10.1002/2013RS005355

Mater, B.D., Venayagamoorthy, S.K.: A unifying framework for parameterizing stably stratified shear-flow turbulence. Phys. Fluids **26**(3), 036601 (2014). doi:10.1063/1.4868142

Mater, B.D., Schaad, S.M., Venayagamoorthy, S.K.: Relevance of the Thorpe length scale in stably stratified turbulence. Phys. Fluids **25**, 076604 (2013). doi:10.1063/1.4813809

Monin, A.S., Obukhov, A.M.: Basic laws of turbulence mixing in the surface layer of the atmosphere. Trudy Geof. Inst. AN SSSR **24**, 163–187 (1954)

Nieuwstadt, F.T.M.: The turbulent structure of the stable, nocturnal boundary layer. J. Atmos. Sci. **41**, 2202–2216 (1984)

Oakey, N.S.: Determination of the rate of dissipation of turbulent energy from simultaneous temperature and velocity shear microstructure measurements. J. Phys. Oceanogr. **12**(3), 256–271 (1982)

Obukhov, A.M.: On the distribution of energy in the spectrum of turbulent flow. Dokl. Akad. Nauk SSSR **32**(1), 22–24 (1941)

Ozmidov, R.V.: On the turbulent exchange in a stably stratified ocean. Izv. Acad. Sci. USSR, Atmos. Oceanic Phys. **1**, 861–871 (1965)

Persson, P.O.G., Fairall, C.W., Andreas, E.L., Guest, P.S., Perovich, D.K.: Measurements near the atmospheric surface flux group tower at SHEBA: near-surface conditions and surface energy budget. J. Geophys. Res. **107**(C10), SHE21.1–SHE21.35 (2002)

Prandtl, L.: Uber die ausgebildete Turbulenz. Proceedings of the Second International Congress for Applied Mechanics, Zürich, pp. 62–74 (1926)

Sorbjan, Z.: Gradient-based similarity in the atmospheric boundary layer. Acta Geophys. **56**(1), 220–233 (2008a)

Sorbjan, Z.: Local scales of turbulence in the stable boundary layer boundary layer. Boundary-Layer Meteorol. **127**(2), 261–271 (2008b)

Sorbjan, Z.: The height correction of the gradient-based and flux-based similarity functions in the stable boundary layer. Boundary-Layer Meteorol. **142**, 21–31 (2012)

Sorbjan, Z.: Gradient-based similarity in the stable atmospheric boundary layer. In: Achievements, History and Challenges in Geophysics. GeoPlanet: Earth and Planetary Sciences pp. 351–375 (2014)

Sorbjan, Z.: Similarity scaling systems for stably stratified turbulent flows. Q. J. R. Meteorol. Soc. **142**(695), 805–810 (2016)

Sorbjan, Z., Balsley, B.B.: Microstructure of turbulence in the nocturnal boundary layer. Boundary-Layer Meteorol. **129**, 191–210 (2009)

Sorbjan, Z., Czerwinska, A.: Statistics of turbulence in the stable boundary layer affected by gravity waves. Boundary-Layer Meteorol. **148**(1), 73–91 (2013)

Sorbjan, Z.: Gradient-based scales and similarity laws in the stable boundary layer. Q. J. R. Meteorol. Soc. **136**, 1243–1254 (2010)

Thorpe, S.A.: Turbulence and mixing in a Scottish Loch. Philos. Trans. Roy. Soc. A **286**, 125–181 (1977). doi:10.1098/rsta.1977.0112

Wyngaard, J.C., Coté, O.R.: Cospectral similarity in the atmospheric surface layer. Q. J. Roy. Meteorol. Soc. **98**, 590–603 (1972)

Part V
Future Opportunities

Chapter 22
Airborne Remote Detection of Turbulence with Forward-Pointing LIDAR

Patrick Sergej Vrancken

Abstract Presently, the airborne remote detection of atmospheric turbulence is limited to radar-visible regions of the sky, i.e., zones that contain hydrometeors like rain or cloud droplets. The bulk of the actual turbulence, possible in clear air at all flight altitudes, evades such a remote detection, though a remote determination of aircraft-relevant physical parameters relevant to turbulence could significantly increase flight safety.

The following chapter reviews possible techniques of remote turbulence detection in clear air and identifies the most promising approaches for future aircraft. These are shown to be optical methods, i.e., LIDAR (Light Detection and Ranging) systems. Principles, as well as pros and cons of some complementary lidar techniques, are discussed.

22.1 Introduction

The forecasting of turbulent-prone zones in the aviation-relevant airspace is steadily making progress as was illustrated in Part III of this volume. Clear-air turbulence (CAT) occurrence is reported verbally by pilots or automatically by onboard in situ detection algorithms and can be used in upper-level turbulence nowcasting and forecasting applications. The understanding of turbulence induced by convective processes is improving, as is the nowcasting of the latter. High-resolution numerical weather prediction (NWP) domains around airports should allow more reliable prognosis of low-level turbulence and wind shear. Wake vortex behavior of nearby aircraft, from drift to dissolution, is increasingly well estimated.

For the coming decades, the aviation sector has set itself the goals of further improving the safety of passengers and crew, enhancing efficiency and timeliness, and increasing environmental sustainability and economic competitiveness (that

P.S. Vrancken (✉)
DLR – German Aerospace Center, Institute of Atmospheric Physics, Oberpfaffenhofen 82234, Germany
e-mail: Patrick.Vrancken@dlr.de

R. Sharman, T. Lane (eds.), *Aviation Turbulence*,
DOI 10.1007/978-3-319-23630-8_22

may both be supported by more lightweight, thus less fuel-consuming aircraft structures). Turbulence in the flightpath of an aircraft is a major obstacle in achieving these challenging goals, in particular in the context of the strong growth rate of global air traffic and climate change (see Chap. 23).

There are at least two aspects that emphasize the need of forecasts and nowcasts to be complemented with onboard remote sensing of turbulence. The first refers to the physical nature of turbulence that inherently shows intermittency and patchiness. Hence, turbulence does not occur at the large scales where it is forecasted, but locally, and temporarily. This results in vast volumes of airspace forecasted to develop turbulence, but it actually occurs in very localized, temporally varying areas wherever the physical conditions are favorable. The second, more operational aspect relates to the circumstance that often aircraft are constrained to follow a certain predefined route, be it through a CAT-prone area on a transatlantic trajectory over Greenland or during a wind shear-threatened approach to a specific airport. An appreciable deviation of a flight route due to a forecast may only be an alternative for very hazardous turbulent conditions since operational and economic aspects play an important role in commercial aviation. This is where the development of new remote sensing techniques comes in. As will be seen in this chapter, where the indispensable workhorse of aviation safety, the onboard weather radar, falls short due to its physical limitations, new optical techniques should hit the mark.

In the next sections we shall explore the two main applications of turbulence remote sensing. In order to understand the technological implications of remote detection of turbulence, we will shortly highlight the possible scope of such detection (Sect. 22.2) and the related physical observables (Sect. 22.3). A brief overview of forthcoming technologies (Sect. 22.4) shall clarify that many of them represent interesting research potential but should retain purely academic relevance. We will see that only active optical remote sensing (i.e., lidar) is a viable solution. In Sect. 22.5 lidar application to the problem is considered in more depth. From this description of lidar principles, we will deduce that there is still a substantial piece of technology and product development work to do that also relies on research into the actual physical peculiarities of turbulence (see Chap. 25). The aim of this chapter though is not to strictly detail lidar or any other technology since this information may be found elsewhere (e.g., Weitkamp 2005; Measures 1984 or cited references).

22.2 Aims of Turbulence Detection

There are two main areas of application of remote turbulence detection within aeronautics, depending on the actual range of detection: on long range to register evidence of turbulence along with information of intensity, i.e., around 15 km distance in order to give the flight crew or some automated system time to respond,

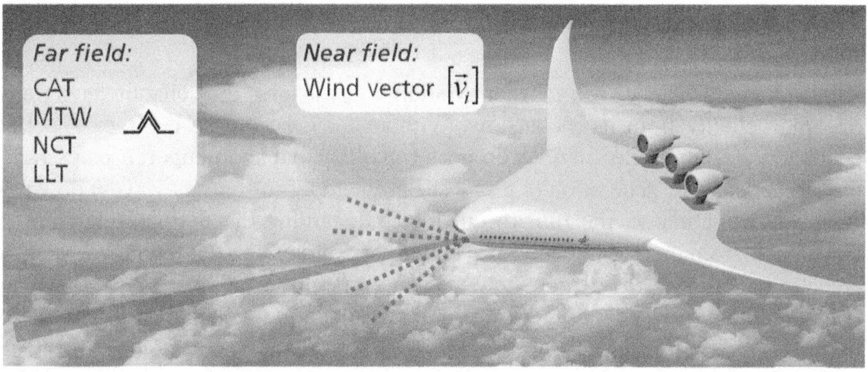

Fig. 22.1 Sketch highlighting the long-range and near-range application of remote turbulence sensing (Image: DLR CC-BY 3.0)

or on short range (some 50–150 m) to directly deliver wind vector information for automated action of the aircraft's control surfaces (see, e.g., Fig. 22.1).

The goal of long-range applications is to remotely detect the severity of turbulent motion of the air ahead with sufficient range to provide warning to the cockpit to take mitigating actions. Within clouds, state-of-the-art radar systems do this job and retrieve a measure of turbulence by analyzing the Doppler-broadened radar return. In clear air, outside clouds, where clear-air, mountain wave, and near-cloud turbulence occur, however, radar is blind to air motion due to the absence of backscattering hydrometeors. Arising in higher altitudes, this class of turbulence concerns the long cruise phase where crew and passengers stroll around the cabin and are thus susceptible to injuries in suddenly occurring turbulence. Further, the aircraft is typically flying at high speed which maximizes the loads and thus the induced bending moments in the wing roots (and consequently their fatigue). Turbulence also has an effect on the onset of high speed buffet and due to flying close to the edge of the envelope, which may lead to high speed stall. Here, a timely warning could alter the situation: from a simple seat belt sign for passenger safety, over mitigation actions such as deceleration, to evasion maneuvers. Anticipating the following section, the actual physical quantity representing the turbulence is of minor importance in this application as long as it reliably determines the presence and "strength" of the turbulent zone ahead, remembering that the most important quantity for an aircraft's disturbance is vertical wind speed.

The short-range application would comprise the determination of an actual wind speed vector, or a vector ensemble, resolved to fit the aircraft and turbulence characteristics. The aircraft's control system could thus determine and set the appropriate control actions of rudders and lift control devices in order to mitigate the induced moments and accelerations. Then, flying even through strong turbulence, the ride would be less bumpy, minimizing passenger and crew discomfort and any potential structural damage.

22.3 Turbulence Observables

Atmospheric turbulence relevant for aviation is characterized by chaotic deviations of wind velocity from its mean flow. Without focusing on the complex spectral characteristics, these fluctuations do occur on all spatial and temporal scales from the larger scales of generation (e.g., by shear due to jets or gravity waves) down to viscous dissipation on the mm to cm level. Regarding the remote detection of turbulence from fast-flying aircraft, and the resulting time interval between a detection ahead and the actual encounter, one may follow Taylor's hypothesis to consider the turbulence as "frozen," i.e., maintaining its characteristics over that time.

The important spatial scales for aviation, though, depend very much on the aircraft characteristics, such as size, weight, speed, and layout. The aircraft acts as a filter, damping high frequencies (by aerodynamics characteristics) and low wavenumbers of the large scales (e.g., through autopilot). From the encountered components[1] of the fluctuating wind velocity (u', v', and w'), the vertical component is the most important, with a tenfold stronger reaction in acceleration compared to the axial/horizontal wind speed fluctuations (Hoblit 1988). The w'- component would thus be the observable of choice for a distant warning, while its relationship to u' (and v') is neither simple nor universal due to anisotropy of the turbulence (e.g., Reiter and Burns 1966; Crooks et al. 1967; Lilly et al. 1973). In contrast we will see throughout the next sections that u' is basically the most accessible quantity (via Doppler shift measurement of backscattered radiation), v' practically immeasurable, and w' can only be determined with indirect methods (Sect. 22.5.3). The three fluctuating velocities may either be seen (directly) in the movement of the air molecules or of suspended particles (i.e., aerosols, ice crystals, water droplets). Over a certain range interval, they may be quantified by a statistical dispersion parameter such as its variance (e.g., $\sigma^2_{u'_{aer}}(R)$ for the fluctuation of head wind speed at distance R, determined from aerosol movement).

The air movement, on its part, generates fluctuations of its state variables temperature, T, and density, ρ, by the displacement and mixing of different atmospheric layers (thus $\sigma_T(R)$ and $\sigma_\rho(R)$). These in turn give rise to a second-order effect on the probing electromagnetic radiation: the variation of the air refractive index which depends on temperature, pressure, and humidity. It is generally quantified by its structure constant $C_n^2(R)$ which is proportional to the variance (multiplied with some length scale characterizing the turbulence).

In addition to the effect of mixing of the turbulent airflow on suspended particles (aerosols), it has a second effect on these passive "scalar" tracers: a certain structuring according to the spatial properties of the turbulent flow. Rather obvious

[1] We will henceforth consider an aircraft-fixed reference system, i.e., $u(R) = \overline{u}(R) + u'(R)$ be the velocity along the aircraft motion axis R, v' the lateral fluctuating component, and w' the vertical component.

and well proven for wake vortices (e.g., Misaka et al. 2012), such a mechanism is less evident for geophysically produced turbulence. The related observable would then be the spatial distribution of the respective scalar atmospheric constituent. However, considering the minute or absent effect on aerodynamics of the latter observables, a quantitative determination of aviation turbulence strength appears fairly precarious.

For now, we may stipulate that for our first aim, the long-range detection of turbulence for aviation the specific physical quantity is of lower importance. This is in contrast to a short-range flight control application where actual three-dimensional wind speed vectors have to be determined. In the following sections we will shortly discuss different technologies in order to determine the most promising ones that may enable future turbulence protection systems.

22.4 Technologies for Remote Detection of Turbulence

This section will give a short review of active remote sensing techniques that may provide information on the previously stated physical observables. Even though passive microwave radiometry had received some attention for airborne CAT detection in the 1960s, its shortcomings for operational use became rapidly evident (e.g., Atlas 1969; Schaffner et al. 2012). However, passive sensing techniques such as air temperature microwave radiometry (e.g., Haggerty et al. 2014) and star scintillation photometry (e.g., Vernin and Pelon 1986) could be valuable methods for determining ancillary scientific data in general experimental research on atmospheric turbulence.

Active remote sensing relies on the emission of electromagnetic radiation into the relevant region and collecting and analyzing the radiation that is backscattered from locally present objects. It generally allows a more robust determination of the aimed observable.

22.4.1 Radar Methods

Electromagnetic radiation in the microwave frequency region is effectively backscattered from hydrometeors such as cloud or rain droplets. The reflectivity is then a measure of liquid water content of the sensed volume ahead. The radar frequencies are chosen as a function of the application: ground-based precipitation radars operate at some GHz (S to C-Band) and modern airborne weather radars employ rather X-band (around 10 GHz). This makes the airborne radars very sensitive to clouds and rain while at the same time turns them essentially "blind" to anything behind very dense clouds (e.g., as occurring in deep convection). The choice of lower frequencies in ground-based systems allows them to more effectively sense through rain and dense clouds.

Scientific (Hamilton et al. 2012; Chap. 5) and state-of-the-art airborne weather radars (Baynes 2014) also exploit the Doppler information in order to determine the turbulence and/or wind shear in or around sufficiently reflecting air volumes (thus detecting in-cloud turbulence). However, such radar systems are not sensitive to anything other than water droplets. Therefore, flow information based on Doppler shift of the backscattered radiation is only available for these water-containing areas.

A different radar technique evaluates the backscatter of clear air. Bragg scatter occurring at successive refractive index and humidity perturbations (Clifford et al. 1994) leads to a coherent adding of the scattered wavefronts at the receiver. Thus, the backscattered radiation, quantified by the reflectivity η, is directly proportional to the refractive index variations $C_{n,\mathrm{Radar}}^2$ (i.e., the radar-optical turbulence intensity) divided by $\lambda_R^{1/3}$, the respective radar wavelength (Good et al. 1982; Ottersten 1969).

The reflectivity η is derived from the signal strength, or more precisely the signal-to-noise ratio (SNR) of the measurement. This technique is used in ground-based research on turbulence in the free atmosphere, employing VHF (30–300 MHz) or UHF (300 MHz–1 GHz) radar systems (e.g., Serafimovich et al. 2005; Luce et al. 2010) due to less atmospheric damping at lower frequencies. The drawback of this approach is the extremely low reflectivity; for instance, in common wind profilers it is typically ten orders of magnitude lower than for S-band weather radars that operate on backscatter on hydrometeors (Clifford et al. 1994). This yields very long integration times (>minutes) and challenges noise-processing algorithms. Further, the extended beam pattern (beam angle equals λ_R/D, with D antenna diameter) does not allow an altitude distinction precise enough for a reliable encounter prediction (of some tens to hundreds meters) in a hypothetic airborne horizontal application. Other inhibiting factors are the necessary size of the antenna, too bulky for aircraft integration, and range limitations due to ground echo. While some experimental effort has been carried out in the 1960s by the Boeing Company (Buehler et al. 1969), a thorough analysis of the relevant quantities' magnitudes (reflectivity, radar wavelength, achievable system parameters) shows the impracticality of a radar approach to airborne turbulence detection in clear air (Atlas et al. 1966; Watkins and Browning 1973).

Summarizing the radar approaches, it can be stated that the actually employed weather radar techniques, with all its possible evolutions, will continue to represent the most important airborne remote sensing instrument due to its ability to detect the chief weather hazards like deep convection along with its embedded wind shear and turbulence. Its limitation to these environments, though, demands a complementary system able to operate in dry and clear air.

22.4.2 THz Remote Sensing

For completeness, we shall mention here another remote sensing technique that is very sensitive to water vapor. THz radiation may effectively be employed to determine humidity and its variability. As conjectured in Sect. 22.3, one could ultimately derive the underlying turbulence from this variable. However, the very strong sensitivity toward water vapor at the same time also limits its applicable range whenever any humidity may be encountered. Therefore, its use in remote turbulence detection may be considered to remain marginal.

22.4.3 LIDAR: Light Detection and Ranging

Light Detection and Ranging techniques were actually already in use before the invention of the laser (at the time using flash lamps). Today lidars use state-of-the-art pulsed (but also continuous wave) lasers being emitted into the atmosphere. These short pulses are backscattered by both air molecules and suspended aerosols. The many different aspects of interaction of light and matter, from Rayleigh, Brillouin, Raman and Mie scattering over depolarization to absorption, give rise to numerous applications in atmospheric physics. Since even the shortest overview of these is already out of the scope of this chapter but may be found in books like Schumann (2012), Weitkamp (2005), and Measures (1984), the following paragraphs shall mention the techniques that possibly may contribute to turbulence detection and characterization. The next section will then describe the techniques relevant to the present aeronautics application in more detail.

Due to the numerous scattering processes, lidar technology may deliver information to nearly all observables indicative of turbulence, as described in Sect. 22.3. To start with, wind speed fluctuations along the line of sight $u'_{mol}(R)$ or its magnitude $\sigma^2_{u'_{mol}}(R)$, directly retrieved from air molecules backscatter, may be measured by determining the Doppler shift in the backscattered signal with interferometric methods. This approach, widely employed in airborne atmospheric research and in preparation for a European satellite mission, is explained in more detail in Sect. 22.5.2. As will be shown in Sect. 22.5.3, the vertical wind speed fluctuations present in clear-air turbulence should be accessible with an indirect method that relies on the adiabatic temperature changes of the air masses being turbulently stirred vertically. As mentioned in Sect. 22.3, the lateral wind v' renounces any measurement, "even" by lidar.

Coherent Doppler wind lidar is a widely used remote sensing technique for determining wind speeds. It is based on the determination of Doppler shift by heterodyne mixing of the narrow spectral line backscattered by aerosols suspended in the turbulent air flow. This technique thus delivers $u'_{aer}(R)$ and $\sigma^2_{u'_{aer}}(R)$. Some more details on this method are given in Sect. 22.5.1. Since it is based on backscatter from aerosols, this method is applicable in regions of sufficient aerosol concentration (or more precisely effective backscatter cross-section). When this is the case, however, it allows a very convenient way of speed measurement along the line of sight.

Both coherent and interferometric methods allow derivation of three-dimensional wind vectors from a certain number of short-range measurements in different directions. Such a setup may thus be designed to provide information about turbulence ahead for input to an automated flight (attitude and lift) control system.

Section 22.3 above noted the air state variables as an important indicator of a turbulent motion of the air. Thus, temperature dispersion measurement with lidar should give the required information. On one hand, lidar technology offers different approaches for the remote measurement of air temperature. The analysis of the signal from different rotational Raman backscatter lines, for instance, allows a powerful technique for temperature determination, without being subject to atmospheric transmission (and thus, e.g., aerosol loading) (Behrendt 2005). Other techniques include the interferometric analysis of the backscattered Rayleigh-Brillouin line itself (Witschas et al. 2014), or the exploiting of the temperature dependence of molecular absorption (DIAL, differential absorption lidar) (Theopold and Bösenberg 1993). However, all these techniques require long integration times (minutes to hours) and rather long range-bins for averaging. This inhibits the measurement of short-scale fluctuations of the temperature $\sigma_T(R)$ as occurring in turbulent events. Moreover, for most Raman approaches daytime operation is not possible. While the temperature distribution (over range) is an important parameter for the understanding of the occurrence of turbulence, the previously mentioned methods do not promise to be applicable for the remote detection of turbulence itself.

The fluctuations of the optical air refractive index n_{opt}, generated by temperature fluctuations, have consequences on the propagation of the laser light: wavefront distortion with beam spread, beam wander, and scintillation. Zilberman and Kopeika (2004) operate a lidar similar to a scintillometer, deriving the refractive index structure constant from the variations of the angle of arrival of the backscatter "image" of the laser beam which is probing aerosol containing atmospheric layers. An application to horizontal airborne sensing is however negated by the resulting poor range resolution and the need for strong aerosol backscatter. More promising approaches have been proposed by Gurvich (2012) and Banakh and Smalikho (2011) relying on the backscatter enhancement effect. The methods employ an angular analysis of the total backscattered radiation, either by transmitting laser pulses in two angularly closely spaced regions or by employing two angularly separated detectors. While, to our knowledge, the proposed schemes have not yet been demonstrated in practice, the analytical considerations show a high potential for a real application (Gurvich 2014).

It was further conjectured that a passive tracer, such as aerosol or to a certain extent water vapor, may be useful for the lidar detection of turbulence in clear air. Among the earliest applications of lidar are the airborne investigations of CAT relying on such a relationship between aerosol load variation and turbulence. Both the University of Michigan (Franken et al. 1966) and NASA (Lawrence et al. 1968; Melfi and Stickle 1969) have been performing flight tests into CAT areas with ruby laser-based lidar systems in the late 1960s. However, the found correlation between turbulence intensity (probed by the aircraft) and measured lidar backscatter variation was rather low in both cases, which was in part attributed to the limited achieved SNR of the systems (ibid.). Abandoned for decades (and replaced by studies with coherent Doppler systems), this relationship between aerosol backscatter fluctuation and turbulence structure -in particular on longitudinal scales- has recently found new interest. Zilberman et al. (2008) retrieve spectral characteristics from aerosol concentration measurements with ground-based lidar and allude to the complex relationship between the aerosol spatial spectrum and the spectrum of the underlying turbulence. Airborne water vapor measurements with lidar (Fischer et al. 2012) display variance spectra in general accordance with turbulence models, at least for the convective troposphere on rather large scales. These studies hint at the potential of passive tracer variance analysis for the determination of turbulence properties. The challenge is then to establish the relationship between the scalar's spectral characteristics and actual strength in terms of fluctuating wind speed.

The above noted observations on the interplay of turbulence sensing techniques and observables highlight the complexity of the underlying phenomenon. Turbulence acts on all dimensions and properties of the air, including the wind speeds relevant for an aircraft, and it generates variations in temperature and density due to these motions. The latter give rise to "optical turbulence," well known to astronomy and optical communication, that acts directly on electromagnetic waves (radio and optical alike). The turbulent air motion also disturbs regions formerly uniformly loaded with aerosols or other passive scalars. An inherent difficulty is represented by the complex relationships between these respective observables. This highlights the need for a deeper understanding and thus research concerning these relationships. Many of these phenomenological expressions of the underlying turbulence may be measured with photonic methods, from low radar frequencies to high ultraviolet laser radiation. As we have seen, probing with laser beams has the biggest potential with many different techniques. The most promising ones will be outlined in the next section.

The previous sections are brief summaries, and some techniques have been omitted due to obvious limitations (such as acoustic methods). Further reviews may be found in, admittedly dated, reviews of Collis (1966) and Atlas (1969).

22.5 Lidar Methods for Airborne Remote Detection of Turbulence

For lidar applications, there are two main approaches to determine the longitudinal wind speed, both relying on the Doppler shift of the backscattered light with respect to the emitted. While both the coherent and the direct detection technique may exploit the narrow backscatter from aerosols, the broad backscatter from air molecules may only be exploited by the direct detection technique. Figure 22.2 depicts a modeled backscatter from "typical" aerosol-laden air at different altitudes according to models. The return from air molecules is hereby broadened by its Brownian motion and thus a function of its temperature. The spectrum's width backscattered from aerosols is only slightly broadened (as compared to the incoming laser line) due to the wind speed fluctuations prevailing in the sensed air volume.

Excellent overviews of the coherent and direct detection lidar techniques may be found in Werner (2005) and Reitebuch (2012); the next sections summarize the approaches and give examples of applications. The third section gives an introduction of an alternative approach aiming the vertical wind speed as the target observable.

22.5.1 *Coherent Doppler Lidar*

The coherent Doppler lidar exploits, as the name implies, the coherent nature of the optical wave by mixing the spectrally narrow aerosol backscatter (left panel in

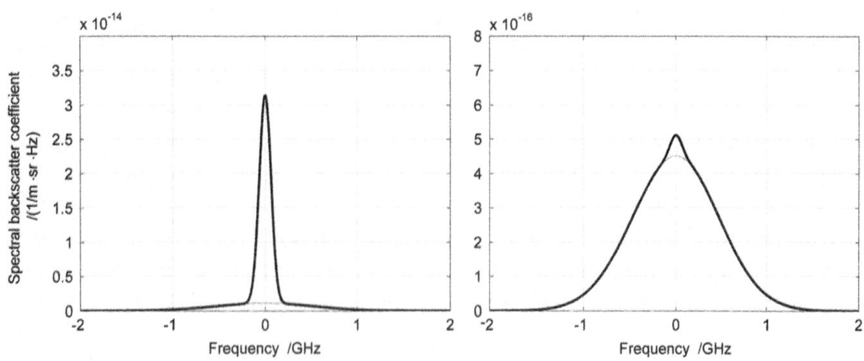

Fig. 22.2 Backscattered spectrum consisting of aerosol and molecular contribution, the latter (*thin lines*) Doppler-broadened due to air temperature. *Thick lines* show the combined spectral response. Two flight altitudes are shown: *left panel* for 500 m, *right panel* for 10,000 m, both for 532 nm irradiation. The lidar backscatter ratios are 3.2 and 0.018, respectively. Note the different scales for the backscatter coefficient [Derived with ESA Reference Atmosphere Model RMA (Vaughan et al. 1995; Vaughan et al. 1998)]

Fig. 22.2) to a local oscillator's optical wave on a fast detector. The frequency difference of the beat signal is then in the range of electronic detectors and circuits, and the Doppler shift may thus be deduced by a spectral analysis of the digitized signal (by Fourier transform, for instance). Further, the spectral width of this signal may be exploited in order to determine the turbulence intensity in terms of longitudinal wind speed variation within the sensed volume (e.g., Frehlich et al. 1998; Smalikho et al. 2005).

The following schematic (Fig. 22.3) depicts a simplified version of such a heterodyne detection lidar system.

The system employs both a pulsed or continuous-wave laser transmitter (LT) and a continuous-wave local oscillator (LO). The actual laser wavelength ν_0 is mainly chosen for a favorable ratio of aerosol to molecular backscatter coefficient, which are functions of wavelength. This aspect and also technological perspectives favor infrared systems with typical wavelengths 1.5–1.6 μm, 2 μm, and 10.6 μm depending on the laser technology. In the course of the development of coherent wind lidars, different laser types have and are being in use: from powerful CO_2 gas lasers over solid state to all fiber-based systems.

The laser LT and the reference laser LO are typically offset one from another by some intermediate frequency and locked with a phase lock loop (PLL). The remaining pulse-to-pulse frequency jitter is monitored by mixing the transmitted pulse with the LO on a reference detector DR. The laser pulse transmitted into the atmosphere is scattered within a range interval $c/2 \cdot \tau$ discriminated by the sample rate $1/\tau$ of the data acquisition's digitizer (for continuous-wave lasers, the focus of the transmit beam defines the single usable range gate). For the applications considered here (that emphasize measurement range), the laser radiation should be scattered from a sufficient number of particles in order to achieve an adequate heterodyne signal. The part scattered into the reverse direction is collected by a telescope, mixed with the LO and guided onto a detector (DA). Its output, similar to the reference DR's output, is subsequently digitized for spectrum analysis. From the shift $\Delta\nu_{aer}$ of the spectra of the atmospheric signal compared to the reference, one

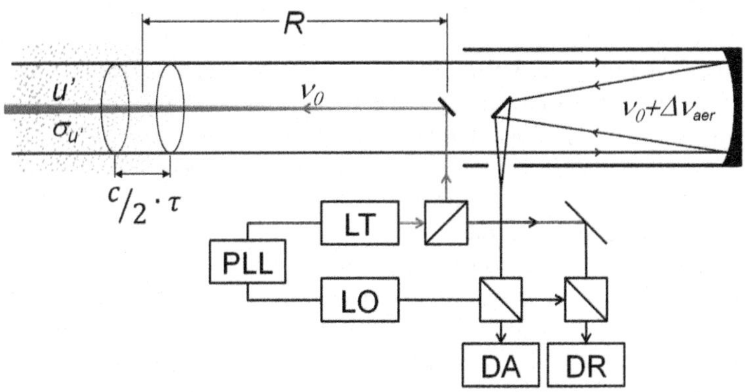

Fig. 22.3 Simplified synopsis of a coherent Doppler lidar. See text for abbreviations

may determine the axial Doppler wind fluctuation $u'_{\mathrm{aer}}(R)$ and from its spectral width retrieve information on the contained turbulence $\sigma^2_{u'_{\mathrm{aer}}}(R)$.

As mentioned, this coherent Doppler scheme may be implemented in various architectures covering near-range air data (true airspeed TAS), determination of the wind speed vector, measurement of wake vortices, and remote detection of clear-air turbulence.

The remote detection of CAT with lidar has long been pioneered by the USA, with activities undertaken by NASA and strong support from the FAA. The research activities on aerosol fluctuations (see above) have been followed by airborne tests of CO_2 coherent wind lidars (Weaver 1971; Huffaker 1975) on NASA's Convair aircraft in 1981 (Bilbro et al. 1984). Subsequent advancements in solid-state laser technology allowed simultaneous comparisons of such systems on NASA's B737 research aircraft in 1994 (Targ et al. 1996). The solid-state technology has been further evaluated by NCAR on the Electra aircraft during the ACLAIM program (Hannon et al. 1999; Soreide et al. 2000) in 1998, in collaboration with NASA. The ACLAIM lidar (build by CTI-Coherent Technologies Inc., now with Lockheed Martin) has further been tested on NASA's DC-8 aircraft within its WxAP project in 2003 (Teets et al. 2006). In 1999, the Japanese JAXA initiated a similar program for the development of an airborne coherent Doppler lidar. Prototypes of this fiber-based lidar system have been flown on JAXA's Beechcraft (for low altitudes research) in 2007 (Inokuchi et al. 2009) and its Gulfstream II aircraft in 2010 (Inokuchi et al. 2010; Inokuchi et al. 2014). All these flight tests of different coherent lidar systems show the high potential of this technology. In particular, wind speed accuracy has been determined to better than 1 m s^{-1}. Turbulence encounters were mostly very well correlated with significant axial wind speed deviations. The coherent systems exhibit a very good performance in low and medium altitudes where backscattering aerosols are abundant. However, for higher altitudes like the typical cruise levels of around 30–40 kft, the attained maximum range was limited throughout the experiments, i.e., too short for a timely turbulence warning. Teets et al. (2006) call for a "significant" increase in laser power for an adequate extension of warning time.

For shorter ranges, however, the ability to remotely detect spatially small phenomena like wake vortices has been demonstrated in an airborne vertical setup during the European AWIATOR project (Rahm et al. 2007) and in an airborne axial setup within the European I-Wake project (Douxchamps et al. 2008), both employing DLR's 2 μm wind lidar. This lidar is based on CTI's transceiver (Henderson et al. 1993; Köpp et al. 2004) and was operated on the German Aerospace Center's Falcon 20 aircraft and on the Dutch Aerospace Laboratory NLR's Cessna Citation 2 aircraft.

Generally, coherent wind lidar technology seems particularly well suited for near- to mid-range applications (of some tens to some hundred meters distance) where the backscatter signal even from low aerosol levels is still sufficient for an adequate signal-to-noise ratio. Accordingly, numerous systems are being developed, often with a focus on reducing size and ruggedizing the setup which favors fiber-based systems. Thus, the French Aerospace Laboratory ONERA has developed a series of coherent lidar systems for near-range turbulence and wake vortex detection (Dolfi-Bouteyre et al. 2009). While the fiber-based technology seems a good candidate for airborne applications, a dedicated flight test campaign has not yet been performed.

Close detection of wake vortices may be used to feed an automated wake identification and alleviation system (Ehlers and Fezans 2015; Ehlers et al. 2015). Such a system then could generate control commands for rudders and lift control devices for counteracting the induced moments and lift forces. The necessary time constant here is dictated by the aircraft actuators, resulting in minimum ranges of around 50 m.

For closer applications, not explicitly covered here, such as the determination of (three-axis) true airspeed (TAS) very close to the fuselage of aircraft or helicopters, coherent Doppler lidar technology appears the method of choice. Some examples of the multitude of activities are the developments of THALES Avionics within the projects NESLIE and DANIELA (Verbeke 2010), JAXA's trial of a helicopter system (Matayoshi et al. 2007), and NCAR's development of an air motion sensor for its HIAPER Gulfstream GV research aircraft (Keeler et al. 1987; Spuler et al. 2011).

22.5.2 Direct Detection Doppler Lidar

As alluded to above, coherent Doppler lidar falls short whenever the aerosol concentration is too low to deliver sufficient backscatter signal-to-noise ratio at the distance of choice. Since the recovered radiation drops with more than the square of the distance, such conditions are usual at higher altitudes, which limits their use for long-range applications (such as turbulence warning), but also at shorter ranges. One way to overcome this difficulty is to determine the axial Doppler shift (or turbulent velocity distribution) from the spectrum of the *molecular* backscatter as opposed to *aerosol* backscatter. Since molecular backscatter is substantially broadened, as compared to a narrow laser line, by the temperature motion of the air (see Fig. 22.2), a heterodyne mixing is not exploitable for realistic pulse powers (Cézard 2008). A shift of this broad backscattered Rayleigh-Brillouin (RB) spectrum, though, may be determined using optical interferometry. Roughly speaking, the frequency information is converted into spatial (or rather angular) information when the received optical wave is brought to interference with itself. The output of the interferometer may then be analyzed radiometrically or spatially in order to derive the associated frequency shift. Accordingly, there exist two main approaches: the edge (or filter) technique and the fringe imaging technique. Actually, both methods also work for the narrow aerosol backscatter, but the optimum design differs depending on the proportion of the respective signals.

Figure 22.4 shows a synopsis of this "direct detection"-type lidar which is very similar for the transmission part and differs for the type of receiver implementation. A pulsed laser LT emits pulses into the atmosphere; in order to maximize the

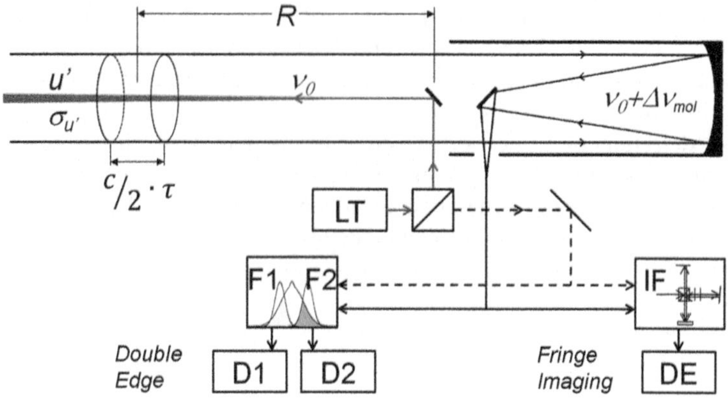

Fig. 22.4 Simplified synopsis of direct detection lidar with *left* double-edge receiver and *right* fringe imaging receiver. See text for abbreviations

backscatter signal, the laser radiation is often converted to ultraviolet wavelengths in order to profit from the high molecular backscatter. The backscattered radiation is collected by a telescope and brought on the receiver setup.

The edge technique (Flesia and Korb 1999) is based on a setup of (at least) two narrow spectral filters (F1 and F2) that are each centered in the flanks of the Rayleigh-Brillouin spectrum as depicted in Fig. 22.2. Part of this spectrum is transmitted through these filters, brought onto two detectors (D1 and D2) and digitized. The Doppler shift $\Delta\nu_{\mathrm{mol}}$ is then derived from the ratio of these two detector signals. This technique may also be implemented with more than two filters, with comparable performance (McGill and Spinhirne 1998). For a large number of filters, this technique basically transitions into the fringe imaging technique (see below). The other extreme is an implementation with only one filter which is also possible and may be realized by implementing an absorptive filter (Liu et al. 2007) (instead of an interference filter). The edge technique has been successfully applied in airborne applications, such as NASA's TWiLiTE lidar, which is autonomously operating on high-altitude research aircraft and the Global Hawk (Gentry et al. 2006), or the airborne demonstrator for Europe's satellite wind lidar ADM-Aeolus (Reitebuch et al. 2009) that is regularly deployed aboard DLR's Falcon 20 aircraft. Though, it has not yet been applied to forward-looking aeronautics applications as discussed herein.

The fringe imaging technique converts, by interferometry, the frequency shift $\Delta\nu_{\mathrm{aer}}$ into an angular shift of the output of the used interferometer (IF). This output is then imaged onto a matrix or line detector (DE). The digitized output is then used for fitting a line-shape model and thus determining the frequency shift as compared to the reference. As for the edge technique, the interferometer has either to be highly stabilized or regularly evaluated by a reference laser beam, derived from the transmitter LT.

Different types of interferometers may be used, each with its own benefits and weaknesses: Bruneau et al. (2004) demonstrated a Mach-Zehnder as spectral analyzer, Cézard (2008) and Liu et al. (2012) a Michelson, whereas the ADM-Aeolus spaceborne instrument will use a Fizeau interferometer (for its aerosol channel). A very widely used device is the Fabry-Perot interferometer. In the European project AWIATOR, a short-range lidar has been tested in forward-pointing configuration for flight control application (Schmitt et al. 2006). The lidar is based on a high repetition rate (18 kHz) solid-state UV laser which is sent, alternatingly, into four directions in order to retrieve a three-dimensional gust vector. The backscattered radiation is collected and fed through a Fabry-Perot etalon, its output intensified and imaged onto a CCD detector. The circular fringe pattern is then analyzed for the radius change due to Doppler shift. During the flight tests on Airbus' A340 test aircraft, wind speed dispersions of around 1.5 m s^{-1} at 50 m range for a flight altitude of 39 kft could be demonstrated (Rabadan et al. 2010), in line with the requirements of automatic control for gust load alleviation.

Whereas the Fabry-Perot etalon has clear advantages (high luminosity, ease of construction), it has the disadvantage of dispersing the received photons circularly on the detector; other interferometers can produce linear fringes (see Fig. 22.5). The latter may be imaged onto linear detector arrays by employing cylindrical lenses, thus enhancing the signal-to-noise ratio. The Fabry-Perot etalon eludes this possibility, thus requiring longer integration times. Optical workarounds have been proposed, such as circle-to-point converters (McGill et al. 1997), but these have other complications.

Generally, direct detection lidar seems a worthy alternative to coherent systems for short-range applications (such as for automatic control), offering the flexibility to operate within aerosol-laden as well as in aerosol-devoid altitudes and conditions.

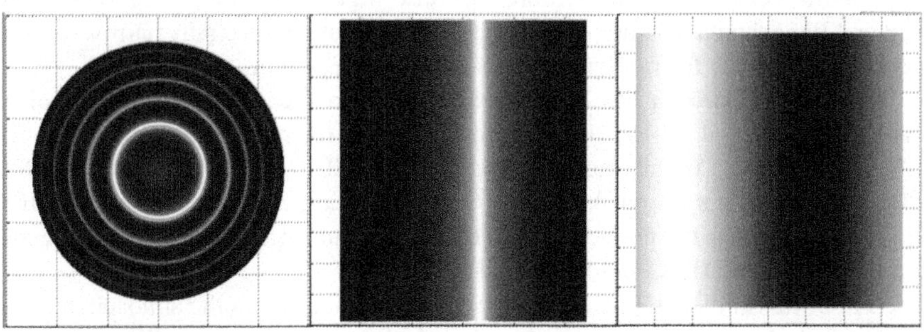

Fig. 22.5 Simulated interference patterns for Fabry-Perot, Fizeau, and Michelson interferometers (*left* to *right*, respectively). The Doppler shift results in a radial or linear shift of the imaged fringe, orders of magnitude smaller than the fringe width itself (around 5 MHz per m/s wind speed compared to the RB spectrum width of 1.7 GHz for UV wavelength at 10 km altitude)

22.5.3 Air Density Method

Another interesting approach for detecting clear-air turbulence was proposed by Feneyrou et al. (2009). It gives access (albeit indirectly) to the vertical wind speed fluctuation $\sigma^2_{w'_{\mathrm{mol}}}(R)$ by relying on the linkage between potential and kinetic energy within turbulence. The approach provides a useful relationship between vertical velocity and air temperature. Air masses stirred up and down within the turbulent region undergo adiabatic cooling and heating, thus temperature fluctuations. The relationship reads (Feneyrou et al. 2009):

$$\sigma_{w'} = \frac{g}{N} \cdot \left(\frac{\partial T_{\mathrm{air}}}{T_{\mathrm{air}}} \right) = \frac{g}{N} \cdot \left(\frac{\partial \rho_{\mathrm{air}}}{\rho_{\mathrm{air}}} \right) = \frac{g}{N} \cdot \left(\frac{\partial \beta_{\mathrm{mol}}}{\beta_{\mathrm{mol}}} \right)$$

with g the gravity acceleration and N the Brunt-Väisälä frequency. N remains between 0.01 and 0.014 rad s^{-1} for typical cruise flight altitudes (around tropopause) (Birner 2003) and may be derived from a numerical weather forecast or in situ during the ascent of the aircraft. The temperature fluctuation measurement $\sigma^2_T(R)$ exceeds the capability limits of lidar detection (see Sect. 22.4.3) but may be replaced by measurement of air density fluctuations $\sigma^2_{\rho_{\mathrm{mol}}}(R)$, which in turn may be determined by the air molecular backscatter fluctuation $\sigma^2_{\beta_{\mathrm{mol}}}(R)$. However, these fluctuations are very minute, amounting to less than the 1 % level. Hence, any aerosol presence hampers this approach because its own backscatter signal may generate sufficient noise to mask the turbulent molecular variability. In order to exploit the stronger molecular backscatter (as with direct detection techniques), such a lidar should operate in the UV region. For an altitude of ≈10,000 m, a typical UV lidar backscatter ratio between aerosol and molecules $\beta_{\mathrm{aer}}/\beta_{\mathrm{mol}}$ amounts to less than 8×10^{-3} most of the time, according to Vaughan et al. (1995). However, this may not be expected to occur during all portions of cruise flight. Therefore, a high-resolution spectral filter should be employed in order to filter out the spectrally narrow aerosol return and use only the molecular wings of the RB spectrum.

The above approach has been tested within the French MMEDTAC project (Hauchecorne et al. 2010; Hauchecorne et al. 2016), relying on aerosol-devoid observation times. Within the European project DELICAT, the DLR built such a UV lidar for operation on the Dutch Cessna Citation 2 (Vrancken et al. 2010). Figure 22.6 depicts the possible layout of such a lidar system and the implementation within MMEDTAC and DELICAT.

A laser transmitter (LT) transmits powerful UV pulses into the atmosphere. The backscattered radiation is sent onto a very narrow spectral filter (SF), such as a Fabry-Perot etalon. In that case, both the transmitted and the reflected part of the spectrum are sent onto detectors DT and DR. Both outputs are used to perform an inversion in order to retrieve the purely molecular signal. Fluctuations of this backscatter may be analyzed by its variance or more advanced methods in order to determine the vertical wind speed fluctuation $\sigma^2_{w'_{\mathrm{mol}}}(R)$. The DELICAT instrument

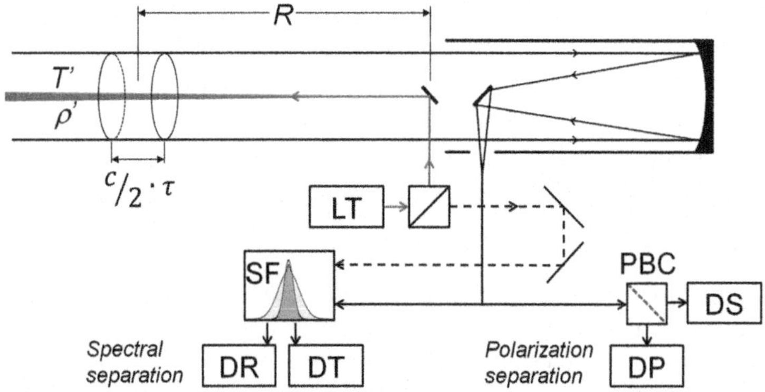

Fig. 22.6 Simplified synopsis of CAT lidar relying on density fluctuations. On the *left* employing spectral separation, on the *right* polarization separation as implemented within the DELICAT project. See text for abbreviations

employed, for these first airborne trials, a polarization separation assembly instead of the spectral filters. A good portion of the aerosols at these altitudes are represented by strongly depolarizing cirrus ice crystals which allow an exclusion of invalid data. Further, the flight campaign was planned according to aerosol content forecasted below a certain threshold. The flight campaign was executed in 2013 (Veerman et al. 2014) and demonstrated that the lidar is able to measure the molecular signal with a dispersion of better than 5×10^{-3} at distances between 10 and 15 km ahead (Vrancken et al. 2015). This allows the resolution of minute density fluctuations from moderate or greater turbulence severity. The instrumentation is subject of evolution within a DLR internal project in order to undergo further flight testing in the future.

This air density approach has the advantage of delivering the turbulence observable that is the most significant to aviation, viz., the vertical wind speed fluctuation $\sigma^2_{w'_{\mathrm{mol}}}(R)$. Even with a noisy signal, and with the uncertainty on the stratification stability N, the derived quantity may be considered a qualified estimator of relevant turbulence intensity as compared to horizontal wind speed derived with Doppler wind lidars.

22.6 The Future

The previous sections highlight the inherent challenge of remote detection and characterization of a phenomenon as complex and capricious as turbulence. Turbulence occurring in clear and clean air, thus comprising CAT, MTW, NCT, and LLT, eludes detection with classical devices like radar. Passive remote sensing instruments, from radio to optical frequencies do not fulfill the tough requirements imposed by the aeronautics applications. Distant detection and near-range quantification (in terms of wind speed) may only be attained by active optical remote sensing, i.e., lidar. Different approaches have been identified and the most important emphasized: the coherent Doppler wind lidar delivering the axial (thus horizontal) speed of aerosols embedded in the turbulent flow, $u'_{\text{aer}}(R)$ and $\sigma^2_{u'_{\text{aer}}}(R)$; the direct detection Doppler wind lidar that provides the axial speed of the air itself, $u'_{\text{mol}}(R)$ and $\sigma^2_{u'_{\text{mol}}}(R)$; and the indirect density method that gives vertical wind speed fluctuation $\sigma^2_{w'_{\text{mol}}}(R)$. One should further mention the potential of passive tracer (i.e., aerosol) structuring and the backscatter enhancement effect that should be explored further. It seems that identifying the optimum instrumentation approach appears (nearly) as difficult as understanding turbulence itself!

Coherent Doppler lidars may be limited in their range due to the low aerosol density typical of normal cruising altitudes. On the other hand, situations with high aerosol content may constrain the density approach. The relationship between the horizontal wind speed and the more important vertical wind speed is not straightforward due to the anisotropy of stratified turbulence. The connection of optical turbulence (refractive index fluctuations) to wind velocity turbulence is even more uncertain. Eye safety issues impede the use of visual and near-visual (as UV and NIR) wavelengths while favoring infrared wavelengths.

In practice, it is worth considering whether a realistic airborne system should be comprised of different technologies or a consolidated method. What about function integration of other hazard detection, such as icing conditions, mineral dust, and volcanic ash? These are the questions that should be answered in the near future by the ongoing instrument and flight testing research.

One thing is for certain, turbulence occurring outside of clouds will remain detectable only with active optical means. Thus, the solution that addresses the issues of air safety, efficiency, and competitiveness in the context of turbulence will be lidar technology of the twenty-first century.

Acknowledgments The author would like to thank Oliver Reitebuch, Todd Lane, and Robert Sharman for the kind review of the present text.

References

Atlas, D.: Clear air turbulence detection methods: a review. In: Pao, Y.-H., Goldburg, A. (eds.) Clear Air Turbulence and Its Detection, pp. 381–401. Plenum Press, New York, NY (1969)

Atlas, D., Hardy, K.R., Naito, K.: Optimizing the radar detection of clear air turbulence. J. Appl. Meteorol. **5**(4), 450–460 (1966)

Banakh, V.A., Smalikho, I.N.: Determination of optical turbulence intensity by atmospheric backscattering of laser radiation. Atmos. Oceanic Opt. **24**(5), 457–465 (2011)

Baynes, J.: Rockwell Collins Unveils New MultiScan ThreatTrack™ Weather Radar. http://investor.rockwellcollins.com/investor-relations/press-releases/press-release-details/2014/Rockwell-Collins-Unveils-New-MultiScan-ThreatTrack-Weather-Radar (2014). Accessed 01 June 2015

Behrendt, A.: Temperature measurements with lidar. In: Weitkamp, C. (ed.) Lidar – Range-Resolved Optical Remote Sensing of the Atmosphere, pp. 273–305. Springer, New York, NY (2005)
Bilbro, J., Fichtl, G., Fitzjarrald, D., Krause, M., Lee, R.: Airborne Doppler lidar wind field measurements. Bull. Am. Meteorol. Soc. **65**(4), 348–359 (1984)
Birner, T.: Die extratropische Tropopausenregion. Dissertation, Ludwig-Maximilans-Universität, München (2003)
Bruneau, D., Garnier, A., Hertzog, A., Porteneuve, J.: Wind-velocity lidar measurements by use of a Mach–Zehnder interferometer, comparison with a Fabry–Perot interferometer. Appl. Opt. **43** (1), 173–182 (2004)
Buehler, W.E., King, C.H., Lunden, C.D.: Radar echoes from clear air inhomogeneities. In: Pao, Y.-H., Goldburg, A. (eds.) Clear Air Turbulence and Its Detection, pp. 425–435. Plenum Press, New York, NY (1969)
Cezard, N.: Etude de faisabilité d'un lidar Rayleigh-Mie pour des mesures à courte portée de la vitesse de l'air, de sa température et de sa densité. Dissertation, Ecole Polytechnique (2008)
Clifford, S.F., Kaimal, J.C., Lataitis, R.J., Strauch, R.G.: Ground-based remote profiling in atmospheric studies: an overview. Proc. IEEE **82**(3), 313–355 (1994)
Collis, R.T.H.: Clear air turbulence detection. Spectr. IEEE **3**(4), 56–61 (1966)
Crooks, W.M., Hoblit, F.M., Prophet, D.T., et al.: Project HICAT, An Investigation of High Altitude Clear Air Turbulence. Technical Report AFFDL-TR-67-123 Vol. 1, Air Force Flight Dynamics Laboratory (1967)
Dolfi-Bouteyre, A., Canat, G., Valla, M., Augere, B., Besson, C., Goular, D., Macq, B.: Pulsed 1.5-μm LIDAR for axial aircraft wake vortex detection based on high-brightness large-core fiber amplifier. IEEE J. Quantum Electron **15**(2), 441–450 (2009)
Douxchamps, D., Lugan, S., Verschueren, Y., Mutuel, L., Macq, B., Chihara, K.: On-board axial detection of wake vortices using a 2 μm LiDAR. IEEE Trans. Aerosp. Electron. Syst. **44**(4), 1276–1290 (2008)
Ehlers, J., Fezans, N.: Airborne doppler LiDAR sensor parameter analysis for wake vortex impact alleviation purposes. In Advances in Aerospace Guidance, Navigation and Control, pp. 433–453. Springer International Publishing (2015)
Ehlers, J., Fischenberg, D., Niedermeier, D.: Wake impact alleviation control based on wake identification. AIAA J. (2015). doi:10.2514/1.C033157
Feneyrou, P., Lehureau, J.-C., Barny, H.: Performance evaluation for long-range turbulence-detection using ultraviolet lidar. Appl. Opt. **48**(19), 3750–3759 (2009)
Fischer, L., Kiemle, Ch., Craig, G.C.: Height-resolved variability of midlatitude tropospheric water vapor measured by an airborne lidar. Geophys. Res. Lett. **39**(6), (2012). doi:10.1029/2011GL050621
Flesia, C., Korb, C.L.: Theory of the double-edge molecular technique for Doppler lidar wind measurement. Appl. Opt. **38**(3), 432–440 (1999)
Franken, P. A., Jenney, J. A., Rank, D. M.: Airborne investigations of clear air turbulence with laser radars (Clear air turbulence detection with laser radar, noting airborne equipment and results). In: 8th Annual Electron and Laser Beam Symposium, University of Michigan, Ann Arbor, pp. 87–103. (1966)
Frehlich, R., Hannon, S.M., Henderson, S.W.: Coherent Doppler lidar measurements of wind field statistics. Boundary-Layer Meteorol. **86**(2), 233–256 (1998)
Gentry, B., McGill, M., Schwemmer, G., Hardesty, M., Brewer, A., Wilkerson, Th., Lindemann, S.: The Tropospheric Wind Lidar Technology Experiment (TWiLiTE): An Airborne Direct Detection Doppler Lidar Instrument Development Program. Presentation to Earth Science Technology Conference, College Park (2006)
Good, R.E., Watkins, B.J., Quesada, A.F., Brown, J.H., Loriot, G.B.: Radar and optical measurements of Cn^2. Appl. Opt. **21**(18), 3373–3376 (1982)
Gurvich, A.S.: Lidar sounding of turbulence based on the backscatter enhancement effect. Izvestiya Atmos. Oceanic Phys. **48**(6), 585–594 (2012)

Gurvich, A.S.: Lidar positioning of higher clear-air turbulence regions. Izvestiya Atmos. Oceanic Phys. **50**(2), 143–151 (2014)

Haggerty, J., Schick, K., Mahoney, M.J., Lim, B.: The NCAR microwave temperature profiler: data applications from recent deployments. In: Microwave Radiometry and Remote Sensing of the Environment (MicroRad), pp. 133–135. 13th Specialist Meeting, Pasadena, CA (2014). doi:10.1109/MicroRad.2014.6878924

Hamilton, D.W., Proctor, F.H., Ahmad, N.N.: Flight Tests of the Turbulence Prediction and Warning System (TPAWS). NASA/TM-2012-217337 (2012)

Hannon, S.M., Bagley, H.R., Bogue, R.K.: Airborne Doppler lidar turbulence detection: ACLAIM flight test results. In: AeroSense'99, pp. 234–241. International Society for Optics and Photonics (1999)

Hauchecorne, A., Cot, Ch., Dalaudier, F., Porteneuve, J., Gaudo, Th., Wilson, R., Besson, C.: Set-up of a ground-based Rayleigh lidar to detect clear air turbulence. In: 25th International Laser Radar Conference (ILRC), pp. 269–272. St. Petersburg (2010)

Hauchecorne, A., Cot, C., Dalaudier, F., Porteneuve, J., Gaudo, T., Wilson, R., Cénac, C., Laqui, C., Keckhut, P., Perrin, J.-M., Dolfi, A., Cézard, N., Lombard, L., Besson, C.: Tentative detection of clear-air turbulence using a ground-based Rayleigh lidar. Appl. Opt. **55**(13), 3420–3428 (2016)

Henderson, S.W., Suni, P.J.M., Hale, C.P., Hannon, S.M., Magee, J.R., Bruns, D.L., Yuen, E.H.: Coherent laser radar at 2 μm using solid-state lasers. IEEE Trans. Geosci. Remote Sens. **31**(1), 4–15 (1993)

Hoblit, F.M.: Gust Loads on Aircraft: Concepts and Applications. AIAA Education Series. AIAA Inc., Washington, DC (1988)

Huffaker, R.M.: CO2 laser Doppler systems for the measurement of atmospheric winds and turbulence. Atmos. Technol. **1**, 71–76 (1975)

Inokuchi, H., Tanaka, H., Ando, T.: Development of an onboard doppler lidar for flight safety. J. Aircraft **46**(4), 1411–1415 (2009)

Inokuchi, H., Tanaka, H., Ando, T.: Development of a long range airborne Doppler LIDAR. In: Proceedings of 27th Congress of International Council of the Aeronautical Sciences (ICAS), **10** (3), (2010)

Inokuchi, H., Furuta, M., Inagaki, T.: High altitude turbulence detection using an airborne Doppler lidar. In: Proceedings of 29th Congress of the International Council of the Aeronautical Sciences (ICAS), St. Petersburg, 7–12 June 2014

Keeler, R.J., Serafin, R.J., Schwiesow, R.L., Lenschow, D.H., Vaughan, J.M., Woodfield, A.A.: An airborne laser air motion sensing system. Part I: Concept and preliminary experiment. J. Atmos. Oceanic Technol. **4**(1), 113–127 (1987)

Köpp, F., Rahm, S., Smalikho, I.: Characterization of aircraft wake vortices by 2 μm pulsed Doppler lidar. J. Atmos. Oceanic Technol. **21**(2), 194–206 (2004)

Lawrence Jr., J.D., McCormick, M.P., Melfi, S.H., Woodman, D.P.: Laser backscatter correlation with turbulent regions of the atmosphere. Appl. Phys. Lett. **12**(3), 72–73 (1968)

Lilly, D.K., Waco, D.E., Adelfang, S.I.: Stratospheric mixing estimated from high-altitude turbulence measurements. J. Appl. Meteorol. **13**(4), 488–493 (1973)

Liu, Z.S., Liu, B.Y., Li, Z.G., Yan, Z.A., Wu, S.H., Sun, Z.B.: Wind measurements with incoherent Doppler lidar based on iodine filters at night and day. Appl. Phys. B **88**(2), 327–335 (2007)

Liu, D., Hostetler, C., Miller, I., Cook, A., Hare, R., Harper, D., Hair, J.: Tilted pressure-tuned field-widened Michelson interferometer for high spectral resolution lidar. In SPIE Photonics Europe, pp. 84390P-84390P. International Society for Optics and Photonics, Apr (2012)

Luce, H., Nakamura, T., Yamamoto, M.K., Yamamoto, M., Fukao, S.: MU radar and lidar observations of clear-air turbulence underneath cirrus. Mon. Weather Rev. **138**(2), 438–452 (2010)

Matayoshi, N., Asaka, K., Okuno, Y.: Flight test evaluation of a helicopter airborne lidar. J. Aircraft **44**(5), 1712–1720 (2007)

McGill, M.J., Spinhirne, J.D.: Comparison of two direct-detection Doppler lidar techniques. Opt. Eng. **37**(10), 2675–2686 (1998)

McGill, M.J., Marzouk, M., Scott, V.S., Spinhirne, J.D.: Holographic circle-to-point converter with particular applications for lidar work. Opt. Eng. **36**(8), 2171–2175 (1997)
Measures, R.M.: Laser Remote Sensing: Fundamentals and Applications. Wiley, New York, NY (1984)
Melfi, S.H., Stickle, J.W.: Airborne laser-radar studies of the lower atmosphere. NASA-TN-D-5558 (1969)
Misaka, T., Holzäpfel, F., Hennemann, I., Gerz, T., Manhart, M., Schwertfirm, F.: Vortex bursting and tracer transport of a counter-rotating vortex pair. Phys. Fluids **24**(2), 025104 (2012)
Ottersten, H.: Atmospheric structure and radar backscattering in clear air. Rad. Sci. **4**(12), 1179–1193 (1969)
Rabadan, G.J., Schmitt, N.P., Pistner, T., Rehm, W.: Airborne lidar for automatic feedforward control of turbulent in-flight phenomena. J. Aircraft **47**(2), 392–403 (2010)
Rahm, S., Smalikho, I., Köpp, F.: Characterization of aircraft wake vortices by airborne coherent Doppler lidar. J. Aircraft **44**(3), 799–805 (2007)
Reitebuch, O.: Wind lidar for atmospheric research. In: Schumann, U. (ed.) Atmospheric Physics, pp. 487–507. Springer, Heidelberg (2012)
Reitebuch, O., Lemmerz, C., Nagel, E., Paffrath, U., Durand, Y., Endemann, M., et al.: The airborne demonstrator for the direct-detection Doppler wind lidar ALADIN on ADM-Aeolus. Part I: Instrument design and comparison to satellite instrument. J. Atmos. Oceanic Technol. **26**(12), 2501–2515 (2009)
Reiter, E.R., Burns, A.: The structure of clear-air turbulence derived from TOPCAT aircraft measurements. J. Atmos. Sci. **23**(2), 206–212 (1966)
Schaffner, Ph.R., Daniels, T.S., West, L.L., Gimmestad, G.G., Lane, S.E., Burdette, E.M., Sharman, R.D.: Experimental validation of a forward looking interferometer for detection of clear air turbulence due to mountain waves. In: 4th AIAA Atmospheric and Space Environments Conference, New Orleans (2012)
Schmitt, N., Rehm, W., Pistner, T., Zeller, P., Diehl, H., Navé, P.: Airborne direct detection UV lidar. In: Proceedings of 23rd International Laser Radar Conference, pp. 167–170. Nara, 24–28 July 2006
Schumann, U. (ed.): Atmospheric physics. Springer, Heidelberg (2012)
Serafimovich, A., Hoffmann, P., Peters, D., Lehmann, V.: Investigation of inertia-gravity waves in the upper troposphere/lower stratosphere over Northern Germany observed with collocated VHF/UHF radars. Atmos. Chem. Phys. **5**(2), 295–310 (2005)
Smalikho, I., Köpp, F., Rahm, S.: Measurement of atmospheric turbulence by 2 μm Doppler lidar. J. Atmos. Oceanic Technol. **22**(11), 1733–1747 (2005)
Soreide, D.C., Bogue, R.K., Ehernberger, J., Hannon, S.M., Bowdle, D.A.: Airborne coherent LIDAR for advanced in-flight measurements (ACLAIM) – flight testing of the LIDAR sensor. NASA-H-2428 (2000)
Spuler, S.M., Richter, D., Spowart, M.P., Rieken, K.: Optical fiber-based laser remote sensor for airborne measurement of wind velocity and turbulence. Appl. Optics **50**(6), 842–851 (2011)
Targ, R., Steakley, B.C., Hawley, J.G., Ames, L.L., Forney, P., Swanson, D., Robinson, P.A.: Coherent lidar airborne wind sensor II: Flight-test results at 2 and 10 μm. Appl. Opt. **35**(36), 7117–7127 (1996)
Teets Jr., E.H., Ashburn, C., Ehernberger, J., Bogue, R.: Turbulence and mountain wave conditions observed with an airborne 2-micron lidar. In: Remote Sensing, pp. 63670O. International Society for Optics and Photonics (2006)
Theopold, F.A., Bösenberg, J.: Differential absorption lidar measurements of atmospheric temperature profiles: theory and experiment. J. Atmos. Oceanic Technol. **10**(2), 165–179 (1993)
Vaughan, J.M., Brown, D.W., Nash, C., Alejandro, S.B., Koenig, G.G.: Atlantic atmospheric aerosol studies: 2. Compendium of airborne backscatter measurements at 10.6 μm. J. Geophys. Res. **100**(D1), 1043–1065 (1995)
Vaughan, J.M., Geddes, N.J., Flamant, P.H., Flesia, C.: Establishment of a backscatter coefficient and atmospheric database. ESA-CR12510 (1998)

Veerman, H.P., Vrancken, P., Lombard, L.: Flight testing DELICAT – a promise for medium-range clear air turbulence protection. In: Proceedings of the 25th SFTE European Chapter Symposium together with SETP European Section, Lulea, 15–18 June 2014

Verbeke, M.: LIDAR Airborne Aerodynamic Sensors – Thales Avionics. Presentation, In WAKENET-3 Europe/Greenwake Workshop, Palaiseau, 29–30 March 2010

Vernin, J., Pelon, J.: Scidar/lidar description of a gravity wave and associated turbulence: preliminary results. Appl. Opt. **25**(17), 2874–2877 (1986)

Vrancken, P., Wirth, M., Rempel, D., Ehret, G., Dolfi-Bouteyre, A., Lombard, L., Rondeau, Ph.: Clear air turbulence detection and characterisation in the DELICAT airborne lidar project. In: Proceedings of the 25th International Laser Radar Conference (ILRC), Saint Petersburg, 5–9 July 2010

Vrancken, P., Wirth, M., Ehret, G., Witschas, B., Veerman, H., Tump, R., Barny, H., Rondeau, Ph., Dolfi-Bouteyre, A., Lombard, L.: Flight tests of the DELICAT airborne LIDAR system for remote clear air turbulence detection. In: Proceedings of the 27th International Laser Radar Conference (ILRC), New York, 5–10 July 2015

Watkins, C.D., Browning, K.A.: The detection of clear air turbulence by radar. Phys. Technol. **4** (1), 28 (1973)

Weaver, E.A.: Clear-air-turbulence detection using lasers. Conf. Proc. NASA Aircraft Saf. Operating Prob. **1**, 89 (1971)

Werner, C.: Doppler wind lidar. In: Weitkamp, pp. 325–354 (2005)

Weitkamp, C. (ed.): Lidar: range-resolved optical remote sensing of the atmosphere. Springer, New York, NY (2005)

Witschas, B., Lemmerz, C., Reitebuch, O.: Daytime measurements of atmospheric temperature profiles (2–15 km) by lidar utilizing Rayleigh–Brillouin scattering. Opt. Lett. **39**(7), 1972–1975 (2014)

Zilberman, A., Kopeika, N.S.: LIDAR measurements of atmospheric turbulence vertical profiles. In Lasers and Applications in Science and Engineering, pp. 288–297. International Society for Optics and Photonics, June (2004)

Zilberman, A., Golbraikh, E., Kopeika, N.S., Virtser, A., Kupershmidt, I., Shtemler, Y.: Lidar study of aerosol turbulence characteristics in the troposphere: Kolmogorov and non-Kolmogorov turbulence. Atmos. Res. **88**(1), 66–77 (2008)

Chapter 23
Clear-Air Turbulence in a Changing Climate

Paul D. Williams and Manoj M. Joshi

Abstract How might the processes generating clear-air turbulence change in a warmer world? We know that observations support an association between clear-air turbulence and shear instability. We also know that the upper atmospheric wind shears are changing in response to greenhouse gas forcing. In particular, theoretical reasoning and climate model simulations both suggest that the vertical shear in horizontal wind is increasing in magnitude at typical aircraft cruising altitudes in the middle latitudes, especially in the winter months in each hemisphere. This increased shearing implies that clear-air turbulence may itself be changing as a consequence of climate change. This chapter reviews the various lines of observational and model-based evidence for trends in clear-air turbulence, by analyzing data from turbulence encounters with aircraft, turbulence diagnosed from reanalysis datasets, passenger injuries caused by turbulence, and turbulence diagnosed from climate models. The possibility of anthropogenic trends in clear-air turbulence opens up a whole new field of academic study, which exists at the interface between the two scientific disciplines of aviation turbulence and climate change. We call for future work to improve our understanding of this poorly understood but potentially important impact of climate change.

23.1 Introduction

Clear-air turbulence is, by definition, atmospheric turbulence on aircraft-affecting length scales that exists outside clouds and thunderstorms and their associated convective updrafts and downdrafts. Aircraft are estimated to spend roughly 3 %

P.D. Williams (✉)
Department of Meteorology, University of Reading, Earley Gate, Reading RG6 6BB, UK
e-mail: p.d.williams@reading.ac.uk

M.M. Joshi
Centre for Ocean and Atmospheric Sciences, University of East Anglia, Norwich Research Park, Norwich NR4 7TJ, UK
e-mail: m.joshi@uea.ac.uk

R. Sharman, T. Lane (eds.), *Aviation Turbulence*,
DOI 10.1007/978-3-319-23630-8_23

of their cruise time in light-or-greater clear-air turbulence (Watkins and Browning 1973) and 1 % of their cruise time in moderate-or-greater clear-air turbulence (Sharman et al. 2006). Observations by Watkins and Browning (1973) support an association between clear-air turbulence and the Kelvin–Helmholtz shear instability. The fluid dynamical theory of this instability applies to stratified shear flows, such as those encountered in the vicinity of the atmospheric jet streams. According to the theory, if the shear is strong enough and the stratification is weak enough, then small, wavelike perturbations are able to grow in amplitude by extracting energy from the background flow. After an initial exponential growth that is governed by linear dynamics, the waves eventually enter the nonlinear regime and break down into turbulence.

The goal of this chapter is to bring together two separate ingredients. The first ingredient is the generally accepted belief, outlined above, that clear-air turbulence is generated by shear instabilities. The second ingredient is the notion that the atmospheric wind shears may be changing because of (or, more precisely, as part of) climate change. The implication of bringing together these two ingredients is that clear-air turbulence itself may be changing as a consequence of climate change. This possibility opens up a whole new field of academic study, which exists at the interface between the two scientific disciplines of aviation turbulence and climate change. This new field is still in its infancy, but our hope in writing this chapter is to spur on the research that will be needed to answer the many open questions.

The outline of the chapter is as follows. Section 23.2 discusses the basic science of climate change, focusing on the response of the upper atmospheric winds to anthropogenic forcing. A mechanism for climate-related trends in wind shear and clear-air turbulence is described, and the role of stratospheric ozone is considered. Section 23.3 discusses the various lines of observational and model-based evidence for trends in clear-air turbulence. Such trends are difficult to detect, but several attempts to do so are described, using data from turbulence encounters with aircraft, turbulence diagnosed from reanalysis datasets, passenger injuries caused by turbulence, and turbulence diagnosed from climate models. Section 23.4 concludes the chapter by calling for future work to improve our understanding of this poorly understood but potentially important impact of climate change.

23.2 Response of Upper Atmospheric Winds to Anthropogenic Forcing

23.2.1 The Changing Climate

It has become apparent from observations in recent decades that Earth's lower atmosphere has been warming globally since the end of the nineteenth century, with the warming trend accelerating in the latter half of the twentieth century (Hartmann et al. 2013). Although some natural factors such as variations in solar radiation have contributed a small amount to the warming, the primary cause is increasing anthropogenic emissions of greenhouse gases, particularly carbon dioxide (CO_2), methane (CH_4), and nitrous oxide (N_2O) (Myhre et al. 2013, Sect. 8.5). Increases in CO_2 mainly come from industrial processes such as fossil fuel combustion and cement manufacture, although other factors such as deforestation are also significant contributors (Ciais et al. 2013, Sect. 6.3.1; Le Quéré et al. 2013). Increases in CH_4 arise from sources such as rice paddies, ruminants, and climate-sensitive ecosystems such as wetlands (Ciais et al. 2013, Sect. 6.3.3), whereas increases in N_2O arise primarily from agriculture (Ciais et al. 2013, Sect. 6.3.4).

The lower atmosphere has warmed by approximately 0.5–1.0 K since the nineteenth century (Hartmann et al. 2013), and projections of future change suggest an additional surface warming of 0.5–2.0 K by the 2050s or 0.5–4.5 K in the absence of reductions in greenhouse gas emissions (Collins et al. 2013, Sect. 12.4). These temperature ranges depend on three major uncertainties. The first uncertainty is the total anthropogenic emissions of greenhouse gases between now and the 2050s, which depends on socioeconomic factors such as economic growth and the development of renewable technologies (e.g., van Vuuren et al. 2011). The second uncertainty is the fraction of emitted carbon that remains in the atmosphere as CO_2, which depends on biogeochemical factors such as uptake of CO_2 by the ocean or the terrestrial biosphere (Friedlingstein et al. 2006). The third uncertainty is the sensitivity of the physical climate system to the CO_2 that remains in the atmosphere, which depends on how a warming world may alter processes such as the seasonal cycle of sea ice or cloud formation (Bony et al. 2006). The relative importance of these three factors is expected to change in time. For instance, changes in anthropogenic emissions of greenhouse gases in the near future will have a much greater bearing on climate change in the latter half of this century than on climate change before the middle of this century (Hawkins and Sutton 2009).

In 1992, the United Nations Framework Convention on Climate Change (UNFCCC) recommended achieving "stabilization of greenhouse gas concentrations in the atmosphere at a level that would prevent dangerous anthropogenic interference with the climate system." The threshold chosen was a 2 K rise from preindustrial levels, although given the uncertainties discussed above, there are significant uncertainties in the greenhouse gas emissions needed to stay within this target (Meinshausen et al. 2009). Although at first sight 2 K seems like a small value, some profound effects are projected from even this small change, because even though climate change will affect the whole of the globe, some parts will warm more than others. For instance, the Arctic is projected to warm more than the tropics or Southern Ocean, and the land is projected to warm more than the ocean (Collins et al. 2013). Associated with such patterns in warming are changes to regional- and continental-scale atmospheric circulations, causing concomitant and potentially profound changes to other atmospheric processes such as winds and rainfall, in addition to changes expected from the direct temperature increase such as sea-level rise or heat waves. Climate model projections suggest that the 2 K rise will happen between 2050 and 2100, depending on future emissions of CO_2 this century (Joshi et al. 2011).

23.2.2 A Mechanism for Climate-Related Trends in Clear-Air Turbulence

A key question of interest to readers of this book will be: How might the processes generating clear-air turbulence change in a warmer world? In addition to spatial gradients in warming, the troposphere and stratosphere warm unevenly in response to climate change (e.g., Collins et al. 2013, their Fig. 12.12). The zonal-mean (i.e., longitudinally averaged) temperature response to climate change is shown in Fig. 23.1. The tropical troposphere is projected to warm more than the tropical surface, because a warmer atmosphere on average is projected to have a higher water vapor concentration, due mostly to the increase in saturation vapor pressure of water vapor with temperature. Associated with the higher concentration of water vapor is more latent heating due to condensation, resulting in the tropical atmosphere being more stable, i.e., having a less negative lapse rate (Bony et al. 2006). In the polar regions, the tropospheric amplification of warming is smaller because there is less water vapor present, but changes in atmospheric heat transport and strong climatic feedbacks associated with changes to sea ice and clouds do result in a strong surface warming (Taylor et al. 2013). The stratosphere, by contrast, cools in response to the addition of anthropogenic greenhouse gases, which is related to

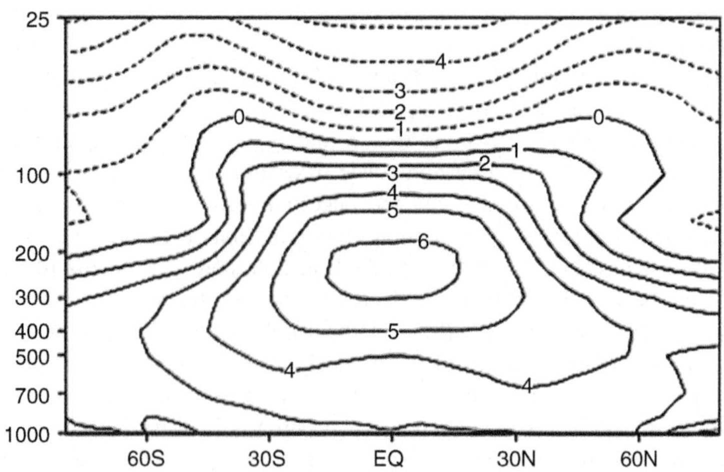

Fig. 23.1 Annual-mean zonal-mean temperature change (K) from 1980–1999 to 2080–2099 averaged over an ensemble of climate models. The horizontal axis is latitude, and the vertical axis is pressure in hPa. *Solid contours* indicate warming, and *dashed contours* indicate cooling. From Lorenz and DeWeaver (2007)

changes in the emission of infrared radiation to space by these increased gases (Fels et al. 1980).

The spatially varying response described above, when allied to the fact that the tropopause, which separates the stratosphere and troposphere, itself decreases in height from equator to pole, gives rise to one key possible change to clear-air turbulence due to the large-scale relationship between wind and temperature. The zonal-mean zonal wind response to climate change is shown in Fig. 23.2. Meridional (north–south) gradients in temperature are related to vertical shear in zonal (east–west) winds, because of geostrophic and hydrostatic balance. Therefore, near the tropopause, which is the approximate cruising altitude for commercial aircraft, climate change causes both the meridional temperature gradient and vertical wind shear to increase in magnitude in the middle latitudes, especially in the winter months in each hemisphere (Lorenz and DeWeaver 2007; Delcambre et al. 2013). This mechanism is at the heart of the projected increases in wind shear and hence clear-air turbulence studied by Williams and Joshi (2013), which will be described in Sect. 23.3.5.

The response described above assumes that both hemispheres warm evenly, which is not the case. In fact, the southern extra-tropics are expected to warm at a

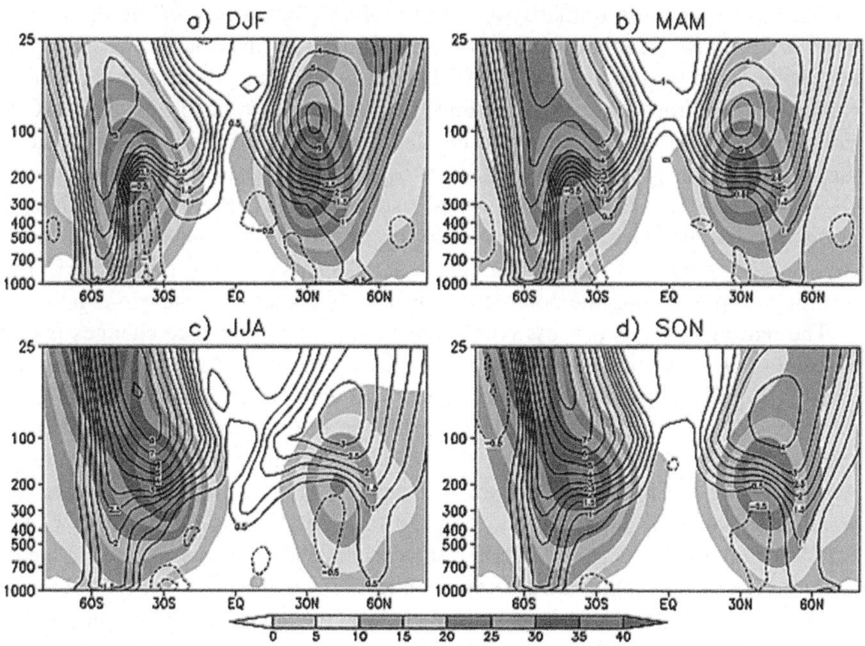

Fig. 23.2 Zonal-mean zonal wind speed for the twentieth century (*shaded*) and change from 1980–1999 to 2080–2099 (*contours*), both averaged over an ensemble of climate models in (**a**) December, January, and February; (**b**) March, April, and May; (**c**) June, July, and August; and (**d**) September, October, and November. The units are m s^{-1}. The horizontal axis is latitude, and the vertical axis is pressure in hPa. *Solid contours* indicate increases, and *dashed contours* indicate decreases. From Lorenz and DeWeaver (2007)

slower rate than the rest of the world, because of the efficiency at which heat is transported away from the surface by the circulation of the Southern Ocean. However, in the upper troposphere, this asymmetry is much less marked, suggesting that the winds in the southern hemisphere's upper troposphere will warm in a similar manner to the northern hemisphere. The zonally averaged picture conceals the fact that the northern hemisphere's upper tropospheric jet stream displays considerable longitudinal variability. The response of the Pacific and Atlantic jet streams will depend to some extent on the differing responses of these respective ocean basins to climate change (Lorenz and DeWeaver 2007), which projections suggest are very different. While climate models project a robust warming of the Pacific Ocean, the response of the North Atlantic Ocean varies more among models (Collins et al. 2013, their Fig. 12.11) because of the added complexity of the ocean circulation's response to climate change (Weaver et al. 2012).

23.2.3 The Role of Stratospheric Ozone

The above picture describes a change to the tropospheric jet stream arising from changes in temperature primarily as a result of CO_2 emissions. While the role of other well-mixed greenhouse gases on the temperature distribution of the upper troposphere and lower stratosphere (UTLS) is smaller, one greenhouse gas whose changes are anthropogenic in origin must be mentioned: stratospheric ozone (O_3). Emissions of chlorofluorocarbons during the twentieth century led to an increase in the concentrations of such gases in the stratosphere. These gases are able to destroy O_3 in a reaction that is catalyzed on surfaces such as cloud particles, which can form in the lower stratosphere in winter and spring, usually in the southern polar lower stratosphere, since this is significantly colder than its northern counterpart. The resulting drop in springtime polar stratospheric O_3 is known as the ozone hole.

The reason why this process is relevant to clear-air turbulence changes is that destroying O_3 cools the polar lower stratosphere (Fels et al. 1980), enhancing the equator-to-pole temperature gradient, associated with which is a strengthening of the tropospheric jet stream wind shear. Depletion of O_3 is thought to have significantly contributed to recent changes in the southern hemisphere's jet stream (e.g., Arblaster and Meehl 2006). However, changes in the northern hemisphere are smaller, so while O_3 depletion is of interest to understanding the behavior of the UTLS region, it is perhaps of less interest to understanding future changes to clear-air turbulence in the regions where aviation traffic is highest.

The above picture also brings into focus how the spread in the projections of regional changes associated with global warming is related to differences between the models. The spread is associated with different model formulations of small-scale processes such as cloud formation and sea ice as described above, as well as differences in the large-scale advection or movement of momentum or tracers in the atmosphere and ocean. These factors also give rise to inter-model spread in projections of exactly how phenomena such as jet streams may change both in position and strength under climate change. Quantifying such uncertainty is a task that is being carried out by the climate research community.

It is perhaps ironic at a time when aviation is being described as part of the climate change problem due to emissions associated with air travel—especially given projected future trends in air traffic—that changes to climate may affect aviation through increases in turbulence. This should not be a surprise though: changes to clear-air turbulence are simply dynamical consequences to changes in the large-scale state of the atmosphere–ocean system, which are to be expected when the system is perturbed significantly over many decades, as humankind has done and is continuing to do.

23.3 Evidence of Trends in Clear-Air Turbulence

23.3.1 The Problem of Detecting Historic Trends in Turbulence

The suggestion of anthropogenic changes in clear-air turbulence according to the mechanisms discussed in Sect. 23.2 naturally leads us to search for turbulence trends in historic data. The detection of historic trends in atmospheric variables such as temperature and precipitation is possible partly because of the existence of high-quality gridded reanalysis data covering at least the past few decades. Reanalysis datasets contain the best estimates of the large-scale state of the three-dimensional global atmosphere, constrained by a wide variety of observations that have been assimilated into a comprehensive general circulation model. In contrast, the detection of historic trends in clear-air turbulence is complicated by a number of problems concerning the quality of the available data. Although global atmospheric reanalysis datasets have finer resolutions than climate models, they are still too coarse in space and time to resolve turbulence on the scales affecting aircraft. The only direct observations of clear-air turbulence are in situ measurements from weather balloons and aircraft.

Radiosondes suspended from weather balloons are widely deployed to obtain operational soundings for weather forecasts. As they ascend, they respond to turbulent eddies that are an order of magnitude smaller than the eddies that affect aircraft. However, under the assumption that the turbulence is in equilibrium, the downscale cascade of three-dimensional turbulence implies that wherever and whenever there is aircraft-affecting turbulence, there will also be balloon-affecting turbulence. Therefore, turbulence measured by weather balloons could offer insights into aviation turbulence. Unfortunately, commercial radiosondes currently have no capability to directly measure and record the atmospheric turbulence they experience.

Motivated by the above capability gap, Harrison and Hogan (2006) proposed a method for adapting conventional meteorological radiosondes to detect atmospheric turbulence. The method involves adding a Hall-effect magnetometer to the instrument package. These inexpensive geomagnetic sensors are able to monitor the terrestrial magnetic field. Rapid fluctuations in the magnetic field measurements are related to the motion of the radiosonde, which is strongly influenced by atmospheric turbulence. Harrison et al. (2007) developed the method, by using three mutually orthogonal Hall-effect magnetometers instead of one, allowing the detection of all three dimensions of the turbulent fluctuations.

Marlton et al. (2015) have recently developed the weather balloon method further, by proposing the use of an accelerometer instead of a geomagnetic sensor. This modification is beneficial because it allows the turbulent motions to be measured in standard units of acceleration. In a series of test flights, strong turbulence was found to induce accelerations of magnitude greater than 5 g, where $g = 9.81\ \mathrm{m\ s^{-2}}$. Calibration of the accelerometer data with a vertically oriented lidar has allowed eddy dissipation rates (ε) of between 10^{-3} and $10^{-2}\ \mathrm{m^2\ s^{-3}}$ to be derived from the acceleration measurements. Data from one of the test flights by Marlton et al. (2015) is shown in Fig. 23.3. In-cloud turbulence, identified as such because the relative humidity is near 100 %, is present at 2–6 km. Clear-air turbulence, identified as such because of the low relative humidity, is present within the fast winds of the jet stream at 8–10 km.

Accelerometer measurements of turbulence have been made routinely by Marlton and colleagues since 2013. However, the launches are made from only a few geographic points, and the record is too short to seek climate-related trends. If inexpensive accelerometers were fitted to the thousands of radiosondes that are launched around the world daily, then the result would be a growing record of direct turbulence measurements with considerable geographic coverage. At the present

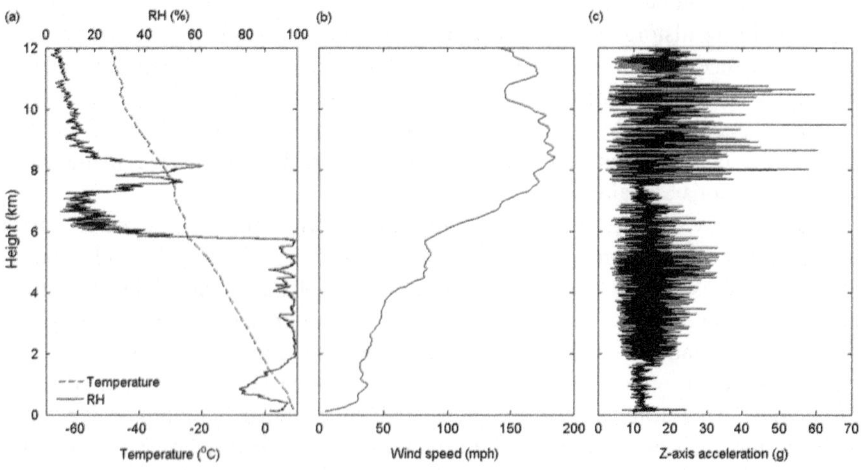

Fig. 23.3 Vertical profiles from a radiosonde balloon flight launched on 5 November 2013 in Reading, UK, showing (**a**) temperature (*dashed*) and relative humidity (*solid*), (**b**) horizontal wind speed, and (**c**) acceleration measured by the vertical axis of the accelerometer. From Marlton et al. (2015)

time, however, datasets of turbulence encounters with aircraft appear to offer the best opportunity for seeking historic trends.

23.3.2 Historic Trends in Turbulence Encounters with Aircraft

Datasets from turbulence encounters with aircraft fall into two categories: automated measurements and pilot reports (PIREPs). In the first category, some aircraft have recently been fitted with appropriate hardware and software to make automated measurements of the atmospheric turbulence through which the aircraft is flying. The measures provide estimates of the eddy dissipation rate (e.g., Sharman et al. 2014) and derived equivalent vertical gust (e.g., Gill 2014). The measures are quantitative and objective and are logged automatically by the aircraft at regular intervals. The eddy dissipation rate is an attractive quantity to measure and analyze, because it is an intrinsic property of the atmosphere and is independent of the specific aircraft flying through it. Unfortunately, these automated measures have only been available for a few years and are not yet amenable to a trend analysis. However, they are likely to become so as more data are acquired over time and as the facility to make automated measurements is installed on more aircraft.

In the second category, PIREPs indicate a turbulence intensity that is estimated by the pilot. The estimates are typically recorded on a calibrated scale in which 0 represents null turbulence, 1 is smooth-to-light, 2 is light, 3 is light-to-moderate, 4 is moderate, 5 is moderate-to-severe, 6 is severe, 7 is severe-to-extreme, and 8 is extreme. PIREPs are known to suffer from a number of limitations (e.g., Schwartz 1996; Sharman et al. 2014). In contrast to automated measurements, PIREPs are only semiquantitative, and, because they inevitably depend upon the experience and knowledge of the pilot, they are also subjective. PIREPs must be interpreted with some caution, because they are a property of the specific aircraft being flown: a small aircraft will experience stronger turbulence than a large aircraft when flying through airspace with a given turbulent eddy dissipation rate. PIREPs can have significant spatial and temporal errors. There may also be cultural biases, with pilots from some countries or airlines less likely to log PIREPs. A final limitation is that aviation hazard forecasts are available to help aircraft avoid flying through suspected regions of strong turbulence, and recent PIREPs from previous aircraft in the same airspace are also available for the same purpose. The usage of this prior information in route planning introduces a bias in PIREPs against the stronger turbulence categories.

Despite the above limitations, PIREPs have been used to create statistical climatologies of clear-air turbulence within given geographic regions. For example, Wolff and Sharman (2008) have constructed a climatology of upper-level turbulence over the contiguous USA. To create their climatology, over 2.3 million PIREPs were analyzed covering the 12-year period from 1994 to 2005. To ensure that the constructed climatologies were as robust as possible, light turbulence reports were ignored, and moderate, severe, and extreme turbulence reports were combined into a single category of moderate-or-greater (MOG) turbulence. The analysis showed that the fraction of PIREPs reporting MOG turbulence increased over this time period, suggesting a trend of increasing turbulence. However, the authors call for a more thorough analysis to verify the existence of the trend because, given that there are only 12 years of PIREP data in the analysis, the statistical significance is unclear.

In a similar manner, Kim and Chun (2011) have analyzed the statistics of PIREPs over South Korea from 2003 to 2008. Korea and eastern Asia have significant potential for turbulence events, because the jet stream there is the strongest on the planet. In the analysis by Kim and Chun (2011), the absolute numbers of PIREPs reporting turbulence are normalized by the total number of PIREPs for each year, to attempt to account for any changes in the volume of air traffic. Their analysis finds that the fraction of PIREPs reporting light-or-greater (LOG) turbulence events increased from around 29 % in 2003 to around 41 % in 2008. In addition, the fraction of PIREPs reporting moderate-or-greater (MOG) turbulence events increased from around 2 % in 2003 to around 6–7 % in 2008. By analyzing jet stream winds, the authors show that the atmospheric conditions over South Korea probably were more conducive for generating turbulence in 2008 than they were in 2003. Although this analysis is objective and the increase in turbulence encounters over the 5-year period appears to be statistically significant, whether the increase reflects inter-annual variability or is indicative of a longer-term climate-related trend is unclear.

23.3.3 Historic Trends in Turbulence Diagnosed from Reanalysis Data

In the absence of multi-decadal records of turbulence encountered directly by aircraft, we must resort to less direct methods to investigate possible trends associated with climate change. As we noted in Sect. 23.3.1, atmospheric reanalysis datasets are too coarse to explicitly resolve turbulence on the scales that are relevant to aircraft. However, this subgrid-scale turbulence can be diagnosed with demonstrable success from the larger-scale synoptic and mesoscale atmospheric flow, which is resolved in reanalysis datasets. Jaeger and Sprenger (2007) exploit this capability to present a 44-year climatology of four clear-air turbulence indicators in the Northern Hemisphere, as diagnosed using ERA-40 reanalysis data from 1958 to 2001. The four turbulence indicators that are calculated are the Richardson number, which diagnoses Kelvin–Helmholtz instability; the Brunt–Väisälä frequency, which diagnoses static instability; the potential vorticity, which diagnoses symmetric instability; and the first Ellrod and Knapp (1992) index, which is an empirical indicator that is commonly used for predictions.

A 44-year climatology of diagnosed clear-air turbulence is long enough to seek climate-related trends. A statistical analysis of all four turbulence indicators by Jaeger and Sprenger (2007) for the North Atlantic sector over the period 1958–2001 found large increases in the frequencies with which thresholds corresponding to significant turbulence were exceeded. The increases were roughly 70 % for turbulence calculated from the Ellrod and Knapp diagnostic, 90 % for the Richardson number diagnostic, 40 % for the Brunt–Väisälä diagnostic, and 60 % for the potential vorticity diagnostic. The trends are shown in Fig. 23.4. Similar trends were found in the US and the European sectors, and the trends were found to be insensitive to the choice of thresholds.

Discussing the positive trends in diagnosed clear-air turbulence, Jaeger and Sprenger (2007) caution that the amount and type of assimilated data were not constant over the 44-year period of study. For example, the sudden onset of assimilation of data from radiosondes and satellites partway through the analysis period may have caused spurious trends. Indeed, Bengtsson et al. (2004) show that the assimilation of additional satellite data produces a kink in approximately 1979 in the time series of total kinetic energy calculated from ERA-40. However,

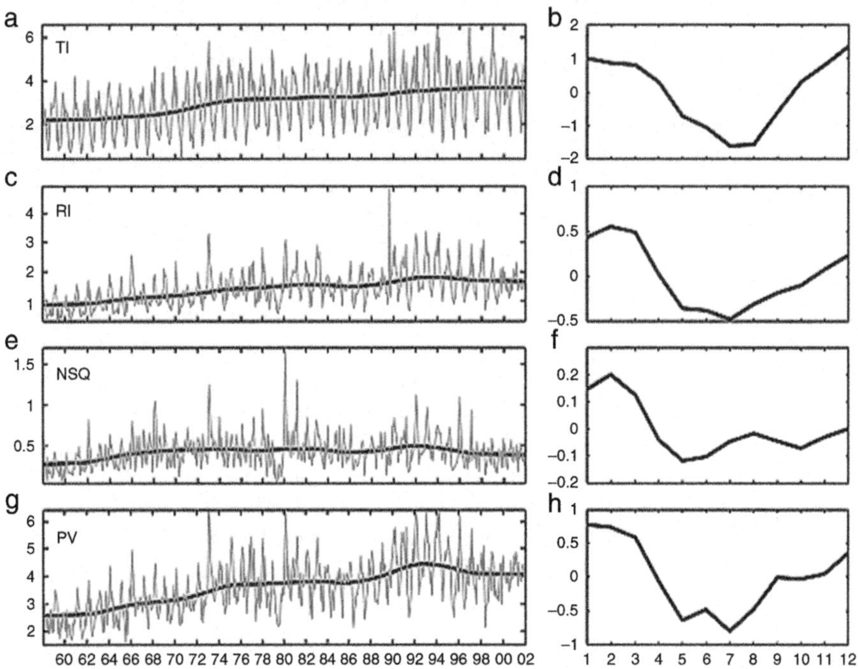

Fig. 23.4 Time series for the period 1958–2001 showing the frequency (%) of clear-air turbulence estimated using the Ellrod and Knapp index (TI), Richardson number (RI), Brunt–Väisälä frequency (NSQ), and potential vorticity (PV). *Gray lines* show the raw data, and *black lines* show the nonlinear trend estimates. The geographic region is the North Atlantic sector from 90°W to 10°E and 30 to 70°N in the tropopause region. The panels on the *right* show the mean seasonal cycle as a deviation from the nonlinear trend. From Jaeger and Sprenger (2007)

because the trends calculated by Jaeger and Sprenger (2007) are reasonably constant throughout the reanalysis period, rather than exhibiting a step change at the start of the satellite era, they appear to be more than merely artifacts of the data assimilation.

23.3.4 Historic Trends in Passenger Injuries Caused by Turbulence

A key source of information on historic trends in passenger injuries caused by turbulence is the US Federal Aviation Administration (FAA) and in particular their advisory circular on preventing injuries caused by turbulence (Ballough 2007). This advisory circular analyzes accident statistics, where an "accident" is defined to be "an occurrence associated with the operation of an aircraft which takes place between the time any person boards the aircraft with the intention of flight and all such persons have disembarked, and in which any person suffers death or serious injury, or in which the aircraft receives substantial damage." In turn, a "serious injury" is defined to be "any injury that (1) requires the individual to be hospitalized for more than 48 hours, commencing within 7 days from the date the injury was received; (2) results in a fracture of any bone (except simple fractures of fingers, toes, or nose); (3) causes severe hemorrhages, nerve, muscle, or tendon damage; (4) involves any internal organ; or (5) involves second- or third-degree burns, or any burns affecting more than 5 per cent of the body surface."

A graph of accident statistics from the FAA report is reproduced in Fig. 23.5. The graph shows that, on average over the 22-year coverage period, in terms of order of magnitude, there was roughly one accident caused by turbulence for every million flight departures by US carriers. Superimposed on this average, however, is a linear trend in which the accident rate more than doubles from 0.5 to 1.2 per million over the 22 years. Ballough (2007) speculates that the controlling factor behind this increase might be load factors. We speculate, however, that at least part of the controlling factor could be an increase in the amount and strength of turbulence in the atmosphere. Incidentally, Ballough (2007) notes that seatbelts play a crucial role in reducing accident rates, stating that in the period 1980–2003, only four people who were seated with seatbelts fastened received serious injuries during turbulence (excluding cases of other people falling onto and injuring properly secured occupants).

Similar data on accident rates from turbulence were analyzed by Kauffmann (2002), except that in his case they were normalized by the number of flight hours rather than the number of flight departures. The use of this normalization makes the trend less pronounced, and consequently calculations by Kauffmann (2002) indicated a lack of statistical significance, although only data up to 1999 were analyzed. Neither the FAA study by Ballough (2007) nor the academic study by Kauffmann

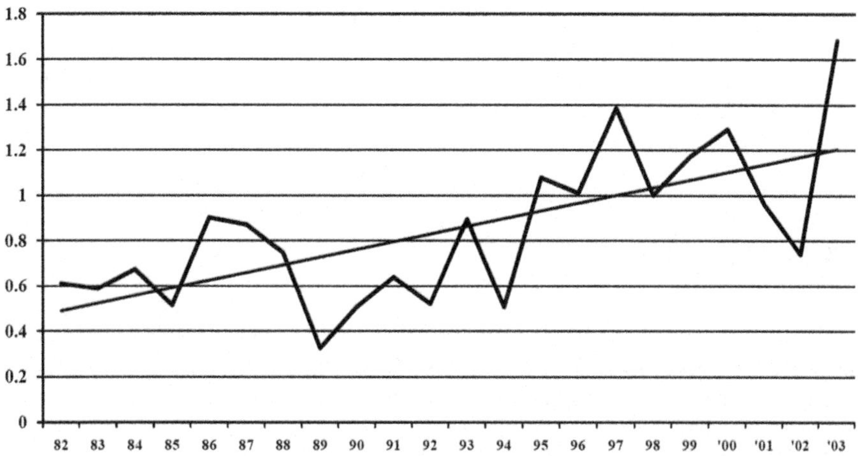

Fig. 23.5 Time series of the number of turbulence accidents per million flight departures. The data are for US air carriers, covering the 22-year period from 1982 to 2003. The data for each year are connected by *lines* and are over-plotted with a *straight line* showing the linear trend. From Ballough (2007)

(2002) appears to have been updated to bring the coverage period up to the present day, despite the present availability of at least a decade's worth of additional data.

23.3.5 Future Trends in Turbulence Diagnosed from Climate Models

The possible existence of historic climate-related trends in turbulence, whether they are inferred from PIREPs (Sect. 23.3.2), reanalysis data (Sect. 23.3.3), or passenger injuries (Sect. 23.3.4), naturally raises the question of what the future holds. To investigate the response of clear-air turbulence to future climate change, Williams and Joshi (2013) used computer simulations from the Geophysical Fluid Dynamics Laboratory (GFDL) CM2.1 coupled atmosphere–ocean model. Twenty years of daily-mean data were analyzed from two model integrations, in which the concentration of atmospheric CO_2 was held constant at its preindustrial level and twice its preindustrial level. The study focused on the 200 hPa pressure level within the busy North Atlantic flight corridor in winter.

Williams and Joshi (2013) used the climate model simulations to calculate a basket of 21 clear-air turbulence diagnostics, including the Richardson number, the relative vorticity advection, the residual of the nonlinear balance equation, the negative absolute vorticity advection, and the horizontal temperature gradient. The central finding of the analysis was that the statistics of the diagnosed clear-air turbulence change significantly when the concentration of carbon dioxide in the atmosphere is doubled. For example, in the doubled-CO_2 integration compared to the preindustrial integration, most of the 21 diagnostics showed a 10–40 % increase in the median strength of turbulence and a 40–170 % increase in the frequency of occurrence of moderate-or-greater turbulence. These results suggest that climate change will lead to bumpier transatlantic flights (along the northern corridor) by the middle of this century.

23.4 Discussion

This chapter has surveyed the current stock of scientific knowledge about the possible long-term trends in clear-air turbulence driven by anthropogenic climate change. Increases to the magnitudes of the jet stream wind shears in the midlatitude upper troposphere and lower stratosphere in each hemisphere are a robust expectation, in the sense that they are not only understood from basic physical principles but also simulated by comprehensive climate models. Given the documented association between wind shears and clear-air turbulence (e.g., Watkins and Browning 1973), it seems inevitable that the result of this process will be an increase in clear-air turbulence. The only study so far to produce a quantitative estimate of the future increase found that the volume of wintertime transatlantic airspace containing moderate-or-greater clear-air turbulence could double by the middle of this century (Williams and Joshi 2013).

Clearly, more work is needed to verify, refine, and extend these predictions and to quantify the range of uncertainty originating from model error and other sources. As this is a multidisciplinary problem, it will require aviation turbulence scientists and climate scientists to work together and collaborate across the usual disciplinary boundaries. It will also require research scientists in academia to collaborate with airline operators. In addition to clear-air turbulence, convective turbulence might also be affected by climate change, but we are not aware of any published research in this area. A key limitation holding back research progress is access to turbulence data collected by the airlines. We call on the airlines to roll out automated turbulence measurements across their fleets of aircraft. Furthermore, in these days of open access to scientific data, we call on the national aviation regulators to make the collected turbulence data available for academic research, to help ensure the long-term safety of air passengers and crew as the climate changes.

References

Arblaster, J., Meehl, G.: Contributions of external forcings to southern annular mode trends. J. Climate **19**, 2896–2905 (2006)

Ballough, J.J.: Advisory Circular 120-88A: Preventing Injuries Caused by Turbulence. U.S. Department of Transportation Federal Aviation Administration, 19 Nov 2007 (2007)

Bengtsson, L., Hagemann, S., Hodges, K.I.: Can climate trends be calculated from reanalysis data? J. Geophys. Res. **109**, D11111 (2004)

Bony, S., et al.: How well do we understand and evaluate climate change feedback processes? J. Climate **19**, 2445–3482 (2006)
Ciais, P., Sabine, C., Bala, G., Bopp, L., Brovkin, V., Canadell, J., Chhabra, A., DeFries, R., Galloway, J., Heimann, M., Jones, C., Le Quéré, C., Myneni, R.B., Piao, S., Thornton, P.: Carbon and other biogeochemical cycles. In: Stocker, T.F., Qin, D., Plattner, G.-K., Tignor, M., Allen, S.K., Boschung, J., Nauels, A., Xia, Y., Bex, V., Midgley, P.M. (eds.) Climate Change 2013: The Physical Science Basis. Contribution of Working Group I to the Fifth Assessment Report of the Intergovernmental Panel on Climate Change. Cambridge University Press, Cambridge (2013)
Collins, M., Knutti, R., Arblaster, J., Dufresne, J.-L., Fichefet, T., Friedlingstein, P., Gao, X., Gutowski, W.J., Johns, T., Krinner, G., Shongwe, M., Tebaldi, C., Weaver, A.J., Wehner, M.: Long-term climate change: projections, commitments and irreversibility. In: Stocker, T.F., Qin, D., Plattner, G.-K., Tignor, M., Allen, S.K., Boschung, J., Nauels, A., Xia, Y., Bex, V., Midgley, P.M. (eds.) Climate Change 2013: The Physical Science Basis. Contribution of Working Group I to the Fifth Assessment Report of the Intergovernmental Panel on Climate Change. Cambridge University Press, Cambridge (2013)
Delcambre, S.C., Lorenz, D.J., Vimont, D.J., Martin, J.E.: Diagnosing northern hemisphere jet portrayal in 17 CMIP3 global climate models: twenty-first-century projections. J. Climate **26**, 4930–4946 (2013)
Ellrod, G., Knapp, D.: An objective clear-air turbulence forecasting technique: verification and operational use. Weather Forecasting **7**, 150–165 (1992)
Fels, S.B., Mahlman, J.D., Schwarzkopf, M.D., Sinclair, R.W.: Stratospheric sensitivity to perturbations in ozone and carbon dioxide: radiative and dynamical response. J. Atmos. Sci. **37**, 2265–2297 (1980)
Friedlingstein, P., et al.: Climate-carbon cycle feedback analysis: results from the C4MIP model intercomparison. J. Climate **19**, 3337–3353 (2006)
Gill, P.G.: Objective verification of World Area Forecast Centre clear air turbulence forecasts. Meteorol. Appl. **21**, 3–11 (2014)
Harrison, R.G., Hogan, R.J.: In situ atmospheric turbulence measurement using the terrestrial magnetic field—a compass for a radiosonde. J. Atmos. Oceanic Technol. **23**, 517–523 (2006)
Harrison, R.G., Rogers, G.W., Hogan, R.J.: A three-dimensional magnetometer for motion sensing of a balloon-carried atmospheric measurement package. Rev. Sci. Instrum. **78**, 124501 (2007)
Hartmann, D.L., Klein Tank, A.M.G., Rusticucci, M., Alexander, L.V., Brönnimann, S., Charabi, F.J., Dentener, E.J., Dlugokencky, D.R., Easterling, A., Kaplan, B.J., Soden, P.W., Thorne, Y., Wild, M., Zhai, P.M.: Observations: atmosphere and surface. In: Stocker, T.F., Qin, D., Plattner, G.-K., Tignor, M., Allen, S.K., Boschung, J., Nauels, A., Xia, Y., Bex, V., Midgley, P.M. (eds.) Climate Change 2013: The Physical Science Basis. Contribution of Working Group I to the Fifth Assessment Report of the Intergovernmental Panel on Climate Change. Cambridge University Press, Cambridge (2013)
Hawkins, E., Sutton, R.: The potential to narrow uncertainty in regional climate predictions. Bull. Am. Meteorol. Soc. **90**, 1095–1107 (2009)
Jaeger, E.B., Sprenger, M.: A northern-hemispheric climatology of indices for clear air turbulence in the tropopause region derived from ERA40 re-analysis data. J. Geophys. Res. **112**, D20106 (2007)
Joshi, M., et al.: Projections of when temperature change will exceed 2°C above pre-industrial levels. Nat. Clim. Change **1**, 407–412 (2011)
Kauffmann, P.: The business case for turbulence sensing systems in the US air transport sector. J. Air Transport Manage. **8**, 99–107 (2002)
Kim, J.-H., Chun, H.-Y.: Statistics and possible sources of aviation turbulence over South Korea. J. Appl. Meteorol. Climatol. **50**(2), 311–324 (2011)
Le Quéré, C., et al.: The global carbon budget 1959–2011. Earth Syst. Sci. Data **5**, 165–186 (2013)
Lorenz, D.J., DeWeaver, E.T.: Tropopause height and zonal wind response to global warming in the IPCC scenario integrations. J. Geophys. Res. **112**, D10119 (2007)

Marlton, G.J., Harrison, R.G., Nicoll, K.A., Williams, P.D.: A balloon-borne accelerometer technique for measuring atmospheric turbulence. Rev. Sci. Instrum. **86**, 016109 (2015)
Meinshausen, M., et al.: Greenhouse-gas emission targets for limiting global warming to 2°C. Nature **458**, 1158–1162 (2009)
Myhre, G., Shindell, D., Bréon, F.-M., Collins, W., Fuglestvedt, J., Huang, J., Koch, D., Lamarque, J.-F., Lee, D., Mendoza, B., Nakajima, T., Robock, A., Stephens, G., Takemura, T., Zhang, H.: Anthropogenic and natural radiative forcing. In: Stocker, T.F., Qin, D., Plattner, G.-K., Tignor, M., Allen, S.K., Boschung, J., Nauels, A., Xia, Y., Bex, V., Midgley, P.M. (eds.) Climate Change 2013: The Physical Science Basis. Contribution of Working Group I to the Fifth Assessment Report of the Intergovernmental Panel on Climate Change. Cambridge University Press, Cambridge (2013)
Schwartz, B.: The quantitative use of PIREPs in developing aviation weather guidance products. Weather Forecasting **11**, 372–384 (1996)
Sharman, R., Tebaldi, C., Wiener, G., Wolff, J.: An integrated approach to mid- and upper-level turbulence forecasting. Weather Forecasting **21**(3), 268–287 (2006)
Sharman, R.D., Cornman, L.B., Meymaris, G., Pearson, J., Farrar, T.: Description and derived climatologies of automated in situ eddy dissipation rate reports of atmospheric turbulence. J. Appl. Meteorol. Climatol. **53**, 1416–1432 (2014)
Taylor, P.C., Cai, M., Hu, A., Meehl, J., Washington, W., Zhang, G.J.: A decomposition of feedback contributions to polar warming amplification. J. Climate **26**, 7023–7043 (2013)
Van Vuuren, D.P., et al.: The representative concentration pathways: an overview. Clim. Change **109**, 5–31 (2011)
Watkins, C.D., Browning, K.A.: The detection of clear air turbulence by radar. Phys. Technol. **4**, 28–61 (1973)
Weaver, A., et al.: Stability of the Atlantic meridional overturning circulation: a model intercomparison. Geophys. Res. Lett. **39**, L20709 (2012)
Williams, P.D., Joshi, M.M.: Intensification of winter transatlantic aviation turbulence in response to climate change. Nat. Clim. Change **3**, 644–648 (2013)
Wolff, J.K., Sharman, R.D.: Climatology of upper-level turbulence over the continental United States. J. Appl. Meteorol. Climatol. **47**, 2198–2214 (2008)

Chapter 24
Application of Aviation Turbulence Information to Air-Traffic Management (ATM)

Jung-Hoon Kim, William N. Chan, and Banavar Sridhar

Abstract Unexpected turbulence, especially in the upper troposphere and lower stratosphere where cabin crews and passengers in cruising aircraft are likely to be unbuckled, causes in-flight injuries, structural damage, and flight delays. Therefore, turbulence information can be used to improve safety while pursuing efficiency in the air-traffic management (ATM). In this chapter, an optimal flight path that minimizes both total flight time (e.g., fuel consumption) and potential encounters of turbulence from departure to arrival airports is derived by combining simple modeling of aircraft flight trajectories with wind and turbulence predictions. In addition, probabilistic ensemble turbulence forecasts, evaluated against in situ eddy dissipation rate turbulence observations from commercial aircraft, are applied to suggest an optimal strategic and tactical ATM route planning for given weather and turbulence conditions in the USA. The variations of long-haul transoceanic flight routes and their turbulence potentials are also investigated using global reanalysis data to understand how the upper-level large-scale flow patterns can affect the long-term ATM planning through the changes of winds and turbulence conditions.

J.-H. Kim (✉)
Cooperative Institute for Research in the Atmosphere/Colorado State University (CIRA/CSU) in affiliation with Aviation Weather Center/National Centers for Environmental Prediction/ National Weather Service/National Oceanic and Atmospheric Administration (AWC/NCEP/ NWS/NOAA), 7220 NW 101st Terrace, Kansas City, MO 64153-2317, USA

NASA Ames Research Center/Oak Ridge Associated Universities (ORAU), Moffett Field, CA, USA
e-mail: jung-hoon.kim@noaa.gov; jhkim99@me.com

W.N. Chan • B. Sridhar
NASA Ames Research Center, Moffett Field, CA, USA
e-mail: william.n.chan@nasa.gov; banavar.sridhar@nasa.gov

R. Sharman, T. Lane (eds.), *Aviation Turbulence*,
DOI 10.1007/978-3-319-23630-8_24

24.1 Introduction

This chapter addresses how turbulence information can be used to improve safety while pursuing improved efficiency in air-traffic management (ATM) operations. In ATM applications, turbulence predictions need to be useful for tactical (0–2-h lead time; also known as nowcasting), strategic (2–8-h lead time; midterm forecasting), and long-term (>8 h lead time) flight planning based on different user demands. The tactical plan is usually necessary when an aircraft is flying along its planned route and needs to make a near-term decision to mitigate potential turbulence encounters ahead. The strategic plan is required to create a weather-related flight plan before departure using a midterm forecast. The long-term plan may be necessary for longer flights, such as trans-oceanic or trans-continent flights, since the optimal route minimizing both fuel consumption (i.e., total flight time) and turbulence potential depends on the large-scale flow patterns, particularly upper-level jet position and strength.

Several researchers have developed strategies for determining optimal flight paths using wind information, called wind-optimal routes (WORs), for the ATM. Ng et al. (2012) developed WORs that minimized total fuel burned by computing minimum-time routes in the presence of winds on multiple flight levels. Palopo et al. (2010) conducted a simulation of WORs and the impact on sector loading, conflicts, and airport arrival rates. And Jardin and Bryson (2012) computed minimum-time flight trajectories using analytical neighboring WOR in the presence of a strong jet stream with winds of up to 80 m s^{-1}. However, these studies did not include turbulence information.

Prior research shows pilots seek to avoid areas of turbulence, and the impact of these avoidance maneuvers on ATM has been well documented. Krozel et al. (2011) studied the maneuvers pilots made when encountering clear-air turbulence (CAT) and showed that the pilot's response to CAT depended on factors such as aircraft type and company policies. Furthermore, since research shows two-thirds of all severe CAT occurs near the jet stream (Lester 1994), ignoring CAT near a jet stream to achieve minimum-time routes may result in fuel savings that cannot be fully realized due to a pilot's unwillingness to traverse turbulent areas to reach the maximum tail winds. Turbulence information can also aid in the development of routes around convective systems. Ng et al. (2009) calculated convective weather avoidance routes considering the probability of pilot deviation using model-based radar data. The model used by them and others, the Convective Weather Avoidance Model (CWAM), uses ground-based radar information to determine areas pilots will likely avoid (Delaura and Evans 2006). CWAM is currently used by NASA's Dynamic Weather Routing tool to create in-flight routing around convective weather and has been evaluated in field studies in collaboration with American Airlines (McNally et al. 2012). Such a model, however, can miss regions of convectively induced turbulence (CIT) outside of convective clouds.

The following sections describe the modeling of aircraft trajectories and application of turbulence information for tactical, strategic, and long-term flight plans. Section 24.2 describes optimization of flight routes in the presence of wind by minimizing the cost function. In Sect. 24.3, examples of strategic and tactical flight plans that minimize both total flight time and potential turbulence encounters for a specific day over the contiguous USA (CONUS), calculated from the ensemble of numerical weather prediction (NWP) model outputs and turbulence diagnostics, are presented. Deviations of long-haul trans-Atlantic flight trajectories and turbulence potentials estimated from the global reanalysis data over the Atlantic Ocean are investigated during two winter seasons that have distinct upper-level weather patterns in Sect. 24.4. Summary and conclusions are discussed in Sect. 24.5.

24.2 Modeling of Aircraft Trajectory

When an aircraft is flying horizontally above the Earth's surface with a true airspeed (V_t) and heading angle (α) during a certain period of time (Δt), as shown in Fig. 24.1, the longitudinal (λ) and latitudinal (ϕ) position changes of the aircraft with the time in the presence of horizontal winds are governed by the following aircraft-motion equations (e.g., Sridhar et al. 2010; Hok et al. 2011; Ng et al. 2011; Kim et al. 2015):

$$\frac{d\lambda(t)}{dt} = \frac{V_t \cos\alpha(t) + u(\lambda, \phi, z)}{R\cos\phi(t)}, \tag{24.1}$$

$$\frac{d\phi(t)}{dt} = \frac{V_t \sin\alpha(t) + v(\lambda, \phi, z)}{R}. \tag{24.2}$$

Here, R is the Earth's radius (the Earth is assumed to be a sphere), z is height above the surface and $R \gg z$, and u and v are the zonal and meridional wind components, respectively. In this literature, it is assumed that V_t is a constant of 250 m s^{-1} once the heading angle (α) is chosen at a given time step.

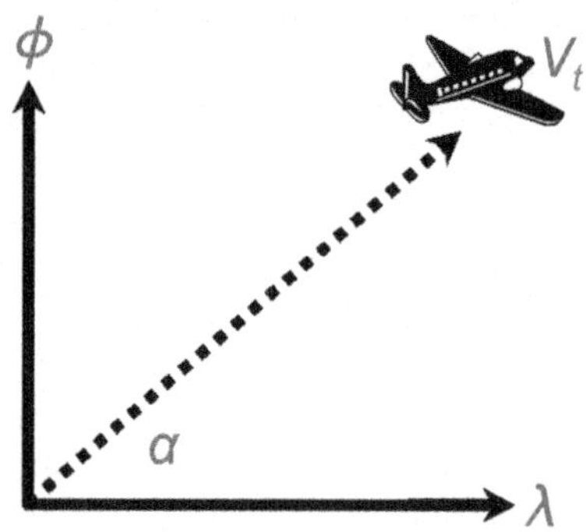

Fig. 24.1 Schematic figure for an aircraft flying horizontally on an Earth's surface with a true airspeed (V_t) and heading angle (α) during a certain period of time (Δt). Here, λ and ϕ are longitudinal and latitudinal directions, respectively

24.2.1 Great Circle Route in the Presence of Winds

To determine the aircraft heading angle (α) at each time step from the departure to the arrival airport in Eqs. (24.1) and (24.2), it is assumed the aircraft follows a great circle route (GCR), which represents the shortest distance between two points on a sphere. In this case, the heading angle between the aircraft position at a given time and the destination is expressed as

$$\alpha(t) = \pi/2 - \tan^{-1}\left\{ \sin\left(\lambda_f - \lambda_t\right)\cos\phi_f,\ \cos\phi_t \sin\phi_f - \sin\phi_t \cos\phi_f \cos\left(\lambda_f - \lambda_t\right)\right\}. \tag{24.3}$$

Here the subscripts t and f denote values at a given time step and the final destination. Note the angle has been converted from the forward azimuth (clockwise from North) to the counterclockwise from East as shown in Fig. 24.1. Starting from the departure airport location, we solve the Eqs. (24.1) and (24.2) using an explicit Euler forward integration scheme, $\left[y(t+1) = y(t) + \Delta t \frac{dy(t)}{dt},\ \text{where}\, y = \lambda\, \text{and}\, \phi\right]$, using the updated information of the great circle heading angle calculated by Eq. (24.3) at each time step. This leads to the GCR trajectory with winds. Note that this trajectory is not the minimum-time path from departure to arrival, because at each time step determination of the great circle heading angle (Eq. 24.3) is independent of the wind. An example will be discussed in Sect. 24.3.

24.2.2 Wind-Optimal Route

In order to maximize the advantage of a tail wind and/or minimize the disadvantage of a head wind in the modeling of an aircraft trajectory, we need to take into account wind variations in the calculation of the heading angle (α) at each time step, which eventually minimizes the total travel time from departure to destination. The minimum-time path in the presence of wind [i.e., wind-optimal route (WOR)] can be obtained by applying Pontryagin's minimum principle (Bryson and Ho 1975) to determine the analytic solution for the control parameter (here, the heading angle of a cruising aircraft, α) that minimizes the cost function (J) defined by

$$J = \int_{t_0}^{t_f} C_t\, dt. \tag{24.4}$$

Here, C_t is the cost coefficient of travel time, and t_0 and t_f are the times at departure and arrival airports, respectively. The analytic solution for the control parameter of heading angle (α) that takes into account the variations of the winds and minimizes the cost function in Eq. (24.4) is

$$\frac{d\alpha(t)}{dt} = -\frac{F_{\text{wind}}(t)}{C_t R \cos\phi(t)}, \text{ where}$$

$$\begin{aligned} F_{\text{wind}}(t) = &- \sin\alpha(t)\cos\alpha(t)\frac{\partial u(\lambda,\phi,z)}{\partial\lambda} + \cos^2\alpha(t)\sin\phi(t)\,u(\lambda,\phi,z) \\ &+ \cos^2\alpha(t)\cos\phi(t)\frac{\partial u(\lambda,\phi,z)}{\partial\phi} - \frac{\partial v(\lambda,\phi,z)}{\partial\lambda} \\ &+ \sin\alpha(t)\cos\alpha(t)\sin\phi(t)\,v(\lambda,\phi,z) \\ &+ \cos\alpha(t)\sin\alpha(t)\cos\phi(t)\frac{\partial v(\lambda,\phi,z)}{\partial\phi} + V_t\cos\alpha(t)\sin\phi(t) \\ &+ \cos^2\alpha(t)\frac{\partial v(\lambda,\phi,z)}{\partial\lambda}. \end{aligned} \tag{24.5}$$

A full derivation of the analytic solution in Eq. (24.5) can be found in previous studies (e.g., Sridhar et al. 2010; Hok et al. 2011; Kim et al. 2015). Next, a shooting method is used to find the optimal initial condition (i.e., heading angle). First, the great circle heading angle (α_{GC}) between the departure and arrival airports, calculated by Eq. (24.3), is used as the first guess for the initial heading angle [$\alpha(t_0)$]. Then, Eqs. (24.1), (24.2), and (24.5) are solved using the explicit Euler forward integration scheme from the departure to the destination. Here, there are two termination conditions for the WOR modeling: (1) the minimum distance between the trajectory and final destination is smaller than 100 km or (2) the distance between the trajectory and initial departure is greater than 1.2× total great circle distance between the departure and arrival destination. This process is iterated with different initial heading angles [$\alpha(t_0)$] between the boundaries of $\alpha_{GC}+90°$ and $\alpha_{GC}-90°$, using an increment of 0.25°. Then, we pick up the candidate trajectories that satisfy the termination condition of (1) the minimum distance. Finally, among these, the one trajectory that arrives at the destination faster than the others is chosen as the WOR.

Figure 24.2 shows an example of the WOR for eastbound (WOREB) and westbound (WORWB) at 250 hPa level (about $z=11$ km) between the John F. Kennedy International Airport (JFK) in New York, USA, and London Heathrow International Airport (LHR) in London, UK, for 18 UTC, 3 January 2005. In this example, the time step (Δt) = 180 s (3 min) and $C_t = 1$ in Eq. (24.5). WOREB (gray lines in Fig. 24.2 upper left) and WORWB (gray lines in Fig. 24.2 upper right) trajectories reach different regions according to the initial heading angles selected [$\alpha(t_0)$] in a given wind situation, which corresponds the minimum distance between each trajectory and the destination (Fig. 24.2 lower). The fastest one to the destination has been picked up as the WOREB (bold black line in Fig. 24.2 upper left) and WORWB (bold black line in Fig. 24.2 upper right) for this wind condition. In this case, the WOREB keeps following the strong westerly and southwesterly jet over North Atlantic Ocean to maximize its tail wind (Fig. 24.2 upper left), while the

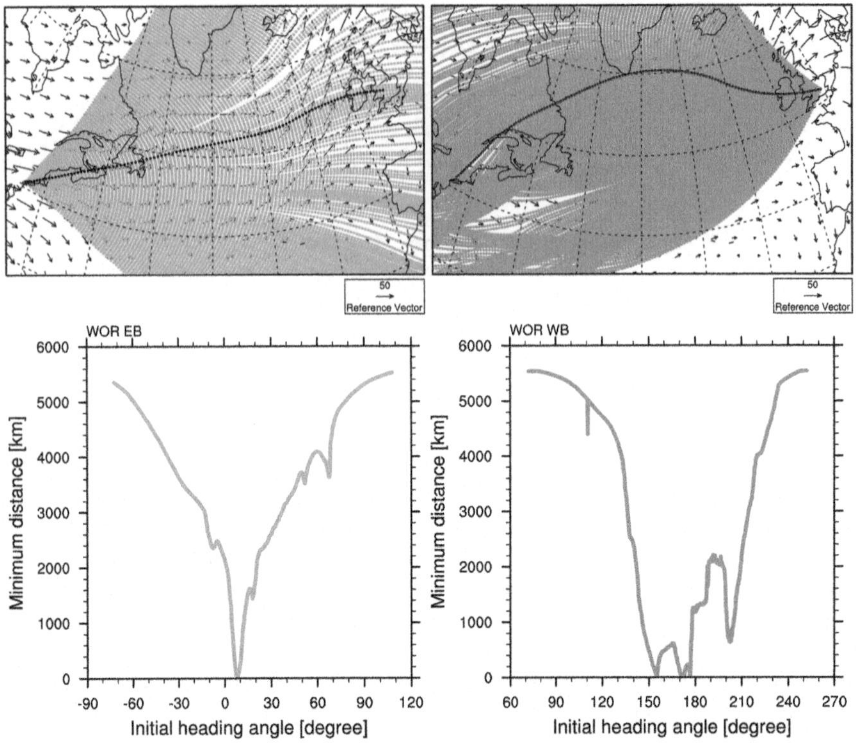

Fig. 24.2 (*Upper*) Trajectories (*gray lines*) for the eastbound (*left*) and westbound (*right*) wind-optimal route (WOR) between John F. Kennedy International Airport (JFK) and London Heathrow International Airport (LHR) with 360 different initial heading angles from the great circle heading angle (α_{GC}) $-90°$ to α_{GC} $+90°$ with 0.25° bins in the presence of wind at 18 UTC, 3 January 2005, and (*lower*) the corresponding minimum distance between the trajectories and destination airport. Reference wind vectors on *bottom right* in *upper panel* are 50 m s^{-1}. The optimal flight routes for eastbound and westbound having the minimum time and distance are depicted as the black lines in *upper-left* and *upper-right* plots

WORWB detours northward near the Southern tip of Greenland to avoid the prevailing westerly jet flow and minimize head winds.

24.2.3 Lateral Turbulence Avoidance Route

The lateral turbulence avoidance route (LTAR) can be determined by following the same approach for the optimization of the WOR in the previous section, but with a different minimization cost function (J):

$$J = \int_{t_0}^{t_f} \{C_t + C_r r(\lambda, \phi, z)\} dt. \tag{24.6}$$

Here, C_t and C_r are the cost coefficients of travel time and turbulence penalty areas along the LTAR, respectively. In this study, $r(\lambda, \phi, z)$ is nonzero (=1) only when the turbulence potential is at or above a chosen threshold. For example, in locations where the probabilistic ensemble turbulence forecast for severe turbulence is greater than 10 %, or the forecasted eddy dissipation rate (EDR) value is greater than 0.4 $m^{2/3}\ s^{-1}$, $r(\lambda, \phi, z) = 1$; elsewhere $r(\lambda, \phi, z) = 0$. Then, the analytic solution for the heading angle taking into account winds and turbulence becomes

$$\begin{aligned}
\frac{d\alpha(t)}{dt} &= -\frac{\{F_{\mathrm{wind}}(t) + F_{\mathrm{turb}}(t)\}}{R \cos\phi(t)\{C_t + C_r r(\lambda, \phi, z)\}}, \text{ where} \\
F_{\mathrm{turb}}(t) &= -\sin\alpha(t)\cos\alpha(t)\sin\phi(t)\, v(\lambda, \phi, z) C_r r(\lambda, \phi, z) \\
&\quad + \cos\phi(t)\cos\alpha(t)\sin\alpha(t) \frac{\partial v(\lambda, \phi, z)}{\partial \phi} C_r r(\lambda, \phi, z) \\
&\quad - \cos\phi(t)\cos\alpha(t)\sin\alpha(t)\, v(\lambda, \phi, z) C_r \frac{\partial r(\lambda, \phi, z)}{\partial \phi} \\
&\quad + V_t \cos\alpha(t)\sin\phi(t) C_r r(\lambda, \phi, z) + V_t \sin\alpha(t) C_r \frac{\partial r(\lambda, \phi, z)}{\partial \lambda} \\
&\quad - \frac{\partial v(\lambda, \phi, z)}{\partial \lambda} C_r r(\lambda, \phi, z) + v(\lambda, \phi, z) C_r \frac{\partial r(\lambda, \phi, z)}{\partial \lambda} \\
&\quad - \sin\alpha(t)\cos\alpha(t) \frac{\partial u(\lambda, \phi, z)}{\partial \lambda} C_r r(\lambda, \phi, z) \\
&\quad + \sin\alpha(t)\cos\alpha(t)\, u(\lambda, \phi, z) C_r \frac{\partial r(\lambda, \phi, z)}{\partial \lambda} \\
&\quad + \cos^2\alpha(t)\sin\phi(t)\, u(\lambda, \phi, z) C_r r(\lambda, \phi, z) \\
&\quad + \cos^2\alpha(t)\cos\phi(t) \frac{\partial u(\lambda, \phi, z)}{\partial \phi} C_r r(\lambda, \phi, z) \\
&\quad - V_t \cos\phi(t)\cos\alpha(t) C_r \frac{\partial r(\lambda, \phi, z)}{\partial \phi} \\
&\quad - \cos\phi(t)\cos^2\alpha(t)\, u(\lambda, \phi, z) C_r \frac{\partial r(\lambda, \phi, z)}{\partial \phi} \\
&\quad + \cos^2\alpha(t) \frac{\partial v(\lambda, \phi, z)}{\partial \lambda} C_r r(\lambda, \phi, z) \\
&\quad - \cos^2\alpha(t)\, v(\lambda, \phi, z) C_r \frac{\partial r(\lambda, \phi, z)}{\partial \lambda}.
\end{aligned} \tag{24.7}$$

Solving Eqs. (24.1), (24.2), and (24.7) using the same integration and shooting methods described in Sect. 24.2.2 for the WOR leads to the LTAR. This LTAR requires information regarding the expected atmospheric turbulence, which is discussed in the next section. A full derivation of the analytic solution in Eq. (24.7) can also be found in previous studies (e.g., Sridhar et al. 2010; Hok et al. 2011; Kim et al. 2015), and LTAR examples will be shown in Fig. 24.4 in Sect. 24.3.

24.3 Example of Turbulence Application to ATM

To address the lack of turbulence information in WOR-based ATM applications, especially for the tactical and strategic plans, a predictive model of aviation-scale turbulence, such as the Graphical Turbulence Guidance (GTG) product (Sharman et al. 2006; Kim et al. 2011, 2015) in which an ensemble of turbulence diagnostics are computed, can be used to modify the ATM plans. From a meteorological perspective, small-scale turbulent eddies that directly affect commercial aircraft at cruising altitudes are generated by a number of possible sources. Well-known turbulence generation mechanisms near an upper-level jet/frontal system include strong vertical shears above and below a jet stream core, inertial instability due to anticyclonic shear and curvature flow, and the gravity wave emissions via geostrophic adjustment in the jet stream exit region (e.g., Lane et al. 2004; Kim and Chun 2010, 2011; Knox et al. 2008; Sharman et al. 2012). Mountain wave breaking frequently causes aviation turbulence over complex terrain (e.g., Lane et al. 2009; Sharman et al. 2011, 2012). Flow deformation, gravity wave breaking, and thermal-shear instability near convective systems are also important sources for aviation turbulence (e.g., Lane et al. 2003, 2012; Lane and Sharman 2008, 2014; Kim and Chun 2012; Kim et al. 2014; Trier and Sharman 2009; Trier et al. 2010; Sharman et al. 2012). To take into account these turbulence generation mechanisms, as well as uncertainties in the NWP model forecasts, a combination of several turbulence metrics from different mechanisms and from different forecasts is essential and is more reliable than using a single diagnostic or simple rule-of-thumb predictor (e.g., Sharman et al. 2006; Kim et al. 2011; Gill 2014; Gill and Stirling 2013). Steiner et al. (2010) reviewed ensemble-based forecasting techniques for ATM and stated that ensemble forecasting can be applied to turbulence. They also pointed out that probabilistic forecasts are appropriate for strategic ATM planning, as they may provide guidance about the uncertainty associated with weather-related phenomena.

In this section, three time-lagged ensemble NWP forecasts are used to derive ensembles of ten turbulence diagnostics to provide probabilistic information about turbulence likelihood. In order to better predict the effects of convection, as well as provide better representation of mountain wave and clear-air turbulence sources, a high-resolution (3 km horizontal grid spacing) NWP model is implemented. Further, each computed turbulence diagnostic is scaled to the energy dissipation rate

(EDR = $\varepsilon^{1/3}$ $m^{2/3}$ s^{-1}) as an aircraft-independent atmospheric turbulence metric (e.g., Cornman et al. 1995; Sharman et al. 2014). EDR is defined as the rate of the turbulent kinetic energy (TKE) transfer from large-scale to small-scale eddies. Large-scale atmospheric eddies can cascade down to smaller-scale eddies until the viscous dissipation becomes dominant and the TKE is converted to heat. The model-derived EDR metric is consistent with in situ EDR estimates currently available from several fleets of commercial airliners including B767s, B757s, and B737s (Cornman et al. 1995; Sharman et al. 2014), which is convenient for forecast verification. The in situ EDR metric can be related to traditional turbulence intensity based on pilot-reported categories of "light (LGT)," "moderate (MOD)," and "severe (SEV)" by appropriate consideration of aircraft type and flight conditions (Sharman et al. 2014). For reasons discussed in Sharman et al. (2014), EDR is the preferred atmospheric turbulence unit for aviation-scale observations and forecasts. For a given valid time, a three-dimensional probabilistic ensemble for severe-or-greater (SOG)-level turbulence areas is calculated by counting how many EDR-scaled individual turbulence diagnostics out of the total 30 diagnostics have EDR values $\geq$0.47 $m^{2/3}$ s^{-1} at each grid point in the model, as shown in Figs. 24.3 and 24.4. Here, the turbulence diagnostics selected are the top 10 diagnostics listed in the operational GTG product for upper levels (e.g., Sharman et al. 2006), and the EDR threshold for SOG level is adapted from the median value of in situ EDR-severe PIREP pairs for longer period over the CONUS (Sharman et al. 2014).

An example eastbound WOR calculated using Eqs. (24.1), (24.2), and (24.5), i.e., including winds but without considering turbulence effects, from Los Angeles International Airport (LAX) to John F. Kennedy International Airport (JFK) is plotted over contours of the probabilistic ensemble EDR for SOG at 35,000 ft (FL350; about 200 hPa level) in Fig. 24.3 (bold black line in upper panel). The probabilistic ensemble EDR uses 3.5–5.5 h forecasts valid at 1730 UTC 7 September 2012. The corresponding vertical cross section is shown in the lower panel (Fig. 24.3). In this case, a flight cruising at FL350 along the WOR would take 238 min (3 h 58 min), which is 2 min less than the elapsed time along the GCR with wind using Eqs. (24.1–24.3) (gray line in Fig. 24.3 upper panel). In Fig. 24.3, however, the WOR would experience a total of 52 min flying in areas with $\geq$10 % probability of encountering SOG-level turbulence over northern Indiana, Ohio, and western Pennsylvania. In the vertical cross section along this WOR (Fig. 24.3 lower panel), the 10 % SOG-level turbulence areas (dark gray) appear to block all possible flight levels from FL260 to FL450 over these regions. This indicates that lateral avoidance turbulence routes (LTARs) would be better suited to avoid turbulence than vertical changes of the flight level in this case.

To demonstrate quantitatively the effects of lateral turbulence avoidance on the WOR routes, we use a probabilistic ensemble EDR $\geq$ 10 % probability of encountering SOG-level turbulence for the calculation of the LTAR using Eqs. (24.1), (24.2), and (24.7), where C_t and $C_r = 1$ and $\Delta t = 60$ s (1 min). The choice of the 10 % SOG probability threshold is arbitrary. But, this is selected because in the aviation community avoiding SOG turbulence is regarded as a hard constraint that should be avoided for safety, while any lower thresholds are a soft constraint that

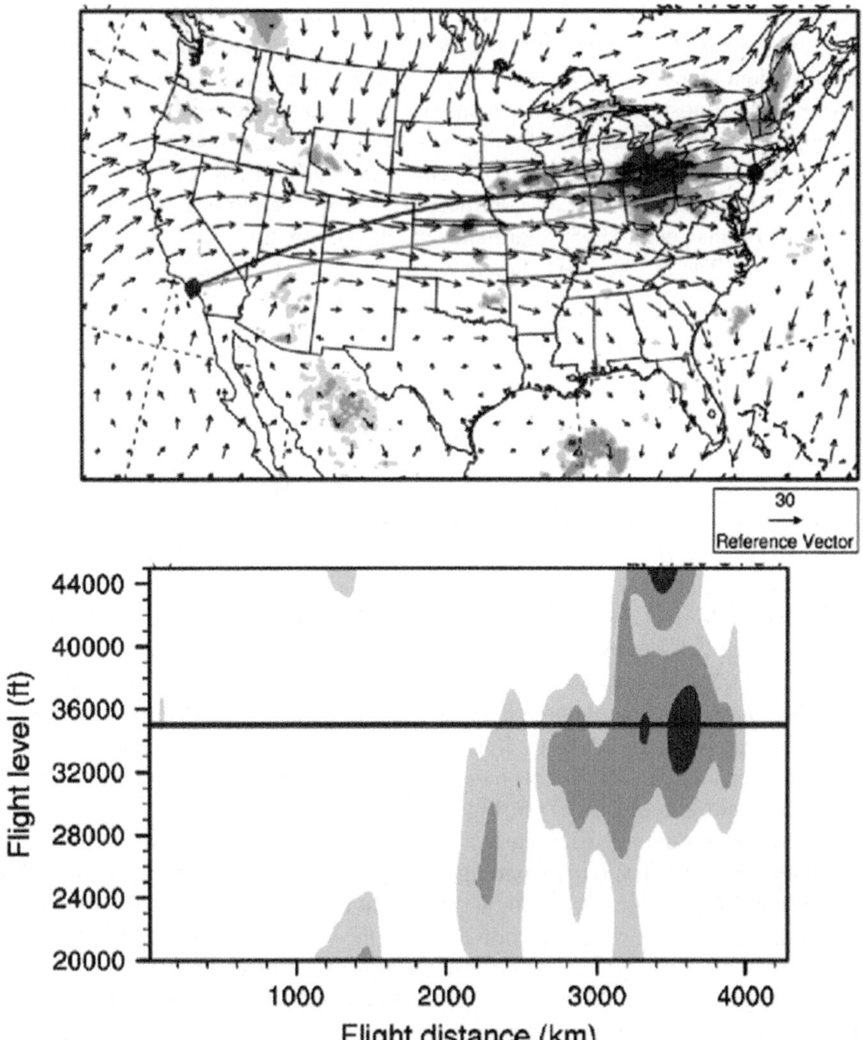

Fig. 24.3 (*Upper*) Probabilistic ensemble forecast for severe-level turbulence (3 %, *light gray*; 10 %, *dark gray*; and 30 %, *black shadings*) with horizontal wind vectors and eastbound wind-optimal route (WOR, *bold black line*) and great circle route with winds (GCR, *gray line*) at a flight level of 35,000 ft from the Los Angeles International Airport (LAX) to John F. Kennedy International Airport (JFK) using 3.5–5.5 h forecasts valid at 1730 UTC, 9 September 2010, and (*lower*) the corresponding vertical (X-Z) cross sections of the severe-level probabilistic turbulence forecast along the WOR. In the *upper panel*, reference wind vector on the *bottom right* is 30 m s^{-1}, and locations of departure (LAX) and arrival (JFK) are also depicted as *dots*

aircraft may penetrate by employing the fasten seatbelt sign. In addition, considering that the background (natural) probability for the SOG-level turbulence encounters in upper troposphere and lower stratosphere (UTLS) is less than 0.1 %

(Sharman et al. 2006, 2014), the forecasted 10 % SOG-level turbulence probability (dark gray shading in Fig. 24.3) is a significantly higher value than the background SOG-level turbulence potential in UTLS.

The LTAR can be initiated at departure as a strategic plan; however, it would be preferable to delay such a maneuver until closer to the forecasted turbulence constraint, because the maneuver decision needs to consider several factors like confidence of the weather forecast in the tactical plan. Therefore, first, in Fig. 24.4 (upper panel), the LTAR trajectory for the 10 % SOG-level turbulence potential using 3.5–5.5 h forecasts initiated from the departure (LAX) is depicted as a red line. The LTAR (red line) takes a total of 254 min flying time and used 6.7 % extra time to entirely avoid the forecasted 10 % SOG-level turbulence areas. From the strategic point of view, this means laterally detouring around these potential areas of the turbulence from the departure airport (LAX), an aircraft would incur 16 min (6.7 %) more travel time to fly to its destination (JFK) (LTAR 1 in Table 24.1). However, two other alternative LTARs were initiated 1.5 h (middle) and 2.5 h (lower) after departing LAX along the WOR (blue lines) with more recently updated forecast data, which is useful for the tactical plan. An aircraft that follows the LTAR 1.5 h after departing LAX (in Fig. 24.4 middle) has a total flying time of 244 min, which saves 10 min more than that if it were to follow the LTAR initiated from LAX (red line in Fig. 24.4 upper). Delaying the horizontal maneuver would result in a savings of 10 min if the maneuver were delayed 1.5 h after leaving LAX (LTAR 2 in Table 24.1). Here, the 10 min time savings can be very significant because this reduction roughly equates to about 160 km less flying distance and about 760 kg of fuel savings, which is a benefit for commercial airline operations. However, if an aircraft follows an LTAR 2.5 h after departing LAX, when it is closer to more recently forecasted turbulence regions (in Fig. 24.4 lower and LTAR 3 in Table 24.1), it takes a total of 256 min of flying time. This is 2 min longer than the LTAR initiated from the departure (red line in Fig. 24.4 upper). Therefore, in this case, the most efficient LTAR is the one that begins its lateral detour 1.5 h after the departure (in Fig. 24.4 middle and LTAR 2 in Table 24.1). This takes 244 min from LAX to JFK in a given weather condition, avoiding entirely all areas of SOG probability >10 %. It is noted that the example of LTARs shown in Fig. 24.4 may not be the most efficient maneuver, because there are several other ways to avoid the potential constraints of turbulence, such as tactical change of flight altitude and route just ahead of turbulence areas.

24.4 Long-Term Variations of the Aircraft Trajectories and Turbulence Potentials

For the long-haul ATM flight plans like trans-oceanic flights, modeling of aircraft trajectories with winds may rely upon the prevailing jet stream position and strength. And the turbulence potential along these trajectories also highly depends

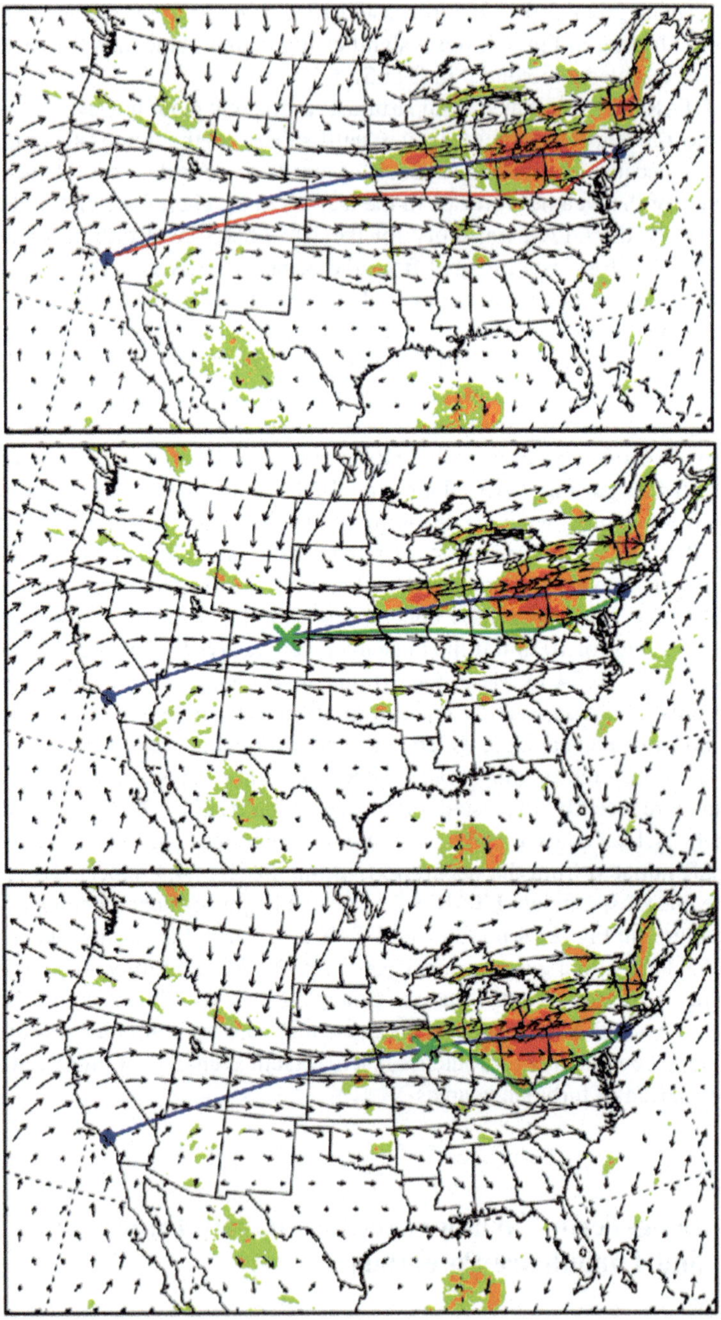

Fig. 24.4 (*Upper*) Probabilistic ensemble forecast for severe-level turbulence (3 %, *green*; 10 %, *orange*; and 30 %, *red shadings*) with horizontal wind vectors and eastbound wind-optimal route (WOR, *blue line*) and lateral turbulence avoidance route (LTAR, *red line*) at a flight level of 35,000 ft from the Los Angeles International Airport (LAX) to John F. Kennedy International Airport (JFK) using 3.5–5.5 h forecasts valid at 1730 UTC, 9 September 2010. *Middle* and *lower*

upon these weather conditions, because local gradients of meteorological variables like horizontal and vertical wind and temperature are generally large within the jet stream (e.g., Jaeger and Sprenger 2007; Williams and Joshi 2013; Kim et al. 2016). In addition, minimizing a head wind in the trajectory modeling can also deviate an aircraft toward the regions like Greenland where mountain waves and associated turbulence are common (e.g., Lane et al. 2009; Sharman et al. 2011). These create a challenge in deciding to maneuver a long-haul aircraft in the hopes of improving safety that may result in higher fuel use. Therefore, variations of the trans-oceanic trajectories and turbulence potentials along these routes are necessary for clustering the large-scale weather patterns responsible for the desired long-term flight route planning.

In this section, variations of the GCRs with winds and WORs and their turbulence potentials between the JFK and LHR are investigated during two distinct winter seasons [December 2004–February 2005 (DJF04-05) and December 2009–February 2010 (DJF09-10)]. The two seasons are selected because of distinct differences in the prevailing westerly flow and jet stream over the Northern Atlantic Ocean. For example, in Fig. 24.5 (upper left), the period of DJF04-05 was in an extremely positive phase of the Northern Atlantic Oscillation (+NAO; Barnston and Livezey 1987). Here the prevailing westerly winds embedded in the jet stream over the Northern Atlantic Ocean were anomalously strong and directed toward the UK and Northwestern Europe, as shown by the thick and long white arrow. This was due to enhanced meridional pressure gradients between the anomalous high-pressure system in the Central Atlantic Ocean and the anomalous low pressure system in Southern Greenland causing strong zonal winds via the geostrophic balance (e.g., Irvine et al. 2013). On the other hand, during DJF09-10 (Fig. 24.5 upper right), the prevailing westerly flow and jet stream are weak and directed toward Southern Europe and the Mediterranean Sea, as shown by the thin and narrow white arrow, due to the weak low- and high-pressure systems over the Atlantic Ocean (negative phase of the NAO).

Figure 24.5a–d (middle and lower panels) show the variability of the eastbound (light green lines) and westbound (dark green lines) GCRs and the eastbound (blue lines) and westbound (red lines) WORs during DJF 04-05 (left) and DJF 09-10 (right). For these plots, modeling of the GCRs with winds and WORs are calculated by Eqs. (24.1–24.3) and (24.5) at 250 hPa level with $V_t = 250\ \mathrm{m\ s^{-1}}$ and $\Delta t = 180$ s from the JFK (LHR) to LHR (JFK) using 6-h Modern Era Retrospective Analysis for Research and Application (MERRA) reanalysis data with 1/2° and 2/3° horizontal grid spacing. The eastbound and westbound trajectories are launched at 0000 (1800) UTC each day during these periods (e.g., Irvine et al. 2013).

Fig. 24.4 (continued) panels are the same as *upper panel* except for the LTARs (*green lines*) initiated after 1.5 h (*middle*) and 2.5 h (*lower*) departing from LAX along the WOR (*blue lines*) between LAX to JFK with winds using 2.5–4.5 h forecasts (*middle*) and using 1.5–3.5 h forecasts (*lower*) valid at 1730 UTC, 9 September 2010

Table 24.1 Minutes of the total travel time (left column), additional flight time along the LTAR compared to wind-optimal route (middle column), and flight time in areas of SOG probability >10 % along the LTARs from the Los Angeles International Airport (LAX) to John F. Kennedy International Airport (JFK)

	Flight time (min)		
Types of the flight routes	Total flight from LAX to JFK	Additional time compared to WOR	Flight time in areas of SOG >10 %
WOR	238	0	52
LTAR1	254	16	0
LTAR2	**244**	**6**	0
LTAR3	256	18	0

Geographical paths of the LTAR1, LTAR2, and LTAR3 are shown as red line in Fig. 24.4 (upper), green line in Fig. 24.4 (middle), and green line in Fig. 24.4 (lower), respectively
Boldface highlights the best LTAR in this case

In Fig. 24.5a, b (middle panel), variability of the GCREBs and GCRWBs is not clearly different from each season, because the heading angle of the GCR trajectory defined by Eq. (24.3) always tries to follow the great circle line between JFK and LHR regardless of the background wind pattern. On the other hand, in Fig. 24.5c, d (lower panel), overall features of the WOREBs (blue lines) and WORWBs (red lines) are obviously different between the two designated winter seasons. The WOREBs (blue lines) from JFK to LHR usually follow the prevailing westerly jet stream to maximize tail winds, reducing total travel time and fuel consumptions. In particular, the WOREBs (blue lines) are direct to Northern Europe following the dominant jet stream in +NAO phase during DJF04-05 (Fig. 24.5c), while the aircraft trajectories deviate southward toward the Southern Europe and Mediterranean Sea, following the dominant jet stream, in −NAO phase during DJF09-10 (Fig. 24.5d).

The westbound flights (WORWBs—red lines) from LHR to JFK, however, try to avoid the prevailing westerly jet stream to minimize head winds. In +NAO phase during DJF04-05 (Fig. 24.5c), the trajectories deviate southward and/or northward to avoid the strong westerly and southwesterly jet stream dominating along the great circle line between the JFK and LHR, so that the envelope of the WORWBs becomes widely spread. However, in the −NAO phase during DJF09-10, in the absence of a strong westerly jet stream, the trajectories mostly follow along the great circle line (i.e., shortest line) between the JFK and LHR, with the exception of some northward deviations (Fig. 24.5d).

Figure 24.6 shows the bar charts of the mean, ±2 standard deviations, and maximum and minimum values of the flight times for the GCREB, GCRWB, WOREB, and WORWB trajectories during the DJF04-05 (left) and DJF09-10 (right). There are several features in this plot. First, eastbound trajectories are faster than westbound trajectories, as expected from Fig. 24.5. Second, WORs are faster than the GCRs with winds in both westbound and eastbound trajectories, which is also somewhat expected from their definitions. Third, eastbound trajectories are faster in +NAO phase during DJF04-05 than those in −NAO phase during DJF09-10,

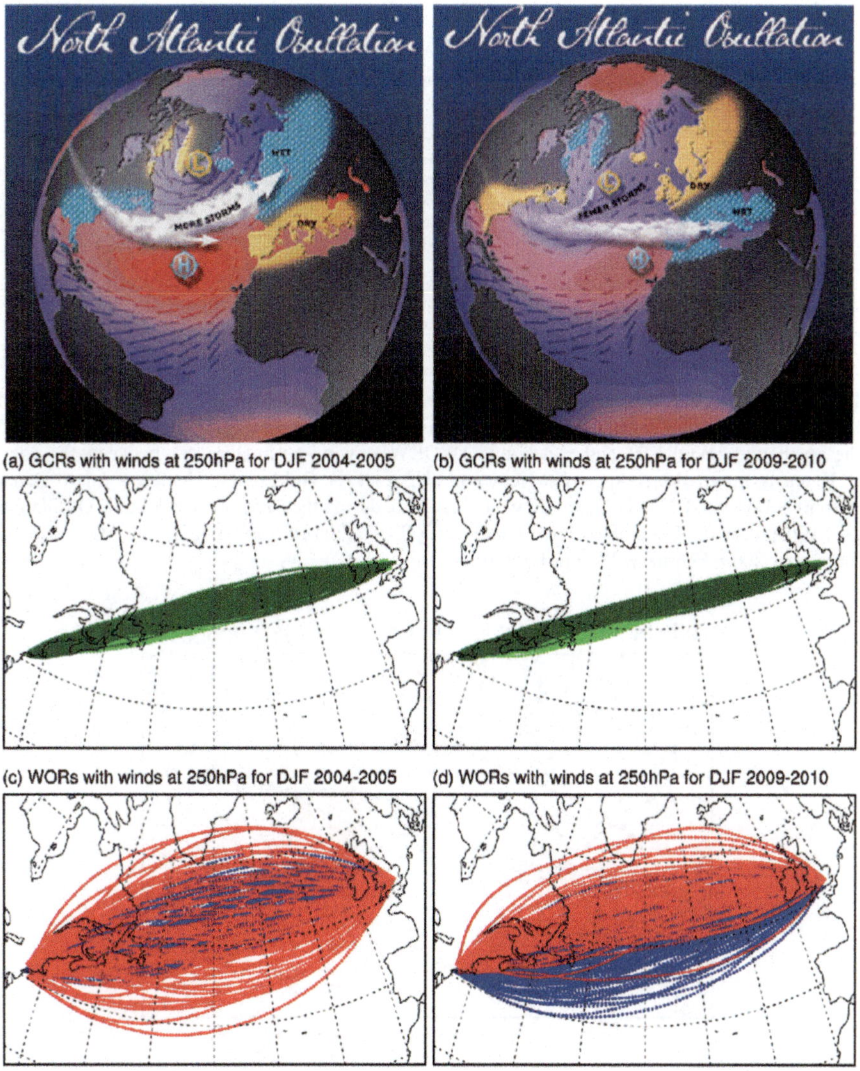

Fig. 24.5 (*Upper*) Schematic features of the prevailing westerly jet position and strength (source from http://www.ldeo.columbia.edu/NAO), (*middle*) variations of the eastbound (*light green lines*) and westbound (*dark green lines*) great circle routes with winds, and (*lower*) variations of the eastbound (*blue lines*) and westbound (*red lines*) wind-optimal routes at 250 hPa between the John F. Kennedy International Airport at New York, USA (JFK), and London Heathrow International Airport at London, UK (LHR), during (*left*) +North Atlantic Oscillation (+NAO) phase on December 2004–February 2005 (DJF04-05) and (*right*) −NAO phase on December 2009–February 2010 (DJF09-10). Note that the *light green lines* in (**a**) and (**b**) and *blue lines* in (**c**) are mostly overlapped with *dark green lines* and *red lines*, respectively

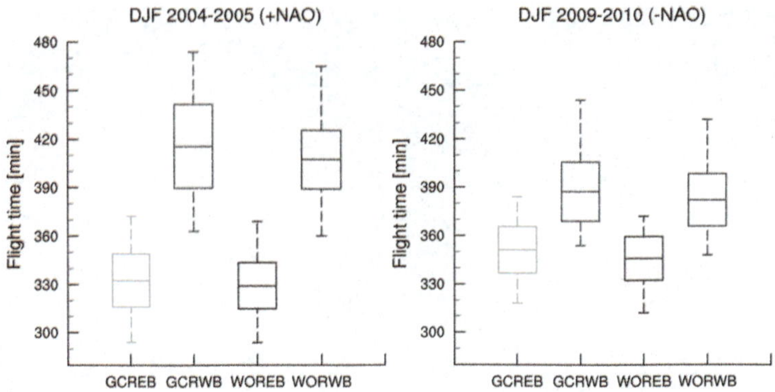

Fig. 24.6 Bar charts of the mean, mean ± 2 stds, and minimum and maximum values of the travel times along the eastbound (GCREB) and westbound (GCRWB) great circle routes with winds and the eastbound (WOREB) and westbound (WORWB) wind-optimal routes between the John F. Kennedy International Airport at New York, USA (JFK), and London Heathrow International Airport at London, UK (LHR), are shown in Fig. 24.5, during the (*left*) +North Atlantic Oscillation (+NAO) phase on December 2004–February 2005 (DJF04-05) and (*right*) −NAO phase on December 2009–February 2010 (DJF09-10)

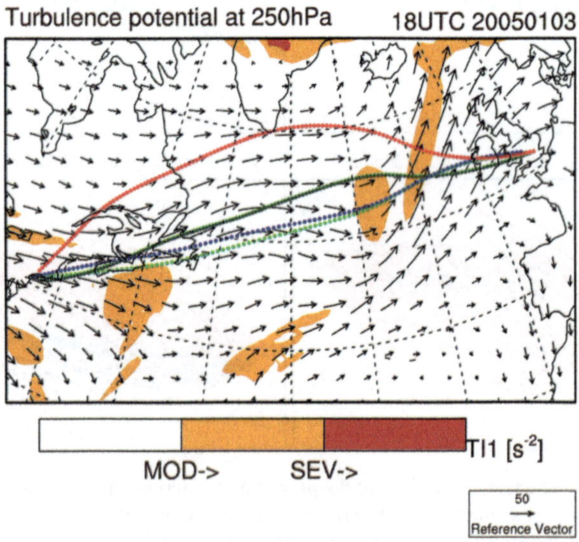

Fig. 24.7 Shadings of turbulence index 1 (TI1; s^{-2}) and horizontal wind vectors with the eastbound (*light green*) and westbound (*dark green*) great circle routes and eastbound (*blue*) and westbound (*red*) wind-optimal routes at 250 hPa level between the John F. Kennedy International Airport at New York, USA (JFK), and London Heathrow International Airport at London, UK (LHR), on 1800 UTC, 3 January 2005. The reference wind vector on the *bottom right* is 50 m s^{-1}

because the prevailing westerly jet stream is faster in +NAO phase over the North Atlantic Ocean. On the other hand, westbound trajectories are faster in −NAO phase than in +NAO phase, because the westerly winds are stronger in +NAO period.

Table 24.2 Relative probability of turbulence encountering time divided by the total flight time in great circle route with winds for eastbound (GCREBs) and westbound (GCRWBs) and wind-optimal route for eastbound (WOREBs) and westbound (WORWBs) during the December, January, and February 2004 and 2005 [+North Atlantic Oscillation (NAO) phase] and 2009 and 2010 (–NAO phase)

	DJF 04-05 (+NAO)		DJF 09-10 (−NAO)	
MOG time/total travel time (%)	GCRs (%)	WORs (%)	GCRs (%)	WORs (%)
Eastbound (EB)	5.6	5.7	4.5	5.1
Westbound (WB)	5.4	4.1	4.3	4.1

Based on the aforementioned variability of the aircraft trajectories in each season, the variability of turbulence potential along these trajectories can be investigated. Figure 24.7 shows the snapshot for the turbulence potential for the moderate-or-greater (MOG) intensity (orange color) estimated by the turbulence index (TI) by Ellrod and Knapp (1992) at 250 hPa at 1800 UTC, 3 January 2005. Here, the threshold of the TI diagnostic for the MOG level is adapted from the operational GTG (Sharman et al. 2006). In this period of time, if an aircraft followed the WORWB, it would have encountered less MOG-level turbulence, because it deviates northward to avoid the prevailing westerly and southwesterly flows. The relative probability of encountering MOG-level turbulence along the trajectories is summarized in Table 24.2. For the GCRs, eastbound and westbound have similar probability in both +NAO and −NAO phases, because the aircraft fly near the great circle route independent of the wind patterns (as shown in Fig. 24.5 middle panel). However, the GCRs in +NAO phase have a higher probability of encountering turbulence due to the stronger westerly jet stream than those in −NAO. For the WORs, eastbound trajectories following the jet stream have a higher turbulence probability than westbound in both the +NAO and −NAO phases. The WOREBs in the +NAO phase have the highest turbulence probability.

24.5 Summary and Conclusions

In this chapter, the application of turbulence information for tactical, strategic, and long-term ATM flight planning was investigated. First, simplified aircraft trajectory models were derived by applying the minimal principal theory to the control parameter of aircraft heading angle in the presence of winds. Second, tactical and strategic aircraft trajectories using the WOR and three LTAR applications for given weather and turbulence conditions over the CONUS showed the utility of this forecast product for route planning applications. The turbulence information was based on the probability of severe turbulence for relatively large aircraft as an ensemble of several EDR-scale turbulence diagnostics at different forecast lead times. Since the potential turbulence areas along the WOR were vertically deep, deviating laterally around the turbulence areas was the best option to avoid turbulence for this case. As a result, delaying calculation of the turbulence avoidance maneuver by 1.5 h from departure is a solution that saves 10-min travel time compared to determining the avoidance strategy initially at departure in this case. Third, the variability of aircraft trajectories and turbulence potentials was investigated during two winter seasons over the Atlantic Ocean with distinctly different upper-level flow patterns, illustrating the usefulness of this technique for making decisions regarding long-term flight plans over the Ocean. Depending on the upper-level winds, the modeled aircraft trajectories had markedly different flight times. eastbound trajectories were faster than westbound ones due to the dominant westerly flow in Northern hemisphere. In addition, flights in +NAO phase were faster, but had a higher probability of encountering turbulence than those in −NAO phase over the North Atlantic Ocean, because prevailing westerly flows and vertical and horizontal wind shears near the jet stream are stronger in +NAO phase than in the −NAO phase.

Future work will use different thresholds instead of the 10 % SOG probability to explore the trade-offs between flight time and fuel consumption used in penetrating certain portions of a turbulence area. In addition, when the fuel consumption model is included in the cost function of Eq. (24.4 or 24.6), the current two-dimensional lateral turbulence avoidance route (LTAR) can be extended to three-dimensional maneuvers that minimize both the fuel consumption and potential of encountering turbulence during the total flight time. The strategic avoidance methodology suggested for turbulence herein can be also applied to other types of weather constraints such as deep convection, icing, volcano ash, wind gust, and the potential for contrail formation. Reducing the run-time would make the new method useful for tactical decisions such as near-term routing around convective weather as well. This can be accomplished by using data from a nowcast version of the GTG or output from a faster-running numerical model.

References

Barnston, A.G., Livezey, R.E.: Classification, seasonality and persistence of low-frequency atmospheric circulation patterns. Mon. Weather Rev. **115**, 1083–1126 (1987)

Bryson, A.E., Ho, Y.C.: Applied Optimal Control. Taylor and Francis, Levittown, PA (1975)

Cornman, L.B., Morse, C.S., Cunning, G.: Real-time estimation of atmospheric turbulence severity from in-situ aircraft measurements. J. Aircraft **32**, 171–177 (1995)

DeLaura, R., Evans, J.: An exploratory study of modeling en route pilot convective storm flight deviation behavior. Preprints, 12th Conference on Aviation, Range, and Aerospace Meteorology, Atlanta, GA. Am. Meteorol. Soc. (2006)

Ellrod, G.P., Knapp, D.I.: An objective clear-air turbulence forecasting technique: verification and operational use. Weather Forecast **7**, 150–165 (1992)

Gill, P.G.: Objective verification of World Area Forecast Centre clear air turbulence forecasts. Meteorol. Appl. **21**, 3–11 (2014). doi:10.1002/met.1288

Gill, P.G., Stirling, A.J.: Including convection in global turbulence forecasts. Meteorol. Appl. **20**, 107–114 (2013). doi:10.1002/met.1315

Hok, H.K., Sridhar, B, Grabbe, S. Chen, N.: Cross-polar aircraft trajectory optimization and the potential climate impact. 30th Digital Avionics Systems Conf. (DASC), Seattle, WA, Institute of Electrical and Electronics Engineers, p 15. [Available online at http://www.aviationsystemsdivision.arc.nasa.gov/ publications/2011/DASC2011_Ng.pdf.] (2011)

Irvine, E.A., Hoskins, B.J., Shine, K.P., Lunnon, R.W., Froemming, C.: Characterizing North Atlantic weather patterns for climate-optimal aircraft routing. Meteorol. Appl. **20**, 80–93 (2013). doi:10.1002/met.1291

Jaeger, E.B., Sprenger, M.: A northern-hemispheric climatology of indices for clear air turbulence in the tropopause region derived from ERA40 re-analysis data. J. Geophys. Res. **112**, D20106 (2007). doi:10.1029/2006JD008189

Jardin, M., Bryson, A.: Methods for computing minimum-time paths in strong winds. J. Aircraft **35** (1), 165–171 (2012)

Kim, J.-H., Chun, H.-Y.: A numerical study of clear-air turbulence (CAT) encounters over South Korea on 2 April 2007. J. Appl. Meteorol. Climatol. **49**(12), 2381–2403 (2010)

Kim, J.-H., Chun, H.-Y.: Statistics and possible sources of aviation turbulence over South Korea. J. Appl. Meteorol. Climatol. **50**(2), 311–324 (2011)

Kim, J.-H., Chun, H.-Y.: A numerical simulation of convectively induced turbulence above deep convection. J. Appl. Meteorol. Climatol. **51**, 1180–1200 (2012)

Kim, J.-H., Chun, H.-Y., Sharman, R.D., Keller, T.L.: Evaluations of upper-level turbulence diagnostics performance using the graphical turbulence guidance (GTG) system and pilot reports (PIREPs) over East Asia. J. Appl. Meteorol. Climatol. **50**, 1936–1951 (2011)

Kim, J.H., Chun, H.-Y., Sharman, R.D., Trier, S.B.: The role of vertical shear on aviation turbulence within cirrus bands of a simulated western Pacific cyclone. Mon. Weather Rev. **142**(8), (2014). doi:10.1175/MWR-D-14-00008.1

Kim, J.-H., Chan, W.N., Banavar, S., Sharman, R.D.: Combined winds and turbulence prediction system for automated Air-Traffic Management applications. J. Appl. Meteorol. Climatol. **54**, 766–784 (2015)

Kim, J.-H., Chan, W.N., Banavar, S., Sharman, R.D., Williams, P.D., Strahan, M.: Impact of the north atlantic oscillation on transatlantic flight routes and clear-air turbulence. J. Appl. Meteorol. Climatol. **55**, 763–771 (2016)

Knox, J.A., McCann, D.W., Williams, P.D.: Application of the Lighthill–Ford theory of spontaneous imbalance to clear-air turbulence forecasting. J. Atmos. Sci. **65**, 3292–3304 (2008)

Krozel, J., Klimenko, V., Sharman, R.D.: Analysis of clear-air turbulence avoidance maneuvers. Air Traffic Control Quart. **4**(2), 147–168 (2011)

Lane, T.P., Sharman, R.D.: Some influences of background flow conditions on the generation of turbulence due to gravity wave breaking above deep convection. J. Appl. Meteorol. Climatol. **47**, 2777–2796 (2008)

Lane, T.P., Sharman, R.D.: Intensity of thunderstorm-generated turbulence revealed by large-eddy simulation. Geophys. Res. Lett. **41**, 2221–2227 (2014). doi:10.1002/2014GL059299

Lane, T.P., Sharman, R.D., Clark, T.L., Hsu, H.-M.: An investigation of turbulence generation mechanisms above deep convection. J. Atmos. Sci. **60**(10), 1297–1321 (2003)

Lane, T.P., Doyle, J.D., Plougonven, R., Shapiro, M.A., Sharman, R.D.: Observations and numerical simulations of inertia-gravity waves and shearing instabilities in the vicinity of a jet stream. J. Atmos. Sci. **61**(22), 2692–2706 (2004)

Lane, T.P., Doyle, J.D., Sharman, R.D., Shapiro, M.A., Watson, C.D.: Statistics and dynamics of aircraft encounters of turbulence over Greenland. Mon. Weather Rev. **137**, 2687–2702 (2009)

Lane, T.P., Sharman, R.D., Trier, S.B., Fovell, R.G., Williams, J.K.: Recent advances in the understanding of near-cloud turbulence. Bull. Am. Meteorol. Soc. **93**, 499–515 (2012). doi:10.1175/BAMS-D-11-00062.1

Lester, P.F.: Turbulence: A New Perspective for Pilots. Jeppesen Sanderson, Englewood, CO (1994)

McNally, D., Sheth, K., Gong, C., Love, J., Lee, C.H., Sahlman, S., Cheng, J.: Dynamic weather routes: a weather avoidance system for near-term trajectory-based operations. 28th International Congress of the Aeronautical Sciences (ICAS), Brisbane, Australia (2012)

Ng, H.K., Grabbe, S., Mukherjee, A.: Design and evaluation of a dynamic programming flight routing algorithm using the Convective Weather Avoidance Model. AIAA-2009-5862, AIAA Guidance, Navigation, and Control Conference, Chicago, IL (2009)

Ng, H.K., Sridhar, B., Grabbe, S., Chen, N.: Cross-polar aircraft trajectory optimization and the potential climate impact. 30th Digital Avionics Systems Conference (DASC), Seattle, WA (2011)

Ng, H.K., Sridhar, B., Grabbe, S., Chen, N.: A practical approach for optimizing aircraft trajectories in winds. 31st Digital Avionics Systems Conference, Institute of Electrical and Electronics Engineers, Williamsburg, VA (2012)

Palopo, K., Windhorst, R.D., Suharwardy, S., Lee, H.-T.: Wind optimal routing in the National Airspace System. J. Aircraft **47**(5), 1584–1592 (2010)

Sharman, R.D., Tebaldi, C., Wiener, G., Wolff, J.: An integrated approach to mid- and upper-level turbulence forecasting. Weather Forecast **21**(3), 268–287 (2006)

Sharman, R.D., Doyle, J.D., Shapiro, M.A.: An investigation of a commercial aircraft encounter with severe clear-air turbulence over western Greenland. J. Appl. Meteorol. Climatol. **51**, 42–53 (2011)

Sharman, R.D., Trier, S.B., Lane, T.P., Doyle, J.D.: Sources and dynamics of turbulence in the upper troposphere and lower stratosphere: a review. Geophys. Res. Lett. **39**, L12803 (2012). doi:10.1029/2012GL051996

Sharman, R.D., Cornman, L.B., Meymaris, G., Pearson, J., Farrar, T.: Description and derived climatologies of automated *in situ* eddy dissipation rate reports of atmospheric turbulence. J. Appl. Meteorol. Climatol. **53**(6), 1416–1432 (2014). doi:10.1175/JAMC-D-13-0329.1

Sridhar, B., Ng, H.K., Chen, N.Y.: Aircraft trajectory optimization and contrails avoidance in the presence of winds. 10th AIAA Aviation Technology, Integration, and Operations (ATIO) Conference, Fort Worth, TX (2010)

Steiner, M., Bateman, R., Megenhardt, D., Liu, Y., Pocernich, M., Krozel, J.: Translation of ensemble weather forecasts into probabilistic air traffic capacity impact. Air Traffic Control Quart. **18**(3), 229–254 (2010)

Trier, S.B., Sharman, R.D.: Convection-permitting simulations of the environment supporting widespread turbulence within the upper-level outflow of a Mesoscale Convective System. Mon. Weather Rev. **137**(6), 1972–1990 (2009)

Trier, S.B., Sharman, R.D., Fovell, R.G., Frehlich, R.G.: Numerical simulation of radial cloud bands within the upper-level outflow of an observed mesoscale convective system. J. Atmos. Sci. **67**(9), 2990–2999 (2010)

Williams, P.D., Joshi, M.M.: Intensification of winter transatlantic aviation turbulence in response to climate change. Nat. Climate Change **3**(7), 644–648 (2013). doi:10.1038/nclimate1866

Chapter 25
Research Needs

Robert Sharman, Todd Lane, and Ulrich Schumann

Abstract The current state of our understanding of aviation-scale turbulence processes and overviews of current detection and forecasting methods were provided in previous chapters of this book. Future progress will require a multipronged multidisciplinary approach from the academic, engineering, and user communities underpinned by advances in our fundamental understanding of turbulence process. The chapter outlines the required research and development, as well as operational needs, required to better understand, observe, and predict turbulence regions.

25.1 Introduction

As described in the previous chapters, great progress has been made in understanding the character of atmospheric turbulence as it affects aircraft, in observational techniques for its detection, and in operational forecasting and nowcasting procedures. However, we are not yet at the point where operators are satisfied with the products currently available to them, including those available for turbulence detection and prediction. Further progress is impaired by several major challenges:

- The nature of aircraft-scale turbulent motion, especially in the upper troposphere and lower stratosphere (UTLS), is still not well-understood.

R. Sharman (✉)
Research Applications Laboratory, National Center for Atmospheric Research, Boulder, CO, USA
e-mail: sharman@ucar.edu

T. Lane
School of Earth Sciences, The University of Melbourne, Melbourne, VIC, Australia
e-mail: tplane@unimelb.edu.au

U. Schumann
Deutsches Zentrum für Luft- und Raumfahrt (DLR), Institut für Physik der Atmosphäre, 82234 Oberpfaffenhofen, Germany
e-mail: ulrich.schumann@dlr.de

R. Sharman, T. Lane (eds.), *Aviation Turbulence*,
DOI 10.1007/978-3-319-23630-8_25

- Aircraft-scale turbulence is too small and transient to actually forecast deterministically because of Numerical Weather Prediction (NWP) model resolution and predictability issues.
- Routine turbulence observations for turbulence forecast verification and nowcasting are lacking in coverage and precision.
- Moderate and severe turbulence is rare, and pilots try to avoid these regions, making the verification of turbulence forecasts difficult.
- Onboard forward-looking turbulence sensors are not yet available for use in all weather conditions, especially clear-air conditions.
- High-quality measurements of turbulence are missing, in particular for low and moderate turbulence, and these are essential for assessing NWP-based turbulence forecasting algorithms.
- From the operational perspective, providing reliable information to the cockpit for real-time avoidance decision-making is moving slowly due to burdensome, yet to some degree necessary, regulations.
- Significant enhancements in pilot training are necessary to facilitate better understanding of the hazards of their environment.
- Current dissemination of information from the research to the operational communities is not timely nor effective (also known as "crossing the valley of death," National Research Council 2000).

Some of these challenges, especially the operational ones, are being addressed by The Next Generation Air Transportation System (NextGen) in the USA and the Single European Sky ATM Research (SESAR) in Europe, but more involvement by government research labs and universities and transition pathways to operations are required to address the more fundamental challenges.

25.2 Research and Development Needs

Some important aviation turbulence research and development areas to meet the challenges listed above are summarized in the following sections, which are organized into observational, forecasting, fundamental understanding, and operational needs. This separation is to some extent arbitrary, and does have some overlap, but most of the highest priority needs should be addressed here.

25.2.1 Need for Enhanced and More Comprehensive Observations of Aircraft-Scale Turbulence

Reliable, routinely available, and systematic aircraft-scale turbulence observations are required for real-time turbulence avoidance, for nowcasting applications, and for verification of turbulence forecasts. Pilot reports (PIREPs) of turbulence have been extremely useful and, to a very large extent, successful, for real-time turbulence monitoring and avoidance, but they have several disadvantages which limit their usefulness not only for real-time avoidance but especially for research applications. The shortcomings of PIREPs for research applications are given in Schwartz (1996) and Sharman et al. (2014), but, beyond that, the reports, whether in the form of PIREPs or automated in situ reports, are not systematic (e.g., commercial aircraft travel only along designated jet routes), so they do not routinely cover the entire airspace all the time. PIREPs in smooth air are particularly underreported (Wolff and Sharman 2008). Further, the pilot is required to initiate the report, and even if the report is provided to Air Traffic Control (ATC), it may not be recorded for later use. Even in the USA where NOAA's Family of Services (FOSs) provide a fairly comprehensive data archive, many reports are still missing due to their perceived proprietary nature by some airlines and by ATC's neglect to provide the information to the database. Automated in situ reports (Chap. 5) are currently augmenting and may someday replace PIREPs, but their coverage has to be expanded by including more airlines, and, to the extent possible, air taxi and general aviation aircraft. More global and nighttime data would be especially valuable. Expansion of data gathering and reporting is inhibited, however, by the cost of transmitting the data and by the treatment of data from some airlines as proprietary. The cost may be decreased by using a combination of fairly sparse routine reporting and event-based reporting (which causes an immediate downlink of data when a certain energy dissipation rate (EDR) threshold is exceeded) and optimizing data reporting strategies (Sharman et al. 2014). In the future, transmission costs may ultimately be avoided entirely when implemented into the Automatic Dependent Surveillance-Broadcast (ADS-B) data stream (Alarcón et al. 2013; Kopeć et al. 2015). The other problem, associated with its proprietary treatment, must be worked out for the common good. There is also the need to standardize EDR implementations to insure that consistent values of EDR estimates are provided (Emanuel et al. 2013). This is most important for short-lived discrete events, because there can be a strong dependence of the peak EDR estimate on the details of the EDR computational algorithm, e.g., the averaging interval (Chap. 5). High-resolution turbulence simulations coupled with aircraft response models and parallel flights with research aircraft may be required to further refine and calibrate such algorithms.

Other currently operational systems could be modified to provide more turbulence observations. One example is EDR maps available to the cockpit from onboard radar (e.g., Chap. 7). This data should be downlinked for dissemination and combined with in situ EDR estimates to provide more comprehensive and timely turbulence maps within the airspace. Ground-based wind profilers, most of which are currently used only in a research mode, could produce vertical profiles of EDR operationally at the location of the profiler for all times the profiler is in operation (e.g., Dehghan et al. 2014). Similarly, high-resolution rawinsonde data (Clayson and Kantha 2008), possibly augmented by other onboard packages such as magnetometers or accelerometers (Marlton et al. 2015; Harrison and Hogan 2006), could be used to provide (nearly) vertical profiles of EDR near the location of the balloon release point and near the time of release. Operationally, rawinsondes are released only twice daily, and the coverage is sparse (roughly 800 measurements globally); however, the vertical resolution is about 25 m and so could provide a useful verification source. Satellite-based feature detectors, although unable to produce turbulence estimates directly, could be used with pattern recognition techniques to provide at least qualitative information about turbulence likelihood on a global basis and would be especially useful over remote regions.

Currently, most forward-looking sensors are either not reliable enough, do not provide all-weather capability, or are not cost-effective for wide-scale operational use (e.g., Kauffmann 2002; Chap. 22). Also, forward-looking sensors obtain information about the longitudinal component (i.e., the component along the direction of travel) of turbulence, not the vertical component which is more important for aircraft response, so knowledge about the anisotropic nature of turbulence is required. However, efforts to develop reliable and cost-effective forward-looking sensors should continue, and these systems should be implemented so that the information can be readily downlinked to the ground.

25.2.2 Need for Improved Nowcasting and Forecasting Products

Efficient strategic avoidance planning requires accurate, timely, and easy-to-understand turbulence forecasts. Besides the expected turbulence level, the forecast should provide the expected duration of turbulence along a user-selected flight path. Moreover, the user needs an indication of how likely it is that the predicted turbulence will actually occur. The users may use this information to take precautionary measures, or try to find alternative routes with less likely turbulence.

The lead time required for mission planning depends on the duration of the flight—international flights would typically require 12–18 h lead times, whereas for regional flights 6–12 h would be adequate. Short-term (less than 1-h) forecasts (nowcasts) provide information to operators during flight and can be used for tactical avoidance decisions if available in the cockpit. Nowcast products would generally provide more accurate information than forecasts, since they can account for highly transient sources such as thunderstorms and can include refinements by integrating various recent observations, either manually or automatically (from, e.g., PIREPs, in situ measurements, and radar EDR estimates), via data assimilation algorithms. In all events, the turbulence forecast or nowcast product would be driven by operational NWP models, which have their own errors and uncertainties, as well as the errors associated with the automated turbulence diagnoses. Thus, improvements require advances in *both* NWP model forecasts *and* turbulence diagnostic algorithms. At the moment it is not clear which of these two sources of error is more significant, although the error partition probably depends on the particular NWP model used and the dynamics of the specific turbulence source being addressed.

Given the total error depends on both NWP and diagnostic model errors, a two-step probabilistic approach could be developed in the future, similar to that developed in Werne et al. (2010). In the first step, the forecast is used to provide the driving forces that have been identified as relevant input for turbulence diagnostics, and in the second step, the turbulence forecast is generated based on diagnoses (or prognostic submodels) for the given input parameters. The approach should be based on ensemble predictions—or should use statistics based on previous predictions—to allow assessment of the likelihood and spread of the outputs from both steps. The probability analysis of the second step depends on the probabilities of the input values and the error covariances of the second step for given input parameters. Examples of driving forces are shear, deformation, frontogenesis, convection updrafts, etc. Examples for turbulence diagnostic parameters used for turbulence prognoses are the subgrid-scale (SGS) turbulence kinetic energy and the Ellrod index, to mention just two of many (Sharman et al. 2006). At present, both steps are intertwined. As a consequence, the verification of the turbulence prognosis is performed by comparing diagnostic output with observations or other information. The advantage of splitting the prediction into two steps would be that one could first assess the validity of the driving force predictions (e.g., by comparing predicted shear with radiosonde observations of shear). In addition to ensemble analysis, this would allow assigning error covariances or probabilities on the input values subsequently used. Second, one could assess the relationship between input parameters and turbulence prognoses in terms of covariances or probabilities.

The verification in the first step may be implemented as part of the automatically running assimilation process in the NWP model. It could encompass a far wider database than the one-step approach by also including regions and periods without flights or without turbulence reports. The verification of the second step would include comparisons of the prognoses with related observations as done so far in the one-step approach, and in addition one may use independent model and observation studies to assess the reliability of the forecast for given inputs.

25.2.2.1 Numerical Weather Prediction Model Forecast Errors

Traditional methods for evaluating NWP forecast model errors have shown significant improvement over the years (e.g., Kalnay 2003; Bauer et al. 2015); however, for forecasting of turbulence, these traditional metrics may not be appropriate. According to our current understanding, what is important, especially for UTLS turbulence, is accurate forecasts of wind shear, stability, frontogenesis, and convection intensity. Yet these properties are not normally examined as part of standard NWP model verification procedures. Verifying these quantities should be emphasized in developing research strategies to improve NWP models for turbulence forecasting purposes.

Accurately simulating vertical and horizontal wind shears can be crucially important to properly represent the turbulence source, both for convection and clear-air turbulence (CAT), but NWP models poorly resolve such shears. As a consequence of numerical smoothing and grid cell resolution, shear variability is strongly underestimated, and consequently Richardson numbers are overestimated in NWP models (e.g., Reiter and Lester 1968; Adelfang 1971; Houchi et al. 2010).

With increases in computational capabilities, NWP models have continually been able to provide higher and higher resolution. This should improve turbulence forecasts. In coarse models, many of the turbulence-generating processes (such as deep convection) occur at scales smaller than the grid scales, requiring approximate SGS models. With increasing numerical resolution, more and more of these processes can be resolved, which avoids such approximations. The higher the resolution, the closer the model comes to being able to properly represent the dynamics of the turbulence sources and consequently the small-scale motions that affect aircraft. However, at intermediate scales, the turbulence is neither fully resolved nor fully parameterized (originally termed "terra incognita" by Wyngaard (2004) or the "gray zone").

Moreover, studies need to be conducted to determine whether turbulence forecasts would benefit more from increases in the horizontal resolution or vertical resolution. The use of nested grids could be included in operational models with the higher-resolution grids incorporated over regions known statistically to be more prone to turbulence or by using higher-resolution grids dynamically in areas that are expected to see enhanced turbulence based on the larger-scale features (e.g., Dietachmayer and Droegemeier 1992, see also Chap. 20). The feasibility of using very high-resolution large-eddy simulation (LES) models operationally over a limited domain but including the UTLS has already been demonstrated (Schalkwijk et al. 2015), and the use of such limited area models should be pursued.

With current generation operational NWP model resolutions, the parameterized SGS turbulence output has not been particularly effective for use in routine aviation-scale turbulence forecasts, especially in the stably stratified shear flow environment characteristic of the UTLS. However, as NWP models' resolution improves, the SGS turbulence information should become more useful. To date most SGS turbulence parameterization development efforts have focused on providing accurate dissipation estimates in the planetary boundary layer (PBL) (e.g., Mellor and Yamada 1974), especially under convectively unstable conditions. For turbulence forecasting in the UTLS, research is required to develop better SGS turbulence parametrizations in stably stratified shear flow and in the presence of gravity waves (Schumann and Gerz 1995; Zilitinkevich et al. 2013; Chap. 21).

Finally, it is clear that the localized and transient nature of turbulence causes it to have low predictability and therefore places limits on the efficacy of deterministic turbulence forecasting. This means that regardless of the resolution of a numerical model or accuracy of its physical parameterizations, turbulence predictions ultimately contain uncertainties that are difficult to quantify. This is a particularly significant issue for convectively induced turbulence because the source of the turbulence, viz., moist convection, has inherently low predictability. Even though CAT is mostly linked to predictable large-scale forcing, the smaller scales of the turbulence-causing instabilities and turbulence itself also lend themselves to low predictability (e.g., Lorenz 1969). Thus, in concert with improvements in NWP models, turbulence forecasting must embrace probabilistic methods. This will likely become especially important when NWP models are at sufficiently high resolution to facilitate explicit predictions of turbulence (e.g., using parameterized SGS turbulent kinetic energy), as well as its most prominent sources. On the other hand, higher-resolution models provide more detailed output, possibly with some stochastic properties, inside a region that was formerly covered with just one grid cell mean value. This may make interpretations more complex and does not necessarily guarantee more accurate conclusions.

25.2.2.2 Turbulence Diagnostic Errors

In the near term, until NWP models can accurately predict aviation-scale turbulence directly, postprocessing algorithms to diagnose turbulence, or simply *diagnostics*, from the routinely available NWP model output quantities such as winds and temperature will continue to be required. This necessarily introduces some empiricism, which may lead to misinterpretation of the results. As Sherlock Holmes said to Dr. Watson in *A Scandal in Bohemia*, "it is a capital mistake to theorize before one has data. Insensibly one begins to twist facts to suit theories, instead of theories to suit facts." So what is really required to develop reliable diagnostics is fundamental research to better understand the characteristics of turbulence-generating sources and the nature of the downscale energy cascade process.

Fundamental research is also required to better predict the vertical scales of turbulence events. In strongly stratified turbulence, the vertical scale is limited by vertical kinetic energy of turbulent motions or turbulent kinetic energy dissipation rate and stratification. The relevant scale [e.g., the Ozmidov scale, Ozmidov 1965, Eq. 6] is often very small (order of 10 m), and only high-resolution models are able to capture these scales (Paoli et al. 2014). Near the tropopause (which is often near commercial airline cruise altitudes), very thin inversion layers can exist (e.g., Birner 2006), making it particularly challenging to design diagnostics on coarser resolution NWP grids.

Many of the turbulence diagnostics in use today were developed decades ago based on NWP models that were much coarser and less sophisticated than those available today. The underlying assumptions behind those diagnostics were linked to the generation of turbulence to aspects of the NWP-resolved synoptic-scale flow. Modern global- and regional-scale NWP models, however, better resolve the turbulence-generating processes, instabilities, and even the inertia-gravity waves that have recently been shown to be important. Thus, the application of diagnostics to these and future models may actually violate many of the original assumptions underlying the diagnostics. As model resolution continues to improve, many of the diagnostics that are well-suited to coarse-scale NWP models may become less useful. This underlines the importance of developing new diagnostics that are better suited to high-resolution NWP and that incorporate some of the recent advances in understanding, such as gravity wave processes. To some extent, the performance of the diagnostics is dependent on the configuration of the NWP model used to drive them, which is usually developed to provide the most robust and accurate weather forecasts. This NWP model configuration may not necessarily be the best for turbulence forecasting applications, however, and research needs to be done to establish the optimal NWP model configuration for that purpose. One example might be to include higher resolution near the tropopause.

From a practical point of view, multiple calibrated diagnostics (i.e., an ensemble) are probably required to ensure that the most important sources are identified in the postprocessing algorithms (Sharman et al. 2006; Gill and Buchanan 2014). But the optimum set of diagnostics used and the settings of any adjustable parameters within the diagnostic algorithms are probably NWP model dependent, thus requiring reconfiguration changes when the underlying NWP model is upgraded. This is typically a human intensive activity, and the capability of autocalibration of diagnostic algorithms should be researched and developed. Further, given a set of turbulence diagnostics, there are many ways in which the diagnostics could be combined to provide the final forecast. Some research into combination strategies has been done (Sharman et al. 2006; Williams 2014), but an optimal combination technique needs to be identified by examining different artificial intelligence methods.

There also remains a significant challenge in the area of determining locations of potential near-cloud turbulence (NCT). Research has shown that many of the current avoidance guidelines are inadequate and in need of updating (see Chaps. 16 and 17). However, updated guidelines would likely involve complicated algorithms that relate atmospheric properties like wind shear curvature and stability to turbulence occurrence. Such algorithms would need to be applied to NWP data and potentially nowcasting products, and to be ultimately useful would need to be uplinked to the cockpit to allow en route avoidance. Thus, new guidelines could be treated as an additional diagnostic, specifically designed for NCT, which could be applied to NWP models. It is likely that such a diagnostic would be best suited to convection-permitting NWP, which should provide a better indication of the characteristics of the convection systems including their organization, depth, and influence on the larger-scale flow.

As mentioned above, given the random nature of turbulent processes, the uncertainty in the underlying NWP model forecasts, coupled with errors in the turbulence diagnostics, a single deterministic forecast should not be provided to end users without some additional information about the expected uncertainty of the result. In the NWP arena ensembles of models and/or various model initializations, configurations and parameterizations have been used to quantify this uncertainty. At current NWP model resolutions, this could be done by using multiple diagnostics, possibly driven by an ensemble of NWP models, to provide probabilistic forecasts. The derivation of probability could be accomplished through assessment of the percentage agreement of ensemble members with some calibration or perhaps by using artificial intelligence techniques that provide probabilistic information.

However, the probability of turbulence at any given place and time is extremely small. For example Sharman et al. (2014), based on climatologies developed from in situ EDR (see Chap. 5), measurements onboard commercial airlines find that the frequency of occurrence (based on one-minute peak data) of EDRs > 0.3 $m^{2/3}\ s^{-1}$ (roughly equivalent to "moderate" turbulence for most commercial aircraft) is $<10^{-3}$ and EDRs > 0.5 $m^{2/3}\ s^{-1}$ ("severe" or greater) have a frequency of occurrence ~10^{-5}. A typical user could be forgiven if they find difficulty in appreciating or understanding these numbers and therefore finds the forecasts useless. Comparisons to climatology may help in understanding the significance of these numbers (Chap. 14). Some of these difficulties in interpretation may be minimized by combining probabilistic information and operational considerations into an automated risk management tool. In all events, user training would be essential, before they could develop appropriate cost-effective route planning strategies.

25.2.2.3 Verification Issues

Verification of turbulence nowcasts and forecasts is a crucial component of turbulence forecasting research and development. Not only is it important to know if one forecasting algorithm or ensemble of algorithms is more skillful than another and under what circumstances, but it is also important for being able to track changes in the underlying NWP model. But levels of moderate or greater (MoG) turbulence (or more correctly peak EDRs above a certain threshold), which are of most importance to aviation, are rare events, and verification is hampered by a dearth of observations. Development of methods to measure the quality of MoG predictions is an ongoing research effort that shares much in common with other extreme but rare events such as strong surface winds, heavy precipitation, or high temperatures, and advantage may be taken of research into verification strategies of these kinds of events.

Compounding the difficulty with MoG turbulence forecasts where observations are rare is the fact that there can be considerable uncertainty in the observational data itself, especially for PIREPs where the location uncertainty can be 10s of km, i.e., several grid points in current generation NWP models. But even in situ EDR data has 1-min or about 15-km resolution, which may also be large compared to grid sizes of regional forecast models. This problem will only become more severe with time as higher and higher-resolution NWP models are brought on line. Thus it becomes particularly important to be able to deal with the uncertainty in observations in the verification process. More sophisticated verification methods, such as object-based or spatial methods (e.g., Warner 2011 his Chap. 9; Ahijevych et al. 2009), may be better suited for turbulence forecast verification purposes.

Verification of probabilistic forecasts requires accurate assessment of reliability, which in turn requires knowledge of the climatology of turbulence. This is difficult due to the nonsystematic nature of observations. Even when pilots are taken out of the reporting loop and replaced by automated in situ turbulence reports, MoG events will be underreported due to intentional turbulence avoidance, making it difficult to determine unbiased climatologies. Thus derivation of actual turbulence climatologies qualifies as another important research need.

Finally, assessments of the operational usefulness of the forecast product must consider not only the forecast quality by whatever measure is appropriate but also the possible economic value (EV) in the daily decision-making process of users. That is, the ultimate value of a forecast can only be evaluated in a decision-making context (see, e.g., Steiner et al. 2010; Chap. 13).

25.2.3 Need for Improved Fundamental Understandings of the Nature and Sources of Aviation-Scale Turbulence

It is important to understand what are the predictability and accuracy limits of turbulence forecasts. This question goes beyond just understanding the sources and relevant pathways that lead to turbulence (or in calm situations without turbulence) and their statistical mean properties. Insight into the nature of initial condition and model errors and their relative contributions to limiting turbulence predictability is needed. Besides fundamental studies on the nature of turbulence, more intense studies comparing predictions to observations is required to better identify research needs.

Improved understanding may be achieved by a combination of theoretical studies, case studies using high-resolution numerical simulations, statistical correlations between large-scale features and turbulence observations, and execution of comprehensive field programs. Theoretical studies may involve idealized very high-resolution (large-eddy or direct numerical) simulations as in, e.g., Fritts et al. (2009), Lane and Sharman (2014), and McHugh and Sharman (2013), and, to the extent possible, analytic or computer-aided investigations. Statistical studies and models may provide correlations that need more examination and motivate fruitful directions for further research.

Case studies have been particularly illuminating in defining the role of convectively generated gravity waves in the near-cloud turbulence generation process (e.g., Lane et al. 2012). Similar studies should be undertaken to better understand cases of enhanced turbulence, based on accident reports, elevated in situ EDR data, and PIREPs. For many cases, PIREPs are simply inadequate due to their large uncertainties in reported time, position, and magnitude. In situ EDR data is more useful because of its 1-min resolution but provides only EDR information. A better strategy would be to identify cases, then request the digital flight data recorder (DFDR) or retrieve quick access data recorder (QAR) information before the data is lost, and use that information to fully characterize the event. This would require cooperation with airlines and the National Safety Transportation Board (NTSB) or similar investigative agencies in other countries to retrieve the high-resolution aircraft data.

This multi-pronged approach should ultimately lead to the formulation and development of improved turbulence forecast algorithms. For example, enhanced understanding of when, where, and why particular diagnostics succeed in terms of the generation of turbulence (e.g., gravity wave generation) could be used to develop situationally dependent turbulence diagnostics. If the diagnostics could be combined in multi-diagnostic approaches in a more optimal way through such research, that should lead to improvements in overall forecasting skill.

25.2.3.1 Fundamental Research Questions

Below is a list of fundamental questions regarding the nature and sources of aviation-scale turbulence that need intensive research investigations to be able to answer them. Many of these are of a very fundamental nature and would have to be addressed through fundamental research sponsoring organizations such as the National Science Foundation in the USA, in addition to continued funding for applied research that has been traditionally done by funding through agencies such as the Federal Aviation Administration and the National Aeronautics and Space Administration. This list is not exhaustive, and other questions might arise as investigations proceed.

- What is the importance of gravity waves and gravity wave breaking in the CAT production process and what are the dominant sources of wave generation? With respect to wave motions, it is important to understand not only the sources of such wave energy but also their sinks (dispersion and dissipation).
- What is the spectrum of gravity waves and how does the energy (in the forms of kinetic and potential energy) cascade either upscale or downscale to the region of wavelengths important to aircraft?
- What are the relative roles of the upscale energy cascade of kinetic and potential energy in stably stratified and nonstratified atmospheres?
- What is the relation of out-of-cloud turbulence to the characteristics of the convective cloud ultimately generating the turbulence?
- What is the connection between upper-level unbalanced flow and upper-level turbulence?
- What processes contribute to the shape of the mesoscale energy spectrum? What is the degree of isotropy (this is important for forward-looking sensors; see Chap. 22)?
- What is the role of the tropopause and tropopause folds? Is turbulence more likely there and if so, why? Does the presence of a tropopause inversion layer influence turbulence production or its life cycle?
- Does Rossby wave amplification and breaking (see McIntyre and Palmer 1983) contribute to turbulence "outbreaks"?
- What conditions lead to the development of upper-level horizontal vortex tubes which are apparently associated with some severe to extreme turbulence encounters (cf. Clark et al. 2000; Chap. 4). What is their frequency of occurrence?

- How do different sources (e.g., convection, baroclinic waves, and mountain waves) interact to produce or damp turbulence?
- How far do cirrus clouds influence the observed widespread turbulence regions in the upper troposphere? Is this influenced by latent heat release, by radiative heating causing local instability, or some other process or processes?
- How prevalent are coherent roll-like structures in the UTLS and what generates them?
- What is the global climatology of turbulence of different sources (clear air, mountain wave, and convectively induced turbulence, i.e., CAT, MWT, and CIT)?
- How large is the mean dissipation rate, and what is its probability density function in various regions of the atmosphere (e.g., middle troposphere, upper troposphere, and lower stratosphere)?
- What is the observational bias associated with routine airborne observations?
- How will climate change impact the intensity and frequency of various turbulence processes in the future (see Chap. 23)? How will this affect operations?
- What is the best method to quantify uncertainty in turbulence forecasts and to verify them?
- Do secondary gravity waves emitted from breaking gravity waves (e.g., in the stratosphere above mountains) play a role in turbulence occurrence at other locations?
- What are the turbulence characteristics of rotors and sub-rotors in the lee of mountain ranges?
- What are the large-scale conditions that promote mountain wave breaking over complex terrain, and how do these vary globally?
- What are the predictability limits of aviation-relevant turbulence using NWP models?
- What observations are needed to improve turbulence forecasts from NWP models, and what are the dominant initial condition model errors that contribute most to limited predictability?

25.2.3.2 Need for Dedicated Field Programs

In the late 1960s and early 1970s, a series of comprehensive field programs were undertaken to measure atmospheric turbulence properties, sponsored mainly by the US Air Force (the Critical Atmospheric Turbulence Program "ALLCAT"). The goals were to develop a statistically significant data set of turbulence properties at all altitudes to be used in aircraft gust design criteria and to obtain data to aid in developing better turbulence forecasting techniques (Loving 1969). Unfortunately most, if not all, of this data seems to be lost, and the supporting meteorological data is probably too coarse to allow serious analysis of the data anyway. There were also a series of field campaigns focused on mountain waves (Lilly and Zipser 1972) in the 1970s, which are among the most complete datasets that document wave breaking. Other, more concentrated programs have been conducted since then.

Examples include the Terrain-Induced Rotor Experiment (T-REX, Grubišić et al. 2008) and the Severe Clear-Air Turbulence Colliding with Aircraft Traffic (SCATCAT; Koch et al. 2005) experiment and programs with armored aircraft to study convectively induced hazards (Honomichl et al. 2013). Nonetheless, the current state of affairs is that as mentioned above, i.e., there are many remaining questions, especially with regard to CAT, that could only be answered with larger-scale field campaigns using state-of-the-art in situ and remote sensors.

To answer these questions, it will be necessary to initiate a set of field campaigns to provide research aircraft flight measurements of turbulence and its environment together with modern state-of-the-art ground-based and remote sensing instrumentation. Specifically, aircraft measurements must be made to provide detailed and continuous records of atmospheric temperature, pressure, and velocity fluctuations (all three components), with a time resolution of tens of Hz. Such flights should be conducted under various conditions of upper-level flow patterns and terrain, to better understand the energy source or sources of the encountered turbulence. This may require both winter and summer field campaigns. In order to address turbulence energetics and the degree of shear and anisotropy, two—preferably three—aircraft should be deployed. The flights must be augmented by simultaneous high-resolution airborne and ground-based remote observations (e.g., by Doppler lidar and microwave temperature profiler, MTP) and rawinsondes (or dropsondes) to define the vertical structure of the atmosphere in the vicinity of the observed turbulence with enough resolution to reliably and accurately compute stability and shear. The use of drones or unmanned aerial systems (UASs) instrumented with high-resolution sensors (e.g., Balsley et al. 2013; Lawrence and Balsley 2013; Elston et al. 2015) may provide a method for obtaining turbulence quantities without putting research aircraft and occupants at risk in sampling high-intensity turbulence events. Incidentally, such drones could also conceivably be used in a more routine measurement mode to get a better picture of the climatology of turbulence characteristics. Once drones are miniaturized, ubiquitous and low cost and the necessary regulations are in place, they could provide a path-finder role, patrolling the major flight paths and sending back turbulence readings to ground-based users.

One ongoing challenge is related to the rarity of enhanced levels of turbulence. Research flights are an expensive endeavor, and some assurance should be provided based on turbulence forecasts, say a day in advance, that encounters with elevated turbulence are likely. Before takeoff, in situ reports of turbulence from commercial airlines can help identify locations where the turbulence is most likely to be found. This, of course, will provide valuable information for verifying the forecasts a posteriori. In fact several models could be run to provide the forecasts and would contribute to a nice intercomparison of model performance under various conditions. Further, whereas the aircraft provides very high-resolution measurements along the direction of flight of the aircraft, the simulation models, if they can adequately represent regions of enhanced turbulence, provide a three-dimensional plus time image of the turbulence and provide some idea of the representativeness of the aircraft measurements.

It should be noted that obtaining accurate turbulence measurements from high-quality instrumented research aircraft is nontrivial. For correct aircraft-measured turbulence, accurate measurement of the aircraft motion is needed, which requires a high-precision combination of an inertial measurement unit (IMU) and a differential Global Positioning System (GPS) navigation system. Both measurement streams have to be optimally combined by forward and backward Kalman filtering (see, e.g., Groves et al. 2008). Also, corrections must be made for aircraft oscillations (phugoid and autopilot-driven motions) and the changes in wind field around the aircraft due to aircraft-induced air circulation (around the wing and fuselage), e.g., the so-called upwash and sidewash (e.g., Crawford et al. 1996). Further, accurate measurements of weak turbulence is far more demanding than measuring strong turbulence because of the relative importance of measurement errors (low signal-to-noise ratios), yet these measurements are critical to defining the turbulence background state. There is a need for direct comparison flights between various turbulence measurement systems under weak and strong turbulence and gravity wave conditions. Moreover, data from various airborne measurement systems and campaigns should be brought together in a unified format and compared to assess accuracies which are achievable today and to provide a standard for future instrument development.

25.3 Operational Needs

Beyond the fundamental research needs listed above to better understand turbulence properties, advances in turbulence nowcasting and forecasting require more operationally-related changes. One major change would be for all commercial airlines to endorse, implement, and share automated onboard in situ turbulence measurements and provide the capability to downlink that information to users on the ground and cross-link the data to other aircraft in flight. This would provide real-time comprehensive maps of EDR in the airspace that could be used for tactical avoidance but also could be used in turbulence forecast algorithm tuning and verification to provide better products. These data would also be valuable for researchers to better define turbulence climatologies and examine in detail large-scale to small-scale connections. Communication costs are an issue, but one way they may be minimized is to use ADS-B (e.g., Alarcón et al. 2013; Kopeć et al. 2015) or other broadband communications networks.

Advances in our understanding of turbulence properties need to be communicated to end users through expanded training programs. These should involve animated graphics output from high-resolution simulation models so that users can start to actually "see" what they are trying to avoid. Training would also involve relating principles of probabilistic forecasting so that users can better appreciate and understand the uncertainty associated with turbulence forecasts, and what it means to them for decision-making. New products are needed to provide situational awareness required by operators and for risk-related decision aids.

Rapid progress is being made to get concise, easy-to-understand weather information (obtained through data-link sources from ground-based and onboard remote and in situ sensors) into the cockpit for enhanced awareness of all types of weather hazards so that the pilot can make informed tactical maneuvers (e.g., Gerz et al. 2012). These may take the form of modified onboard displays or data-link displays on carry-on laptops and tablets. Electronic flight bags (EFBs), which replace paper-based reference material found in the pilot's carry-on flight bag, are becoming more common in the cockpits of general and commercial aviation aircraft. Weather-in-the-cockpit displays on tablets are currently in test and evaluation phases by various US, European (the FLYSAFE project, Mirza et al. 2008), and Asian air carriers, and display standards are being developed for the NextGen and the European ATC modernization program, SESAR. However, the usefulness of this cockpit display information does depend on the pilot's experience (Ambs 2014).

On the aircraft side, gust alleviation strategies, already deployed with various sophistications on most commercial aircraft, should continue to be refined by airframe manufacturers. Most of the details of these technologies are proprietary and not available in the public domain, but an overview of strategies used can be found in Regan and Jutte (2012). These gust alleviation systems may in the future be coupled with short-range look-ahead sensors to automatically modify autopilot controls in preparation for detected turbulence ahead (e.g., Sordeide et al. 1996).

25.4 Summary

A multi-pronged approach consisting of establishing better and more comprehensive operational turbulence sensing, performing theoretical and numerical studies of observed turbulence events, developing improved turbulence forecasting techniques, engineering improved gust alleviation strategies, and broadcasting information to the flight deck for real-time avoidance will undoubtedly reduce the number of moderate, severe, and extreme turbulence encounters in the future. Impressive progress has been made in recent years through enhanced understanding and better observational and forecasting techniques. However, given the complexity of the turbulence problem, improvements are likely to be manifested as only incremental steps toward fulfilling the operational goals of strategic and tactical turbulence avoidance.

Acknowledgments The authors gratefully acknowledge the reviews and suggestions supplied by John Knox, Paul Williams, Wiebke Deierling, Stan Trier, and Jim Doyle. Todd Lane is supported by the Australian Research Council's Centres of Excellence Scheme (CE110001028).

References

Adelfang, S.I.: On the relations between wind shears over various altitude intervals. J. Appl. Meteorol. **10**(2), 156–159 (1971)

Ahijevych, D., Gilleland, E., Brown, B.G., Ebert, E.E.: Application of spatial verification methods to idealized and NWP-gridded precipitation forecasts. Weather Forecasting **24**, 1485–1497 (2009). doi:10.1175/2009WAF2222298.1

Alarcón, J.F.A., Nieto, F.J.S., Carretero, J.G.-H.: Aircraft used as a sensor for atmospheric behaviour determination. Practical case: pressure estimation using automatic dependent surveillance-broadcast. Proc. Inst. Mech. Eng. G: J. Aerospace Eng. **227**(5), 778–797 (2013). doi:10.1177/0954410012442044

Ambs, K.: The influence of cockpit weather automation on pilot perception and decision-making in severe weather conditions. McNair Scholars Res. J. **7**(1), (2014). http://commons.emich.edu/mcnair/vol7/iss1/4

Balsley, B.B., Lawrence, D.A., Woodman, R.F., Fritts, D.C.: Fine-Scale characteristics of temperature, wind, and turbulence in the lower atmosphere (0–1,300 m) over the South Peruvian Coast. Boundary-Layer Meteorol. **147**, 165–178 (2013)

Bauer, P., Thorpe, A., Brunet, G.: The quiet revolution of numerical weather prediction. Nature. **525**, 47–55 (2015). doi:10.1038/nature14956

Birner, T.: Fine-scale structure of the extratropical tropopause region. J. Geophys. Res. **111**(D4), D04104 (2006). doi:10.1029/2005JD006301

Clark, T.L., Hall, W.D., Kerr, R.M., Middleton, D., Radke, L., Ralph, F.M., Nieman, P.J., Levinson, D.: Origins of aircraft-damaging clear air turbulence during the 9 December 1992 Colorado downslope windstorm: numerical simulations and comparison to observations. J. Atmos. Sci. **57**(8), 1105–1131 (2000)

Clayson, C.A., Kantha, L.: On turbulence and mixing in the free atmosphere inferred from high-resolution soundings. J. Atmos. Oceanic Technol. **25**(6), 833–852 (2008)

Crawford, T.L., Dobosy, R.J., Dumas, E.J.: Aircraft wind measurement considering lift-induced upwash. Boundary-Layer Meteorol. **80**, 79–94 (1996)

Dehghan, A., Hocking, W.R., Srinivasan, R.: Comparisons between multiple in-situ aircraft measurements and radar in the troposphere. J. Atmos. Sol. Terr. Phys. **118**, 64–77 (2014)

Dietachmayer, G.S., Droegemeier, K.K.: Application of continuous dynamic grid adaptation techniques to meteorological modeling. Part I. Mon. Weather Rev. **120**, 1675–1706 (1992)

Elston, J., Argrow, B., Stachura, M., Weibel, D., Lawrence, D., Pope, D.: Overview of small fixed-wing unmanned aircraft for meteorological sampling. J. Atmos. Oceanic Technol. **32**, 97–115 (2015). doi:10.1175/JTECH-D-13-00236.1

Emanuel, M., Sherry, J., Catapano, S., Cornman, L., Robinson, P.: *In situ* performance standard for eddy dissipation rate. 16th Conference of Aviation, Range and Aerospace Meteorology, Austin, TX. Paper 11.3 (2013)

Fritts, D.C., Wang, L., Werne, J., Lund, T., Wan, K.: Gravity wave instability dynamics at high Reynolds numbers. Part I: Wave field evolution at large amplitudes and high frequencies. J. Atmos. Sci. **66**, 1126–1148 (2009)

Gerz, T., Forster, C., Tafferner, A.: Mitigating the impact of adverse weather on aviation. In: Schumann, U. (ed.) Atmospheric Physics, Research Topics in Aerospace, pp. 645–659. Springer, Berlin (2012). doi:10.1007/978-3-642-30183-4_39

Gill, P.G., Buchanan, P.: An ensemble based turbulence forecast system. Meteorol. Appl. **21**, 12–19 (2014). doi:10.1002/met.1373

Groves, P.D.: Principles of GNSS, Inertial, and Multisensor Integrated Navigation Systems. ARTECH House, Boston, MA, 518 pp. (2008)

Grubišić, V., Doyle, J.D., Kuettner, J., Mobbs, S., Smith, R.B., Whiteman, C.D., Dirks, R., Czyzyk, S., Cohn, S.A., Vosper, S., Weissman, M., Haimov, S., De Wekker, S., Pan, L., Chow, F.K.: The Terrain-Induced Rotor Experiment: an overview of the field campaign and some highlights of special observations. Bull. Am. Meteorol. Soc. **89**, 1513–1533 (2008)

Harrison, R.G., Hogan, R.J.: In situ atmospheric turbulence measurement using the terrestrial magnetic field—a compass for a radiosonde. J. Atmos. Oceanic Technol. **23**, 517–523 (2006)

Honomichl, S.B., Detwiler, A.G., Smith, P.L.: Observed hazards to aircraft in deep summertime convective clouds from 4-7 km. J. Aircraft **50**(3), 926–935 (2013). doi:10.2514/1.C032057

Houchi, K., Stoffelen, A., Marseille, G.J., De Kloe, J.: Comparison of wind and wind shear climatologies derived from high-resolution radiosondes and the ECMWF model. J. Geophys. Res. **115**, D22123 (2010). doi:10.1029/2009JD013196

Kalnay, E.: Atmospheric Modeling, Data Assimilation and Predictability. Cambridge University Press, Cambridge (2003)

Kauffmann, P.: The business case for turbulence sensing systems in the US air transport sector. J. Air Transport Manage. **8**, 99–107 (2002)

Koch, S.E., Jamison, B.D., Lu, C., Smith, T.L., Tollerud, E.L., Girz, C., Wang, N., Lane, T.P., Shapiro, M.A., Parrish, D.D., Cooper, O.R.: Turbulence and gravity waves within an upper-level front. J. Atmos. Sci. **62**(11), 3885–3908 (2005)

Kopeć, J.M., Kwiatkowski, K., de Haan, S., Malinowski, S.P.: Retrieving clear-air turbulence information from regular commercial aircraft using Mode-S and ADS-B broadcast. Atmos. Meas. Tech. Discuss. **8**, 11817–11852 (2015). doi:10.5194/amtd-8-11817-2015

Lane, T.P., Sharman, R.D.: Intensity of thunderstorm-generated turbulence revealed by large-eddy simulation. Geophys. Res. Lett. **41**, 2221–2227 (2014). doi:10.1002/2014GL059299

Lane, T.P., Sharman, R.D., Trier, S.B., Fovell, R.G., Williams, J.K.: Recent advances in the understanding of near-cloud turbulence. Bull. Am. Meteorol. Soc. **93**(4), 499–515 (2012). doi:10.1175/BAMS-D-11-00062.1

Lawrence, D.A., Balsley, B.B.: High-resolution atmospheric sensing of multiple atmospheric variables using the DataHawk small airborne measurement system. J. Atmos. Oceanic Technol. **30**, 2352–2366 (2013). doi:10.1175/JTECH-D-12-00089.1

Lilly, D.K., Zipser, E.J.: The Front Range windstorm of 11 January 1972: a meteorological narrative. Weatherwise **25**, 56–63 (1972)

Lorenz, E.N.: The predictability of a flow which possesses many scales of motion. Tellus **21**, 289–307 (1969)

Loving, N.V.: Technical and meteorological planning to meet the ALLCAT program objectives. In: Pao, Y.-H., Goldburg, A. (eds.) Clear Air Turbulence and Its Detection, pp. 127–143. Plenum Press, New York, NY (1969)

Marlton, G.J., Harrison, R.G., Nicoll, K.A., Williams, P.D.: A balloon-borne accelerometer technique for measuring atmospheric turbulence. Rev. Sci. Instrum. **86**, 016109 (2015). doi:10.1063/1.4905529

McHugh, J., Sharman, R.: Generation of mountain wave-induced mean flows and turbulence near the tropopause. Q. J. Roy. Meteorol. Soc. **139**, 1632–1642 (2013). doi:10.1002/qj.2035

McIntyre, M.E., Palmer, T.N.: Breaking planetary waves in the stratosphere. Nature **305**, 593–600 (1983)

Mellor, G.L., Yamada, T.: A hierarchy of turbulence closure models for planetary boundary layers. J. Atmos. Sci. **31**, 1791–1806 (1974)

Mirza, A., Pagé, C., Geindre, S.: FLYSAFE – an approach to flight safety – using GML/XML objects to define hazardous volumes space for aviation. 13th Conference on Aviation, Range and Aerospace Meteorology, American Meteorological Society, New Orleans, USA (2008)

National Research Council, Board on Atmospheric Sciences and Climate: From Research to Operations in Weather Satellites and Numerical Weather Prediction: Crossing the *Valley of Death*. National Academic Press, Washington, DC (2000)

Ozmidov, R.V.: On the turbulent exchange in a stably stratified ocean. Izv. Acad. Sci. USSR Atmos. Oceanic Phys. **1**, 853–860 (1965)

Paoli, R., Thouron, O., Escobar, J., Picot, J., Cariolle, D.: High-resolution large-eddy simulations of stably stratified flows: application to subkilometer-scale turbulence in the upper troposphere–lower stratosphere. Atmos. Chem. Phys. **14**, 5037–5055 (2014)

Regan, C.D., Jutte, C.V.: Survey of applications of active control technology for gust alleviation and new challenges for lighter-weight aircraft. NASA TM-2012-216008 (2012)
Reiter, E.R., Lester, P.F.: Richardson's Number in the free atmosphere. Arch. Met. Geoph. Biokl. A **17**, 1–7 (1968)
Schalkwijk, J., Jonker, H.J.J., Siebesma, A.P., Van Meijgaard, E.: Weather forecasting using GPU-based large-eddy simulations. Bull. Am. Meteorol. Soc. **96**, 715–723 (2015). doi:10.1175/BAMS-D-14-00114.1
Schumann, U., Gerz, T.: Turbulent mixing in stably stratified shear flows. J. Appl. Meteorol. **34**(1), 33–48 (1995)
Schwartz, B.: The quantitative use of PIREPs in developing aviation weather guidance products. Weather Forecasting **11**, 372–384 (1996)
Sharman, R., Tebaldi, C., Wiener, G., Wolff, J.: An integrated approach to mid- and upper-level turbulence forecasting. Weather Forecasting **21**(3), 268–287 (2006). doi:10.1175/WAF924.1
Sharman, R.D., Cornman, L.B., Meymaris, G., Pearson, J., Farrar, T.: Description and derived climatologies of automated in situ eddy dissipation rate reports of atmospheric turbulence. J. Appl. Meteorol. Climatol. **53**(6), 1416–1432 (2014). doi:10.1175/JAMC-D-13-0329.1
Sordeide, D., Bogue, R.K., Ehernberger, L.J., Bagley, H.: Coherent lidar turbulence measurement for gust load alleviation. NASA TM 104318 (1996)
Steiner, M., Bateman, R., Megenhardt, D., Liu, Y., Pocernich, M., Krozel, J.: Translation of ensemble weather forecasts into probabilistic air traffic capacity impact. Air Traffic Control Quart. **18**(3), 229–254 (2010)
Warner, T.T.: Numerical Weather and Climate Prediction. Cambridge University Press, Cambridge (2011)
Werne, J., Fritts, D., Wang, L., Lund, T., Wan, K.: Atmospheric turbulence forecasts for Air Force and missile defense applications. 20th DoD HPC User Group Conference, Schaumburg, IL, 14–17 June (2010). doi:10.1109/HPCMP-UGC.2010.75
Williams, J.K.: Using random forests to diagnose aviation turbulence. Mach. Learn. **95**, 51–70 (2014). doi:10.1007/s10994-013-5346-7
Wolff, J.K., Sharman, R.D.: Climatology of upper-level turbulence over the continental Unites States. J. Appl. Meteorol. Climatol. **47**, 2198–2214 (2008). doi:10.1175/2008JAMC1799.1
Wyngaard, J.C.: Toward numerical modeling in the "terra incognita". J. Atmos. Sci. **61**(14), 1816–1826 (2004). doi:10.1175/1520-0469(2004)061<1816:TNMITT>2.0.CO;2
Zilitinkevich, S.S., Elperin, T., Kleeorin, N., Rogachevskii, I., Esau, I.: A hierarchy of energy- and flux-budget (EFB) turbulence closure models for stably-stratified geophysical flows. Boundary-Layer Meteorol. **146**, 341–373 (2013). doi:10.1007/s10546-012-9768-8

Index

R. Sharman, T. Lane (eds.), *Aviation Turbulence*,
DOI 10.1007/978-3-319-23630-8

Printed by Printforce, the Netherlands